NUCLEAR FACTOR κB

Nuclear Factor κB

Regulation and Role in Disease

Edited by

Rudi Beyaert

Ghent University – VIB,
Ghent, Belgium

SPRINGER SCIENCE+BUSINESS MEDIA, B.V.

A C.I.P. Catalogue record for this book is available from the Library of Congress.

ISBN 978-94-010-3983-3 ISBN 978-94-010-0163-2 (eBook)
DOI 10.1007/978-94-010-0163-2

Printed on acid-free paper

CONTENTS

CONTRIBUTORS

ALNEMRI, Emad S.
Thomas Jefferson University, Kimmel Cancer Institute, Center for Apoptosis Research and Department of Microbiology and Immunology, 233 South 10th Street, Philadelphia, PA 19107, USA.

ANDREAKOS, Evangelos
Imperial College of Science, Technology and Medicine, Faculty of Medicine, Kennedy Institute of Rheumatology, 1 Aspenlea Road, Hammersmith, London W6 8LH, UK.

BACHER, Susanne
German Cancer Research Center, Division of Immunochemistry, Im Neuenheimer Feld 280, 69120 Heidelberg, Germany.

BALDWIN, Albert S. Jr.
University of North Carolina, Lineberger Comprehensive Cancer Center, School of Medicine, Chapel Hill, NC 27599, USA.

BEYAERT, Rudi
Ghent University and Flanders Interuniversity Institute for Biotechnology, Department for Molecular Biomedical Research, Unit for Molecular Signal Transduction in Inflammation, K.L. Ledeganckstraat 35, 9000 Gent, Belgium.

BONNARD, Madeleine
University of Toronto, Department of Medical Biophysics, and Advanced Medical Discovery Institute, University Health Network, 620 University Avenue, Suite 706, Toronto M5G 2C1, Canada.

CARPENTIER, Isabelle
Ghent University and Flanders Interuniversity Institute for Biotechnology, Department for Molecular Biomedical Research, Unit for Molecular Signal Transduction in Inflammation, K.L. Ledeganckstraat 35, 9000 Gent, Belgium.

CHEN, Lin-Feng
University of California, Gladstone Institute of Virology and Immunology, POB 419100, San Francisco, CA 94143, USA.

CHEN, Zhijian J.
University of Texas Southwestern Medical Center, Department of Molecular Biology, 5323 Harry Hines Boulevard, Dallas, TX 75390, USA.

COURTOIS, Gilles
Institut Pasteur, Unité de Biologie Moléculaire de l'Expression Génique, FRE2364 CNRS, 26 rue du Dr. Roux, 75724 Paris, France.

DE BOSSCHER, Karolien
Ghent University and Flanders Interuniversity Institute for Biotechnology, Department for Molecular Biomedical Research, Unit of Eukaryotic Expression and Signal Transduction, K.L. Ledeganckstraat 35, 9000 Gent, Belgium.

DE MARTIN, Rainer
University of Vienna, Department of Vascular Biology and Thrombosis Research, Brunnerstraße 59, 1235 Vienna, Austria.
DENG, Li
University of Texas Southwestern Medical Center, Department of Molecular Biology, 5323 Harry Hines Boulevard, Dallas, TX 75390, USA.
DIENZ, Oliver
German Cancer Research Center, Division of Immunochemistry, Im Neuenheimer Feld 280, 69120 Heidelberg, Germany.
FOXWELL, Brian M.
Imperial College of Science, Technology and Medicine, Faculty of Medicine, Kennedy Institute of Rheumatology, 1 Aspenlea Road, Hammersmith, London W6 8LH, UK.
GAYNOR, Richard B.
University of Texas Southwestern Medical Center, Department of Medicine, Division of Hematology-Oncology, Harold Simmons Cancer Center, 5323 Harry Hines Boulevard, Dallas, TX 75390, USA.
GEORGEL, Philippe
Centre National de la Recherche Scientifique, Institut de Biologie Moléculaire et Cellulaire, UPR 9022, 15 rue Descartes, 67000 Strasbourg, France.
GHOSH, Gourisankar
University of California, San Diego, Department of Chemistry and Biochemistry, 9500 Gilman Drive, MC 0359, La Jolla, CA 92093, USA.
GILMORE, Thomas D.
Boston University, Biology Department, 5 Cummington Street, Boston, MA 02215, USA.
GREENE, Warner C.
University of California, Gladstone Institute of Virology and Immunology, Departments of Medicine, and Microbiology and Immunology, POB 419100, San Francisco, CA 94143, USA.
HAEGEMAN, Guy
Ghent University and Flanders Interuniversity Institute for Biotechnology, Department for Molecular Biomedical Research, Unit of Eukaryotic Expression and Signal Transduction, K.L. Ledeganckstraat 35, 9000 Gent, Belgium.
HEYNINCK, Karen
Ghent University and Flanders Interuniversity Institute for Biotechnology, Department for Molecular Biomedical Research, Unit for Molecular Signal Transduction in Inflammation, K.L. Ledeganckstraat 35, 9000 Gent, Belgium.
HUXFORD, Tom
University of California, San Diego, Department of Chemistry and Biochemistry, 9500 Gilman Drive, MC 0359, La Jolla, CA 92093, USA.

ISRAËL, Alain
Institut Pasteur, Unité de Biologie Moléculaire de l'Expression Génique, FRE2364 CNRS, 26 rue du Dr. Roux, 75724 Paris, France.

JANSSENS, Sophie
Ghent University and Flanders Interuniversity Institute for Biotechnology, Department for Molecular Biomedical Research, Unit for Molecular Signal Transduction in Inflammation, K.L. Ledeganckstraat 35, 9000 Gent, Belgium.

KALTSCHMIDT, Barbara
University of Witten/Herdecke, Institute of Neurobiochemistry, Stockumer Straße 10, 58448 Witten, Germany.

KALTSCHMIDT, Christian
University of Witten/Herdecke, Institute of Neurobiochemistry, Stockumer Straße 10, 58448 Witten, Germany.

KARIN, Michael
University of California, San Diego, Department of Pharmacology, Laboratory of Gene Regulation and Signal Transduction, 9500 Gilman Drive, La Jolla, CA 92093, USA.

LI, Xiaoxia
The Cleveland Clinic Foundation, Lerner Research Institute, Department of Immunology, 9500 Euclid Avenue, Cleveland, OH 44195, USA.

LOOP, Torsten
University Hospital of Freiburg, Department of Anesthesiology and Intensive Care Medicine, Breisacher Straße 66, 79106 Freiburg, Germany.

MOSIALOS, George
Biomedical Sciences Research Center — Alexander Fleming, Institute of Immunology, 14-16 Alexander Fleming Street, 16672 Vari, Greece.

O'MAHONY, Alison
University of California, Gladstone Institute of Virology and Immunology, POB 419100, San Francisco, CA 94143, USA.

PAHL, Heike L.
University Hospital of Freiburg, Department of Anesthesiology and Intensive Care Medicine, Breisacher Straße 66, 79106 Freiburg, Germany.

POYET, Jean-Luc
Thomas Jefferson University, Kimmel Cancer Institute, Center for Apoptosis Research and Department of Microbiology and Immunology, 233 South 10th Street, Philadelphia, PA 19107, USA.

SACRE, Sandra
Imperial College of Science, Technology and Medicine, Faculty of Medicine, Kennedy Institute of Rheumatology, 1 Aspenlea Road, Hammersmith, London W6 8LH, UK.

SCHMITZ, M. Lienhard
University of Bern, Department for Chemistry and Biochemistry, Freiestrasse 3, 3012 Bern, Switzerland.
SENFTLEBEN, Uwe
University of Ulm, Clinic for Anesthesiology, Steinhövelstraße 9, 89075 Ulm, Germany.
STARK, George R.
The Cleveland Clinic Foundation, Lerner Research Institute, Department of Molecular Biology, 9500 Euclid Avenue, Cleveland, OH 44195, USA.
SUZUKI, Nobutaka
University of Toronto, Department of Medical Biophysics, and Advanced Medical Discovery Institute, University Health Network, 620 University Avenue, Suite 706, Toronto M5G 2C1, Canada.
SUZUKI, Shinobu
University of Toronto, Department of Medical Biophysics, and Advanced Medical Discovery Institute, University Health Network, 620 University Avenue, Suite 706, Toronto M5G 2C1, Canada.
UDALOVA, Irina
The John Radcliffe Hospital, University Department of Paediatrics, Headley Way, Headington, Oxford OX3 9DU, UK.
VANDEN BERGHE, Wim
Ghent University and Flanders Interuniversity Institute for Biotechnology, Department for Molecular Biomedical Research, Unit of Eukaryotic Expression and Signal Transduction, K.L. Ledeganckstraat 35, 9000 Gent, Belgium.
VERMA, Udit N.
University of Texas Southwestern Medical Center, Department of Medicine, Division of Hematology-Oncology, Harold Simmons Cancer Center, 5323 Harry Hines Boulevard, Dallas, TX 75390, USA.
YAMAMOTO, Yumi
University of Texas Southwestern Medical Center, Department of Medicine, Division of Hematology-Oncology, Harold Simmons Cancer Center, 5323 Harry Hines Boulevard, Dallas, TX 75390, USA.
YEH, Wen-Chen
University of Toronto, Department of Medical Biophysics, and Advanced Medical Discovery Institute, University Health Network, 620 University Avenue, Suite 706, Toronto M5G 2C1, Canada.

PREFACE

Seventeen years after its initial description, nuclear factor-κB (NF-κB) endures as one of the most studied transcription factors. NF-κB has attracted widespread interest based on the variety of stimuli that activate it, the diverse genes and biological responses that it controls, the striking evolutionary conservation of structure and function among species, and its involvement in a variety of human diseases. The biochemical basis by which several stimuli converge to activate NF-κB has been largely elucidated during recent years. While first discovered as a key regulatory factor of the immune system, NF-κB is now recognized as an important player in the functioning of many organs and cell types. The ongoing examination of NF-κB signaling has revealed its ever expanding role in immune and inflammatory responses, but also in cancer and development. For this reason, numerous efforts are underway to develop safe inhibitors of NF-κB to be used in the treatment of both chronic and acute disease situations.

The present book is the first to review and synthesize our knowledge of this interesting transcription factor. As such, the choice of subjects to review was daunting. To set the stage, an introductory chapter on activators and target genes, as well as the role they play in several responses, has been included. The first part of the book then focuses on intracellular signaling pathways leading to NF-κB activation in response to selected stimuli, with some chapters zooming in on specific signaling steps or molecules. Also information on the structure of NF-κB family members, as well as an overview of several strategies that allow to interfere at several steps of the NF-κB-signaling pathway is provided. Two chapters deal with forward genetics and gene knockout as tools to characterize the role of specific molecules in the activation of NF-κB. The second part of the book describes the role of NF-κB in development and disease, with chapters on inflammatory disease, genetic disorders, immunity, cancer and neurobiology. One of these chapters also describes the work done in the model system of Drosophila.

In the forthcoming years we will witness fabulous changes in the way patients are treated. If we can understand the biology of the fundamental cellular mechanisms in which NF-κB is involved, as well as the NF-κB-signaling pathways operating under normal and pathological conditions, it will be possible to take advantage of the powerful therapeutic potential of NF-κB.

I would like to thank the many colleagues for contributing to this book, and apologize to any researchers who feel that their work was overlooked in this collection of chapters. I also acknowledge the members of the MSTI unit at the DMBR for their continuous effort to dissect the NF-κB puzzle. I am particularly grateful to Maurits Vandecasteele, who did a great job to improve and finalize the book. I thank the Ghent University and VIB, as well as several granting institutions (IUAP, FWO, IWT) for their continuous support of research on NF-κB in my laboratory.

R.B., April 2003

T. LOOP & H.L. PAHL

ACTIVATORS AND TARGET GENES OF Rel/NF-κB TRANSCRIPTION FACTORS

From Bench to Bedside

1. INTRODUCTION

Sixteen years have passed since the description of the nuclear factor-κB (NF-κB) as a regulator of κ light-chain gene expression in murine B lymphocytes (Sen & Baltimore, 1986a). During that time, over 4,000 publications have appeared, characterizing the family of Rel/NF-κB transcription factors involved in the control of a large number of normal and pathological cellular processes. The physiological functions of NF-κB proteins include immunological and inflammatory responses, developmental processes, cellular growth and modulating effects on apoptosis. In addition, these factors are activated in a number of diseases, including cancer, arthritis, acute and chronic inflammatory states, asthma, as well as neurodegenerative and heart diseases.

Research on NF-κB proteins has spanned the bridge from basic sciences to applied clinical investigations. The fact that several large pharmaceutical companies are rapidly developing NF-κB inhibitory agents attests to the potential of these proteins as therapeutic targets. In order to understand why NF-κB proteins have attracted such interest both from basic and clinical scientists, we have compiled a list of extracellular κB activators as well as the NF-κB target genes known to date (see Section 6).

2. THE Rel/NF-κB FAMILY OF TRANSCRIPTION FACTORS

The Rel/NF-κB transcription factor family comprises a collection of homo- and heterodimeric proteins binding to related 10 bp DNA-binding sites, with a consensus sequence of 5´-GGGPuNNPyPyCC-3´. Individual stimuli can induce different NF-κB/Rel proteins, resulting in the activation of varying hetero- or homodimeric NF-κB complexes. Depending on the subunit composition, different DNA-binding sites are bound with varying affinities, leading to the activation of different target genes. The commonest form of active NF-κB consists of a heterodimer, containing the p65 (also called RelA) and the p50 subunit. Other heterodimers have been described in which RelA, RelB and c-Rel can substitute

R. Beyaert (ed.), Nuclear Factor κB. Regulation and Role in Disease, 1-48.

for p65 while Bcl-3, p100 and p105 can take the place of p50. Among these subunits only RelA, RelB and c-Rel contain a transactivation domain and are able to induce NF-κB-mediated gene transcription. Thus some NF-κB proteins, such as the p50/p50 homodimers, do not activate gene transcription. For the purposes of this review, "NF-κB" will be used to refer to any hetero- or homodimeric complex of NF-κB/Rel proteins that binds to κB sites.

3. DEMONSTRATION OF NF-κB ACTIVATION AND DISCOVERY OF TARGET GENES

How were the data in Section 6 obtained? The simplest and most reliable assay to demonstrate NF-κB activation is the electrophoretic mobility shift assay (EMSA). When combined with the appropriate controls, i.e. unstimulated control cells, as well as competition and supershift assays, EMSA is sensitive and accurate. Transient transfection experiments using a κB motiv-driven reporter gene are also frequently used. The appropriate controls include a reporter gene lacking the κB sites or, more specific and therefore better, a reporter gene preceded by mutated κB sites. Cotransfection of IκB, the inhibitory subunit, provides additional specificity. All inducers listed were demonstrated to activate NF-κB using one of these assays. What makes a gene an NF-κB target gene? Not all genes listed in Section 6.2 are bona fide NF-κB target genes. Genes denoted with an asterisk only contain a putative NF-κB-binding site in their promoter/enhancer region. To establish that a gene is truly regulated by NF-κB, these experiments must be performed: (i) NF-κB binding to the putative DNA site must be shown in EMSA, preferably using cell extracts from a tissue usually expressing the gene investigated; (ii) the promoter/enhancer region must be cloned before a reporter gene and functional importance of the κB-site must be demonstrated by mutagenesis. These experiments are required to consider a gene an NF-κB target. The data may be misleading if, for example, NF-κB can bind to a DNA sequence, but this site or adjacent sequences are occupied by other proteins in vivo. In such a scenario, NF-κB would bind the isolated site in vitro in EMSA. Mutation of this site would result in a loss of function in reporter gene assays, but only because binding of the adjacent unidentified protein is lost. For final proof of NF-κB involvement, in vivo DNA footprinting of the κB-binding site should be performed. This, however, has not been done for most of the target genes listed in Section 6.2, although the majority is expected to be true NF-κB target genes.

4. THE ROLE OF NF-κB ACTIVATION

While it has been recognized for some time that NF-κB is a central mediator of the human immune response (Pahl, 1999), it is now clear that NF-κB has an important role in initiating, perpetuating and potentiating pathophysiological processes (Section 6). The next sections illustrate our broadening understanding of the role played by this transcription factor in human diseases and disorders.

4.1. Lung

Increasing evidence shows an important role for NF-κB in the pathogenesis of acute lung inflammation, a potentially fatal complication that can occur during the course of various serious illnesses. Previously published findings, such as increased levels of TNFα or IL-8 in serum and bronchoalveolar lavage fluid of patients with acute lung inflammation (Hyers et al., 1991; Miller et al., 1992), implicated the NF-κB pathway in the pathogenesis of acute lung inflammation. Although the importance of NF-κB in transcription of these cytokines has been established, only two in vivo studies have directly demonstrated enhanced NF-κB activation in alveolar macrophages from rats or humans with acute lung inflammation (Manning et al., 1995; Schwartz et al., 1996). Interestingly, macrophage depletion largely abrogated activation of NF-κB in whole-lung tissue in a rat model of lung inflammation (Lentsch et al., 1999). In parallel, upregulation of lung vascular adhesion molecules (MIP-2, TNFα and CXC chemokine) were substantially reduced (Lentsch et al., 1999). However, lung instillation of TNFα into macrophage-depleted rat lungs nonetheless induced NF-κB activation in whole-lung tissue (Lentsch et al., 1999). These data suggest that, in this model, initial NF-κB activation occurred in alveolar macrophages, but propagation of the inflammation occurs in parenchymal lung cells as a response of NF-κB target proteins. In this way, the parenchymal lung cells themselves also contribute to the inflammatory process, since the activated endothelium participates in the inflammatory response (Lentsch & Ward, 2001). NF-κB regulates the expression of several genes that encode adhesion molecules such as ICAM-1, ELAM-1 and VCAM-1 (Section 6.2). These findings suggest a central role of NF-κB in the activation of genes, which promote adhesion and extravasation of leukocytes across the endothelial barrier.

Acute lung inflammation is characterized by the accumulation of neutrophils and a significant rise in IL-8 content in the lungs of these patients (Miller et al., 1992; Goodman et al., 1996). This is associated with activated NF-κB, anti-IL-8 antibodies/IL-8 complexes in bronchoalveolar lavage fluid from patients (Miller et al., 1992; Goodman et al., 1996; Kurdowska et al., 1996a; Kurdowska et al., 1997b). In a cohort of 64 patients, high levels of NF-κB activation and anti-IL-8 antibodies/IL-8 complexes coincided with the occurrence of lung failure (Kurdowska et al., 2001; Kurdowska et al., 2002). Moreover, the level of transcription factor activation and cytokine/antibody complexes correlated with the severity of lung dysfunction.

Based on these data, several studies were initiated to investigate the effect of NF-κB inhibition in acute lung inflammation. Inhibition of endotoxin-induced NF-κB activation in rat lungs by N-acetylcysteine or pyrrolidine dithiocarbamate leads to diminished cytokine-induced neutrophil chemoattractant (CINC) mRNA expression and decreased iNOS expression in lung tissue. Moreover, these antioxidants suppressed the alveolitis (Blackwell et al., 1996a; Liu et al., 1997a-b; Liu et al., 1999). Exogenously administered IL-10 and IL-13 greatly attenuated lung injury in an alveolitis model of rat lungs and prevented IκBα degradation (Lentsch et al., 1997). In rats, treatment of lipopolysaccharide (LPS)-induced lung injury with genistein, a specific protein tyrosine inhibitor, inhibited the induction of NF-κB activation and prevented CINC and matrix metalloproteinase (MMP-9) expression (Kang et al., 2001). Exogenous administration of a secretory leukocyte protease inhibitor (SLPI) greatly reduced NF-κB activation in rat lung cells in vivo (Lentsch et al., 1999). These different therapeutic strategies of modulating NF-κB activation may not only reduce the expression of proinflammatory mediators, but also speed the resolution of lung injury by decreasing the numbers of activated resident neutrophils (Kang et al., 2001).

The pathogenesis of asthma involves the overexpression of a distinct profile of inflammatory proteins that are regulated by NF-κB (Section 6.1) (Barnes, 1996). Increased NF-κB-binding activity has been identified in alveolar macrophages recovered from sputum from asthmatic patients and in airway epithelial cells from bronchial mucosal biopsies (Hart et al., 1998). The NF-κB-activating stimuli in asthmatic airways have not been identified so far, but several agents that cause asthma exacerbations activate NF-κB. For example, allergens such as Derp1 or ozone have been shown in vitro to activate NF-κB in asthmatic bronchial epithelial cells (Stacey et al., 1997). Exposure to aerosolized ovalbumin has been demonstrated to result in profound activation of NF-κB in the airways of rats (Liu et al., 1997). Rhinovirus infections of the upper respiratory tract may exacerbate asthma by activation of NF-κB and increase of ICAM-1 gene expression in bronchial epithelial cells (Papi & Johnston, 1999a-b). Respiratory syncytial virus (RSV) has been shown to activate NF-κB and to result in IL-8 and IL-11 gene expression in human type-II alveolar epithelial cells (A549 cells) (Garofalo et al., 1996; Mastronarde et al., 1996). Interestingly, mice lacking p50 or c-Rel develop less airway inflammation on antigen challenge, thus demonstrating a causal role of NF-κB in airway inflammation (Yang et al., 1998; Donovan et al., 1999; Das et al., 2001). Strikingly, $p50^{-/-}$ mice failed to express IL-5, which is important for the proliferation and differentiation of eosinophils, or eotaxin, which is important for the recruitment of eosinophils into the asthmatic airways (Yang et al., 1998). Immuno-histochemistry identified bronchial epithelial cell as the source of NF-κB-induced inflammatory protein expression, including chemokines, MIP-2 and eotaxin. These agents mediate the influx of neutrophils and eosinophils into the lung tissue (Poynter et al., 2002). Since NF-κB targets, such as TNFα and IL-1β, can also activate NF-κB expression, an autocrine loop may form the basis for the persistence of the chronic inflammatory process in asthma patients (Bureau et al., 2000).

Activation of NF-κB has been demonstrated in different models suffering from asbestosis (Janssen et al., 1995; Simeonova & Luster, 1996; Janssen et al., 1997; Simeonova et al., 1997; Luster & Simeonova, 1998). It has been shown that iron-containing asbestos fibers induce NF-κB in bronchial epithelial cells (Simeonova & Luster, 1996; Luster & Simeonova, 1998). In other mineral dust diseases, for example, a rat model of silica dioxide inhalation, NF-κB activation correlates with the intensity of alveolitis (Sacks et al., 1998). A wide array of air pollutants results in NF-κB activation in animal models, including residual oil fly ash (Quay et al., 1998), copper-ion (Kennedy et al., 1998) and diesel exhaust articles (Takizawa et al., 1999). Although human data are lacking, these data suggest that NF-κB induction could contribute to the pathogenesis of asbestosis, silicosis and other occupational lung diseases.

Constitutive NF-κB activation and IκB degradation have previously been observed in cystic fibrosis cells (DiMango et al., 1998; Tabary et al., 2000). These observations can be explained by the accumulation of ΔF508-CFTR in the endoplasmic reticulum, inducing NF-κB activity via the endoplasmic overload response (Knorre et al., 2002). Inhibition of NF-κB activation has been suggested as a method of decreasing sputum viscosity in cystic fibrosis as well as an effective treatment for reducing airway inflammation and attenuating detrimental changes in lung function (Ghio et al., 1996; Knorre et al., 2002).

4.2. Bowel

Increases in the production of proinflammatory cytokines have also been implicated in the pathogenesis of both acute and chronic inflammatory bowel diseases, including appendicitis, pancreatitis, Crohn´s disease and ulcerative colitis (Podolsky, 1991; Neurath et al., 1998; Rogler et al., 1998; Schreiber et al., 1998; Pennington et al., 2000). In patients with acute appendicitis, NF-κB-binding activity is increased in extracts from mononuclear cells and the level of activation correlates with the length of time of symptoms before operation (Pennington et al., 2000). This suggests that NF-κB represents a molecular indicator in predicting and staging surgical inflammatory conditions.

It has recently been found that activation of NF-κB occurs within pancreatic acinar cells early in the course of experimental, cerulein-induced acute pancreatitis in rats (Gukovsky et al., 1998; Steinle et al., 1999) and is correlated with the expression of TNFα, IL-6, KC (murine GROα/IL-8 analogue) and MCP-1 in this model (Grady et al., 1997; Gukovskaya et al., 1997; Gukovsky et al., 1998). Direct activation of NF-κB by adenoviral-mediated intraductal gene transfer of the RelA subunit led to infection within the acinar cells of the pancreas and a profound increase in the level of activity of myeloperoxidase and amylase in the organ (Chen et al., 2002). In addition, activation of NF-κB has been demonstrated in the intestine of patients with Crohn´s disease, ulcerative colitis and self-limited colitis (Neurath et al., 1996; Rogler et al., 1998; Schreiber et al., 1998). Immuno-histochemistry demonstrated that the amount of activated NF-κB correlates with the degree of mucosal inflammation. The transcription factor is induced in the intestinal epithelial cells located in the crypts, but not in cells at the surface (Rogler et al., 1998).

The intestine comprises a pleated layer of epithelial cells whose basal membranes form the lamina propria. Underneath this layer immunologically active cells reside in clusters termed Peyer´s patches. Thus, in a healthy intestinal mucosa and homeostasis, commensal bacteria only interact with intestinal epithelial cells. NF-κB activation in intestinal epithelial cells is thus only observed in response to pathogenic species such as enteropathogenic Escherichia coli (EPEC). Consequently, nonpathogenic E. coli or purified E. coli (LPS) have no effect in an intestinal epithelial cell model (Savkovic et al., 1997). Deletion of specific EPEC virulence genes abrogated NF-κB activation and IL-8 gene expression (Savkovic et al., 1997). Infection of intestinal epithelial cells with other bacteria, such as Salmonella or Yersinia, also triggers IKK activation, partial IκBα degradation and NF-κB activation (Hobbie et al., 1997; Elewaut et al., 1999). It is not clear whether bacterial invasion is required for NF-κB activation, since Salmonella strains can also trigger NF-κB activation in cultured intestinal epithelial cells when invasion is prevented by the placement of a membrane, which still allows bacterial adhesion (Eaves-Pyles et al., 1999; Gewirtz et al., 2000). Helicobacter pylori-associated gastritis is also marked by increased NF-κB activity in gastric epithelial cells (van den Brink et al., 2000). The number of NF-κB-positive cells correlates with the degree of gastritis.

Surprisingly, nonpathogenic Lactobacillus rhamnosus activates NF-κB-signaling pathways in human macrophages in vitro (Miettinen et al., 2000). These authors suggest that transcription factor activation by this commensal bacterium is a time-dependent direct effect, occurring without mediation by cytokines and hypothetically mediated by human Toll-like receptors. Physiologically, NF-κB activation by Lactobacillus rhamnosus may take place when the bacterium crosses the lamina propria and encounters mucosal macrophages, a pathologic process called translocation.

4.3. Rheuma

Activation of the NF-κB pathway occurs in patients with rheumatoid arthritis, a chronic inflammatory disease of unknown cause, which is characterized by synovial inflammation and joint destruction. Recently, high NF-κB DNA-binding activity has been found in the synovium of patients with rheumatoid arthritis compared to healthy synovium and animals with induced osteoarthritis (Miagkov et al., 1998; Miyazawa et al., 1998). Likewise, cytokines such as TNFα and IL-1, which activate NF-κB, are elevated in the synovial fluid of patients (Feldmann et al., 1996). In fibroblast-like synoviocytes, an in vitro model for rheumatoid arthritis, cytokine-induced NF-κB activation is regulated by IκB kinase-β (Aupperle et al., 2001). This enzyme appears to play a critical part in the disease process, because gene therapy transferring an IKKβ kinase-inactive mutant to rat joints ameliorated the severity of arthritis (Tak et al., 2001). Clinical studies have demonstrated that the administration of antibodies against TNFα (Infliximab [Remicade]) or a truncated TNFα receptor that binds to TNFα, thereby inactivating the cytokine (Etanercept [Enbrel]) can markedly improve the symptoms of patients with rheumatoid arthritis. The same effect is seen in inflammatory ankylosing spondylitis (morbus Bechterew) (Maini et al., 1999; Braun et al., 2002; Gorman et al., 2002).

4.4. Acute phase response

The acute phase response is a systemic alteration in normal biological set points, which is induced in response to pathophysiological insults, such as trauma, surgery (Hawkey et al., 1983) or myocardial infarction (Rennie & Ogston, 1976). It is mediated by cytokines including IL-1, IL-6 and TNFα, which activate NF-κB by itself (Section 6.1). During the acute phase, coagulation is activated and fever is induced. Both changes are accompanied with increased levels of various circulating plasma proteins (IL-1, IL-6, TNFα, factor VIII; see Section 6.2). It has recently been demonstrated that an NF-κB response element is contained in the promoter of the factor VIII gene (Begbie et al., 2000).

4.5. Systemic inflammatory response syndrome (SIRS) and sepsis

SIRS is defined as a clinical state resembling sepsis, but without the provable presence of infection. Both SIRS and sepsis can lead to multiple organ dysfunction, a potentially lethal situation. In a clinical setting, it is often difficult to distinguish a single etiology leading to the development of SIRS or sepsis. For example, patients who initially suffered blood loss may subsequently become infected. While both hemorrhage and endotoxemia can individually lead to the development of SIRS/sepsis, the relative contributions in a patient who suffers both conditions are unclear. Data in sepsis models of rats exposed to bacteria or endotoxin suggest a link between in vivo activation of NF-κB and general inflammatory response (Blackwell et al., 1994; Blackwell et al., 1996a; Blackwell et al., 1997b; Chen et al., 2001). In a transgenic mouse model of systemic inflammation, intraperitoneal injection of E. coli LPS displayed a time- and dose-dependent, organ specific pattern of NF-κB activation (Koay et al., 2001; Sadikot et al., 2001). Studies on the DNA-binding activity of NF-κB in nuclear extracts from peripheral blood mononuclear cells of septic patients revealed that increased NF-κB activation predicts mortality (Bohrer et al., 1997; Adib-Conquy et al., 2000). Specifically, all patients in these studies who died from sepsis, had increased NF-κB activation (greater than twice baseline) in the first 6 days, whereas all surviving patients had NF-κB activation less than twice baseline at each time point during the 14-day study period (Bohrer et al., 1997).

Other pathophysiological and proinflammatory states leading to SIRS, such as hemorrhage and resuscitation, also result in a dramatic increase in NF-κB activation in rat hearts (Meldrum et al., 1997a-b) and mice lung neutrophils (Shenkar & Abraham, 1997; Shenkar & Abraham, 1999; Abraham et al., 2000; Shenkar et al., 2001). It is well known that hemorrhagic shock initiates inflammatory cascades characterized by increased cytokine production, leukocyte adhesion molecule expression and neutrophil infiltration into tissues (Hierholzer et al., 1997). NF-κB as well as another transcription factor, the cAMP response element-binding protein (CREB), are activated in the lung after hemorrhage (Shenkar & Abraham, 1999). The mechanisms leading to acute inflammatory lung injury after hemorrhage differ from endotoxemia, since in the former xanthin oxidase has an important role in increasing proinflammatory cytokines (Shenkar & Abraham, 1999). Our present knowledge would suggest that patterns of NF-κB and CREB activation could serve to distinguish between groups of distinct intracellular mechanisms producing lung injury or other organ failure. Because NF-κB activation constitutes an important determinant of systemic inflammation, interventions causing NF-κB inhibition could prove beneficial in SIRS or sepsis. Agents which have previously been described to inhibit NF-κB, such as antioxidants, proteasome inhibitors, corticosteroids or specific kinase inhibitors, might be effective in the treatment of SIRS and sepsis. To date, no clinical study has investigated NF-κB inhibition as a target of therapy.

4.6. Muscle wasting

Muscle wasting is associated with disease states such as cancer, sepsis, HIV infection and end-stage heart failure (Lecker et al., 1999). It is also a result of chronic disuse, for example due to bed rest, sedentary lifestyles, exposure to space flights and muscle denervation (Booth et al., 1996; Yan et al., 1996). There is evidence that disuse muscle atrophy is associated with activation of an alternative NF-κB pathway, distinct from that used by cytokines, which involves the activation of p50 and its interaction with bcl-3, but not p65 (Hunter et al., 2002). However, in other models of muscle wasting, induction of NF-κB seems to represent a survival mechanism. For instance, muscular biopsies of type 2A muscular dystrophy patients contain apoptotic myonuclei at much higher incidence than that recorded from patients with a different type of myopathy (i.e. Duchenne´s muscular dystrophy) (Baghdiguian et al., 1999). This is associated with the accumulation of IκBα and the absence of NF-κB in type 2A dystrophic myonuclei (Baghdiguian et al., 2001). Calpain 3, present in healthy muscles, may control IκBα degradation, since calpain 3-deficient knockout mice displayed a myonuclear apoptotic state associated with a persistence of IκBα protein and the ability to activate NF-κB (Richard et al., 2000). These results suggest that type 2A muscular dystrophy is a consequence of a failure to prevent apoptosis caused by a lack of calpain 3-dependent IκBα degradation/NF-κB activation.

The target genes of NF-κB during the various types of muscle wasting are unknown. TNF-R-associated factor 2 (TRAF2) is an antiapoptotic target of TNFα-induced NF-κB activation in cultured muscle cells (MacLachlan & Giordano, 1998). Members of the antiapoptotic Bcl-2 family are known targets of NF-κB in non-muscle cells (Tamatani et al., 1999; Chen et al., 2000; Tamatani et al., 2000). Another survival gene, HO-1, has been shown to be a target of NF-κB in the liver and its promoter region contains multiple κB-binding sites (Lavrovsky et al., 1994). Since it has recently been shown that HO-1 is upregulated by disuse atrophy of skeletal muscle, this gene could be a potential target of NF-κB in atrophying muscles (Hunter et al., 2001). Additional cytoprotective, beneficial genes, such as heat shock proteins, inducible nitric oxide synthase and cyclooxygenase-2 (COX-2), are also activated by NF-κB in muscle cells.

4.7. Apoptosis

The role played by NF-κB activation in apoptosis appears to be complex since it has been found to depend on the cell type. Antiapoptotic genes directly activated by NF-κB include the cellular inhibitors of apoptosis (c-IAP1, c-IAP2 and IXAP), the TNF-R-associated factors (TRAF1 and TRAF2), the Bcl-2 homologue A1/Bfl-1 and IEX-IL (Section 6.2) (Wang et al., 1998; Wu et al., 1998). NF-κB also acts in controlling the cell cycle by inducing cyclin D1 expression (Guttridge et al., 1999). Insulin triggers antiapoptotic signaling by activating its cognate receptor and activates NF-κB in Chinese hamster ovary cells (Bertrand et al., 1998).

In addition, several investigations provided evidence that NF-κB acts as a survival signal in T- and B-cells (Wu et al., 1996; Mayo et al., 1997). In contrast, treatment of human immature thymocytes or promyelocytic leukemia cells with etoposide activates NF-κB and induces apoptosis (Bessho et al., 1994). NF-κB is concomitantly activated with TNF-induced apoptosis in osteoblasts (Kitajima et al., 1996). Proapoptotic effects are also mediated by pro-inflammatory stimuli via the TNF-R1. Moreover, cross-linking of the Fas receptor by anti-Fas antibodies and binding of the Trail receptors activate NF-κB. Fas receptor and its ligand are encoded by NF-κB target genes. How can NF-κB mediate both life and death signals? It is possible that the role of NF-κB activation in apoptosis depends on the cell type or the nature of the stimulus applied, which is still elusive.

5. CONCLUSIONS

NF-κB is activated by over 150 substances, organisms or conditions which constitute inflammatory, physiological and stressful processes (Section 6.1). They result in a selective pattern of gene expression, which is specific for both the inducing agent and the cell type responding. This specificity is achieved through the different availability of receptors, signal transduction molecules and additional transcription factors required for expression of NF-κB target genes. Because such a large variety of physiological stimuli induces NF-κB, it should come as no surprise that NF-κB is also a primary effector of pathophysiological processes and human diseases. Thus NF-κB proteins and their activators have become prime targets for therapeutic modulation. Indeed, many common synthetic drugs, for example aspirin, as well as traditional remedies such as green tea and curcumin were found to modulate NF-κB activity. If empirically developed therapies are found to exert their effect via NF-κB, it is likely that our knowledge of the molecular details of this pathway will enable us to develop more specific and potent strategies.

6. OVERVIEWS

6.1. Activators of NF-κB activity

Conditions	*References*
—Bacteria	
Bordetella pertussis	Belcher et al., 2000
Enteropathogenic E. coli	Savkovic et al., 1997
Gardnerella vaginalis	Al-Harthi et al., 1998
Helicobacter pylori	Munzenmaier et al., 1997
Lactobacilli	Klebanoff et al., 1999
Listeria monocytogenes	Hauf et al., 1994
Mycoplasma fermentans	Marie et al., 1999
Mycobacteria tuberculosis	Zhang et al., 1994
Neisseria gonorrhoeae	Naumann et al., 1997
Rickettsia rickettsii	Sporn et al., 1997
Salmonella dublin	Eaves-Pyles et al., 1999
Salmonella typhimurium	Hobbie et al., 1997
Shigella flexneri	Dyer et al., 1993
Staphylococcus aureus	Busam et al., 1992
Streptococcus group B (GBS)	Glibetic et al., 2001
Yersinia enterocolitica	Elewaut et al., 1999
—Bacterial products	
CpG	Krieg, 2000
Cytolysin (Vibrio vulnificus)	Kang et al., 2002
Diphosphoryl lipid A (Rhodobacter sphaeroides)	Lawrence et al., 1995
Exotoxin B	Busam et al., 1992
G(Anh) M tetra	Dokter et al., 1994
Lipoteichoic acid (Listeria)	Hauf et al., 1997
LPS	Sen & Baltimore, 1986b
Membrane lipoproteins (Mycoplasma fermentans)	Garcia et al., 1998
Muramyl peptides	Schreck et al., 1992
PlcA (phospholipase) (Listeria)	Hauf et al., 1997
PlcB (phospholipase) (Listeria)	Hauf et al., 1997
Staphylococcus enterotoxin A/B (superantigen)	Trede et al., 1993 Busam et al., 1992
Toxic shock syndrome toxin 1	Trede et al., 1993
—Viruses	
Adenovirus	Shurman et al., 1989
Cytomegalovirus	Sambucetti et al., 1989
Epstein-Barr virus (EBV)	Hammarskjold & Simurda, 1992
Hepatitis B virus	Siddiqui et al., 1989
Herpesvirus saimiri	Yao et al., 1995
Human herpesvirus 6	Ensoli et al., 1989

(6.1. cont.) *Conditions*	*References*
HIV-1	Bachelerie et al., 1991
Herpex simplex virus -1	Gimble et al., 1988
HTLV-I	Ballard et al., 1988
	Leung & Nabel, 1988
Influenza virus	Ronni et al., 1997
Measles virus	Harcourt et al., 1999
Molony murine leukemia virus	Pak & Faller, 1996
Newcastle disease virus	Ten et al., 1993
Respiratory syncytial virus	Mastronarde et al., 1996
	Garofalo et al., 1996
Rhinovirus	Zhu et al., 1996a-b
Sendai paramyxovirus	Hiscott et al., 1989
Sindbis virus	Lin et al., 1995
—Viral products	
Adenovirus 5 E1A	Shurman et al., 1989
Adenovirus E3/19K	Pahl et al., 1996
CMV ie1	Sambucetti et al., 1989
Double-stranded RNA	Visvanathan & Goodbourn, 1989
EBV EBNA-2	Scala et al., 1993
EBV LMP	Hammarskjold & Simurda, 1992
HBV HBx	Twu et al., 1989
HBV LHBs	Hildt et al., 1996
HBV MHBs[t]	Meyer et al., 1992
HCV core protein	You et al., 1999
Herpes saimiri HVS13	Yao et al., 1995
HIV-1 gp160	Chirmule et al., 1994
HIV-1 Tat	Westendorp et al., 1995
HTLV-I Tax1	Ballard et al., 1988
	Leung & Nabel, 1988
HTLV-II Tax2	Tanaka et al., 1996
Influenza virus hemagglutinin	Pahl & Baeuerle, 1995b
Parvovirus B19 NSI	Moffatt et al., 1996
—Eukaryotic parasite	
Theileria parva	Ivanov et al., 1989
—Cytokines	
Cardiotrophin-like cytokine	Shi et al., 1999
IL-1	Osborn et al., 1989
IL-2	Hazan et al., 1990
IL-12	Grohmann et al., 1998
IL-15	McDonald et al., 1998
IL-17	Shalom-Barak et al., 1998
IL-18	Matsumoto et al., 1997
p43	Ko et al., 2001
S100B	Lam et al., 2001
THANK	Mukhopadhyay et al., 1999

(6.1. cont.) Conditions	*References*
TNFα	Osborn et al., 1989
	Israel et al., 1989a
TNFβ	Messer et al., 1990
	Paul et al., 1990
—Pathophysiological conditions and diseases	
Acute lung injury/respiratory distress syndrome	Manning et al., 1995
	Schwartz et al., 1996
Adhesion	Lin et al., 1995
Angina pectoris (unstable)	Valen et al., 2001
Appendicitis	Pennington et al., 2000
Asthma	Hart et al., 1998
	Zhang et al., 1999
	Ray & Cohn, 1999
Cecal ligation and puncture (murine sepsis model)	Williams et al., 1999
Crohn's disease/ulcerative colitis	Rogler et al., 1998
Depolarization	Kaltschmidt et al., 1995
	Pahl & Baeuerle, 1995a
Hemorrhage	Shenkar et al., 1996
Hypercholesterolemia	Wilson et al., 2000
Hyperglycemia	Yerneni et al., 1999
Hyperosmotic shock	Courtois et al., 1997
Hyperoxia	Shea et al., 1996
Hypoxia/reoxygenation	Tamatani et al., 2000
Ischemia (transient, focal)	Gabriel et al., 1999
	Li et al., 1999
Ischemia/reperfusion injury	Zwacka et al., 1998
Liver regeneration	Tewari et al., 1992
	Cressman et al., 1994
Mechanical ventilation (in vitro)	Pugin et al., 1998
Muscle disuse	Hunter et al., 2002
Muscle dystrophy (type 2A)	Baghdiguian et al., 2001
Neuronal firing	Lerner-Natoli et al., 2000
Pancreatitis (acute/experimental)	Gukovsky et al., 1998
Rheumatoid arthritis/osteoarthritis	Miyazawa et al., 1998
	Miagkov et al., 1998
Reoxygenation	Rupec & Baeuerle, 1995
Shear stress	Lan et al., 1994
T-cell selection	Moore et al., 1995
—Physical stress	
PPME photosensitization	Legrand-Poels et al., 1995
Ultraviolet irradiation (A, B and C)	Stein et al., 1989
Wounding combined with HeNe irradiation	Haas et al., 1998
Wounding combined with thermal ablation	Nomoto et al., 2001
Gamma radiation	Brach et al., 1991a

(6.1. cont.) *Conditions*	*References*
—Oxidative stress	
Butyl peroxide	Munroe et al., 1995
Glutathione (prooxidative shift)	Hehner et al., 2000
Hydrogen peroxide	Schreck et al., 1991
Ozone	Haddad et al., 1996
Pervanadate	Imbert et al., 1996
Reoxygenation	Rupec & Baeuerle, 1995
—Environmental hazards	
3,3',4,4'-Tetrachlorobiphenyl (PCB77)	Hennig et al., 1999
Chromium	Ye et al., 2000
Cigarette smoke	Nishikawa et al., 1999
Cobalt	Goebeler et al., 1995
Copper iron	Kennedy et al., 1998
Crocidolite asbestos fibers	Janssen et al., 1995
Dicamba (herbicide, peroxisome proliferator)	Espandiari et al., 1998
Diesel exhaust articles	Takizawa et al., 1999
Lead	Ramesh et al., 1999
Nickel	Goebeler et al., 1995
Oil fly ash	Quay et al., 1998
Silica particles	Chen et al., 1995
Wood smoke	Leonard et al., 2000
—Therapeutically used drugs	
1-β-D-arabinofuranosyl cytosine (ara-C)	Strum et al., 1994
Anthralin	Schmidt et al., 1996
Azidothymidine (AZT)	Kurata, 1994
Camptothecin/7-ethyl-10-hydroxycamptothecin	Piret & Piette, 1996
(active metabolite)	Bottero et al., 2001
Ciprofibrate	Li et al., 1996a
Cisplatin	Nie et al., 1998
Daunomycin	Das & White, 1997
	Hellin et al., 1998
Daunorubicin	Wang et al., 1996
5,6-Dimethylxanthenone-4-acetic acid	Ching et al., 1999
Doxorubicin	Das & White, 1997
Etoposide	Bessho et al., 1994
Flavone-8-acetic acid	Ching et al., 1999
Haloperidol	Post et al., 1998
Lithium	Nemeth et al., 2002
Methamphetamine	Asanuma & Cadet, 1998
Mitoxantrone	Boland et al., 2000
Norepinephrine	Minneman et al., 2000
Phenobarbital	Li et al., 1996b
Tamoxifen	Ferlini et al., 1999
Taxol (paclitaxel)	Hwang & Ding, 1995
Vinblastine	Rosette & Karin, 1995

(6.1. cont.) *Conditions*	*References*
Vincristine	Das & White, 1997
WR1065 (free thiol form of amifostine)	Grdina et al., 2002
—Modified proteins	
Advanced glycated end products (AGEs)	Yan et al., 1994
	Wautier et al., 1994
Amyloid protein fragment (βA4)	Behl et al., 1994
Amyloid β-peptide	Kaltschmidt et al., 1997
Anti-PR3	Mayet et al., 1999
Maleylated BSA	Misra et al., 1996
Modified (oxidized) LDL	Rajavashisth et al., 1995
	Andalibi et al., 1993
—Overexpressed proteins (ER overload)	
CFTR	Knorre et al., 2002
Erythropoietin receptor	Knorre & Pahl, unpublished
Hematopoietic progenitor kinase 1	Arnold et al., 2001
Heat shock proteins	Srivastava, 2002
Ig heavy chain	Pahl & Baeuerle, 1995a
MHC class I	Pahl & Baeuerle, 1995a
—Receptor ligands	
Antigen (IgM ligand)	Marcuzzi et al., 1989
β-D-glucan ligand	Battle et al., 1998
CD11b/CD18 ligand (complement)	Thieblemont et al., 1995
CD28 ligand (B7-1)	Verweij et al., 1991
CD2 ligand	Bressler et al., 1991
CD35 ligand (complement)	Thieblemont et al., 1995
CD3 ligand	Tong-Starksen et al., 1990
CD25 ligand	Vockerodt et al., 2001
CD40 ligand	Berberich et al., 1994
CD43 ligand	Santana et al., 2000
CD4 ligand (gp120)	Chirmule et al., 1994
CD66a ligand	Donda et al., 2000
Fc-2a receptor ligand (IgG2a)	Muroi et al., 1994
Flt-1 ligand	Reikerstorfer et al., 1995
Ly6A/E ligand	Ivanov et al., 1994
N-CAM	Krushel et al., 1999
Trail receptor-1 ligand (Trail)	Schneider et al., 1997
Trail receptor-2 ligand (Trail)	Schneider et al., 1997
Trail receptor-4 ligand (Trail)	Degli-Esposti et al., 1997
—Apoptotic mediators	
Anti-Fas/Apo-1	Rensing-Ehl et al., 1995
Glucan	Battle et al., 1998
Trail	Schneider et al., 1997
—Mitogens, growth factors and hormones	
Angiotensin II	Li & Brasier, 1996
Bone morphogenic protein 2	Mohan et al., 1998

(6.1. cont.) *Conditions*	*References*
Bone morphogenic protein 4	Mohan et al., 1998
Epidermal growth factor	Biswas et al., 2000
Folicle stimulating hormone	Delfino & Walker, 1998
Human growth hormone	Shen et al., 1997
Insulin	Bertrand et al., 1995
Insulin-like growth factor I	Liu et al., 2001
Luteinizing hormone-releasing hormone	Grundker et al., 2000
M-CSF	Brach et al., 1991b
Müllerian inhibiting substance	Segev et al., 2001
Nerve growth factor	Wood, 1995
	Carter et al., 1996
Pigment epithelium-derived factor	Yabe et al., 2001
Platelet-derived growth factor	Olashaw et al., 1992
Serum	Baldwin et al., 1991
TGFα	Lee et al., 1995
—*Physiological mediators*	
12(R)-hydroxyeicosatrienoic acid	Laniado-Schwartzman et al., 1994
Amino acid analogues	Kretz-Remy et al., 1998
Anaphylatoxin C3a	Pan, 1998
Anaphylatoxin C5a	Pan, 1998
Angiotensin II	Li & Brasier, 1996
Basic calcium phosphate crystals	McCarthy et al., 1998
Bradykinin	Pan et al., 1996
C2 ceramide (N-acetyl sphingosine)	Andrieu et al., 1995
Cerulein	Gukovsky et al., 1998
	Steinle et al., 1999
Collagen lattice	Xu et al., 1998
Collagen type I	Lee et al., 1995
Des-arg10-kallidin (B1 receptor agonist)	Schanstra et al., 1998
Double-stranded polynucleotides	Suzuki et al., 1999
f-Met-Leu-Phe	Browning et al., 1997
Heat shock protein 60 (HSP 60)	Kol et al., 1999
Hemoglobin	Simoni et al., 1998
Homocysteine	Mercie et al., 2000
Hyaluronan	Noble et al., 1996
Kaianic acid (kainate)	Kaltschmidt et al., 1995
Leukotriene B4	Brach et al., 1992
L-glutamate	Guerrini et al., 1995
Lysophosphatidylcholine (LysoPC)	Zhu et al., 1997
Neutrophil elastase	Walsh et al., 2001
Platelet-activating factor	Smith & Shearer, 1994
	Mutoh et al., 1994
Polyunsaturated fatty acids	Maziere et al., 1999
Potassium	Kaltschmidt et al., 1995
Prostaglandin A1	Cornelussen et al., 2001

(6.1. cont.)

Conditions	*References*
Thioredoxin	Hirota et al., 1999
Thrombin	Mari et al., 1994
Trypsin/SLIGRL	Bretschneider et al., 1999
—Chemical agents	
2-Deoxyglucose	Pahl & Baeuerle, 1995a
Anisomycin	Sen & Baltimore, 1986b
Arsenic	Huang et al., 2001
Benzyl isothiocyanate	Patten & DeLong, 1999
Brefeldin A	Pahl & Baeuerle, 1995a
Calchicine	Rosette & Karin, 1995
Calcium ionophores	Novak et al., 1990
Calyculin A	Pahl & Baeuerle, 1995a
	Suzuki et al., 1994
Cobalt chloride	Sultana et al., 1999
ConA	Rattner et al., 1991
Cycloheximide	Sen & Baltimore, 1986b
Cyclopiazonic acid	Pahl et al., 1996
Ethanol	Lehmann et al., 2000
Forskolin	Delfino & Walker, 1998
Gadolinium chloride	Rose et al., 2001
Glass fibers	Ye et al., 1999
Kainic acid	Blondeau et al., 2001
Linoleic acid	Hennig et al., 1996
L-NMA	Peng et al., 1995
Lysophosphatidic acid	Shahrestanifar et al., 1999
Malondialdehyde	Oh et al., 2001
Monensin	Pahl & Baeuerle, unpublished
Nafenopin	West et al., 1999
N-methyl-D-aspartate	Guerrini et al., 1995
Nocodazol	Rosette & Karin, 1995
Okadaic acid	Thevenin et al., 1990
PHA	Sen & Baltimore, 1986b
Phorbol ester	Sen & Baltimore, 1986b
Podophyllotoxin	Rosette & Karin, 1995
Pyrogallol	Adcock et al., 1994
Quinolinic acid	Qin et al., 1998
Staurosporine	Chabot-Fletcher & Breton, 1998
Sanglifehrin A	Zhang et al., 2001
Thapsigargin	Adcock et al., 1994
	Pahl et al., 1996
Tunicamycin	Adcock et al., 1994
	Pahl & Baeuerle, 1995a
Vinblastine	Rosette & Karin, 1995

6.2. Target genes of NF-κB

Genes	*Functions*	*References*
—Cytokines/chemokines and their modulators		
CINC	Cytokine-induced neutrophil chemoattractant	Blackwell et al., 1994 Ohtsuka et al., 1996
*CXCL 11	Chemokine ligand for CXCR3	Tensen et al., 1999
Eotaxin	β-Chemokine, eosinophil-specific	Hein et al., 1997
Gro α-γ	Melanoma growth-stimulating activity	Anisowicz et al., 1991
IFNγ	Macrophage activator	Sica et al., 1992 Sica et al., 1997
IL-1α	Synthesis of other mediators	Mori & Prager, 1997
IL-1β	Synthesis of other mediators	Hiscott et al., 1993
IL-1 receptor antagonist	Inhibitor of IL-1 activity	Smith et al., 1994
IL-2	T-cell proliferation	Serfling et al., 1989 Hoyos et al., 1989 Lai et al., 1995
IL-6	Acute-phase reaction mediator	Libermann & Baltimore, 1990 Shimizu et al., 1990
IL-8	Neutrophil granulocyte activator	Kunsch & Rosen, 1993
*IL-9	Th-cell stimulator	Zhu et al., 1996
IL-11	Antigen/body reaction promoter	Bitko et al., 1997
IL-12 (p40)	Lymphocyte proliferation	Murphy et al., 1995
*IL-15	Mast-cell proliferation	Azimi et al., 1998
IFNβ	Unspecific humoral immune response	Hiscott et al., 1989 Lenardo et al., 1989
IP-10	α-Chemokine	Ohmori & Hamilton, 1993
KC	α-Chemokine	Ohmori & Hamilton, 1995
Lymphotoxin-α	—	Worm et al., 1998
Lymphotoxin-β	Anchors TNF to cell surface	Kuprash et al., 1996
MCP-1/JE	Macrophage chemotactic protein, β-chemokine	Ueda et al., 1994
MIP-1α,β	Macrophage inflammatory protein-1, β-chemokine	Grove & Plumb, 1993 Widmer et al., 1993
MIP-2	Macrophage inflammatory protein-1, β-chemokine	Widmer et al., 1993
TCA3	T-cell activation gene 3, β-chemokine	Oh & Metcalfe, 1994
RANTES	Regulated upon activation normal T-lymphocyte expressed and secreted, β-chemokine	Moriuchi et al., 1997
Peptide-78	Neutrophil activator	Keates et al.,2001

(6.2. cont.) *Genes*	*Functions*	*References*
TNFα	Tumor necrosis factor-α	Drouet et al., 1991 Collart et al., 1990
TNFβ	Tumor necrosis factor-β	Messer et al., 1990
—Immunoreceptors		
B7.1 (CD80)	Costimulation of T-cells via CD28 binding	Fong et al., 1996 Zhao et al., 1996
BRL-1	B-cell homing receptor	Wolf et al., 1998
CCR5	Chemokine receptor	Liu et al., 1998
CD154	CD40 ligand	Srahna et al., 2001
CD48	Antigen of stimulated lymphocytes	Klaman & Thorley-Lawson, 1995
$F_c\epsilon$ receptor II (CD23)	Receptor for IgE	Richards & Katz, 1997
IL-2 receptor α-chain	IL-2 receptor subunit	Ballard et al., 1988
Ig Cγ1	IgG heavy chain	Lin & Stavnezer, 1996
Ig ϵ heavy chain	IgE heavy chain	Iciek et al., 1997
Ig κ light chain	Antibody light chain	Sen & Baltimore, 1986b
Invariant chain I_I	Antigen presentation	Pessara & Koch, 1990
MHC class I (H-2K^b)	Mouse histocompatibility Ag	Israel et al., 1989a-b
MHC class I HLA-B7	Mouse histocompatibility Ag	Johnson & Pober, 1994
β_2 microglobulin	Binds MHC class I	Israel et al., 1989a-b
TNF receptor-associated factor 2 (TRAF2)	Antiapoptotic factor	MacLachlan & Giordano, 1998 MacLachlan et al., 1998
T-cell receptor β-chain	T-cell receptor subunit	Jamieson et al., 1989
*TNF receptor, p75/80	High-affinity TNF receptor	Santee & Owen-Schaub, 1996
—Proteins involved in antigen presentation		
Proteasome subunit LMP2	Subunit of 26S proteasome, cysteine protease	Wright et al., 1995
Peptide transporter TAP1	Peptide transporter for ER	Wright et al., 1995
—Cell adhesion molecules		
ELAM-1	E-selectin, endothelial cell leukocyte adhesion molecule	Whelan et al., 1991
ICAM-1	Intracellular adhesion mol. 1	van de Stolpe et al., 1994
MadCAM-1	Mucosal addressin cell adhesion molecule	Takeuchi & Baichwal, 1995
P-selectin	Platelet adhesion receptor	Pan & McEver, 1995
Tenascin-C	ECM protein (controls cell attachment and migration, cell growth)	Mettouchi et al., 1997
VCAM-1	Vascular cell adhesion molecule	Iademarco et al., 1992
—Acute phase proteins		
Angiotensinogen	Angiotensin precursor (regulates blood pressure)	Brasier et al., 1990 Ron et al., 1991

(6.2. cont.)

Genes	*Functions*	*References*
β-Amyloid (Aβ)	Activated glia component	Akama et al., 1998
β-Defensin 2	Antimicrobial peptide	Tsutsumi & Nagaoka, 2002
C4b-binding protein	Complement-binding protein	Moffat & Tack, 1992
Complement factor B	Complement factor	Nonaka & Huang, 1990
Complement factor C4	Activates extrinsic pathway of complement activation	Yu et al., 1989
C-reactive protein	Pentraxin	Zhang et al., 1995
Factor VIII	Coagulation factor	Begbie et al., 2000
LPS-binding protein	Binds to LPS receptor (CD14) with LPS	Schumann, 1995
PTX3	Pentraxin	Basile et al., 1997
Serum amyloid A precursor	Serum component	Edbrooke et al., 1991 Li & Liao, 1991
Serum amyloid A3	Type I acute phase protein	Bing et al., 2000
Tissue factor-1	Activates extrinsic pathway of complement activation	Mackman et al., 1991
Urokinase-type plasminogen activator	Activates fibrinogen for fibrin clot lysis	Novak et al., 1991
—Stress response genes		
Angiotensin II	Peptide hormone	Brasier et al., 1990
COX-2	Cyclooxygenase, prostaglandin endoperoxide synthase	Yamamoto et al., 1995
Ferritin H chain	Iron storage protein	Kwak et al., 1995
*5-Lipoxygenase	Arachidonic acid metabolic enzyme, leukotriene synthesis	Chopra et al., 1992
12-Lipoxygenase	Arach. acid metabolic enzyme	Arakawa et al., 1995
Inducible NO synthase	NO synthesis	Geller et al., 1993
MnSOD	Superoxide dismutase	Das et al., 1995
NAD(P)H quinone oxido-reductase (DT-diaphorase)	Bioreductive enzyme	Yao & O'Dwyer, 1995
Phospholipase A2	Fatty acid metabolism	Morri et al., 1994
—Cell surface receptors		
A1 adenosine receptor	Pleiotropic physiological effects	Nie et al., 1998
Bradykinin B1 receptor	idem	Nie et al., 1998
*CD23	Cell-surface molecule	Tinnell et al., 1998
CD69	Lectin mainly on activated T-cells	Lopez-Cabrera et al., 1995
Gall receptor	Galanine receptor, neuroendocrine peptide	Lorimer et al., 1997
Lox-1	Receptor for oxidized low-density lipoprotein	Nagase et al., 1998
Mdr1	Multiple drug resistance mediator (P glycoprotein)	Zhou & Kuo, 1997

(6.2. cont.) Genes	Functions	References
PAF receptor 1	Platelet activator receptor	Mutoh et al., 1994
Neuropeptide Y-Y1 receptor	Pleiotropic physiological effects	Musso et al., 1997
RAGE receptor	Receptor for advanced glycation end products	Li & Schmidt, 1997
Toll-like receptor 2	Pathogen recognition molecule	Wang et al., 2001
—Regulators of apoptosis or differentiation		
Bfl1/A1	Prosurvival Bcl-2 homologue	Grumont et al., 1999 Zong et al., 1999
Bcl-w		Bottero et al., 2001
Bcl-xL	Prosurvival Bcl-2 homologue	Chen et al., 1999 Lee et al., 1999
Bcl-2	Prosurvival factor	Catz & Johnson, 2001
c-IAP-1, -2, XIAP	Inhibitor of apoptosis	Bottero et al., 2001
Nr13	Prosurvival Bcl-2 homologue	Lee et al., 1999
c-FLIP	Prosurvival factor	Kreuz et al., 2001
CD95 (Fas)	Proapoptotic receptor	Chan et al., 1999
*Fas-associated phosphatase-1	Protein phosphatase	Mizuno et al., 2001
Fas ligand	Inducer of apoptosis	Matsui et al., 1998
IAPs	Inhibitors of apoptosis	You et al., 1997 Stehlik et al., 1998
IEX-1L	Immediate early gene	Wu et al., 1998
Hes1	Basic helix-loop-helix gene	Espinosa et al., 2002
Mcl-1	—	Bottero et al., 2001
Wnt-1	Prosurvival oncogene	Bournat et al., 2000
—Growth factors and their modulators		
G-CSF	Granulocyte colony stimulating factor	Nishizawa & Nagata, 1990
GM-CSF	Granulocyte macrophage colony stimulating factor	Schreck et al., 1990
*IGFBP-1	Insulin-like growth factor-binding protein-1	Lang et al., 1999
IGFBP-2	Insulin-like growth factor-binding protein-2	Cazals et al., 1999
M-CSF (CSF-1)	Macrophage colony stimulating factor	Brach et al., 1991b
PDGF B chain	Platelet-derived growth factor	Khachigian et al., 1995
Proenkephalin	Hormone	Rattner et al., 1991
*Thrombospondin	Matrix glycoprotein t	Adolph et al., 1997
VEGF C	Vascular endothelial growth factor	Chilov et al., 1997

(6.2. cont.) Genes	*Functions*	*References*
—Early response genes		
p22/PRG1	Rat homology of IEX	Schafer et al., 1998
*p62	Nonproteasomal multi-ubiquitin chain-binding protein	Vadlamudi & Shin, 1998
—Transcription factors		
A20	TNF-inducible zinc finger	Krikos et al., 1992
c-myb	Protooncogene	Toth et al., 1995
c-myc	Protooncogene	Duyao et al., 1992
c-rel	Protooncogene	Hannink & Temin, 1990
IRF-1	IFN regulatory factor-1	Harada et al., 1994
IRF-2	IFN regulatory factor-2	Harada et al., 1994
IRF-4	IFN regulatory factor-4	Grumont & Gerondakis, 2000
IκBα	Inhibitor of Rel/NF-κB	Haskill et al., 1991 Sun et al., 1993 de Martin et al., 1993
JunB	Protooncogene	Brown et al., 1995
nfκb2	NF-κB p100 precursor	Lombardi et al., 1995
nfκb1	NF-κB p105 precursor	Ten et al., 1992
p53	Tumor suppressor	Wu & Lozano, 1994
Relb	Transcription factor	Bren et al., 2001
—Viruses		
Adenovirus (E3 region)	Adenovirus	Williams et al., 1990
Avian leukosis virus	Causes avian leukosis	Bowers et al., 1996
Bovine leukemia virus	Causes bovine leukemia	Brooks et al., 1998
CMV	Cytomegalovirus	Sambucetti et al., 1989
EBV (Wp promoter)	Epstein-Barr virus	Sugano et al., 1997
HIV-1	Human immunodeficiency virus	Nabel & Baltimore, 1987 Griffin et al., 1989
HSV	Herpes simplex virus	Rong et al., 1992
JC virus	Polyomavirus	Ranganathan & Khalili, 1993
HPV type 16	Human papillomavirus	Fontaine et al., 2000
SIV	Simian immunodef. virus	Bellas et al., 1993
SV40	Simian virus 40	Kanno et al., 1989
—Enzymes		
*Ceramide glycosyl-transferase	Glycosphingolipid	Ichikawa et al., 1998
Collagenase 1	Matrix metalloproteinase	Vincenti et al., 1998
*Dihydrodiol-dehydrogenase	Oxidoreductase, oxidation of trans-hydrodiols	Ciaccio et al., 1996
*GAD67	Glutamic acid decarboxylase	Szabo et al., 1996
Gelatinase B	Matrix metalloproteinase	He, 1996
GSTP1-1	Glutathione transferase	Xia et al., 1996
*Glucose-6-phosphate-dehydrogenase	Hexose monophosphate	Garcia-Nogales et al., 1999
*HO-1	Heme oxygenase	Lavrovsky et al., 1994

(6.2. cont.) Genes	*Functions*	*References*
Glutamate-cysteine ligase (modifier)	Enzyme in glutathione (GSH) synthesis	Yang et al., 2001a-b
Hyaluronan synthase	Synthesizes hyaluronic acid	Ohkawa et al., 1999
Lysozyme	Hydrolyzes bacterial cell walls	Phi, 1996
MMP-9, matrix metallo-proteinase-9	Secreted gelatinase involved in metastasis	Bond et al., 1998 Farina et al., 1999
PIM-1	Ser/Thr kinase	Zhu et al., 2002
Protein kinase Cδ (PKCδ)	Calcium-independent PKC isozyme	Kurkinen et al., 2000
*PTGIS	Prostaglandin synthase	Yokoyama et al., 1996
Transglutaminase	Forms isopeptide bonds	Mirza et al., 1997
Type II-secreted phospholipase A_2	Proinflammatory	Couturier et al., 1999
*Xanthine oxidase	Oxidative metabolism of purines	Xu et al., 1996
—Miscellanea		
α_1-Acid glycoprotein	Serum protein	Mejdoubi et al., 1999
α_1-Antitrypsin	Protease inhibitor	Ray et al., 1995
Apolipoprotein C III	Apoprotein of HDL	Gruber et al., 1994
*Biglycan	Connective tissue proteoglycan	Ungefroren & Krull, 1996
BRCA2	Tumor suppressor gene	Wu et al., 2000
Cyclin D1	Cell-cycle regulation	Guttridge et al., 1999 Hinz et al., 1999
*Cyclin D3	Cell-cycle regulation	Wang et al., 1996
Fibronectin	Extracellular matrix	Ritchie et al., 2000
Gadd45β	DNA repair/cell cycle	De Smaele et al., 2001
Galectin 3	β-Galactosidase-binding lectin	Hsu et al., 1996
HMG14	High-mobility group 14	Walker & Enrietto, 1996
Human serotonin transporter	—	Flattem & Blakely, 2000
K3 keratin	Intermediate filament protein	Wu et al., 1994
K6 keratin	Cytoskeletal protein	Komine et al., 2000
Laminin B2 chain	Basement membrane protein	Richardson et al., 1995
Mts1	Multiple tumor suppressor	Tulchinsky et al., 1997
μ-Opioid receptor	Analgesic, euphoric and immune effects	Borner et al., 2002
*Pax8	Paired box gene	Okladnova et al., 1997
Pregnancy-specific glycoprotein rnCGM3	Placental expression	Wang et al., 2001
Prostate-specific antigen	Serum protein prostate cancer	Chen & Sawyers, 2002
Smad7	Inhibitor of TGFβ	Patil et al., 2000
Sox9	Essential for chondrogenesis	Murakami et al., 2000
*UCP-2	Uncoupling protein-2	Lee et al., 1999
Vimentin	Intermediate filament protein	Lilienbaum et al., 1990

(6.2. cont.) *Genes*	*Functions*	*References*
von Willebrand factor	—	Keightley et al., 1999
Wilm's tumor suppressor gene	Tumor suppressor	Dehbi et al., 1998

7. REFERENCES

Abraham E, Carmody A, Shenkar R & Arcaroli J (2000) Neutrophils as early immunologic effectors in hemorrhage- or endotoxemia-induced acute lung injury. Am. J. Physiol. Lung Cell. Mol. Physiol. 279: L1137-L1145.

Adcock IM, Brown CR, Kwon O & Barnes PJ (1994) Oxidative stress induces NFκB DNA binding and inducible NOS mRNA in human epithelial cells. Biochem. Biophys. Res. Commun. 199: 1518-1524.

Adib-Conquy M, Adrie C, Moine P, Asehnoune K, Fitting C, Pinsky MR, Dhainaut JF & Cavaillon JM (2000) NF-κB expression in mononuclear cells of patients with sepsis resembles that observed in lipopolysaccharide tolerance. Am. J. Respir. Crit. Care Med. 162: 1877-1883.

Adolph KW, Liska DJ & Bornstein P (1997) Analysis of the promoter and transcription start sites of the human thrombospondin 2 gene (THBS2). Gene 193: 5-11.

Akama KT, Albanese C, Pestell RG & Van Eldik LJ (1998) Amyloid β-peptide stimulates nitric oxide production in astrocytes through an NFκB-dependent mechanism. Proc. Natl. Acad. Sci. USA 95: 5795-5800.

Al-Harthi L, Spear GT, Hashemi FB, Landay A, Sha BE & Roebuck KA (1998) A human immunodeficiency virus (HIV)-inducing factor from the female genital tract activates HIV-1 gene expression through the κB enhancer. J. Infect. Dis. 178: 1343-1351.

Andalibi A, Liao F, Imes S, Fogelman AM & Lusis AJ (1993) Oxidized lipoproteins influence gene expression by causing oxidative stress and activating the transcription factor NF-κB. Biochem. Soc. Trans. 21: 651-655.

Andrieu N, Salvayre R, Jaffrezou JP & Levade T (1995) Low temperatures and hypertonicity do not block cytokine-induced stimulation of the sphingomyelin pathway but inhibit nuclear factor-κB activation. J. Biol. Chem. 270: 24518-24524.

Anisowicz A, Messineo M, Lee SW & Sager R (1991) An NF-κB-like transcription factor mediates IL-1/TNF-α induction of gro in human fibroblasts. J. Immunol. 147: 520-527.

Arakawa T, Nakamura M, Yoshimoto T & Yamamoto S (1995) The transcriptional regulation of human arachidonate 12-lipoxygenase gene by NFκB/Rel. FEBS Lett. 363: 105-110.

Arnold R, Liou J, Drexler HC, Weiss A & Kiefer F (2001) Caspase-mediated cleavage of hematopoietic progenitor kinase 1 (HPK1) converts an activator of NFκB into an inhibitor of NFκB. J. Biol. Chem. 276: 14675-14684.

Asanuma M & Cadet JL (1998) Methamphetamine-induced increase in striatal NF-κB DNA-binding activity is attenuated in superoxide dismutase transgenic mice. Brain Res. Mol. Brain Res. 60: 305-309.

Aupperle K, Bennett B, Han Z, Boyle D, Manning A & Firestein G (2001) NF-κB regulation by IκB kinase-2 in rheumatoid arthritis synoviocytes. J. Immunol. 166: 2705-2711.

Azimi N, Brown K, Bamford RN, Tagaya Y, Siebenlist U & Waldmann TA (1998) Human T cell lymphotropic virus type I Tax protein trans-activates interleukin 15 gene transcription through an NF-κB site. Proc. Natl. Acad. Sci. USA 95: 2452-2457.

Bachelerie F, Alcami J, Arenzana-Seisdedos F & Virelizier JL (1991) HIV enhancer activity perpetuated by NF-κB induction on infection of monocytes. Nature 350: 709-712.

Baghdiguian S, Martin M, Richard I, Pons F, Astier C, Bourg N, Hay RT, Chemaly R, Halaby G, Loiselet J, Anderson LV, Lopez dM, Fardeau M, Mangeat P, Beckmann JS & Lefranc G (1999) Calpain 3 deficiency is associated with myonuclear apoptosis and profound perturbation of the

IκBα/NF-κB pathway in limb-girdle muscular dystrophy type 2A. Nat. Med. 5: 503-511.

Baghdiguian S, Richard I, Martin M, Coopman P, Beckmann JS, Mangeat P & Lefranc G (2001) Pathophysiology of limb girdle muscular dystrophy type 2A: hypothesis and new insights into the IκBα/NF-κB survival pathway in skeletal muscle. J. Mol. Med. 79: 254-261.

Baldwin AS, Azizkhan JC, Jensen DE, Beg AA & Coodly LR (1991) Induction of NF-κB DNA-binding activity during the G0-to-G1 transition in mouse fibroblasts. Mol. Cell. Biol. 11: 4943-4951.

Ballard DW, Bohnlein E, Lowenthal JW, Wano Y, Franza BR & Greene WC (1988) HTLV-I tax induces cellular proteins that activate the κB element in the IL-2 receptor α gene. Science 241: 1652-1655.

Barnes PJ (1996) Pathophysiology of asthma. Br. J. Clin. Pharmacol. 42: 3-10.

Basile A, Sica A, d'Aniello E, Breviario F, Garrido G, Castellano M, Mantovani A & Introna M (1997) Characterization of the promoter for the human long pentraxin PTX3. Role of NF-κB in tumor necrosis factor-α and interleukin-1β regulation. J. Biol. Chem. 272: 8172-8178.

Battle J, Ha T, Li C, Della B, V, Rice P, Kalbfleisch J, Browder W & Williams D (1998) Ligand binding to the (1→3)-β-D-glucan receptor stimulates NFκB activation, but not apoptosis in U937 cells. Biochem. Biophys. Res. Commun. 249: 499-504.

Begbie M, Notley C, Tinlin S, Sawyer L & Lillicrap D (2000) The factor VIII acute phase response requires the participation of NFκB and C/EBP. Thromb. Haemost. 84: 216-222.

Behl C, Davis JB, Lesley R & Schubert D (1994) Hydrogen peroxide mediates amyloid β protein toxicity. Cell 77: 817-827.

Belcher CE, Drenkow J, Kehoe B, Gingeras TR, McNamara N, Lemjabbar H, Basbaum C & Relman DA (2000) From the cover: the transcriptional responses of respiratory epithelial cells to Bordetella pertussis reveal host defensive and pathogen counter-defensive strategies. Proc. Natl. Acad. Sci. USA 97: 13847-13852.

Bellas RE, Hopkins N & Li Y (1993) The NF-κB binding site is necessary for efficient replication of simian immunodeficiency virus of macaques in primary macrophages but not in T cells in vitro. J. Virol. 67: 2908-29013.

Berberich I, Shu GL & Clark EA (1994) Cross-linking CD40 on B cells rapidly activates nuclear factor-κB. J. Immunol. 153: 4357-4366.

Bertrand F, Philippe C, Antoine PJ, Baud L, Groyer A, Capeau J & Cherqui G (1995) Insulin activates nuclear factor κB in mammalian cells through a Raf-1-mediated pathway. J. Biol. Chem. 270: 24435-24441.

Bertrand F, Atfi A, Cadoret A, L'Allemain G, Robin H, La cols O, Capeau J & Cherqui G (1998) A role for nuclear factor κB in the antiapoptotic function of insulin. J. Biol. Chem. 273: 2931-2938.

Bessho R, Matsubara K, Kubota M, Kuwakado K, Hirota H, Wakazono Y, Lin YW, Okuda A, Kawai M & Nishikomori R (1994) Pyrrolidine dithiocarbamate, a potent inhibitor of nuclear factor κB (NF-κB) activation, prevents apoptosis in human promyelocytic leukemia HL-60 cells and thymocytes. Biochem. Pharmacol. 48: 1883-1889.

Bing Z, Huang JH & Liao WS (2000) NFκB interacts with serum amyloid A3 enhancer factor to synergistically activate mouse serum amyloid A3 gene transcription. J. Biol. Chem. 275: 31616-31623.

Biswas DK, Cruz AP, Gansberger E & Pardee AB (2000) Epidermal growth factor-induced nuclear factor κB activation: A major pathway of cell-cycle progression in estrogen-receptor negative breast cancer cells. Proc. Natl. Acad. Sci. USA 98: 8542-8547.

Bitko V, Velazquez A, Yang L, Yang YC & Barik S (1997) Transcriptional induction of multiple cytokines by human respiratory syncytial virus requires activation of NF-κB and is inhibited by sodium salicylate and aspirin. Virology 232: 369-378.

Blackwell TS, Holden EP, Blackwell TR, DeLarco JE & Christman JW (1994) Cytokine-induced neutrophil chemoattractant mediates neutrophilic alveolitis in rats: association with nuclear factor κB activation. Am. J. Respir. Cell. Mol. Biol. 11: 464-472.

Blackwell TS, Blackwell TR, Holden EP, Christman BW & Christman JW (1996) In vivo antioxidant treatment suppresses nuclear factor-κB activation and neutrophilic lung inflammation. J. Immunol.

157: 1630-1637.
Blackwell TS, Blackwell TR & Christman JW (1997) Impaired activation of nuclear factor-κB in endotoxin-tolerant rats is associated with down-regulation of chemokine gene expression and inhibition of neutrophilic lung inflammation. J. Immunol. 158: 5934-5940.
Blondeau N, Widmann C, Lazdunski M & Heurteaux C (2001) Activation of the nuclear factor-κB is a key event in brain tolerance. J. Neurosci. 21: 4668-4677.
Bohrer H, Qiu F, Zimmermann T, Zhang Y, Jllmer T, Mannel D, Bottiger BW, Stern DM, Waldherr R, Saeger HD, Ziegler R, Bierhaus A, Martin E & Nawroth PP (1997) Role of NFκB in the mortality of sepsis. J. Clin. Invest. 100: 972-985.
Boland MP, Fitzgerald KA & O'Neill LA (2000) Topoisomerase II is required for mitoxantrone to signal nuclear factor κB activation in HL60 cells. J. Biol. Chem. 275: 25231-25238.
Bond M, Fabunmi RP, Baker AH & Newby AC (1998) Synergistic upregulation of metalloproteinase-9 by growth factors and inflammatory cytokines: an absolute requirement for transcription factor NF-κB. FEBS Lett. 435: 29-34.
Booth FW, Lou W, Hamilton MT & Yan Z (1996) Cytochrome c mRNA in skeletal muscles of immobilized limbs. J. Appl. Physiol. 81: 1941-1945.
Borner C, Hollt V & Kraus J (2002) Involvement of activator protein-1 in transcriptional regulation of the human mu-opioid receptor gene. Mol. Pharmacol. 61: 800-805.
Bottero V, Rossi F, Samson M, Mari M, Hofman P & Peyron JF (2001) IκB-α, the NF-κB inhibitory subunit, interacts with ANT, the mitochondrial ATP/ADP translocator. J. Biol. Chem. 276: 21317-21324.
Bournat JC, Brown AM & Soler AP (2000) Wnt-1 dependent activation of the survival factor NF-κB in PC12 cells. J. Neurosci. Res. 61: 21-32.
Bowers WJ, Baglia LA & Ruddel A (1996) Regulation of avian leukosis virus long terminal repeat-enhanced transcription by C/EBP-Rel interactions. J. Virol. 70: 3051-3059.
Brach MA, Hass R, Sherman ML, Gunji H, Weichselbaum R & Kufe D (1991a) Ionizing radiation induces expression and binding activity of the nuclear factor κB. J. Clin. Invest. 88: 691-695.
Brach MA, Henschler R, Mertelsmann RH & Herrmann F (1991b) Regulation of M-CSF expression by M-CSF: role of protein kinase C and transcription factor NFκB. Pathobiology 59: 284-288.
Brach MA, de Vos S, Arnold C, Gruss HJ, Mertelsmann R & Herrmann F (1992) Leukotriene B4 transcriptionally activates interleukin-6 expression involving NK-chi B and NF-IL6. Eur. J. Immunol. 22: 2705-2711.
Brasier AR, Ron D, Tate JE & Habener JF (1990) A family of constitutive C/EBP-like DNA binding proteins attenuate the IL-1α induced, NFκB mediated trans-activation of the angiotensinogen gene acute-phase response element. EMBO J. 9: 3933-3944.
Braun J, Brandt J, Listing J, Zink A, Alten R, Golder W, Gromnica-Ihle E, Kellner H, Krause A, Schneider M, Sorensen H, Zeidler H, Thriene W & Sieper J (2002) Treatment of active ankylosing spondylitis with infliximab: a randomised controlled multicentre trial. Lancet 359: 1187-1193.
Bren GD, Solan NJ, Miyoshi H, Pennington KN, Pobst LJ & Paya CV (2001) Transcription of the RelB gene is regulated by NF-κB. Oncogene 20: 7722-7733.
Bressler P, Pantaleo G, Demaria A & Fauci AS (1991) Anti-CD2 receptor antibodies activate the HIV long terminal repeat in T lymphocytes. J. Immunol. 147: 2290-2294.
Bretschneider E, Kaufmann R, Braun M, Wittpoth M, Glusa E, Nowak G & Schror K (1999) Evidence for proteinase-activated receptor-2 (PAR-2)-mediated mitogenesis in coronary artery smooth muscle cells. Br. J. Pharmacol. 126: 1735-1740.
Brooks PA, Cockerell GL & Nyborg JK (1998) Activation of BLV transcription by NF-κB and Tax. Virology 243: 94-98.
Brown RT, Ades IZ & Nordan RP (1995) An acute phase response factor/NF-κB site downstream of the junB gene that mediates responsiveness to interleukin-6 in a murine plasmacytoma. J. Biol. Chem. 270: 31129-31135.
Browning DD, Pan ZK, Prossnitz ER & Ye RD (1997) Cell type- and developmental stage-specific activation of NF-κB by fMet-Leu-Phe in myeloid cells. J. Biol. Chem. 272: 7995-8001.
Bureau F, Delhalle S, Bonizzi G, Fievez L, Dogne S, Kirschvink N, Vanderplasschen A, Merville

MP, Bours V & Lekeux P (2000) Mechanisms of persistent NF-κB activity in the bronchi of an animal model of asthma. J. Immunol. 165: 5822-5830.
Busam K, Gieringer C, Freudenberg M & Hohmann HP (1992) Staphylococcus aureus and derived exotoxins induce nuclear factor κB-like activity in murine bone marrow macrophages. Infect. Immun. 60: 2008-2015.
Carter BD, Kaltschmidt C, Kaltschmidt B, Offenhauser N, Bohm-Matthaei R, Baeuerle PA & Barde YA (1996) Selective activation of NF-κB by nerve growth factor through the neurotrophin receptor p75. Science 272: 542-545.
Catz SD & Johnson JL (2001) Transcriptional regulation of bcl-2 by nuclear factor κB and its significance in prostate cancer. Oncogene 20: 7342-7351.
Cazals V, Nabeyrat E, Corroyer S, de Keyzer Y & Clement A (1999) Role for NF-κB in mediating the effects of hyperoxia on IGF-binding protein 2 promoter activity in lung alveolar epithelial cells. Biochim. Biophys. Acta 1448: 349-362.
Chabot-Fletcher M & Breton JJ (1998) Effect of staurosporine on transcription factor NF-κB in human keratinocytes. Biochem. Pharmacol. 56: 71-78.
Chan H, Bartos DP & Owen-Schaub LB (1999) Activation-dependent transcriptional regulation of the human Fas promoter requires NF-κB p50-p65 recruitment. Mol. Cell. Biol. 19: 2098-2108.
Chen C, Edelstein LC & Gelinas C (2000) The Rel/NF-κB family directly activates expression of the apoptosis inhibitor Bcl-x_L. Mol. Cell. Biol. 20: 2687-2695.
Chen CD & Sawyers CL (2002) NF-κB activates prostate-specific antigen expression and is upregulated in androgen-independent prostate cancer. Mol. Cell. Biol. 22: 2862-2870.
Chen F, Sun SC, Kuh DC, Gaydos LJ & Demers LM (1995) Essential role of NF-κB activation in silica-induced inflammatory mediator production in macrophages. Biochem. Biophys. Res. Commun. 214: 985-992.
Chen F, Demers LM, Vallyathan V, Lu Y, Castranova V & Shi X (1999) Involvement of 5'-flanking κB-like sites within bcl-x gene in silica-induced Bcl-x expression. J. Biol. Chem. 274: 35591-35595.
Chen J, Raj N, Kim P, Andrejko KM & Deutschman CS (2001) Intrahepatic nuclear factor-κB activity and α1-acid glycoprotein transcription do not predict outcome after cecal ligation and puncture in the rat. Crit. Care Med. 29: 589-596.
Chen X, Ji B, Han B, Ernst SA, Simeone D & Logsdon CD (2002) NF-κB activation in pancreas induces pancreatic and systemic inflammatory response. Gastroenterology 122: 448-457.
Chilov D, Kukk E, Taira S, Jeltsch M, Kaukonen J, Palotie A, Joukov V & Alitalo K (1997) Genomic organization of human and mouse genes for vascular endothelial growth factor C. J. Biol. Chem. 272: 25176-25183.
Ching LM, Young HA, Eberly K & Yu CR (1999) Induction of STAT and NFκB activation by the antitumor agents 5,6-dimethylxanthenone-4-acetic acid and flavone acetic acid in a murine macrophage cell line. Biochem. Pharmacol. 58: 1173-1181.
Chirmule N, Kalyanaraman VS & Pahwa S (1994) Signals transduced through the CD4 molecule on T lymphocytes activate NF-κB. Biochem. Biophys. Res. Commun. 203: 498-505.
Chopra A, Ferreira-Alves DL, Sirois P & Thirion JP (1992) Cloning of the guinea pig 5-lipoxygenase gene and nucleotide sequence of its promoter. Biochem. Biophys. Res. Commun. 185: 489-495.
Ciaccio PJ, Walsh ES & Tew KD (1996) Promoter analysis of a human dihydrodiol dehydrogenase. Biochem. Biophys. Res Commun. 228: 524-529.
Collart MA, Baeuerle P & Vassalli P (1990) Regulation of tumor necrosis factor α transcription in macrophages: involvement of four κB-like motifs and of constitutive and inducible forms of NF-κB. Mol. Cell. Biol. 10: 1498-1506.
Cornelussen RN, Gupta S & Knowlton AA (2001) Regulation of prostaglandin A1-induced heat shock protein expression in isolated cardiomyocytes. J. Mol. Cell. Cardiol. 33: 1447-1454.
Courtois G, Whiteside ST, Sibley CH & Israel A (1997) Characterization of a mutant cell line that does not activate NF-κB in response to multiple stimuli. Mol. Cell. Biol. 17: 1441-1449.
Couturier C, Brouillet A, Couriaud C, Koumanov K, Bereziat G & Andreani M (1999) Interleukin 1β induces type II-secreted phospholipase A_2 gene in vascular smooth muscle cells by a nuclear factor

κB and peroxisome proliferator-activated receptor-mediated process. J. Biol. Chem. 274: 23085-23093.

Cressman DE, Greenbaum LE, Haber BA & Taub R (1994) Rapid activation of post-hepatectomy factor/nuclear factor κB in hepatocytes, a primary response in the regenerating liver. J. Biol. Chem. 269: 30429-30435.

Das J, Chen CH, Yang L, Cohn L, Ray P & Ray A (2001) A critical role for NF-κB in GATA3 expression and TH2 differentiation in allergic airway inflammation. Nat. Immunol. 2: 45-50.

Das KC, Lewis-Molock Y & White CW (1995) Activation of NF-κB and elevation of MnSOD gene expression by thiol reducing agents in lung adenocarcinoma (A549) cells. Am. J. Physiol. 269: L588-L602.

Das KC & White CW (1997) Activation of NF-κB by antineoplastic agents. Role of protein kinase C. J. Biol. Chem. 272: 14914-14920.

de Martin R, Vanhove B, Cheng Q, Hofer E, Csizmadia V, Winkler H & Bach FH (1993) Cytokine-inducible expression in endothelial cells of an IκBα-like gene is regulated by NFκB. EMBO J. 12: 2773-2779.

De Smaele E, Zazzeroni F, Papa S, Nguyen DU, Jin R, Jones J, Cong R & Franzoso G (2001) Induction of gadd45β by NF-κB downregulates pro-apoptotic JNK signalling. Nature 414: 308-313.

Degli-Esposti MA, Dougall WC, Smolak PJ, Waugh JY, Smith CA & Goodwin RG (1997) The novel receptor TRAIL-R4 induces NF-κB and protects against TRAIL-mediated apoptosis, yet retains an incomplete death domain. Immunity 7: 813-820.

Dehbi M, Hiscott J & Pelletier J (1998) Activation of the wt1 Wilms' tumor suppressor gene by NF-κB. Oncogene 16: 2033-2039.

Delfino F & Walker WH (1998) Stage-specific nuclear expression of NF-κB in mammalian testis. Mol. Endocrinol. 12: 1696-1707.

DiMango E, Ratner AJ, Bryan R, Tabibi S & Prince A (1998) Activation of NF-κB by adherent Pseudomonas aeruginosa in normal and cystic fibrosis respiratory epithelial cells. J. Clin. Invest. 101: 2598-2605.

Dokter WH, Dijkstra AJ, Koopmans SB, Stulp BK, Keck W, Halie MR & Vellenga E (1994) G(Anh)MTetra, a natural bacterial cell wall breakdown product, induces interleukin-1β and interleukin-6 expression in human monocytes. A study of the molecular mechanisms involved in inflammatory cytokine expression. J. Biol. Chem. 269: 4201-4206.

Donda A, Mori L, Shamshiev A, Carena I, Mottet C, Heim MH, Beglinger C, Grunert F, Rochlitz C, Terracciano L, Jantscheff P & De Libero G (2000) Locally inducible CD66a (CEACAM1) as an amplifier of the human intestinal T cell response. Eur. J. Immunol. 30: 2593-2603.

Donovan CE, Mark DA, He HZ, Liou HC, Kobzik L, Wang Y, De Sanctis GT, Perkins DL & Finn PW (1999) NF-κB/Rel transcription factors: c-Rel promotes airway hyperresponsiveness and allergic pulmonary inflammation. J. Immunol. 163: 6827-6833.

Drouet C, Shakhov AN & Jongeneel CV (1991) Enhancers and transcription factors controlling the inducibility of the tumor necrosis factor-α promoter in primary macrophages. J. Immunol. 147: 1694-1700.

Duyao MP, Kessler DJ, Spicer DB, Bartholomew C, Cleveland JL, Siekevitz M & Sonenshein GE (1992) Transactivation of the c-myc promoter by human T cell leukemia virus type 1 tax is mediated by NFκB. J. Biol. Chem. 267: 16288-16291.

Dyer RB, Collaco CR, Niesel DW & Herzog NK (1993) Shigella flexneri invasion of HeLa cells induces NF-κB DNA-binding activity. Infect. Immun. 61: 4427-4433.

Eaves-Pyles T, Szabo C & Salzman AL (1999) Bacterial invasion is not required for activation of NF-κB in enterocytes. Infect. Immun. 67: 800-804.

Edbrooke MR, Foldi J, Cheshire JK, Li F, Faulkes DJ & Woo P (1991) Constitutive and NF-κB-like proteins in the regulation of the serum amyloid A gene by interleukin 1. Cytokine 3: 380-388.

Elewaut D, DiDonato JA, Kim JM, Truong F, Eckmann L & Kagnoff MF (1999) NF-κB is a central regulator of the intestinal epithelial cell innate immune response induced by infection with enteroinvasive bacteria. J. Immunol. 163: 1457-1466.

Ensoli B, Lusso P, Schachter F, Josephs SF, Rappaport J, Negro F, Gallo RC & Wong-Staal F

(1989) Human herpes virus-6 increases HIV-1 expression in co-infected T cells via nuclear factors binding to the HIV-1 enhancer. EMBO J. 8: 3019-3037.

Espandiari P, Ludewig G, Glauert HP & Robertson LW (1998) Activation of hepatic NF-κB by the herbicide Dicamba (2-methoxy-3,6-dichlorobenzoic acid) in female and male rats. J. Biochem. Mol. Toxicol. 12: 339-344.

Espinosa L, Santos S, Ingles-Esteve J, Munoz-Canoves P & Bigas A (2002) p65-NFκB synergizes with Notch to activate transcription by triggering cytoplasmic translocation of the nuclear receptor corepressor N-CoR. J. Cell. Sci. 115: 1295-1303.

Farina AR, Tacconelli A, Vacca A, Maroder M, Gulino A & Mackay AR (1999) Transcriptional up-regulation of matrix metalloproteinase-9 expression during spontaneous epithelial to neuroblast phenotype conversion by SK-N-SH neuroblastoma cells, involved in enhanced invasivity, depends upon GT-box and nuclear factor κB elements. Cell. Growth Differ. 10: 353-367.

Feldmann M, Brennan FM & Maini RN (1996) Role of cytokines in rheumatoid arthritis. Annu. Rev. Immunol. 14: 397-440.

Ferlini C, Scambia G, Marone M, Distefano M, Gaggini C, Ferrandina G, Fattorossi A, Isola G, Benedetti PP & Mancuso S (1999) Tamoxifen induces oxidative stress and apoptosis in oestrogen receptor-negative human cancer cell lines. Br. J. Cancer 79: 257-263.

Flattem NL & Blakely RD (2000) Modified structure of the human serotonin transporter promoter. Mol. Psychiatry 5: 110-115.

Fong TC, Wu Y & Kipps TJ (1996) Identification of a promoter element that regulates tissue-specific expression of the human CD80 (B7.1) gene. J. Immunol. 157: 4442-4450.

Fontaine V, van der Meijden E, de Graaf J, ter Schegget J & Struyk L (2000) A functional NF-κB binding site in the human papillomavirus type 16 long control region. Virology 272: 40-49.

Gabriel C, Justicia C, Camins A & Planas AM (1999) Activation of nuclear factor-κB in the rat brain after transient focal ischemia. Brain Res. Mol. Brain Res. 65: 61-69.

Garcia-Nogales P, Almeida A, Fernandez E, Medina JM & Bolanos JP (1999) Induction of glucose-6-phosphate dehydrogenase by lipopolysaccharide contributes to preventing nitric oxide-mediated glutathione depletion in cultured rat astrocytes. J. Neurochem. 72: 1750-1758.

Garcia J, Lemercier B, Roman-Roman S & Rawadi G (1998) A Mycoplasma fermentans-derived synthetic lipopeptide induces AP-1 and NF-κB activity and cytokine secretion in macrophages via the activation of mitogen-activated protein kinase pathways. J. Biol. Chem. 273: 34391-34398.

Garofalo R, Sabry M, Jamaluddin M, Yu RK, Casola A, Ogra PL & Brasier AR (1996) Transcriptional activation of the interleukin-8 gene by respiratory syncytial virus infection in alveolar epithelial cells: nuclear translocation of the RelA transcription factor as a mechanism producing airway mucosal inflammation. J. Virol. 70: 8773-8781.

Geller DA, Nussler AK, Di Silvio M, Lowenstein CJ, Shapiro RA, Wang SC, Simmons RL & Billiar TR (1993) Cytokines, endotoxin, and glucocorticoids regulate the expression of inducible nitric oxide synthase in hepatocytes. Proc. Natl. Acad. Sci. USA 90: 522-526.

Gewirtz AT, Rao AS, Simon POJ, Merlin D, Carnes D, Madara JL & Neish AS (2000) Salmonella typhimurium induces epithelial IL-8 expression via Ca^{2+}-mediated activation of the NF-κB pathway. J. Clin. Invest. 105: 79-92.

Ghio AJ, Marshall BC, Diaz JL, Hasegawa T, Samuelson W, Povia D, Kennedy TP & Piantodosi CA (1996) Tyloxapol inhibits NF-κB and cytokine release, scavenges HOCI, and reduces viscosity of cystic fibrosis sputum. Am. J. Resp. Crit. Care Med. 154: 783-788.

Gimble JM, Duh E, Ostrove JM, Gendelman HE, Max EE & Rabson AB (1988) Activation of the human immunodeficiency virus long terminal repeat by herpes simplex virus type 1 is associated with induction of a nuclear factor that binds to the NF-κB/core enhancer sequence. J. Virol. 62: 4104-4112.

Glibetic M, Samlalsingh-Parker J, Raykova V, Ofenstein J & Aranda JV (2001) Group B Streptococci and inducible nitric oxide synthase: modulation by nuclear factor κB and ibuprofen. Sem. Perinatol. 25: 65-69.

Goebeler M, Roth J, Brocker EB, Sorg C & Schulze-Osthoff K (1995) Activation of nuclear factor-κB and gene expression in human endothelial cells by the common haptens nickel and cobalt. J.

Immunol. 155: 2459-2467.
Goodman RB, Strieter RM, Martin DP, Steinberg KP, Milberg JA, Maunder RJ, Kunkel SL, Walz A, Hudson LD & Martin TR (1996) Inflammatory cytokines in patients with persistence of the acute respiratory distress syndrome. Am. J. Respir. Crit. Care Med. 154: 602-611.
Gorman JD, Sack KE & Davis JC (2002) Treatment of ankylosing spondylitis by inhibition of tumor necrosis factor α. N. Engl. J. Med. 346: 1349-1356.
Grady T, Liang P, Ernst SA & Logsdon CD (1997) Chemokine gene expression in rat pancreatic acinar cells is an early event associated with acute pancreatitis. Gastroenterology 113: 1966-1975.
Grdina DJ, Murley JS, Kataoka Y & Calvin DP (2002) Differential activation of nuclear transcription factor κB, gene expression, and proteins by amifostine's free thiol in human microvascular endothelial and glioma cells. Semin. Radiat. Oncol. 12: 103-111.
Griffin GE, Leung K, Folks TM, Kunkel S & Nabel GJ (1989) Activation of HIV gene expression during monocyte differentiation by induction of NF-κB. Nature 339: 70-73.
Grohmann U, Belladonna ML, Bianchi R, Orabona C, Ayroldi E, Fioretti MC & Puccetti P (1998) IL-12 acts directly on DC to promote nuclear localization of NF-κB and primes DC for IL-12 production. Immunity 9: 315-323.
Grove M & Plumb M (1993) C/EBP, NF-κB, and c-Ets family members and transcriptional regulation of the cell-specific and inducible macrophage inflammatory protein 1α immediate-early gene. Mol. Cell. Biol. 13: 5276-5289.
Gruber PJ, Torres-Rosado A, Wolak ML & Leff T (1994) Apo CIII gene transcription is regulated by a cytokine inducible NF-κB element. Nucleic Acids Res. 22: 2417-2422.
Grumont RJ, Rourke IJ & Gerondakis S (1999) Rel-dependent induction of A1 transcription is required to protect B cells from antigen receptor ligation-induced apoptosis. Genes Dev. 13: 400-411.
Grumont RJ & Gerondakis S (2000) Rel induces interferon regulatory factor 4 (IRF-4) expression in lymphocytes: modulation of interferon-regulated gene expression by rel/nuclear factor κB. J. Exp. Med. 191: 1281-1292.
Grundker C, Schulz K, Gunthert AR & Emons G (2000) Luteinizing hormone-releasing hormone induces nuclear factor κB-activation and inhibits apoptosis in ovarian cancer cells. J. Clin. Endocrinol. Metab. 85: 3815-3820.
Guerrini L, Blasi F & Denis-Donini S (1995) Synaptic activation of NF-κB by glutamate in cerebellar granule neurons in vitro. Proc. Natl. Acad. Sci. USA 92: 9077-9081.
Gukovskaya AS, Gukovsky I, Zaninovic V, Song M, Sandoval D, Gukovsky S & Pandol SJ (1997) Pancreatic acinar cells produce, release, and respond to tumor necrosis factor-α. Role in regulating cell death and pancreatitis. J. Clin. Invest. 100: 1853-1862.
Gukovsky I, Gukovskaya AS, Blinman TA, Zaninovic V & Pandol SJ (1998) Early NF-κB activation is associated with hormone-induced pancreatitis. Am. J. Physiol. 275: G1402-G1414.
Guttridge DC, Albanese C, Reuther JY, Pestell RG & Baldwin AS (1999) NF-κB controls cell growth and differentiation through transcriptional regulation of cyclin D1. Mol. Cell. Biol. 19: 5785-5799.
Haas AF, Wong JW, Iwahashi CK, Halliwell B, Cross CE & Davis PA (1998) Redox regulation of wound healing? NF-κB activation in cultured human keratinocytes upon wounding and the effect of low energy HeNe irradiation. Free Radic. Biol. Med. 25: 998-1005.
Haddad EB, Salmon M, Koto H, Barnes PJ, Adcock I & Chung KF (1996) Ozone induction of cytokine-induced neutrophil chemoattractant (CINC) and nuclear factor-κB in rat lung: inhibition by corticosteroids. FEBS Lett. 379: 265-268.
Hammarskjold ML & Simurda MC (1992) Epstein-Barr virus latent membrane protein transactivates the human immunodeficiency virus type 1 long terminal repeat through induction of NF-κB activity. J. Virol. 66: 6496-6501.
Hannink M & Temin HM (1990) Structure and autoregulation of the c-rel promoter. Oncogene 5: 1843-1850.
Harada H, Takahashi E, Itoh S, Harada K, Hori TA & Taniguchi T (1994) Structure and regulation of the human interferon regulatory factor 1 (IRF-1) and IRF-2 genes: implications for a gene network in the interferon system. Mol. Cell. Biol. 14: 1500-1509.

Harcourt BH, Rota PA, Hummel KB, Bellini WJ & Offermann MK (1999) Induction of intercellular adhesion molecule 1 gene expression by measles virus in human umbilical vein endothelial cells. J. Med. Virol. 57: 9-16.

Hart LA, Krishnan VL, Adcock IM, Barnes PJ & Chung KF (1998) Activation and localization of transcription factor, nuclear factor-κB, in asthma. Am. J. Respir. Crit. Care Med. 158: 1585-1592.

Haskill S, Beg AA, Tompkins SM, Morris JS, Yurochko AD, Sampson-Johannes A, Mondal K, Ralph P & Baldwin AS (1991) Characterization of an immediate-early gene induced in adherent monocytes that encodes IκB-like activity. Cell 65: 1281-1289.

Hauf N, Goebel W, Serfling E & Kuhn M (1994) Listeria monocytogenes infection enhances transcription factor NF-κB in P388D1 macrophage-like cells. Infect. Immun. 62: 2740-2747.

Hauf N, Goebel W, Fiedler F, Sokolovic Z & Kuhn M (1997) Listeria monocytogenes infection of P388D1 macrophages results in a biphasic NF-κB (RelA/p50) activation induced by lipoteichoic acid and bacterial phospholipases and mediated by IκBα and IκBβ degradation. Proc. Natl. Acad. Sci. USA 94: 9394-9399.

Hawkey CJ, Stirling Y, Chakrabarti R, Brozovic M, Cox AG & Meade TW (1983) Haemostatic changes following surgery. Thromb. Res. 32: 223-227.

Hazan U, Thomas D, Alcami J, Bachelerie F, Israel N, Yssel H, Virelizier JL & Arenzana-Seisdedos F (1990) Stimulation of a human T-cell clone with anti-CD3 or tumor necrosis factor induces NF-κB translocation but not human immunodeficiency virus 1 enhancer-dependent transcription. Proc. Natl. Acad. Sci. USA 87: 7861-7865.

He C (1996) Molecular mechanism of transcriptional activation of human gelatinase B by proximal promoter. Cancer Lett 106: 185-191.

Hehner SP, Breitkreutz R, Shubinsky G, Unsoeld H, Schulze-Osthoff K, Schmitz ML & Droge W (2000) Enhancement of T cell receptor signaling by a mild oxidative shift in the intracellular thiol pool. J. Immunol. 165: 4319-4328.

Hein H, Schluter C, Kulke R, Christophers E, Schroder JM & Bartels J (1997) Genomic organization, sequence, and transcriptional regulation of the human eotaxin gene. Biochem. Biophys. Res. Commun. 237: 537-542.

Hellin AC, Calmant P, Gielen J, Bours V & Merville MP (1998) Nuclear factor-κB-dependent regulation of p53 gene expression induced by daunomycin genotoxic drug. Oncogene 16: 1187-1195.

Hennig B, Toborek M, Joshi-Barve S, Barger SW, Barve S, Mattson MP & McClain CJ (1996) Linoleic acid activates nuclear transcription factor-κB (NF-κB) and induces NF-κB-dependent transcription in cultured endothelial cells. Am. J. Clin. Nutr. 63: 322-328.

Hennig B, Slim R, Toborek M & Robertson LW (1999) Linoleic acid amplifies polychlorinated biphenyl-mediated dysfunction of endothelial cells. J. Biochem. Mol. Toxicol. 13: 83-91.

Hierholzer C, Kelly E, Tsukada K, Loeffert E, Watkins S, Billiar TR & Tweardy DJ (1997) Hemorrhagic shock induces G-CSF expression in bronchial epithelium. Am. J. Physiol. 273: L1058-L1064.

Hildt E, Saher G, Bruss V & Hofschneider PH (1996) The hepatitis B virus large surface protein (LHBs) is a transcriptional activator. Virology 225: 235-239.

Hinz M, Krappmann D, Eichten A, Heder A, Scheidereit C & Strauss M (1999) NF-κB function in growth control: regulation of cyclin D1 expression and G0/G1-to-S-phase transition. Mol. Cell. Biol. 19: 2690-2698.

Hirota K, Murata M, Sachi Y, Nakamura H, Takeuchi J, Mori K & Yodoi J (1999) Distinct roles of thioredoxin in the cytoplasm and in the nucleus. A two-step mechanism of redox regulation of transcription factor NF-κB. J. Biol. Chem. 274: 27891-27897.

Hiscott J, Alper D, Cohen L, Leblanc JF, Sportza L, Wong A & Xanthoudakis S (1989) Induction of human interferon gene expression is associated with a nuclear factor that interacts with the NF-κB site of the human immunodeficiency virus enhancer. J. Virol. 63: 2557-2566.

Hiscott J, Marois J, Garoufalis J, D'Addario M, Roulston A, Kwan I, Pepin N, Lacoste J, Nguyen H & Bensi G (1993) Characterization of a functional NF-κB site in the human interleukin 1β promoter: evidence for a positive autoregulatory loop. Mol. Cell. Biol. 13: 6231-6240.

Hobbie S, Chen LM, Davis RJ & Galan JE (1997) Involvement of mitogen-activated protein kinase pathways in the nuclear responses and cytokine production induced by Salmonella typhimurium in cultured intestinal epithelial cells. J. Immunol. 159: 5550-5559.

Hoyos B, Ballard DW, Bohnlein E, Siekevitz M & Greene WC (1989) κB-specific DNA binding proteins: role in the regulation of human interleukin-2 gene expression. Science 244: 457-460.

Hsu DK, Hammes SR, Kuwabara I, Greene WC & Liu FT (1996) Human T lymphotropic virus-I infection of human T lymphocytes induces expression of the β-galactoside-binding lectin, galectin-3. Am. J. Pathol. 148: 1661-1670.

Huang C, Li J, Ding M, Wang L, Shi X, Castranova V, Vallyathan V, Ju G & Costa M (2001) Arsenic-induced NFκB transactivation through Erks- and JNKs-dependent pathways in mouse epidermal JB6 cells. Mol. Cell. Biochem. 222: 29-34.

Hunter RB, Mitchell-Felton H, Essig DA & Kandarian SC (2001) Expression of endoplasmic reticulum stress proteins during skeletal muscle disuse atrophy. Am. J. Physiol. Cell. Physiol. 281: C1285-C1290.

Hunter RB, Stevenson E, Koncarevic A, Mitchell-Felton H, Essig DA & Kandarian SC (2002) Activation of an alternative NF-κB pathway in skeletal muscle during disuse atrophy. FASEB J. 16: 529-538.

Hwang S & Ding A (1995) Activation of NF-κB in murine macrophages by taxol. Cancer Biochem. Biophys. 14: 265-272.

Hyers TM, Tricomi SM, Dettenmeier PA & Fowler AA (1991) Tumor necrosis factor levels in serum and bronchoalveolar lavage fluid of patients with the adult respiratory distress syndrome. Am. Rev. Respir. Dis. 144: 268-271.

Iademarco MF, McQuillan JJ, Rosen GD & Dean DC (1992) Characterization of the promoter for vascular cell adhesion molecule-1 (VCAM-1). J. Biol. Chem. 267: 16323-16329.

Ichikawa S, Ozawa K & Hirabayashi Y (1998) Molecular cloning and characterization of the mouse ceramide glucosyltransferase gene. Biochem. Biophys. Res. Commun. 253: 707-711.

Iciek LA, Delphin SA & Stavnezer J (1997) CD40 cross-linking induces Igε germline transcripts in B cells via activation of NF-κB. Synergy with IL-4 induction. J. Immunol. 158: 4769-4779.

Imbert V, Rupec RA, Livolsi A, Pahl HL, Traenckner EB, Mueller-Dieckmann C, Farahifar D, Rossi B, Auberger P, Baeuerle PA & Peyron JF (1996) Tyrosine phosphorylation of IκB-α activates NF-κB without proteolytic degradation of IκB-α. Cell 86: 787-798.

Israel A, Le Bail O, Hatat D, Piette J, Kieran M, Logeat F, Wallach D, Fellous M & Kourilsky P (1989a) TNF stimulates expression of mouse MHC class I genes by inducing an NFκB-like enhancer binding activity which displaces constitutive factors. EMBO J. 8: 3793-3800.

Israel A, Yano O, Logeat F, Kieran M & Kourilsky P (1989b) Two purified factors bind to the same sequence in the enhancer of mouse MHC class I genes: one of them is a positive regulator induced upon differentiation of teratocarcinoma cells. Nucleic Acids Res. 17: 5245-5257.

Ivanov V, Stein B, Baumann I, Dobbelaere DA, Herrlich P & Williams RO (1989) Infection with the intracellular protozoan parasite Theileria parva induces constitutively high levels of NF-κB in bovine T lymphocytes. Mol. Cell. Biol. 9: 4677-4686.

Ivanov V, Fleming TJ & Malek TR (1994) Regulation of nuclear factor-κB and activator protein-1 activities after stimulation of T cells via glycosylphosphatidylinositol-anchored Ly-6A/E. J. Immunol. 153: 2394-2406.

Jamieson C, Mauxion F & Sen R (1989) Identification of a functional NF-κB binding site in the murine T cell receptor β2 locus. J. Exp. Med. 170: 1737-1743.

Janssen YM, Barchowsky A, Treadwell M, Driscoll KE & Mossman BT (1995) Asbestos induces nuclear factor κB (NF-κB) DNA-binding activity and NF-κB-dependent gene expression in tracheal epithelial cells. Proc. Natl. Acad. Sci. USA 92: 8458-8462.

Janssen YM, Driscoll KE, Howard B, Quinlan TR, Treadwell M, Barchowsky A & Mossman BT (1997) Asbestos causes translocation of p65 protein and increases NF-κB DNA binding activity in rat lung epithelial and pleural mesothelial cells. Am. J. Pathol. 151: 389-401.

Johnson DR & Pober JS (1994) HLA class I heavy-chain gene promoter elements mediating synergy between tumor necrosis factor and interferons. Mol. Cell. Biol. 14: 1322-1332.

Kaltschmidt C, Kaltschmidt B & Baeuerle PA (1995) Stimulation of ionotropic glutamate receptors activates transcription factor NF-κB in primary neurons. Proc. Natl. Acad. Sci. USA 92: 9618-9622.

Kaltschmidt B, Uherek M, Volk B, Baeuerle PA & Kaltschmidt C (1997) Transcription factor NF-κB is activated in primary neurons by amyloid β peptides and in neurons surrounding early plaques from patients with Alzheimer disease. Proc. Natl. Acad. Sci. USA 94: 2642-2647.

Kang JL, Lee HW, Lee HS, Pack IS, Chong Y, Castranova V & Koh Y (2001) Genistein prevents nuclear factor-κB activation and acute lung injury induced by lipopolysaccharide. Am. J. Respir. Crit. Care Med. 164: 2206-2212.

Kang MK, Jhee EC, Koo BS, Yang JY, Park BH, Kim JS, Rho HW, Kim HR & Park JW (2002) Induction of nitric oxide synthase expression by Vibrio vulnificus cytolysin. Biochem. Biophys. Res. Commun. 290: 1090-1095.

Kanno M, Fromental C, Staub A, Ruffenach F, Davidson I & Chambon P (1989) The SV40 TC-II(κB) and the related H-2Kb enhansons exhibit different cell type specific and inducible proto-enhancer activities, but the SV40 core sequence and the AP-2 binding site have no enhanson properties. EMBO J. 8: 4205-4214.

Keates AC, Keates S, Kwon JH, Arseneau KO, Law DJ, Bai L, Merchant JL, Wang TC & Kelly CP (2001) ZBP-89, Sp1, and nuclear factor-κB regulate epithelial neutrophil-activating peptide-78 gene expression in Caco-2 human colonic epithelial cells. J. Biol. Chem. 276: 43713-43722.

Keightley AM, Lam YM, Brady JN, Cameron CL & Lillicrap D (1999) Variation at the von Willebrand factor (vWF) gene locus is associated with plasma vWF:Ag levels: identification of three novel single nucleotide polymorphisms in the vWF gene promoter. Blood 93: 4277-4283.

Kennedy T, Ghio AJ, Reed W, Samet J, Zagorski J, Quay J, Carter J, Dailey L, Hoidal JR & Devlin RB (1998) Copper-dependent inflammation and nuclear factor-κB activation by particulate air pollution. Am. J. Resp. Cell Mol. Biol. 19: 366-378.

Khachigian LM, Resnick N, Gimbrone MAJ & Collins T (1995) Nuclear factor-κB interacts functionally with the platelet-derived growth factor B-chain shear-stress response element in vascular endothelial cells exposed to fluid shear stress. J. Clin. Invest. 96: 1169-1175.

Kitajima I, Soejima Y, Takasaki I, Beppu H, Tokioka T & Maruyama I (1996) Ceramide-induced nuclear translocation of NF-κB is a potential mediator of the apoptotic response to TNF-α in murine clonal osteoblasts. Bone 19: 263-270.

Klaman LD & Thorley-Lawson DA (1995) Characterization of the CD48 gene demonstrates a positive element that is specific to Epstein-Barr virus-immortalized B-cell lines and contains an essential NF-κB site. J. Virol. 69: 871-881.

Klebanoff SJ, Watts DH, Mehlin C & Headley CM (1999) Lactobacilli and vaginal host defense: activation of the human immunodeficiency virus type 1 long terminal repeat, cytokine production, and NF-κB. J. Infect. Dis. 179: 653-660.

Knorre A, Wagner M, Schaefer HE, Colledge WH & Pahl HL (2002) ΔF508-CFTR causes constitutive NF-κB activation through an ER-overload response in cystic fibrosis lungs. Biol. Chem. 383: 271-282.

Ko YG, Park H, Kim T, Lee JW, Park SG, Seol W, Kim JE, Lee WH, Kim SH, Park JE & Kim S (2001) A cofactor of tRNA synthetase, p43, is secreted to up-regulate proinflammatory genes. J. Biol. Chem. 276: 23028-23033.

Koay MA, Christman JW, Segal BH, Venkatakrishnan A, Blackwell TR, Holland SM & Blackwell TS (2001) Impaired pulmonary NF-κB activation in response to lipopolysaccharide in NADPH oxidase-deficient mice. Infect. Immun. 69: 5991-5996.

Kol A, Bourcier T, Lichtman AH & Libby P (1999) Chlamydial and human heat shock protein 60s activate human vascular endothelium, smooth muscle cells, and macrophages. J. Clin. Invest. 103: 571-577.

Komine M, Rao LS, Kaneko T, Tomic-Canic M, Tamaki K, Freedberg IM & Blumenberg M (2000) Inflammatory versus proliferative processes in epidermis. Tumor necrosis factor α induces K6b keratin synthesis through a transcriptional complex containing NFκB and C/EBPβ. J. Biol. Chem. 275: 32077-32088.

Kretz-Remy C, Bates EE & Arrigo AP (1998) Amino acid analogs activate NF-κB through redox-dependent IκB-α degradation by the proteasome without apparent IκB-α phosphorylation. Consequence on HIV-1 long terminal repeat activation. J. Biol. Chem. 273: 3180-31891.

Kreuz S, Siegmund D, Scheurich P & Wajant H (2001) NF-κB inducers upregulate cFLIP, a cycloheximide-sensitive inhibitor of death receptor signaling. Mol. Cell. Biol. 21: 3964-3973.

Krieg AM (2000) The role of CpG motifs in innate immunity. Curr. Opin. Immunol. 12: 35-43.

Krikos A, Laherty CD & Dixit VM (1992) Transcriptional activation of the tumor necrosis factor α-inducible zinc finger protein, A20, is mediated by κB elements. J. Biol. Chem. 267: 17971-17976.

Krushel LA, Cunningham BA, Edelman GM & Crossin KL (1999) NF-κB activity is induced by neural cell adhesion molecule binding to neurons and astrocytes. J. Biol. Chem. 274: 2432-2439.

Kunsch C & Rosen CA (1993) NF-κB subunit-specific regulation of the interleukin-8 promoter. Mol. Cell. Biol. 13: 6137-6146.

Kuprash DV, Osipovich OA, Pokholok DK, Alimzhanov MB, Biragyn A, Turetskaya RL & Nedospasov SA (1996) Functional analysis of the lymphotoxin-β promoter. Sequence requirements for PMA activation. J. Immunol. 156: 2465-2472.

Kurata S (1994) Potential of azidothymidine to activate the HIV-1 promoter. J. Biol. Chem. 269: 24553-24556.

Kurdowska A, Miller EJ, Noble JM, Baughman RP, Matthay MA, Brelsford WG & Cohen AB (1996) Anti-IL-8 autoantibodies in alveolar fluid from patients with the adult respiratory distress syndrome. J. Immunol. 157: 2699-2706.

Kurdowska A, Carr FK, Stevens MD, Baughman RP & Martin TR (1997) Studies on the interaction of IL-8 with human plasma α2-macroglobulin: evidence for the presence of IL-8 complexed to α2-macroglobulin in lung fluids of patients with adult respiratory distress syndrome. J. Immunol. 158: 1930-1940.

Kurdowska A, Noble JM, Steinberg KP, Ruzinski JT, Hudson LD & Martin TR (2001) Anti-interleukin 8 autoantibody: interleukin 8 complexes in the acute respiratory distress syndrome. Relationship between the complexes and clinical disease activity. Am. J. Respir. Crit. Care Med. 163: 463-468.

Kurdowska AK, Geiser TK, Alden SM, Dziadek BR, Noble JM, Nuckton TJ & Matthay MA (2002) Activity of pulmonary edema fluid interleukin-8 bound to α_2-macroglobulin in patients with acute lung injury. Am. J. Physiol. Lung Cell Mol. Physiol. 282: L1092-L1098.

Kurkinen KM, Keinanen RA, Karhu R & Koistinaho J (2000) Genomic structure and chromosomal localization of the rat protein kinase Cδ-gene. Gene 242: 115-123.

Kwak EL, Larochelle DA, Beaumont C, Torti SV & Torti FM (1995) Role for NF-κB in the regulation of ferritin H by tumor necrosis factor-α. J. Biol. Chem. 270: 15285-15293.

Lai JH, Horvath G, Subleski J, Bruder J, Ghosh P & Tan TH (1995) RelA is a potent transcriptional activator of the CD28 response element within the interleukin 2 promoter. Mol. Cell. Biol. 15: 4260-4271.

Lam AG, Koppal T, Akama KT, Guo L, Craft JM, Samy B, Schavocky JP, Watterson DM & Van Eldik LJ (2001) Mechanism of glial activation by S100B: involvement of the transcription factor NFκB. Neurobiol. Aging 22: 765-772.

Lan Q, Mercurius KO & Davies PF (1994) Stimulation of transcription factors NFκB and AP1 in endothelial cells subjected to shear stress. Biochem. Biophys. Res. Commun. 201: 950-956.

Lang CH, Nystrom GJ & Frost RA (1999) Regulation of IGF binding protein-1 in hep G2 cells by cytokines and reactive oxygen species. Am. J. Physiol. 276: G719-G727.

Laniado-Schwartzman M, Lavrovsky Y, Stoltz RA, Conners MS, Falck JR, Chauhan K & Abraham NG (1994) Activation of nuclear factor κB and oncogene expression by 12(R)-hydroxyeicosatrienoic acid, an angiogenic factor in microvessel endothelial cells. J. Biol. Chem. 269: 24321-24327.

Lavrovsky Y, Schwartzman ML, Levere RD, Kappas A & Abraham NG (1994) Identification of binding sites for transcription factors NF-κB and AP-2 in the promoter region of the human heme oxygenase 1 gene. Proc. Natl. Acad. Sci. USA 91: 5987-5991.

Lawrence O, Rachie N, Qureshi N, Bomsztyk K & Sibley CH (1995) Diphosphoryl lipid A from Rhodobacter sphaeroides transiently activates NF-κB but inhibits lipopolysaccharide induction of κ light chain and Oct-2 in the B-cell lymphoma line 70Z/3. Infect. Immun. 63: 1040-1046.

Lecker SH, Solomon V, Mitch WE & Goldberg AL (1999) Muscle protein breakdown and the critical role of the ubiquitin-proteasome pathway in normal and disease states. J. Nutr. 129: 227S-237S.

Lee FY, Li Y, Zhu H, Yang S, Lin HZ, Trush M & Diehl AM (1999) Tumor necrosis factor increases mitochondrial oxidant production and induces expression of uncoupling protein-2 in the regenerating mice [correction of rat] liver. Hepatology 29: 677-687.

Lee HH, Dadgostar H, Cheng Q, Shu J & Cheng G (1999) NF-κB-mediated up-regulation of Bcl-x and Bfl-1/A1 is required for CD40 survival signaling in B lymphocytes. Proc. Natl. Acad. Sci. USA 96: 9136-9141.

Lee KS, Buck M, Houglum K & Chojkier M (1995) Activation of hepatic stellate cells by TGFα and collagen type I is mediated by oxidative stress through c-myb expression. J. Clin. Invest. 96: 2461-2468.

Lee RM, Gillet G, Burnside J, Thomas SJ & Neiman P (1999) Role of Nr13 in regulation of programmed cell death in the bursa of Fabricius. Genes Dev. 13: 718-728.

Legrand-Poels S, Bours V, Piret B, Pflaum M, Epe B, Rentier B & Piette J (1995) Transcription factor NF-κB is activated by photosensitization generating oxidative DNA damages. J. Biol. Chem. 270: 6925-6934.

Lehmann TG, Wheeler MD, Schwabe RF, Connor HD, Schoonhoven R, Bunzendahl H, Brenner DA, Jude SR, Zhong Z & Thurman RG (2000) Gene delivery of Cu/Zn-superoxide dismutase improves graft function after transplantation of fatty livers in the rat. Hepatology 32: 1255-1264.

Lenardo MJ, Fan CM, Maniatis T & Baltimore D (1989) The involvement of NF-κB in β-interferon gene regulation reveals its role as widely inducible mediator of signal transduction. Cell 57: 287-294.

Lentsch AB, Shanley TP, Sarma V & Ward PA (1997) In vivo suppression of NF-κB and preservation of IκBα by interleukin-10 and interleukin-13. J. Clin. Invest. 100: 2443-2448.

Lentsch AB, Jordan JA, Czermak BJ, Diehl KM, Younkin EM, Sarma V & Ward PA (1999) Inhibition of NF-κB activation and augmentation of IκBβ by secretory leukocyte protease inhibitor during lung inflammation. Am. J. Pathol. 154: 239-247.

Lentsch AB & Ward PA (2001) Regulation of experimental lung inflammation. Respir. Physiol. 128: 17-22.

Leonard SS, Wang S, Shi X, Jordan BS, Castranova V & Dubick MA (2000) Wood smoke particles generate free radicals and cause lipid peroxidation, DNA damage, NFκB activation and TNF-α release in macrophages. Toxicology 150: 147-157.

Lerner-Natoli M, Montpied P, Rousset MC, Bockaert J & Rondouin G (2000) Sequential expression of surface antigens and transcription factor NFκB by hippocampal cells in excitotoxicity and experimental epilepsy. Epilepsy Res. 41: 141-154.

Leung K & Nabel GJ (1988) HTLV-1 transactivator induces interleukin-2 receptor expression through an NF-κB-like factor. Nature 333: 776-778.

Li C, Browder W & Kao RL (1999) Early activation of transcription factor NF-κB during ischemia in perfused rat heart. Am. J. Physiol. 276: H543-H552.

Li J & Brasier AR (1996) Angiotensinogen gene activation by angiotensin II is mediated by the rel A (nuclear factor-κB p65) transcription factor: one mechanism for the renin angiotensin system positive feedback loop in hepatocytes. Mol. Endocrinol. 10: 252-264.

Li J & Schmidt AM (1997) Characterization and functional analysis of the promoter of RAGE, the receptor for advanced glycation end products. J. Biol. Chem. 272: 16498-16506.

Li XX & Liao WS (1991) Expression of rat serum amyloid A1 gene involves both C/EBP-like and NFκB-like transcription factors. J. Biol. Chem. 266: 15192-15201.

Li Y, Leung LK, Glauert HP & Spear BT (1996a) Treatment of rats with the peroxisome proliferator ciprofibrate results in increased liver NF-κB activity. Carcinogenesis 17: 2305-2309.

Li Y, Leung LK, Spear BT & Glauert HP (1996b) Activation of hepatic NF-κB by phenobarbital in rats. Biochem. Biophys. Res. Commun. 229: 982-989.

Libermann TA & Baltimore D (1990) Activation of interleukin-6 gene expression through the NF-κB transcription factor. Mol. Cell. Biol. 10: 2327-2334.

Lilienbaum A, Duc DM, Alexandre C, Gazzolo L & Paulin D (1990) Effect of human T-cell leukemia virus type I tax protein on activation of the human vimentin gene. J. Virol. 64: 256-263.

Lin KI, Lee SH, Narayanan R, Baraban JM, Hardwick JM & Ratan RR (1995) Thiol agents and Bcl-2 identify an alphavirus-induced apoptotic pathway that requires activation of the transcription factor NF-κB. J. Cell. Biol. 131: 1149-1161.

Lin SC & Stavnezer J (1996) Activation of NF-κB/Rel by CD40 engagement induces the mouse germ line immunoglobulin Cγ1 promoter. Mol. Cell. Biol. 16: 4591-4603.

Lin TH, Rosales C, Mondal K, Bolen JB, Haskill S & Juliano RL (1995) Integrin-mediated tyrosine phosphorylation and cytokine message induction in monocytic cells. A possible signaling role for the Syk tyrosine kinase. J. Biol. Chem. 270: 16189-16197.

Liu R, Zhao X, Gurney TA & Landau NR (1998) Functional analysis of the proximal CCR5 promoter. AIDS Res. Hum. Retro. 14: 1509-1519.

Liu SF, Haddad EB, Adcock I, Salmon M, Koto H, Gilbey T, Barnes PJ & Chung KF (1997a) Inducible nitric oxide synthase after sensitization and allergen challenge of Brown Norway rat lung. Br. J. Pharmacol. 121: 1241-1246.

Liu SF, Ye X & Malik AB (1997b) In vivo inhibition of nuclear factor-κB activation prevents inducible nitric oxide synthase expression and systemic hypotension in a rat model of septic shock. J. Immunol. 159: 3976-3983.

Liu SF, Ye X & Malik AB (1999) Pyrrolidine dithiocarbamate prevents I-κB degradation and reduces microvascular injury induced by lipopolysaccharide in multiple organs. Mol. Pharmacol. 55: 658-667.

Liu W, Liu Y & Lowe JWJ (2001) The role of phosphatidylinositol 3-kinase and the mitogen-activated protein kinases in insulin-like growth factor-I-mediated effects in vascular endothelial cells. Endocrinology 142: 1710-1719.

Lombardi L, Ciana P, Cappellini C, Trecca D, Guerrini L, Migliazza A, Maiolo AT & Neri A (1995) Structural and functional characterization of the promoter regions of the NFκB2 gene. Nucleic Acids Res. 23: 2328-2336.

Lopez-Cabrera M, Munoz E, Blazquez MV, Ursa MA, Santis AG & Sanchez-Madrid F (1995) Transcriptional regulation of the gene encoding the human C-type lectin leukocyte receptor AIM/CD69 and functional characterization of its tumor necrosis factor-α-responsive elements. J. Biol. Chem. 270: 21545-21551.

Lorimer DD, Matkowskj K & Benya RV (1997) Cloning, chromosomal location, and transcriptional regulation of the human galanin-1 receptor gene (GALN1R). Biochem. Biophys. Res. Commun. 241: 558-564.

Luster MI & Simeonova PP (1998) Asbestos induces inflammatory cytokines in the lung through redox sensitive transcription factors. Toxicol. Lett. 102-103: 271-275.

Mackman N, Brand K & Edgington TS (1991) Lipopolysaccharide-mediated transcriptional activation of the human tissue factor gene in THP-1 monocytic cells requires both activator protein 1 and nuclear factor κB binding sites. J. Exp. Med. 174: 1517-1526.

MacLachlan TK & Giordano A (1998) TRAF2 expression in differentiated muscle. J. Cell. Biochem. 71: 461-466.

MacLachlan TK, Sang N, De Luca A, Puri PL, Levrero M & Giordano A (1998) Binding of CDK9 to TRAF2. J. Cell. Biochem. 71: 467-478.

Maini R, St Clair EW, Breedveld F, Furst D, Kalden J, Weisman M, Smolen J, Emery P, Harriman G, Feldmann M & Lipsky P (1999) Infliximab (chimeric anti-tumour necrosis factor α monoclonal antibody) versus placebo in rheumatoid arthritis patients receiving concomitant methotrexate: a randomised phase III trial. ATTRACT Study Group. Lancet 354: 1932-1939.

Manning AM, Bell FP, Rosenbloom CL, Chosay JG, Simmons CA, Northrup JL, Shebuski RJ, Dunn CJ & Anderson DC (1995) NF-κB is activated during acute inflammation in vivo in association with elevated endothelial cell adhesion molecule gene expression and leukocyte recruitment. J. Inflamm. 45: 283-296.

Marcuzzi A, Van Ness B, Rouse T & Lafrenz D (1989) Effects of anti-IgM suppression on polyclonally activated murine B cells: analysis of immunoglobulin mRNA, gene specific nuclear factors and cell cycle distribution. Nucleic Acids Res. 17: 10455-10472.

Mari B, Imbert V, Belhacene N, Far DF, Peyron JF, Pouyssegur J, Van Obberghen-Schilling E, Rossi B & Auberger P (1994) Thrombin and thrombin receptor agonist peptide induce early events of T cell activation and synergize with TCR cross-linking for CD69 expression and interleukin 2 production. J. Biol. Chem. 269: 8517-8523.

Marie C, Roman-Roman S & Rawadi G (1999) Involvement of mitogen-activated protein kinase pathways in interleukin-8 production by human monocytes and polymorphonuclear cells stimulated with lipopolysaccharide or Mycoplasma fermentans membrane lipoproteins. Infect. Immun. 67: 688-693.

Mastronarde JG, He B, Monick MM, Mukaida N, Matsushima K & Hunninghake GW (1996) Induction of interleukin (IL)-8 gene expression by respiratory syncytial virus involves activation of nuclear factor (NF)-κB and NF-IL-6. J. Infect. Dis. 174: 262-267.

Matsui K, Fine A, Zhu B, Marshak-Rothstein A & Ju ST (1998) Identification of two NF-κB sites in mouse CD95 ligand (Fas ligand) promoter: functional analysis in T cell hybridoma. J. Immunol. 161: 3469-3473.

Matsumoto S, Tsuji-Takayama K, Aizawa Y, Koide K, Takeuchi M, Ohta T & Kurimoto M (1997) Interleukin-18 activates NF-κB in murine T helper type 1 cells. Biochem. Biophys. Res. Commun. 234: 454-457.

Mayet W, Schwarting A, Barreiros AP, Schlaak J & Neurath M (1999) Anti-PR-3 antibodies induce endothelial IL-8 release. Eur. J. Clin. Invest. 29: 973-979.

Mayo MW, Wang CY, Cogswell PC, Rogers-Graham KS, Lowe SW, Der CJ & Baldwin AS (1997) Requirement of NF-κB activation to suppress p53-independent apoptosis induced by oncogenic Ras. Science 278: 1812-1815.

Maziere C, Conte MA, Degonville J, Ali D & Maziere JC (1999) Cellular enrichment with polyunsaturated fatty acids induces an oxidative stress and activates the transcription factors AP1 and NFκB. Biochem. Biophys. Res. Commun. 265: 116-122.

McCarthy GM, Augustine JA, Baldwin AS, Christopherson PA, Cheung HS, Westfall PR & Scheinman RI (1998) Molecular mechanism of basic calcium phosphate crystal-induced activation of human fibroblasts. Role of nuclear factor κB, activator protein 1, and protein kinase c. J. Biol. Chem. 273: 35161-35169.

McDonald PP, Russo MP, Ferrini S & Cassatella MA (1998) Interleukin-15 (IL-15) induces NF-κB activation and IL-8 production in human neutrophils. Blood 92: 4828-4835.

Mejdoubi N, Henriques C, Bui E & Porquet D (1999) NF-κB is involved in the induction of the rat hepatic α1-acid glycoprotein gene by phenobarbital. Biochem. Biophys Res. Commun. 254: 93-99.

Meldrum DR, Cleveland JC, Sheridan BC, Rowland RT, Selzman CH, Banerjee A & Harken AH (1997a) α-Adrenergic activation of myocardial NFκB during hemorrhage. J. Surg. Res. 69: 268-276.

Meldrum DR, Shenkar R, Sheridan BC, Cain BS, Abraham E & Harken AH (1997b) Hemorrhage activates myocardial NFκB and increases TNF-α in the heart. J. Mol. Cell. Cardiol. 29: 2849-2854.

Mercie P, Belloc F, Bihlou-Nabera C, Barthe C, Pruvost A, Renard M, Seigneur M, Bernard P, Marit G & Boisseau MR (2000) Comparative methodologic study of NFκB activation in cultured endothelial cells. J. Lab. Clin. Med. 136: 402-411.

Messer G, Weiss EH & Baeuerle PA (1990) Tumor necrosis factor β (TNF-β) induces binding of the NF-κB transcription factor to a high-affinity κB element in the TNF-β promoter. Cytokine 2: 389-397.

Mettouchi A, Cabon F, Montreau N, Dejong V, Vernier P, Gherzi R, Mercier G & Binetruy B (1997) The c-Jun-induced transformation process involves complex regulation of tenascin-C expression. Mol. Cell. Biol. 17: 3202-3209.

Meyer M, Caselmann WH, Schluter V, Schreck R, Hofschneider PH & Baeuerle PA (1992) Hepatitis B virus transactivator MHBst: activation of NF-κB, selective inhibition by antioxidants and integral membrane localization. EMBO J. 11: 2991-3001.

Miagkov AV, Kovalenko DV, Brown CE, Didsbury JR, Cogswell JP, Stimpson SA, Baldwin AS & Makarov SS (1998) NF-κB activation provides the potential link between inflammation and hyperplasia in the arthritic joint. Proc. Natl. Acad. Sci. USA 95: 13859-13864.

Miettinen M, Lehtonen A, Julkunen I & Matikainen S (2000) Lactobacilli and Streptococci activate

NF-κB and STAT signaling pathways in human macrophages. J. Immunol. 161: 3733-3740.
Miller EJ, Cohen AB, Nagao S, Griffith D, Maunder RJ, Martin TR, Weiner-Kronish JP, Sticherling M, Christophers E & Matthay MA (1992) Elevated levels of NAP-1/interleukin-8 are present in the airspaces of patients with the adult respiratory distress syndrome and are associated with increased mortality. Am. Rev. Respir. Dis. 146: 427-432.
Minneman KP, Lee D, Zhong H, Berts A, Abbott KL & Murphy TJ (2000) Transcriptional responses to growth factor and G protein-coupled receptors in PC12 cells: comparison of α_1-adrenergic receptor subtypes. J. Neurochem. 74: 2392-2400.
Mirza A, Liu SL, Frizell E, Zhu J, Maddukuri S, Martinez J, Davies P, Schwarting R, Norton P & Zern MA (1997) A role for tissue transglutaminase in hepatic injury and fibrogenesis, and its regulation by NF-κB. Am. J. Physiol. 272: G281-G288.
Misra UK, Shackelford RE, Florine-Casteel K, Thai SF, Alford PB, Pizzo SV & Adams DO (1996) Maleylated-BSA induces hydrolysis of PIP2, fluxes of Ca^{2+}, NF-κB binding, and transcription of the TNF-α gene in murine macrophages. J. Leukocyte Biol. 60: 784-792.
Miyazawa K, Mori A, Yamamoto K & Okudaira H (1998) Constitutive transcription of the human interleukin-6 gene by rheumatoid synoviocytes: spontaneous activation of NF-κB and CBF1. Am. J. Pathol. 152: 793-803.
Mizuno K, Irie S & Sato TA (2001) Overexpression of EXTL3/EXTR1 enhances NF-κB activity induced by TNF-α. Cell. Signal. 13: 125-130.
Moffat GJ & Tack BF (1992) Regulation of C4b-binding protein gene expression by the acute-phase mediators tumor necrosis factor-α, interleukin-6, and interleukin-1. Biochemistry 31: 12376-12384.
Moffatt S, Tanaka N, Tada K, Nose M, Nakamura M, Muraoka O, Hirano T & Sugamura K (1996) A cytotoxic nonstructural protein, NS1, of human parvovirus B19 induces activation of interleukin-6 gene expression. J. Virol. 70: 8485-8491.
Mohan RR, Kim WJ, Chen L & Wilson SE (1998) Bone morphogenic proteins 2 and 4 and their receptors in the adult human cornea. Invest. Ophthalmol. Vis. Sci. 39: 2626-2636.
Moore NC, Girdlestone J, Anderson G, Owen JJ & Jenkinson EJ (1995) Stimulation of thymocytes before and after positive selection results in the induction of different NF-κB/Rel protein complexes. J. Immunol. 155: 4653-4660.
Mori N & Prager D (1997) Activation of the interleukin-10 gene in the human T lymphoma line HuT 78: identification and characterization of NF-κB binding sites in the regulatory region of the interleukin-10 gene. Eur. J. Haematol. 84: 162-170.
Moriuchi H, Moriuchi M & Fauci AS (1997) Nuclear factor-κB potently up-regulates the promoter activity of RANTES, a chemokine that blocks HIV infection. J. Immunol. 158: 3483-3491.
Morri H, Ozaki M & Watanabe Y (1994) 5'-flanking region surrounding a human cytosolic phospholipase A2 gene. Biochem. Biophys. Res. Commun. 205: 6-11.
Mukhopadhyay A, Ni J, Zhai Y, Yu GL & Aggarwal BB (1999) Identification and characterization of a novel cytokine, THANK, a TNF homologue that activates apoptosis, nuclear factor-κB, and c-Jun NH2-terminal kinase. J. Biol. Chem. 274: 15978-15981.
Munroe DG, Wang EY, MacIntyre JP, Tam SS, Lee DH, Taylor GR, Zhou L, Plante RK, Kazmi SM & Bauerle PA (1995) Novel intracellular signaling function of prostaglandin H synthase-1 in NF-κB activation. J. Inflamm. 45: 260-268.
Munzenmaier A, Lange C, Glocker E, Covacci A, Moran A, Bereswill S, Baeuerle PA, Kist M & Pahl HL (1997) A secreted/shed product of Helicobacter pylori activates transcription factor nuclear factor-κB. J. Immunol. 159: 6140-6147.
Murakami S, Lefebvre V & de Crombrugghe B (2000) Potent inhibition of the master chondrogenic factor Sox9 gene by interleukin-1 and tumor necrosis factor-α. J. Biol. Chem. 275: 3687-3692.
Muroi M, Muroi Y & Suzuki T (1994) The binding of immobilized IgG2a to Fcγ 2a receptor activates NF-κB via reactive oxygen intermediates and tumor necrosis factor-α 1. J. Biol. Chem. 269: 30561-30568.
Murphy TL, Cleveland MG, Kulesza P, Magram J & Murphy KM (1995) Regulation of interleukin 12 p40 expression through an NF-κB half-site. Mol. Cell. Biol. 15: 5258-5267.
Musso R, Grilli M, Oberto A, Gamalero SR & Eva C (1997) Regulation of mouse neuropeptide Y

Y1 receptor gene transcription: a potential role for nuclear factor-κB/Rel proteins. Mol. Pharmacol. 51: 27-35.

Mutoh H, Ishii S, Izumi T, Kato S & Shimizu T (1994) Platelet-activating factor (PAF) positively auto-regulates the expression of human PAF receptor transcript 1 (leukocyte-type) through NF-κB. Biochem. Biophys. Res. Commun. 205: 1137-1142.

Nabel G & Baltimore D (1987) An inducible transcription factor activates expression of human immunodeficiency virus in T cells. Nature 326: 711-713.

Nagase M, Abe J, Takahashi K, Ando J, Hirose S & Fujita T (1998) Genomic organization and regulation of expression of the lectin-like oxidized low-density lipoprotein receptor (LOX-1) gene. J. Biol. Chem. 273: 33702-33707.

Naumann M, Wessler S, Bartsch C, Wieland B & Meyer TF (1997) Neisseria gonorrhoeae epithelial cell interaction leads to the activation of the transcription factors nuclear factor κB and activator protein 1 and the induction of inflammatory cytokines. J. Exp. Med. 186: 247-258.

Nemeth ZH, Deitch EA, Szabo C, Fekete Z, Hauser CJ & Hasko G (2002) Lithium induces NF-κB activation and interleukin-8 production in human intestinal epithelial cells. J. Biol. Chem. 277: 7713-7719.

Neurath MF, Pettersson S, Meyer ZBK & Strober W (1996) Local administration of antisense phosphorothioate oligonucleotides to the p65 subunit of NF-κB abrogates established experimental colitis in mice. Nat. Med. 2: 998-1004.

Neurath MF, Becker C & Barbulescu K (1998) Role of NF-κB in immune and inflammatory responses in the gut. Gut 43: 856-860.

Nie Z, Mei Y, Ford M, Rybak L, Marcuzzi A, Ren H, Stiles GL & Ramkumar V (1998) Oxidative stress increases A1 adenosine receptor expression by activating nuclear factor κB. Mol. Pharmacol. 53: 663-669.

Nishikawa M, Kakemizu N, Ito T, Kudo M, Kaneko T, Suzuki M, Udaka N, Ikeda H & Okubo T (1999) Superoxide mediates cigarette smoke-induced infiltration of neutrophils into the airways through nuclear factor-κB activation and IL-8 mRNA expression in guinea pigs in vivo. Am. J. Respir. Cell Mol. Biol. 20: 189-198.

Nishizawa M & Nagata S (1990) Regulatory elements responsible for inducible expression of the granulocyte colony-stimulating factor gene in macrophages. Mol. Cell. Biol. 10: 2002-2011.

Noble PW, McKee CM, Cowman M & Shin HS (1996) Hyaluronan fragments activate an NF-κB/I-κBα autoregulatory loop in murine macrophages. J. Exp. Med. 183: 2373-2378.

Nomoto Y, Yamamoto M, Fukushima T, Kimura H, Ohshima K & Tomonaga (2001) Expression of nuclear factor κB and tumor necrosis factor α in the mouse brain after experimental thermal ablation injury. Neurosurgery 48: 158-166.

Nonaka M & Huang ZM (1990) Interleukin-1-mediated enhancement of mouse factor B gene expression via NFκB-like hepatoma nuclear factor. Mol. Cell. Biol. 10: 6283-6289.

Novak TJ, Chen D & Rothenberg EV (1990) Interleukin-1 synergy with phosphoinositide pathway agonists for induction of interleukin-2 gene expression: molecular basis of costimulation. Mol. Cell. Biol. 10: 6325-6334.

Novak U, Cocks BG & Hamilton JA (1991) A labile repressor acts through the NFκB-like binding sites of the human urokinase gene. Nucleic Acids Res. 19: 3389-3393.

Oh CK & Metcalfe DD (1994) Transcriptional regulation of the TCA3 gene in mast cells after FcεRI cross-linking. J. Immunol. 153: 325-332.

Oh GS, Pae HO, Seo WG, Kim NY, Pyun KH, Kim IK, Shin M & Chung HT (2001) Capsazepine, a vanilloid receptor antagonist, inhibits the expression of inducible nitric oxide synthase gene in lipopolysaccharide-stimulated RAW264.7 macrophages through the inactivation of nuclear transcription factor-κB. Int. Immunopharmacol. 1: 777-784.

Ohkawa T, Ueki N, Taguchi T, Shindo Y, Adachi M, Amuro Y, Hada T & Higashino K (1999) Stimulation of hyaluronan synthesis by tumor necrosis factor-α is mediated by the p50/p65 NF-κB complex in MRC-5 myofibroblasts. Biochim. Biophys. Acta 1448: 416-424.

Ohmori Y & Hamilton TA (1993) Cooperative interaction between interferon (IFN) stimulus response element and κB sequence motifs controls IFNγ- and lipopolysaccharide-stimulated transcription

from the murine IP-10 promoter. J. Biol. Chem. 268: 6677-6688.
Ohmori Y & Hamilton TA (1995) The interferon-stimulated response element and a κB site mediate synergistic induction of murine IP-10 gene transcription by IFN-γ and TNF-α. J. Immunol. 154: 5235-5244.
Ohtsuka T, Kubota A, Hirano T, Watanabe K, Yoshida H, Tsurufuji M, Iizuka Y, Konishi K & Tsurufuji S (1996) Glucocorticoid-mediated gene suppression of rat cytokine-induced neutrophil chemoattractant CINC/gro, a member of the interleukin-8 family, through impairment of NF-κB activation. J. Biol. Chem. 271: 1651-1659.
Okladnova O, Poleev A, Fantes J, Lee M, Plachov D & Horst J (1997) The genomic organization of the murine Pax 8 gene and characterization of its basal promoter. Genomics 42: 452-461.
Olashaw NE, Kowalik TF, Huang ES & Pledger WJ (1992) Induction of NF-κB-like activity by platelet-derived growth factor in mouse fibroblasts. Mol. Biol. Cell 3: 1131-1139.
Osborn L, Kunkel S & Nabel GJ (1989) Tumor necrosis factor α and interleukin 1 stimulate the human immunodeficiency virus enhancer by activation of the nuclear factor κB. Proc. Natl. Acad. Sci. USA 86: 2336-2340.
Pahl HL (1999) Activators and target genes of Rel/NF-κB transcription factors. Oncogene 18: 6853-6866.
Pahl HL & Baeuerle PA (1995a) A novel signal transduction pathway from the endoplasmic reticulum to the nucleus is mediated by transcription factor NF-κB. EMBO J. 14: 2580-2588.
Pahl HL & Baeuerle PA (1995b) Expression of influenza virus hemagglutinin activates transcription factor NF-κB. J. Virol. 69: 1480-1484.
Pahl HL, Sester M, Burgert HG & Baeuerle PA (1996) Activation of transcription factor NF-κB by the adenovirus E3/19K protein requires its ER retention. J. Cell. Biol. 132: 511-522.
Pak J & Faller DV (1996) Moloney murine leukemia virus activates NF-κB. J. Virol. 70: 4167-4172.
Pan J & McEver RP (1995) Regulation of the human P-selectin promoter by Bcl-3 and specific homodimeric members of the NF-κB/Rel family. J. Biol. Chem. 270: 23077-23083.
Pan ZK, Zuraw BL, Lung CC, Prossnitz ER, Browning DD & Ye RD (1996) Bradykinin stimulates NF-κB activation and interleukin 1β gene expression in cultured human fibroblasts. J. Clin. Invest. 98: 2042-2049.
Pan ZK (1998) Anaphylatoxins C5a and C3a induce nuclear factor κB activation in human peripheral blood monocytes. Biochim. Biophys. Acta 1443: 90-98.
Papi A & Johnston SL (1999a) Rhinovirus infection induces expression of its own receptor intercellular adhesion molecule 1 (ICAM-1) via increased NF-κB-mediated transcription. J. Biol. Chem. 274: 9707-9720.
Papi A & Johnston SL (1999b) Respiratory epithelial cell expression of vascular cell adhesion molecule-1 and its up-regulation by rhinovirus infection via NF-κB and GATA transcription factors. J. Biol. Chem. 274: 30041-30051.
Patil S, Wildey GM, Brown TL, Choy L, Derynck R & Howe PH (2000) Smad7 is induced by CD40 and protects WEHI 231 B-lymphocytes from transforming growth factor-β-induced growth inhibition and apoptosis. J. Biol. Chem. 275: 38363-38370.
Patten EJ & DeLong MJ (1999) Temporal effects of the detoxification enzyme inducer, benzyl isothiocyanate: activation of c-Jun N-terminal kinase prior to the transcription factors AP-1 and NFκB. Biochem. Biophys. Res. Commun. 257: 149-155.
Paul NL, Lenardo MJ, Novak KD, Sarr T, Tang WL & Ruddle NH (1990) Lymphotoxin activation by human T-cell leukemia virus type I-infected cell lines: role for NF-κB. J. Virol. 64: 5412-5419.
Peng HB, Libby P & Liao JK (1995) Induction and stabilization of IκBα by nitric oxide mediates inhibition of NF-κB. J. Biol. Chem. 270: 14214-14219.
Pennington C, Dunn J, Li C, Ha T & Browder W (2000) Nuclear factor κB activation in acute appendicitis: a molecular marker for extent of disease? Am. Surg. 66: 914-918.
Pessara U & Koch N (1990) Tumor necrosis factor α regulates expression of the major histocompatibility complex class II-associated invariant chain by binding of an NF-κB-like factor to a promoter element. Mol. Cell. Biol. 10: 4146-4154.
Phi VL (1996) Transcriptional activation of the chicken lysozyme gene by NF-κB p65 (RelA) and c-

Rel, but not by NF-κB p50. Biochem. J. 313: 39-44.
Piret B & Piette J (1996) Topoisomerase poisons activate the transcription factor NF-κB in ACH-2 and CEM cells. Nucleic Acids Res. 24: 4242-4248.
Podolsky DK (1991) Inflammatory bowel disease. N. Engl. J. Med. 325: 1008-1016.
Post A, Holsboer F & Behl C (1998) Induction of NF-κB activity during haloperidol-induced oxidative toxicity in clonal hippocampal cells: suppression of NF-κB and neuroprotection by antioxidants. J. Neurosci. 18: 8236-8246.
Poynter ME, Irvin CG & Janssen-Heininger YM (2002) Rapid activation of nuclear factor-κB in airway epithelium in a murine model of allergic airway inflammation. Am. J. Pathol. 160: 1325-1334.
Pugin J, Dunn I, Jolliet P, Tassaux D, Magnenat JL, Nicod LP & Chevrolet JC (1998) Activation of human macrophages by mechanical ventilation in vitro. Am. J. Physiol. 275: L1040-L1050.
Qin ZH, Wang Y, Nakai M & Chase TN (1998) Nuclear factor-κB contributes to excitotoxin-induced apoptosis in rat striatum. Mol. Pharmacol. 53: 33-42.
Quay JL, Reed W, Samet J & Devlin RB (1998) Air pollution particles induce IL-6 gene expression in human airway epithelial cells via NF-κB activation. Am. J. Resp. Cell Mol. Biol. 19: 98-106.
Rajavashisth TB, Yamada H & Mishra NK (1995) Transcriptional activation of the macrophage-colony stimulating factor gene by minimally modified LDL. Involvement of nuclear factor-κB. Arterioscl. Thromb. Vasc. Biol. 15: 1591-1598.
Ramesh GT, Manna SK, Aggarwal BB & Jadhav AL (1999) Lead activates nuclear transcription factor-κB, activator protein-1, and amino-terminal c-Jun kinase in pheochromocytoma cells. Toxicol. Appl. Pharmacol. 155: 280-286.
Ranganathan PN & Khalili K (1993) The transcriptional enhancer element, κB, regulates promoter activity of the human neurotropic virus, JCV, in cells derived from the CNS. Nucleic Acids Res. 21: 1959-1964.
Rattner A, Korner M, Rosen H, Baeuerle PA & Citri Y (1991) Nuclear factor κB activates proenkephalin transcription in T lymphocytes. Mol. Cell. Biol 12: 1017-1022.
Ray A & Cohn L (1999) Th2 cells and GATA-3 in asthma: new insights into the regulation of airway inflammation. J. Clin. Invest. 104: 985-993.
Ray A, Gao X & Ray BK (1995) Role of a distal enhancer containing a functional NF-κB-binding site in lipopolysaccharide-induced expression of a novel α1-antitrypsin gene. J. Biol. Chem. 270: 29201-29208.
Reikerstorfer A, Holz H, Stunnenberg HG & Busslinger M (1995) Low affinity binding of interleukin-1β and intracellular signaling via NF-κB identify Fit-1 as a distant member of the interleukin-1 receptor family. J. Biol. Chem. 270: 17645-17648.
Rennie JA & Ogston D (1976) Changes in coagulation factors following acute myocardial infarction in man. Haemostasis 5: 258-264.
Rensing-Ehl A, Hess S, Ziegler-Heitbrock HW, Riethmuller G & Engelmann H (1995) Fas/Apo-1 activates nuclear factor κB and induces interleukin-6 production. J. Inflamm. 45: 161-174.
Richard I, Roudaut C, Marchand S, Baghdiguian S, Herasse M, Stockholm D, Ono Y, Suel L, Bourg N, Sorimachi H, Lefranc G, Fardeau M, Sebille A & Beckmann JS (2000) Loss of calpain 3 proteolytic activity leads to muscular dystrophy and to apoptosis-associated IκBα/nuclear factor κB pathway perturbation in mice. J. Cell. Biol. 151: 1583-1590.
Richards ML & Katz DH (1997) Analysis of the promoter elements necessary for IL-4 and anti-CD40 antibody induction of murine FcεRII (CD23). Comparison with the germline ε promoter. J. Immunol. 158: 263-272.
Richardson CA, Gordon KL, Couser WG & Bomsztyk K (1995) IL-1β increases laminin B2 chain mRNA levels and activates NF-κB in rat glomerular epithelial cells. Am. J. Physiol. 268: F273-F278.
Ritchie CK, Giordano A & Khalili K (2000) Integrin involvement in glioblastoma multiforme: possible regulation by NF-κB. J. Cell. Physiol. 184: 214-221.
Rogler G, Brand K, Vogl D, Page S, Hofmeister R, Andus T, Knuechel R, Baeuerle PA, Scholmerich J & Gross V (1998) Nuclear factor κB is activated in macrophages and epithelial cells of inflamed intestinal mucosa. Gastroenterology 115: 357-369.

Ron D, Brasier AR & Habener JF (1991) Angiotensinogen gene-inducible enhancer-binding protein 1, a member of a new family of large nuclear proteins that recognize nuclear factor κB-binding sites through a zinc finger motif. Mol. Cell. Biol. 11: 2887-2895.

Rong BL, Libermann TA, Kogawa K, Ghosh S, Cao LX, Pavan-Langston D & Dunkel EC (1992) HSV-1-inducible proteins bind to NF-κB-like sites in the HSV-1 genome. Virology 189: 750-756.

Ronni T, Matikainen S, Sareneva T, Melen K, Pirhonen J, Keskinen P & Julkunen I (1997) Regulation of IFN-α/β, MxA, 2',5'-oligoadenylate synthetase, and HLA gene expression in influenza A-infected human lung epithelial cells. J. Immunol. 158: 2363-2374.

Rose ML, Bradford BU, Germolec DR, Lin M, Tsukamoto H & Thurman RG (2001) Gadolinium chloride-induced hepatocyte proliferation is prevented by antibodies to tumor necrosis factor α. Toxicol. Appl. Pharmacol. 170: 39-45.

Rosette C & Karin M (1995) Cytoskeletal control of gene expression: depolymerization of microtubules activates NF-κB. J. Cell. Biol. 128: 1111-1119.

Rupec RA & Baeuerle PA (1995) The genomic response of tumor cells to hypoxia and reoxygenation. Differential activation of transcription factors AP-1 and NF-κB. Eur. J. Biochem. 234: 632-640.

Sacks M, Gordon J, Bylander J, Porter D, Shi XL, Castranova V, Kaczmarczyk W, Van Dyke K & Reasor MJ (1998) Silica-induced pulmonary inflammation in rats: activation of NF-κB and its suppression by dexamethasone. Biochem. Biophys. Res. Commun. 253: 181-184.

Sadikot RT, Jansen ED, Blackwell TR, Zoia O, Yull F, Christman JW & Blackwell TS (2001) High-dose dexamethasone accentuates nuclear factor-κB activation in endotoxin-treated mice. Am. J. Respir. Crit. Care Med. 164: 873-878.

Sambucetti LC, Cherrington JM, Wilkinson GW & Mocarski ES (1989) NF-κB activation of the cytomegalovirus enhancer is mediated by a viral transactivator and by T cell stimulation. EMBO J. 8: 4251-4258.

Santana MA, Pedraza-Alva G, Olivares-Zavaleta N, Madrid-Marina V, Horejsi V, Burakoff SJ & Rosenstein Y (2000) CD43-mediated signals induce DNA binding activity of AP-1, NF-AT, and NFκB transcription factors in human T lymphocytes. J. Biol. Chem. 275: 31460-31468.

Santee SM & Owen-Schaub LB (1996) Human tumor necrosis factor receptor p75/80 (CD120b) gene structure and promoter characterization. J. Biol. Chem. 271: 21151-21159.

Savkovic SD, Koutsouris A & Hecht G (1997) Activation of NF-κB in intestinal epithelial cells by enteropathogenic Escherichia coli. Am. J. Physiol. 273: C1160-C1167.

Scala G, Quinto I, Ruocco MR, Mallardo M, Ambrosino C, Squitieri B, Tassone P & Venuta S (1993) Epstein-Barr virus nuclear antigen 2 transactivates the long terminal repeat of human immunodeficiency virus type 1. J. Virol. 67: 2853-2861.

Schafer H, Diebel J, Arlt A, Trauzold A & Schmidt WE (1998) The promoter of human p22/PACAP response gene 1 (PRG1) contains functional binding sites for the p53 tumor suppressor and for NFκB. FEBS Lett. 436: 139-143.

Schanstra JP, Bataille E, Marin CM, Barascud Y, Hirtz C, Pesquero JB, Pecher C, Gauthier F, Girolami JP & Bascands JL (1998) The B1-agonist [des-Arg10]-kallidin activates transcription factor NF-κB and induces homologous upregulation of the bradykinin B1-receptor in cultured human lung fibroblasts. J. Clin. Invest. 101: 2080-2091.

Schmidt KN, Podda M, Packer L & Baeuerle PA (1996) Anti-psoriatic drug anthralin activates transcription factor NF-κB in murine keratinocytes. J. Immunol. 156: 4514-4519.

Schneider P, Thome M, Burns K, Bodmer JL, Hofmann K, Kataoka T, Holler N & Tschopp J (1997) TRAIL receptors 1 (DR4) and 2 (DR5) signal FADD-dependent apoptosis and activate NF-κB. Immunity 7: 831-836.

Schreck R, Zorbas H, Winnacker EL & Baeuerle PA (1990) The NF-κB transcription factor induces DNA bending which is modulated by its 65-kD subunit. Nucleic Acids Res. 18: 6497-6502.

Schreck R, Rieber P & Baeuerle PA (1991) Reactive oxygen intermediates as apparently widely used messengers in the activation of the NF-κB transcription factor and HIV-1. EMBO J. 10: 2247-2258.

Schreck R, Bevec D, Dukor P, Baeuerle PA, Chedid L & Bahr GM (1992) Selection of a muramyl peptide based on its lack of activation of nuclear factor-κB as a potential adjuvant for AIDS vaccines. Clin. Exp. Immunol. 90: 188-193.

Schreiber S, Nikolaus S & Hampe J (1998) Activation of nuclear factor κB inflammatory bowel disease. Gut 42: 477-484.

Schumann RR (1995) Mechanisms of transcriptional activation of lipopolysaccharide binding protein (LBP). Prog. Clin. Biol. Res. 392: 297-304.

Schwartz MD, Moore EE, Moore FA, Shenkar R, Moine P, Haenel JB & Abraham E (1996) Nuclear factor-κB is activated in alveolar macrophages from patients with acute respiratory distress syndrome. Crit. Care Med. 24: 1285-1292.

Segev DL, Hoshiya Y, Stephen AE, Hoshiya M, Tran TT, MacLaughlin DT, Donahoe PK & Maheswaran S (2001) Mullerian inhibiting substance regulates NFκB signaling and growth of mammary epithelial cells in vivo. J. Biol. Chem. 276: 26799-26806.

Sen R & Baltimore D (1986a) Multiple nuclear factors interact with the immunoglobulin enhancer sequences. Cell 46: 705-716.

Sen R & Baltimore D (1986b) Inducibility of κ immunoglobulin enhancer-binding protein NF-κB by a posttranslational mechanism. Cell 47: 921-928.

Serfling E, Barthelmas R, Pfeuffer I, Schenk B, Zarius S, Swoboda R, Mercurio F & Karin M (1989) Ubiquitous and lymphocyte-specific factors are involved in the induction of the mouse interleukin 2 gene in T lymphocytes. EMBO J. 8: 465-473.

Shahrestanifar M, Fan X & Manning DR (1999) Lysophosphatidic acid activates NF-κB in fibroblasts. A requirement for multiple inputs. J. Biol. Chem. 274: 3828-3833.

Shalom-Barak T, Quach J & Lotz M (1998) Interleukin-17-induced gene expression in articular chondrocytes is associated with activation of mitogen-activated protein kinases and NF-κB. J. Biol. Chem. 273: 27467-27473.

Shea LM, Beehler C, Schwartz M, Shenkar R, Tuder R & Abraham E (1996) Hyperoxia activates NF-κB and increases TNF-α and IFN-γ gene expression in mouse pulmonary lymphocytes. J. Immunol. 157: 3902-3908.

Shen BJ, Chang CJ, Lee HS, Tsai WH, Miau LH & Lee SC (1997) Transcriptional induction of the agp/ebp (c/ebpβ) gene by hepatocyte growth factor. DNA Cell Biol. 16: 703-711.

Shenkar R & Abraham E (1997) Hemorrhage induces rapid in vivo activation of CREB and NF-κB in murine intraparenchymal lung mononuclear cells. Am. J. Respir. Cell Mol. Biol. 16: 145-152.

Shenkar R & Abraham E (1999) Mechanisms of lung neutrophil activation after hemorrhage or endotoxemia: roles of reactive oxygen intermediates, NF-κB, and cyclic AMP response element binding protein. J. Immunol. 163: 954-962.

Shenkar R, Schwartz MD, Terada LS, Repine JE, McCord J & Abraham E (1996) Hemorrhage activates NF-κB in murine lung mononuclear cells in vivo. Am. J. Physiol. 270: L729-L735.

Shenkar R, Yum HK, Arcaroli J, Kupfner J & Abraham E (2001) Interactions between CBP, NF-κB, and CREB in the lungs after hemorrhage and endotoxemia. Am. J. Physiol. Lung Cell Mol. Physiol. 281: L418-L426.

Shi Y, Wang W, Yourey PA, Gohari S, Zukauskas D, Zhang J, Ruben S & Alderson RF (1999) Computational EST database analysis identifies a novel member of the neuropoietic cytokine family. Biochem. Biophys. Res. Commun. 262: 132-138.

Shimizu H, Mitomo K, Watanabe T, Okamoto S & Yamamoto K (1990) Involvement of a NF-κB-like transcription factor in the activation of the interleukin-6 gene by inflammatory lymphokines. Mol. Cell. Biol. 10: 561-568.

Shurman L, Sen R & Bergman Y (1989) Adenovirus E1A products activate the Ig k-chain enhancer in fibroblasts. A possible involvement of the NF-κB binding site. J. Immunol. 143: 3806-3812.

Sica A, Tan TH, Rice N, Kretzschmar M, Ghosh P & Young HA (1992) The c-rel protooncogene product c-Rel but not NF-κB binds to the intronic region of the human interferon-γ gene at a site related to an interferon-stimulable response element. Proc. Natl. Acad. Sci. USA 89: 1740-1744.

Sica A, Dorman L, Viggiano V, Cippitelli M, Ghosh P, Rice N & Young HA (1997) Interaction of NF-κB and NFAT with the interferon-γ promoter. J. Biol. Chem. 272: 30412-30420.

Siddiqui A, Gaynor R, Srinivasan A, Mapoles J & Farr RW (1989) trans-activation of viral enhancers including long terminal repeat of the human immunodeficiency virus by the hepatitis B virus X protein. Virology 169: 479-484.

Simeonova PP & Luster MI (1996) Asbestos induction of nuclear transcription factors and interleukin 8 gene regulation. Am. J. Respir. Cell Mol. Biol. 15: 787-795.

Simeonova PP, Toriumi W, Kommineni C, Erkan M, Munson AE, Rom WN & Luster MI (1997) Molecular regulation of IL-6 activation by asbestos in lung epithelial cells: role of reactive oxygen species. J. Immunol. 159: 3921-3928.

Simoni J, Simoni G, Martinez-Zaguilan R, Wesson DE, Lox CD, Prien SD & Kumar RV (1998) Improved blood substitute: evaluation of its effects on human endothelial cells. ASAIO J. 44: M356-M367.

Smith CS & Shearer WT (1994) Activation of NF-κB and immunoglobulin expression in response to platelet-activating factor in a human B cell line. Cell. Immunol. 155: 292-303.

Smith MFJ, Eidlen D, Arend WP & Gutierrez-Hartmann A (1994) LPS-induced expression of the human IL-1 receptor antagonist gene is controlled by multiple interacting promoter elements. J. Immunol. 153: 3584-3593.

Sporn LA, Sahni SK, Lerner NB, Marder VJ, Silverman DJ, Turpin LC & Schwab AL (1997) Rickettsia rickettsii infection of cultured human endothelial cells induces NF-κB activation. Infect. Immun. 65: 2786-2791.

Srahna M, Remacle JE, Annamalai K, Pype S, Huylebroeck D, Boogaerts MA & Vandenberghe P (2001) NF-κB is involved in the regulation of CD154 (CD40 ligand) expression in primary human T cells. Clin. Exp. Immunol. 125: 229-236.

Srivastava P (2002) Roles of heat-shock proteins in innate and adaptive immunity. Nat. Rev. Immunol. 2: 185-194.

Stacey MA, Sun G, Vassalli G, Marini M, Bellini A & Mattoli S (1997) The allergen Der p1 induces NF-κB activation through interference with IκBα function in asthmatic bronchial epithelial cells. Biochem. Biophys. Res. Commun. 236: 522-526.

Stehlik C, de Martin R, Binder BR & Lipp J (1998) Cytokine induced expression of porcine inhibitor of apoptosis protein (iap) family member is regulated by NF-κB. Biochem. Biophys. Res. Commun. 243: 827-832.

Stein B, Kramer M, Rahmsdorf HJ, Ponta H & Herrlich P (1989) UV-induced transcription from the human immunodeficiency virus type 1 (HIV-1) long terminal repeat and UV-induced secretion of an extracellular factor that induces HIV-1 transcription in nonirradiated cells. J. Virol. 63: 4540-4544.

Steinle AU, Weidenbach H, Wagner M, Adler G & Schmid RM (1999) NF-κB/Rel activation in cerulein pancreatitis. Gastroenterology 116: 420-430.

Strum JC, Small GW, Pauig SB & Daniel LW (1994) 1-β-D-Arabinofuranosylcytosine stimulates ceramide and diglyceride formation in HL-60 cells. J. Biol. Chem. 269: 15493-15497.

Sugano N, Chen W, Roberts ML & Cooper NR (1997) Epstein-Barr virus binding to CD21 activates the initial viral promoter via NF-κB induction. J. Exp. Med. 186: 731-737.

Sultana C, Shen Y, Johnson C & Kalra VK (1999) Cobalt chloride-induced signaling in endothelium leading to the augmented adherence of sickle red blood cells and transendothelial migration of monocyte-like HL-60 cells is blocked by PAF-receptor antagonist. J. Cell. Physiol. 179: 67-78.

Sun SC, Ganchi PA, Ballard DW & Greene WC (1993) NF-κB controls expression of inhibitor IκBα: evidence for an inducible autoregulatory pathway. Science 259: 1912-1915.

Suzuki K, Mori A, Ishii KJ, Saito J, Singer DS, Klinman DM, Krause PR & Kohn LD (1999) Activation of target-tissue immune-recognition molecules by double-stranded polynucleotides. Proc. Natl. Acad. Sci. USA 96: 2285-2290.

Suzuki YJ, Mizuno M & Packer L (1994) Signal transduction for nuclear factor-κB activation. Proposed location of antioxidant-inhibitable step. J. Immunol. 153: 5008-5015.

Szabo G, Katarova Z, Kortvely E, Greenspan RJ & Urban Z (1996) Structure and the promoter region of the mouse gene encoding the 67-kD form of glutamic acid decarboxylase. DNA Cell Biol. 15: 1081-1091.

Tabary O, Escotte S, Couetil JP, Hubert D, Dusser D, Puchelle E & Jacquot J (2000) High susceptibility for cystic fibrosis human airway gland cells to produce IL-8 through the IκB kinase α pathway in response to extracellular NaCl content. J. Immunol. 164: 3377-3384.

Tak PP, Gerlag DM, Aupperle KR, van de Geest DA, Overbeek M, Bennett BL, Boyle DL, Manning

AM & Firestein GS (2001) Inhibitor of nuclear factor κB kinase β is a key regulator of synovial inflammation. Arth. Rheum. 44: 1897-1907.

Takeuchi M & Baichwal VR (1995) Induction of the gene encoding mucosal vascular addressin cell adhesion molecule 1 by tumor necrosis factor α is mediated by NF-κB proteins. Proc. Natl. Acad. Sci. USA 92: 3561-3565.

Takizawa H, Ohtoshi T, Kawasaki S, Kohyama T, Desaki M, Kasama T, Kobayashi K, Nakahara K, Yamamoto K, Matsushima K & Kudoh S (1999) Diesel exhaust particles induce NF-κB activation in human bronchial epithelial cells in vitro: importance in cytokine transcription. J. Immunol. 162: 4705-4711.

Tamatani M, Che YH, Matsuzaki H, Ogawa S, Okado H, Miyake S, Mizuno T & Tohyama M (1999) Tumor necrosis factor induces Bcl-2 and Bcl-x expression through NFκB activation in primary hippocampal neurons. J. Biol. Chem. 274: 8531-8538.

Tamatani M, Mitsuda N, Matsuzaki H, Okado H, Miyake S, Vitek MP, Yamaguchi A & Tohyama M (2000) A pathway of neuronal apoptosis induced by hypoxia/reoxygenation: roles of nuclear factor-κB and Bcl-2. J. Neurochem. 75: 683-693.

Tanaka Y, Hayashi M, Takagi S & Yoshie O (1996) Differential transactivation of the intercellular adhesion molecule 1 gene promoter by Tax1 and Tax2 of human T-cell leukemia viruses. J. Virol. 70: 8508-8517.

Ten RM, Paya CV, Israel N, Le Bail O, Mattei MG, Virelizier JL, Kourilsky P & Israel A (1992) The characterization of the promoter of the gene encoding the p50 subunit of NF-κB indicates that it participates in its own regulation. EMBO J. 11: 195-203.

Ten RM, Blank V, Le Bail O, Kourilsky P & Israel A (1993) Two factors, IRF1 and KBF1/NF-κB, cooperate during induction of MHC class I gene expression by interferon α/β or Newcastle disease virus. Compt. Rend. Acad. Sci. III, Sci. Vie 316: 496-501.

Tensen CP, Flier J, Rampersad SS, Sampat-Sardjoepersad S, Scheper RJ, Boorsma DM & Willemze R (1999) Genomic organization, sequence and transcriptional regulation of the human CXCL 11_1 gene. Biochim. Biophys. Acta 1446: 167-172.

Tewari M, Dobrzanski P, Mohn KL, Cressman DE, Hsu JC, Bravo R & Taub R (1992) Rapid induction in regenerating liver of RL/IF-1 (an IκB that inhibits NF-κB, RelB-p50, and c-Rel-p50) and PHF, a novel κB site-binding complex. Mol. Cell. Biol. 12: 2898-2908.

Thevenin C, Kim SJ, Rieckmann P, Fujiki H, Norcross MA, Sporn MB, Fauci AS & Kehrl JH (1990) Induction of nuclear factor-κB and the human immunodeficiency virus long terminal repeat by okadaic acid, a specific inhibitor of phosphatases 1 and 2A. New Biologist 2: 793-800.

Thieblemont N, Haeffner-Cavaillon N, Haeffner A, Cholley B, Weiss L & Kazatchkine MD (1995) Triggering of complement receptors CR1 (CD35) and CR3 (CD11b/CD18) induces nuclear translocation of NF-κB (p50/p65) in human monocytes and enhances viral replication in HIV-infected monocytic cells. J. Immunol. 155: 4861-4867.

Tinnell SB, Jacobs-Helber SM, Sterneck E, Sawyer ST & Conrad DH (1998) STAT6, NF-κB and C/EBP in CD23 expression and IgE production. Int. Immunol. 10: 1529-1538.

Tong-Starksen SE, Welsh TM & Peterlin BM (1990) Differences in transcriptional enhancers of HIV-1 and HIV-2. Response to T cell activation signals. J. Immunol. 145: 4348-4354.

Toth CR, Hostutler RF, Baldwin AS & Bender TP (1995) Members of the nuclear factor κB family transactivate the murine c-myb gene. J. Biol. Chem. 270: 7661-7671.

Trede NS, Castigli E, Geha RS & Chatila T (1993) Microbial superantigens induce NF-κB in the human monocytic cell line THP-1. J. Immunol. 150: 5604-5613.

Tsutsumi-Ishii Y & Nagaoka I (2002) NF-κB-mediated transcriptional regulation of human β-defensin-2 gene following lipopolysaccharide stimulation. J. Leukocyte Biol. 71: 154-162.

Tulchinsky E, Prokhortchouk E, Georgiev G & Lukanidin E (1997) A κB-related binding site is an integral part of the mts1 gene composite enhancer element located in the first intron of the gene. J. Biol. Chem. 272: 4828-4835.

Twu JS, Chu K & Robinson WS (1989) Hepatitis B virus X gene activates κB-like enhancer sequences in the long terminal repeat of human immunodeficiency virus 1. Proc. Natl. Acad. Sci. USA 86: 5168-5172.

Ueda A, Okuda K, Ohno S, Shirai A, Igarashi T, Matsunaga K, Fukushima J, Kawamoto S, Ishigatsubo Y & Okubo T (1994) NF-κB and Sp1 regulate transcription of the human monocyte chemoattractant protein-1 gene. J. Immunol. 153: 2052-2063.

Ungefroren H & Krull NB (1996) Transcriptional regulation of the human biglycan gene. J. Biol. Chem. 271: 15787-15795.

Vadlamudi RK & Shin J (1998) Genomic structure and promoter analysis of the p62 gene encoding a non-proteasomal multiubiquitin chain binding protein. FEBS Lett. 435: 138-142.

Valen G, Yan ZQ & Hansson GK (2001) Nuclear factor κB and the heart. J. Am. Coll. Cardiol. 38: 307-314.

van de Stolpe A, Caldenhoven E, Stade BG, Koenderman L, Raaijmakers JA, Johnson JP & van der Saag PT (1994) 12-O-tetradecanoylphorbol-13-acetate- and tumor necrosis factor α-mediated induction of intercellular adhesion molecule-1 is inhibited by dexamethasone. Functional analysis of the human intercellular adhesion molecular-1 promoter. J. Biol. Chem. 269: 6185-6192.

van den Brink GR, ten Kate FJ, Ponsioen CY, Rive MM, Tytgat GN, van Deventer SJ & Peppelenbosch MP (2000) Expression and activation of NF-κB in the antrum of the human stomach. J. Immunol. 164: 3353-3359.

Verweij CL, Geerts M & Aarden LA (1991) Activation of interleukin-2 gene transcription via the T-cell surface molecule CD28 is mediated through an NF-κB-like response element. J. Biol. Chem. 266: 14179-14182.

Vincenti MP, Coon CI & Brinckerhoff CE (1998) Nuclear factor κB/p50 activates an element in the distal matrix metalloproteinase 1 promoter in interleukin-1β-stimulated synovial fibroblasts. Arth. Rheum. 41: 1987-1994.

Visvanathan KV & Goodbourn S (1989) Double-stranded RNA activates binding of NF-κB to an inducible element in the human β-interferon promoter. EMBO J. 8: 1129-1138.

Vockerodt M, Tesch H & Kube D (2001) Epstein-Barr virus latent membrane protein-1 activates CD25 expression in lymphoma cells involving the NFκB pathway. Genes Immun. 2: 433-441.

Walker AK & Enrietto PJ (1996) Analysis of the role of v-rel in transcriptional regulation of high mobility group 14. Oncogene 12: 2515-2525.

Walsh DE, Greene CM, Carroll TP, Taggart CC, Gallagher PM, O'Neill SJ & McElvaney NG (2001) Interleukin-8 up-regulation by neutrophil elastase is mediated by MyD88/IRAK/TRAF-6 in human bronchial epithelium. J. Biol. Chem. 276: 35494-35499.

Wang CD, Chang GD, Lee YK & Chen H (2001) A functional composite cis-element for NFκB and RBJκ in the rat pregnancy-specific glycoprotein gene. Biol. Reprod. 65: 1437-1443.

Wang CY, Mayo MW & Baldwin AS (1996) TNF- and cancer therapy-induced apoptosis: potentiation by inhibition of NF-κB. Science 274: 784-787.

Wang CY, Mayo MW, Korneluk RG, Goeddel DV & Baldwin AS (1998) NF-κB antiapoptosis: induction of TRAF1 and TRAF2 and c-IAP1 and c-IAP2 to suppress caspase-8 activation. Science 281: 1680-1683.

Wang T, Lafuse WP & Zwilling BS (2001) NFκB and Sp1 elements are necessary for maximal transcription of toll-like receptor 2 induced by Mycobacterium avium. J. Immunol. 167: 6924-6932.

Wang Z, Sicinski P, Weinberg RA, Zhang Y & Ravid K (1996) Characterization of the mouse cyclin D3 gene: exon/intron organization and promoter activity. Genomics 35: 156-163.

Wautier JL, Wautier MP, Schmidt AM, Anderson GM, Hori O, Zoukourian C, Capron L, Chappey O, Yan SD & Brett J (1994) Advanced glycation end products (AGEs) on the surface of diabetic erythrocytes bind to the vessel wall via a specific receptor inducing oxidant stress in the vasculature: a link between surface-associated AGEs and diabetic complications. Proc. Natl. Acad. Sci. USA 91: 7742-7746.

West DA, James NH, Cosulich SC, Holden PR, Brindle R, Rolfe M & Roberts RA (1999) Role for tumor necrosis factor α receptor 1 and interleukin-1 receptor in the suppression of mouse hepatocyte apoptosis by the peroxisome proliferator nafenopin. Hepatology 30: 1417-1424.

Westendorp MO, Shatrov VA, Schulze-Osthoff K, Frank R, Kraft M, Los M, Krammer PH, Droge W & Lehmann V (1995) HIV-1 Tat potentiates TNF-induced NF-κB activation and cytotoxicity by altering the cellular redox state. EMBO J. 14: 546-554.

Whelan J, Ghersa P, Hooft vH, Gray J, Chandra G, Talabot F & DeLamarter JF (1991) An NFκB-like factor is essential but not sufficient for cytokine induction of endothelial leukocyte adhesion molecule 1 (ELAM-1) gene transcription. Nucleic Acids Res. 19: 2645-2653.

Widmer U, Manogue KR, Cerami A & Sherry B (1993) Genomic cloning and promoter analysis of macrophage inflammatory protein (MIP)-2, MIP-1α, and MIP-1β, members of the chemokine superfamily of proinflammatory cytokines. J. Immunol. 150: 4996-5012.

Williams DL, Ha T, Li C, Kalbfleisch JH & Ferguson DA (1999) Early activation of hepatic NFκB and NF-IL6 in polymicrobial sepsis correlates with bacteremia, cytokine expression, and mortality. Ann. Surg. 230: 95-104.

Williams JL, Garcia J, Harrich D, Pearson L, Wu F & Gaynor R (1990) Lymphoid specific gene expression of the adenovirus early region 3 promoter is mediated by NF-κB binding motifs. EMBO J. 9: 4435-4442.

Wilson SH, Caplice NM, Simari RD, Holmes DRJ, Carlson PJ & Lerman A (2000) Activated nuclear factor-κB is present in the coronary vasculature in experimental hypercholesterolemia. Atherosclerosis 148: 23-30.

Wolf I, Pevzner V, Kaiser E, Bernhardt G, Claudio E, Siebenlist U, Forster R & Lipp M (1998) Downstream activation of a TATA-less promoter by Oct-2, Bob1, and NF-κB directs expression of the homing receptor BLR1 to mature B cells. J. Biol. Chem. 273: 28831-28836.

Wood JN (1995) Regulation of NF-κB activity in rat dorsal root ganglia and PC12 cells by tumour necrosis factor and nerve growth factor. Neurosci. Lett. 192: 41-44.

Worm MM, Tsytsykova A & Geha RS (1998) CD40 ligation and IL-4 use different mechanisms of transcriptional activation of the human lymphotoxin α promoter in B cells. Eur. J. Immunol. 28: 901-906.

Wright KL, White LC, Kelly A, Beck S, Trowsdale J & Ting JP (1995) Coordinate regulation of the human TAP1 and LMP2 genes from a shared bidirectional promoter. J. Exp. Med. 181: 1459-1471.

Wu H & Lozano G (1994) NF-κB activation of p53. A potential mechanism for suppressing cell growth in response to stress. J. Biol. Chem. 269: 20067-20074.

Wu K, Jiang SW, Thangaraju M, Wu G & Couch FJ (2000) Induction of the BRCA2 promoter by nuclear factor-κB. J. Biol. Chem. 275: 35548-35556.

Wu M, Lee H, Bellas RE, Schauer SL, Arsura M, Katz D, FitzGerald MJ, Rothstein TL, Sherr DH & Sonenshein GE (1996) Inhibition of NF-κB/Rel induces apoptosis of murine B cells. EMBO J. 15: 4682-4690.

Wu MX, Ao Z, Prasad KV, Wu R & Schlossman SF (1998) IEX-1L, an apoptosis inhibitor involved in NF-κB-mediated cell survival. Science 281: 998-1001.

Wu RL, Chen TT & Sun TT (1994) Functional importance of an Sp1- and an NFκB-related nuclear protein in a keratinocyte-specific promoter of rabbit K3 keratin gene. J. Biol. Chem. 269: 28450-28459.

Xia C, Hu J, Ketterer B & Taylor JB (1996) The organization of the human GSTP1-1 gene promoter and its response to retinoic acid and cellular redox status. Biochem. J. 313: 155-161.

Xu J, Zutter MM, Santoro SA & Clark RA (1998) A three-dimensional collagen lattice activates NF-κB in human fibroblasts: role in integrin α2 gene expression and tissue remodeling. J. Cell Biol. 140: 709-719.

Xu P, Huecksteadt TP & Hoidal JR (1996) Molecular cloning and characterization of the human xanthine dehydrogenase gene (XDH). Genomics 34: 173-180.

Yabe T, Wilson D & Schwartz JP (2001) NFκB activation is required for the neuroprotective effects of pigment epithelium-derived factor (PEDF) on cerebellar granule neurons. J. Biol. Chem. 276: 43313-43319.

Yamamoto K, Arakawa T, Ueda N & Yamamoto S (1995) Transcriptional roles of nuclear factor κB and nuclear factor-interleukin-6 in the tumor necrosis factor α-dependent induction of cyclooxygenase-2 in MC3T3-E1 cells. J. Biol. Chem. 270: 31315-31320.

Yan SD, Schmidt AM, Anderson GM, Zhang J, Brett J, Zou YS, Pinsky D & Stern D (1994) Enhanced cellular oxidant stress by the interaction of advanced glycation end products with their receptors/binding proteins. J. Biol. Chem. 269: 9889-9897.

Yan Z, Salmons S, Dang YI, Hamilton MT & Booth FW (1996) Increased contractile activity decreases RNA-protein interaction in the 3'-UTR of cytochrome c mRNA. Am. J. Physiol. 271: C1157-C1166.

Yang H, Wang J, Huang ZZ, Ou X & Lu SC (2001a) Cloning and characterization of the 5'-flanking region of the rat glutamate-cysteine ligase catalytic subunit. Biochem. J. 357: 447-455.

Yang H, Wang J, Ou X, Huang ZZ & Lu SC (2001b) Cloning and analysis of the rat glutamate-cysteine ligase modifier subunit promoter. Biochem. Biophys. Res. Commun. 285: 476-482.

Yang L, Cohn L, Zhang DH, Homer R, Ray A & Ray P (1998) Essential role of nuclear factor κB in the induction of eosinophilia in allergic airway inflammation. J. Exp. Med. 188: 1739-1750.

Yao KS & O'Dwyer PJ (1995) Involvement of NF-κB in the induction of NAD(P)H:quinone oxidoreductase (DT-diaphorase) by hypoxia, oltipraz and mitomycin C. Biochem. Pharmacol. 49: 275-282.

Yao Z, Fanslow WC, Seldin MF, Rousseau AM, Painter SL, Comeau MR, Cohen JI & Spriggs MK (1995) Herpesvirus Saimiri encodes a new cytokine, IL-17, which binds to a novel cytokine receptor. Immunity 3: 811-821.

Ye J, Shi X, Jones W, Rojanasakul Y, Cheng N, Schwegler-Berry D, Baron P, Deye GJ, Li C & Castranova V (1999) Critical role of glass fiber length in TNF-α production and transcription factor activation in macrophages. Am. J. Physiol. 276: L426-L434.

Ye J, Xie X, Tarassishin L & Horwitz MS (2000) Regulation of the NF-κB activation pathway by isolated domains of FIP3/IKKγ, a component of the IκB-α kinase complex. J. Biol. Chem. 275: 9882-9889.

Yerneni KK, Bai W, Khan BV, Medford RM & Natarajan R (1999) Hyperglycemia-induced activation of nuclear transcription factor κB in vascular smooth muscle cells. Diabetes 48: 855-864.

Yokoyama C, Yabuki T, Inoue H, Tone Y, Hara S, Hatae T, Nagata M, Takahashi EI & Tanabe T (1996) Human gene encoding prostacyclin synthase (PTGIS): genomic organization, chromosomal localization, and promoter activity. Genomics 36: 296-304.

You LR, Chen CM & Lee YH (1999) Hepatitis C virus core protein enhances NF-κB signal pathway triggering by lymphotoxin-β receptor ligand and tumor necrosis factor α. J. Virol. 73: 1672-1681.

You M, Ku PT, Hrdlickova R & Bose HRJ (1997) ch-IAP1, a member of the inhibitor-of-apoptosis protein family, is a mediator of the antiapoptotic activity of the v-Rel oncoprotein. Mol. Cell. Biol. 17: 7328-7341.

Yu DY, Huang ZM, Murakami S, Takahashi M & Nonaka M (1989) Specific binding of a hepatoma nuclear factor to the NFκB/H2TF1 recognition motif found in the C4 promoter, but not in the Slp promoter. J. Immunol. 143: 2395-2400.

Zhang D, Jiang SL, Rzewnicki D, Samols D & Kushner I (1995) The effect of interleukin-1 on C-reactive protein expression in Hep3B cells is exerted at the transcriptional level. Biochem. J. 310: 143-148.

Zhang DH, Yang L, Cohn L, Parkyn L, Homer R, Ray P & Ray A (1999) Inhibition of allergic inflammation in a murine model of asthma by expression of a dominant-negative mutant of GATA-3. Immunity 11: 473-482.

Zhang LH, Youn HD & Liu JO (2001) Inhibition of cell cycle progression by the novel cyclophilin ligand sanglifehrin A is mediated through the NFκB-dependent activation of p53. J. Biol. Chem. 276: 43534-43540.

Zhang Y, Broser M & Rom WN (1994) Activation of the interleukin 6 gene by Mycobacterium tuberculosis or lipopolysaccharide is mediated by nuclear factors NF-IL6 and NF-κB. Proc. Natl. Acad. Sci. USA 91: 2225-2229.

Zhao J, Freeman GJ, Gray GS, Nadler LM & Glimcher L (1996) A cell type-specific enhancer in the human B7.1 gene regulated by NF-κB. J. Exp. Med. 183: 777-789.

Zhou G & Kuo MT (1997) NF-κB-mediated induction of mdr1b expression by insulin in rat hepatoma cells. J. Biol. Chem. 272: 15174-15183.

Zhu N, Ramirez LM, Lee RL, Magnuson NS, Bishop GA & Gold MR (2002) CD40 signaling in B cells regulates the expression of the Pim-1 kinase via the NF-κB pathway. J. Immunol. 168: 744-754.

Zhu Y, Lin JH, Liao HL, Verna L & Stemerman MB (1997) Activation of ICAM-1 promoter by lysophosphatidylcholine: possible involvement of protein tyrosine kinases. Biochim. Biophys. Acta 1345: 93-98.

Zhu YX, Kang LY, Luo W, Li CC, Yang L & Yang YC (1996a) Multiple transcription factors are required for activation of human interleukin 9 gene in T cells. J. Biol. Chem. 271: 15815-15822.

Zhu Z, Tang W, Ray A, Wu Y, Einarsson O, Landry ML, Gwaltney JJ & Elias JA (1996b) Rhinovirus stimulation of interleukin-6 in vivo and in vitro. Evidence for nuclear factor κB-dependent transcriptional activation. J. Clin. Invest. 97: 421-430.

Zong WX, Edelstein LC, Chen C, Bash J & Gelinas C (1999) The prosurvival Bcl-2 homolog Bfl-1/A1 is a direct transcriptional target of NF-κB that blocks TNFα-induced apoptosis. Genes Dev. 13: 382-387.

Zwacka RM, Zhang Y, Zhou W, Halldorson J & Engelhardt JF (1998) Ischemia/reperfusion injury in the liver of BALB/c mice activates AP-1 and nuclear factor κB independently of IκB degradation. Hepatology 28: 1022-1030.

8. AFFILIATION

University Hospital of Freiburg, Department of Anesthesiology and Intensive Care Medicine, Breisacher Straße 66, 79106 Freiburg, Germany.

I. CARPENTIER, S. JANSSENS, K. HEYNINCK
& R. BEYAERT

NF-κB ACTIVATION BY TUMOR NECROSIS FACTOR AND INTERLEUKIN-1

1. INTRODUCTION

Tumor necrosis factor (TNF) and interleukin-1 (IL-1) are two important proinflammatory cytokines with overlapping and often synergizing activities. Both proteins are produced mainly by monocytes/macrophages, but also by other cell types of the immune system in response to inflammation, infection and other stress-inducing factors. Although TNF was originally identified on the basis of its antitumor activity and as a mediator of cachexia, it is a pluripotent cytokine involved in cell growth modulation, inflammation, tumorigenesis, tumor metastasis, viral replication, septic shock and several auto-immune diseases (Aggarwal et al., 2000). IL-1 has been found to be one of the most powerful endogenous pyrogens involved in the same physiological phenomena as TNF, except for tumor modulation (Dinarello et al., 2000a-b).

The pleiotropic character of TNF and IL-1 is reflected by triggering of several signal transduction pathways. Whereas the cytotoxic effects of TNF result from the induction of a cell death program, the proinflammatory function of both TNF and IL-1 is the consequence of transcriptional gene activation. TNF- and IL-1-activated mitogen-activated protein kinase (MAPK) p38 and c-Jun NH_2-terminal kinase (JNK) modulate TNF- and IL-1-induced gene expression (Kyriakis & Avruch, 2001). Both cytokines can also activate the IκB kinase (IKK) complex leading to translocation of the transcription factor NF-κB. This chapter will focus on proteins involved in the regulation of TNF- and IL-1-induced signal transduction towards activation of the IKK complex.

2. TNF-MEDIATED SIGNAL TRANSDUCTION

2.1. TNF induces TNF-receptor trimerization

TNF, the founder of a still growing trimeric cytokine family called the TNF superfamily, is expressed as a type II transmembrane precursor protein that is converted to mature secreted TNF (Kriegler et al., 1988). Both secreted and membrane-bound TNF have biological activity. TNF exerts its effects by binding

R. Beyaert (ed.), Nuclear Factor κB. Regulation and Role in Disease, 49-67.

to two distinct receptor types called TNF-R1 (p55) and TNF-R2 (p75) (Vandenabeele et al., 1992). These receptors are the prototype of a large family of cell surface receptors called the TNF-R superfamily. Their extracellular domains are characterized by the presence of a variable number of cysteine-rich domains (Smith et al., 1994). These receptors are subdivided into two classes based on their cytoplasmic tails: those containing a death domain (DD), such as TNF-R1 (Tartaglia et al., 1993), and those with a TNF-R-associated factor (TRAF)-binding domain such as TNF-R2 (Rothe et al., 1994). The 80 amino acids large DD was originally identified in apoptosis-inducing proteins, but was later recognized as a general protein interaction motif also found in proteins not involved in cell death (Feinstein et al., 1995). The TRAF-binding motif has no recognizable secondary motif (Ye et al., 1999). In most cell types, a large number of TNF effects is mediated via TNF-R1 (for example NF-κB activation, cytotoxicity), whereas in T-lymphocytes TNF signaling mainly occurs via TNF-R2 (Tartaglia & Goeddel, 1992; Vandenabeele et al., 1995). Both receptors can cooperate with each other and share common adaptors (Declercq et al., 1998). Pre-ligand assembly domains (PLADs) were identified in the extracellular part of the TNF-Rs responsible for oligomerization of an unliganded complex (Chan et al., 2000). The ligand TNF binds to its cognate receptor showing 3/3 stoichiometry in which a single TNF trimer binds three receptors (Banner et al., 1993). This event triggers these pre-associated oligomers and induces conformational changes in the cytoplasmic tails of the receptors and subsequent signal transduction (Chan et al., 2000).

2.2. Adaptors of the TNF-R complex

2.2.1. TRADD

Under unstimulated conditions, spontaneous signaling of TNF-R1 is blocked by DD-DD interaction with the negative regulator called "silencer-of-death domain" (SODD) (Jiang et al., 1999). After ligand binding, SODD dissociates from the receptor to allow recruitment of other TNF-R-associated proteins (Fig. 1). Immediately after receptor triggering, the TNF-R1-associated DD protein (TRADD) is recruited to TNF-R1 via DD-DD interaction (Jones et al., 1999). TRADD functions as a platform to recruit three additional mediators: Fas-associated DD protein (FADD) and receptor-interacting protein (RIP) through its C-terminal DD, and TRAF2 through its N-terminal domain (Hsu et al., 1996a-b). Whereas FADD is not involved in NF-κB activation, both RIP and TRAF2 are needed for TNF-R1-induced NF-κB activation. TRADD, however, does not directly bind TNF-R2.

2.2.2. TRAF2

TRAF2 (and the related protein TRAF1) was first identified biochemically by its interaction with the cytoplasmic tail of TNF-R2 (Rothe et al., 1994). Since then, the TRAF family has expanded to six related proteins, but only TRAF1 and

TRAF2 have been shown to be involved in TNF signal transduction (Chung et al., 2002). TRAF2, like most of the TRAF proteins, holds an N-terminal RING finger, a zinc finger domain and a C-terminal TRAF domain (Rothe et al., 1994). The latter domain mediates interaction with TNF-R2, TRADD and RIP (Rothe et al., 1994; Hsu et al., 1996b). Structural analysis of TRAF2 revealed a trimeric self-association through this TRAF domain (Park et al., 1999).

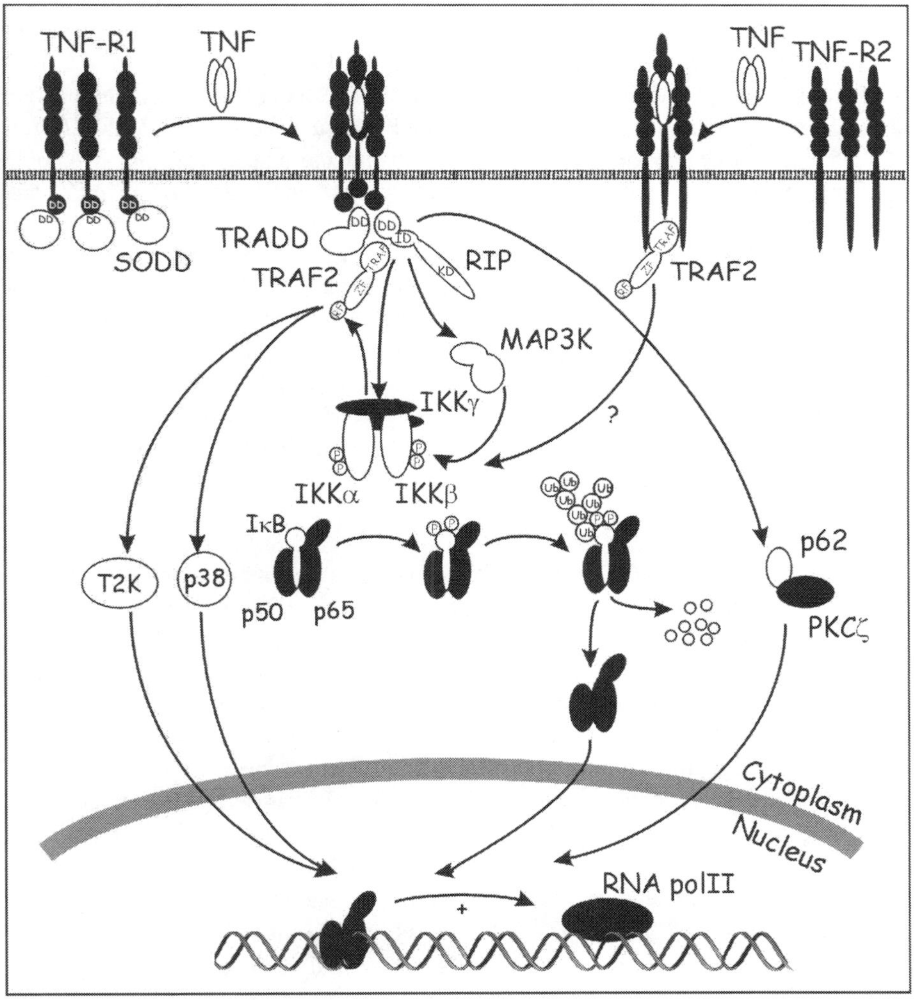

Fig. 1. Overview of TNF-induced activation of NF-κB (DD, death domain; ID, intermediate domain; KD, kinase domain; RF, RING finger; TRAF, TRAF domain; ZF, zinc finger).

Studies on cell lines derived from TRAF2-deficient mice demonstrated only slightly reduced and delayed TNF-induced NF-κB nuclear translocation (Yeh et al., 1997), but almost completely abolished MAPK activation of JNK and p38 (I. Carpentier, unpublished data). Nevertheless, IKK activity is severely reduced in TRAF2$^{-/-}$ MEF cells after TNF stimulation due to a lack of recruitment of IKK to the activated TNF-R1 complex (Devin et al., 2000). Introduction of TRAF2.DN (a dominant-negative mutant of TRAF2 lacking the RING finger) in MEF-TRAF2$^{-/-}$ cells could not reconstitute this recruitment, suggesting a requisite for the RING finger. Indeed, enforced multimerization of the N-terminal effector domain of TRAF2 consisting of the RING finger and the zinc fingers is sufficient for the activation of the IKK complex (Baud et al., 1999). In contrast to TRAF2$^{-/-}$ MEF cells, TRAF2/TRAF5 double-knockout MEF cells exhibit a complete defect in nuclear translocation of NF-κB upon TNF stimulation (Tada et al., 2001). TRAF5$^{-/-}$ T-thymocytes or MEF cells show no physiological role for TRAF5 in TNF-mediated NF-κB activation, implying redundant roles for TRAF2 and TRAF5 in TNF-mediated signaling.

In the case of TNF-R2, TRAF1 and TRAF2 form a heteromeric complex with the TRAF-binding domain of TNF-R2 (Rothe et al., 1995). Overexpression of TRAF2.DN inhibits TNF-R2-induced NF-κB gene expression (Rothe et al., 1994). Remarkably, structural-biochemical analysis of the TRAF2-TRADD interaction revealed a higher affinity of TRAF2 for TRADD than for TNF-R2 (Park et al., 2000). The degradation of TRAF2 after TNF-R2 signaling might serve as a negative feedback loop for TNF signaling towards NF-κB activation (Li et al., 2002b).

2.2.3. RIP

Triggering of TNF-R1 also leads to recruitment of another DD-containing protein called RIP (Stanger et al., 1995; Hsu et al., 1996a). RIP consists of an N-terminal kinase domain (KD), a central intermediate domain (ID) and a C-terminal DD (Hsu et al., 1996a). Studies on RIP$^{-/-}$ MEF or RIP-deficient Jurkat cells showed that RIP is essential for TNF-induced NF-κB activation, reflected by a highly reduced IKK activity (Ting et al., 1996; Kelliher et al., 1998). However, IKK recruitment to the receptor complex is unaffected in these cells (Devin et al., 2000). Reconstitution experiments in RIP-deficient Jurkat cells demonstrated that the charge-rich element within the ID of RIP is sufficient for RIP-mediated IKK activation. Indeed, after TNF stimulation RIP interacts through its ID with IKKγ, the regulatory component of the IKK complex (Zhang et al., 2000). Enforced oligomerization of the ID (which naturally occurs through the DD–DD interactions of RIP) induces oligomerization of IKKγ, resulting in activation of the IKK complex (Poyet et al., 2000). It is possible that RIP-IKKγ interaction induces conformational changes in IKKα and IKKβ, the catalytic subunits of the IKK complex, leading to their activation via trans-autophosphorylation (Delhase et al., 1999; Poyet et al., 2000). Another possibility may involve RIP-mediated recruitment and/or activation of an upstream IKK kinase, most probably a MAP3K.

Despite its well-studied role and function in TNF-R1 signal transduction, less is known about the involvement of RIP in TNF-R2 signal transduction towards

NF-κB. Interestingly, in T-thymocytes where TNF activities seem to be mediated by TNF-R2 (Tartaglia et al., 1991; Vandenabeele et al., 1992), it has been shown that in the presence of RIP, TNF-R2 triggers cell death, whereas in the absence of RIP, TNF-R2 activates NF-κB (Pimentel-Muinos & Seed, 1999).

2.2.4. IKK activation by IKK kinases

Although TRAF2 and RIP are recruited independently to TNF-R1, both are needed for full activation of the IKK complex: TRAF2 is responsible for the recruitment of the IKK complex to the activated TNF-R1, whereas RIP mediates IKK oligomerization. IKK oligomerization through IKKγ might induce trans-autophosphorylation of the activation loop serines of IKKα and IKKβ, resulting in their activation (Poyet et al., 2000). However, the possibility that another upstream kinase, which might be recruited and/or activated by RIP-IKKγ, is responsible for the activation of IKKα/β cannot be excluded.

Several MAP3Ks (upstream activators of the MAPK pathway) such as NF-κB-inducing kinase (NIK) and MAPK/extracellular signal-regulated kinase kinase kinase (MEKK)-1, 2 and 3 have been proposed as regulators of IKK activation (Woronicz et al., 1997; Nakano et al., 1998; Zhao & Lee, 1999). Both IKKα and IKKβ contain a canonical MAP2K activation loop required for their activation, which might explain why so many MAP3K can phosphorylate and subsequently activate IKKα and IKKβ after overexpression (Mercurio et al., 1997). Gene disruption studies clearly rejected a physiological role for NIK or MEKK1 in TNF-induced IKK activation and TNF-induced NF-κB DNA binding (Yujiri et al., 2000; Yin et al., 2001). It has to be noted, however, that NIK is required (together with IKKα) for processing of the NF-κB precursor p100/NF-κB2 to the mature p52 form and in this way possibly may affect the expression of a limited set of p52-dependent genes (Senftleben et al., 2001). In contrast, TNF-induced IKK activation is reduced in MEKK3$^{-/-}$ MEF cells, which could only be restored by a kinase-active MEKK3 (Yang et al., 2001). However, residual NFκB activity even in these cells suggests that other MEKKs may be involved in serving a redundant function. In addition, also the non-MAP3K protein kinase Cζ (PKCζ) has been implicated in TNF-mediated IKK activation. PKCζ can be recruited to TNF-R1 via an indirect interaction with RIP mediated by the adaptor protein p62. Antisense p62 impaired RIP-induced NF-κB-dependent gene expression (Sanz et al., 1999). Furthermore, genetic disruption of PKCζ resulted in abrogation of IKK activation after TNF injection. However, this could only be observed in lung where this kinase is abundantly expressed (Leitges et al., 2001).

3. IL-1-MEDIATED SIGNAL TRANSDUCTION

3.1. IL-1 exerts its effects via a heterodimeric receptor complex

The term IL-1 refers to two molecules, namely IL-1α and IL-1β, which are the oldest representatives of a recently expanded family of IL-1-like ligands, now comprising ten different family members (Lin et al., 2001). IL-1 mediates its multiple effects via a heterodimeric receptor complex consisting of IL-1 receptor 1 (IL-1R1) and IL-1 receptor accessory protein (IL-1RAcP) (Sims, 2002). Both belong to the IL-1 receptor family, whose members are defined by the presence of three Ig domains in the extracellular part and a Toll/IL-1 receptor-like (TIR) domain in the intracellular part (O'Neill, 2000). The TIR domain is a region of high similarity between the human IL-1 receptor and the related Drosophila Toll protein and is characteristic for the Toll/IL-1R superfamily of proteins. All members of this family are implicated in host defence responses against pathogens and can be found in organisms as diverse as plants, insects, nematodes or even bacteria, indicating that this motif is extremely conserved and one of the oldest signaling motifs in evolution (O'Neill, 2000).

Knockout studies of the IL-1R1 and studies in EL-4D6/76 murine thymoma cells lacking expression of the IL-1RAcP revealed that in both cases all IL-1-induced responses were abrogated, showing that IL-1R1 as well as IL-1RAcP are required and that there is no redundancy with other receptors of the IL-1R family (Labow et al., 1997; Wesche et al., 1997b). Crystal structure studies revealed that IL-1 contacts the IL-1R1 extracellular domain both at the tightly linked Ig domains 1 and 2 and at the flexible Ig domain 3. This provokes a movement of the Ig domain 3, which enables recruitment of IL-1RAcP, resulting in the formation of a functional receptor complex (Vigers et al., 1997) and activation of downstream signaling pathways.

3.2. IL-1-induced signal transduction cascade to NF-κB

Triggering of the IL-1R complex results in activation of the IKK complex, leading to phosphorylation of IκB and translocation of NF-κB. Neither IL-1R1 nor IL-1RAcP possesses intrinsic kinase activity, so that they rely on the recruitment of adaptor molecules to transduce their signals. Knockout studies revealed that activation of NF-κB by IL-1 requires at least three signaling intermediates: myeloid differentiation factor 88 (MyD88), IL-R-associated kinase (IRAK) and TRAF6 (Adachi et al., 1998; Kanakaraj et al., 1998; Lomaga et al., 1999). Reconstitution experiments in a cell-free system showed that TGFβ-activated kinase 1 (TAK1), a MAP3K, is essential for TRAF6-mediated activation of the IKK complex (Wang et al., 2001).

3.2.1. MyD88

The adaptor protein MyD88 was originally identified in $M1D^+$ myeloid precursors as a novel myeloid differentiation primary response gene, activated upon induction of terminal differentiation and growth arrest by IL-6 (Lord et al., 1990). Later on, MyD88 was purified from IL-1-stimulated 293 cells by immunoprecipitation of the IL-1R complex (Wesche et al., 1997a). It consists of an N-terminal DD, a short ID and a C-terminal TIR domain. Upon IL-1 stimulation, a functional IL-1R1/IL-1RAcP receptor complex is formed, generating a docking site for MyD88 homodimers, which are recruited via their TIR domains (Fig. 2). With its dual domain organization, MyD88 was originally considered an essential adaptor for DD-mediated recruitment of the downstream molecule IRAK to the activated receptor complex (Wesche et al., 1997a).

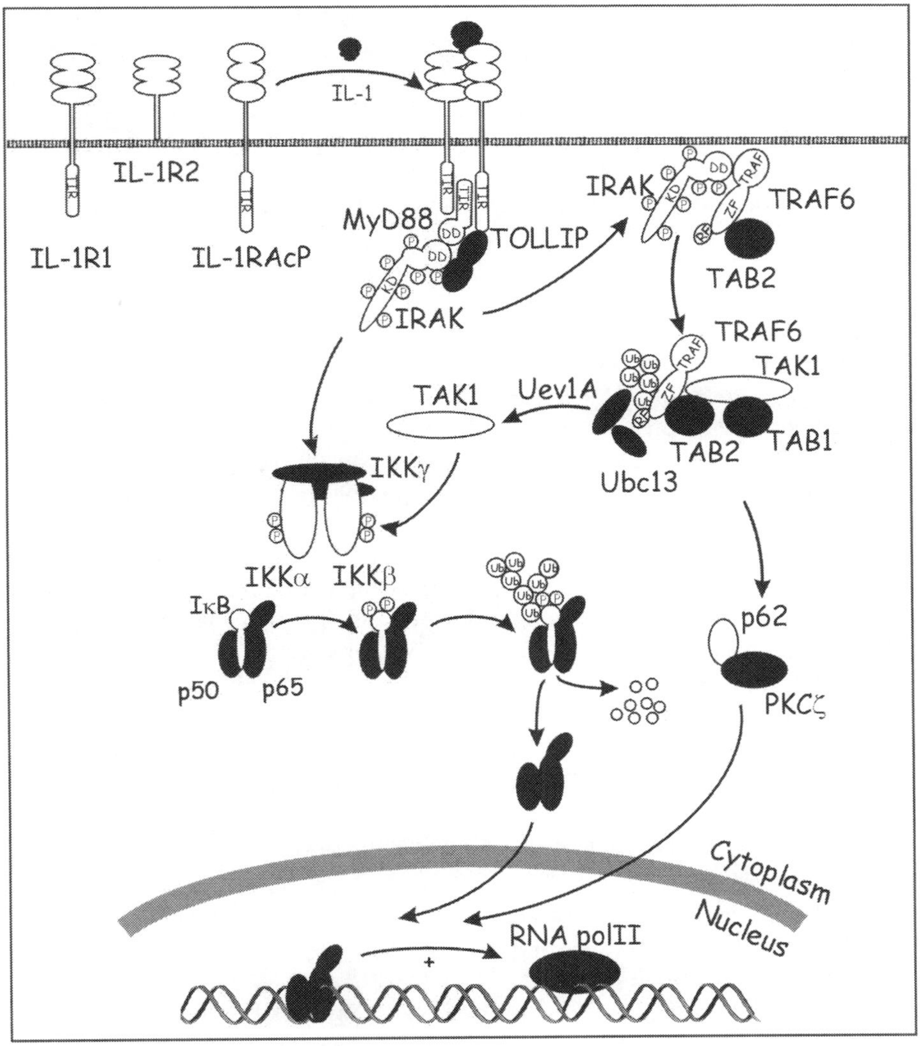

Fig. 2. Overview of IL-1-induced activation of NF-κB (DD, death domain; ID, intermediate domain; KD, kinase domain; RF, RING finger; TIR, Toll-IL-1R homology domain; TRAF, TRAF domain; ZF, zinc finger).

However, it was reported that IRAK can also be recruited in an MyD88-independent manner, probably through binding with another adaptor protein, viz. toll-interacting protein (Tollip) (Burns et al., 2000). Tollip was identified in a yeast-two-hybrid screening with IL-1RAcP as bait. In unstimulated cells, Tollip was shown to form heterodimers with IRAK, preventing IRAK from being autophosphorylated. Upon IL-1 induction, Tollip/IRAK complexes translocate to the IL-1R complex, providing an alternative for MyD88-mediated recruitment of IRAK. However, overexpression of Tollip dose-dependently inhibits IL-1-mediated signaling, implying the need for knockout studies to reveal if Tollip acts as an adaptor or an inhibitor of the IL-1-signaling pathway (Burns et al., 2000). In contrast, knockout studies revealed that MyD88 is indispensable for IL-1-mediated NF-κB and JNK activation (Adachi et al., 1998). Recently, MyD88 adaptor-like molecule (Mal) has been identified which also contains a TIR domain, but does not seem to be implicated in IL-1 signaling (Fitzgerald et al., 2001).

3.2.2. IRAK

IRAK consists of an N-terminal DD, an undetermined domain (UD), a KD and a C-terminal stretch, and shows significant homology to the Drosophila DD protein kinase pelle (Cao et al., 1996a). IRAK was originally identified by its protein kinase activity that coprecipitated with the IL-1R1 upon IL-1 stimulation (Croston et al., 1995). Indeed, IRAK is rapidly recruited to the IL-1R complex (through interaction with MyD88 or Tollip), becomes phosphorylated and dissociates from the receptor complex (Cao et al., 1996a).

The mechanism of IRAK activation is still controversial. Many authors believe that DD-DD interaction with MyD88 at the receptor complex is sufficient to trigger IRAK autophosphorylation. However, in IRAK$^{-/-}$ cells, IRAK(K239A), a kinase-dead version of IRAK which completely restores IL-1-induced NF-κB activation, still becomes phosphorylated upon IL-1 stimulation, showing that phosphorylation must be due to another kinase (Li et al., 1999). We showed that a splice variant of MyD88, which lacks the ID, is still able to bind IRAK at the receptor complex. However, this interaction is not sufficient to induce IRAK phosphorylation (Janssens et al., 2002). Probably, the ID of MyD88 is involved in recruiting and/or activating an as yet unidentified kinase, which then induces IRAK phosphorylation.

Although IRAK kinase activity is not required for NF-κB signaling, phosphorylation of IRAK is correlated with IL-1-mediated NF-κB activation (Li et al., 1999). Recently, a homologue of IRAK, called IRAK-4, has been proposed as an IRAK kinase (Li et al., 2002a). Presumably, phosphorylation of IRAK triggers its dissociation from MyD88 and/or Tollip and allows interaction with the downstream signaling intermediate TRAF6 at the membrane. Like RIP in the TNF-signaling cascade, IRAK also interacts with IKK-γ, although the implications of this interaction have not been revealed yet (Cooke et al., 2001).

At the moment, four different IRAK family members have been identified in humans (Muzio et al., 1997; Wesche et al., 1999; Suzuki et al., 2002), but their expression patterns and physiological relevance have not been established yet. IRAK$^{-/-}$ MEF cells showed a strong defect in IL-1-mediated JNK and p38 activation, but only a slight defect in NF-κB activation, suggesting that IRAK proteins might have redundant functions (Kanakaraj et al., 1998). In contrast, IRAK-$4^{-/-}$ MEF cells are completely deficient in IL-1-mediated signaling to NF-κB and JNK (Suzuki et al., 2002). Indeed, expression of IRAK, IRAK-2 or IRAK-3/-M restores the IL-1-mediated signaling pathway to NF-κB in IRAK-deficient 293 cells but not in IRAK-$4^{-/-}$ MEF cells (Li et al., 1999; Suzuki et al., 2002). However, studies with IRAK-3/-M-deficient cells recently showed a negative regulatory function for IRAK-3/-M in TLR/IL-1R-signaling (Kobayashi et al., 2002a). The reason for these controversial findings is still unclear, but may be due to overexpression.

3.2.3. TRAF6 and activation of the IKK complex

TRAF6 was identified by blast searches with the TRAF domain of TRAF2. Conform to the architecture of other TRAF proteins, TRAF6 consists of a conserved TRAF domain (although it is the most divergent of all TRAF proteins), an N-terminal RING finger domain and five zinc-finger structures (Cao et al., 1996b). Knockout studies revealed that TRAF6 is indispensable for IL-1-mediated NF-κB and JNK activation (Lomaga et al., 1999). Unlike TRAF2 in the TNF-signaling pathway, TRAF6 is not recruited to the IL-1R, but becomes activated at the membrane by receptor-dissociated IRAK (Qian et al., 2001). In analogy with TRAF2, the N-terminal part of TRAF6 is the effector domain and the C-terminal TRAF domain is required for oligomerization (Baud et al., 1999). This suggests that an upstream activator like IRAK, which binds to the TRAF-domain, induces activation of TRAF6 by clustering its effector domain. Furthermore, IRAK also mediates the translocation of TRAF6 together with another adaptor protein TAK1-binding protein 2 (TAB2) from the membrane to the cytosol (Takaesu et al., 2000; Qian et al., 2001). There TRAF6 and TAB2 form a complex with the downstream signaling molecules TAK1 and TAK1-binding protein 1 (TAB1) (Qian et al., 2001).

TAK1 is a MAP3K, originally identified in the TGFβ-signaling pathway (Yamaguchi et al., 1995), but also implicated in the phosphorylation of the IKK complex (Ninomiya-Tsuji et al., 1999). Indeed, in a cell-free system it was demonstrated that the only two complexes needed for TRAF6 to activate the IKK complex are the complex consisting of TAK1, TAB1 and TAB2 and the dimeric E2 ubiquitin-conjugating enzyme complex Ubc13/Uev1A (Deng et al., 2000; Wang et al., 2001). Furthermore, upon IL-1 stimulation or after enforced clustering, TRAF6 is conjugated by a K63-linked polyubiquitin chain. Mutants of TRAF6 that cannot be ubiquitinated, as well as the addition of methylated ubiquitin block TRAF6-mediated activation of IKK, suggesting that TAK1 is a ubiquitin-dependent kinase of IKK, representing a new mode of MAP3K activation (Wang et al., 2001). These authors showed that the RING finger is essential for polyubiquitination, indicating that, like in many other RING finger proteins, this structure enables TRAF6 to function as an E3 ubiquitin ligase (Lorick et al., 1999). Remarkably, deletion mutants of TRAF6 lacking the RING finger were capable of restoring the NF-κB pathway in TRAF6$^{-/-}$ cells, although these deletion constructs did not activate TAK1 (Kobayashi et al., 2001). It might be that in MEF cells the contribution of TAK1 to IL-1-induced NF-κB is negligible and that other IKK activation pathways might exist.

Other TRAF6-binding proteins indeed suggest the existence of multiple pathways downstream of TRAF6. Via the adaptor protein p62, TRAF6 is linked to PKCζ (Sanz et al., 2000). Knockout studies of PKCζ showed a partial inhibition of IL-1-induced NF-κB activation, but no correlation to a defect in IKK activation (Leitges et al., 2001). Another TRAF6-binding protein is the RIP-related protein RIP2, a serine/threonine kinase that carries a C-terminal caspase-recruitment domain (CARD) instead of a DD in RIP (Thome et al., 1998). RIP2$^{-/-}$ MEF cells were shown to be hyporesponsive to IL-1-mediated IL-6 production or thymocyte proliferation, indicating that RIP2 also contributes to optimal IL-1-induced NF-κB signaling (Kobayashi et al., 2002b). Finally, TRAF6 also binds to the conserved adaptor protein called "evolutionarily conserved signaling intermediate in Toll pathways" (ECSIT), through which it is linked with MEKK1 (Kopp et al., 1999). Although MEKK1 was proposed to be an activator of the IKK complex, MEKK1$^{-/-}$ cells showed no defect in IL-1-induced NF-κB activation (Yujiri et al., 2000), undermining the physiological relevance of this interaction to NF-κB activation. Hence the signaling pathway from TRAF6 to IKK activation probably involves TAK1, though redundancy with other unidentified MAP3K might exist.

4. NF-κB TRANSACTIVATION BY TNF AND IL-1

Much attention has been paid to the identification of the IκB kinase leading to phosphorylation and degradation of IκBα, thereby allowing nuclear translocation of NF-κB. In addition, other TNF- and IL-1-initiated signaling pathways enhance the transcriptional activity of nuclear NF-κB, a phenomenon called transactivation (Bergmann et al., 1998). For several kinases a role in transactivation has been deduced from knockout studies (TRAF2-associated kinase, glycogen synthase kinase 3β, PKCζ) or inhibitor studies (p38, extracellular regulated kinase, phosphatidylinositol 3-kinase), demonstrating normal nuclear NF-κB translocation but impaired NF-κB-dependent gene expression (Beyaert et al., 1996; Vanden Berghe et al., 1998; Sizemore et al., 1999; Bonnard et al., 2000; Hoeflich et al., 2000; Leitges et al., 2001).

In general, transactivation can be achieved by posttranslational modifications of NF-κB itself, coactivators, histones or cooperative transcription factors. Casein kinase II (CKII) and IKKs themselves have been shown to phosphorylate serine residues in the transactivation domain of p65, while PKCζ phosphorylates the Rel homology domain of p65 (Sakurai et al., 1999; Wang et al., 2000; Leitges et al., 2001). Phosphorylated sites on nuclear p65 can serve as docking sites to recruit histone acetyl transferases (HATs) (Gerritsen et al., 1997; Sheppard et al., 1999). HATs are coactivators that function by facilitating or bridging NF-κB to the basal transcription machinery and altering chromatin structure.

The second type of chromatin modification, viz. phosphorylation of histone 3 (H3), involves p38 and ERK. This is an indirect event requiring a downstream kinase called 'mitogen- and stress-activated kinase 1' (MSK1) (Deak et al., 1998). Chromatin immunoprecipitation assays showed that p38-dependent phosphorylation of H3 affects the chromatin structure and enhances the accessibility of cryptic NF-κB-binding sites in a subset of NF-κB-dependent promoters (Saccani et al., 2002). Otherwise, p38 might also regulate NF-κB-driven gene expression by phosphorylation of the core element of the TATA box-binding protein TFIID (Carter et al., 1999). In conclusion, the activity of nuclear NF-κB is also a regulated process adding another level of complexity to the activation of NF-κB.

5. NEGATIVE REGULATION OF TNF- AND IL-1-MEDIATED ACTIVATION OF NF-κB BY CELLULAR PROTEINS

Under physiological conditions, TNF activity is regulated extracellularly by constitutive proteolytical cleavage of both membrane-bound TNF-R1 and TNF-R2, inducing enhanced levels of soluble TNF-Rs, also called TNF-binding proteins (Pinckard et al., 1997). These soluble receptors compete with membrane-bound TNF-R for binding of TNF, resulting in negative regulation of TNF activities (Ythier et al., 1993). Alternatively, it has also been suggested that soluble TNF-Rs form a stable complex with TNF, thereby keeping TNF for longer periods in its active form (Aderka et al., 1992).

Regulation of IL-1-mediated signal transduction at the receptor level occurs by counteracting homologues of both IL-1 and IL-1R. The IL-1R antagonist IL-1ra binds with high affinity to IL-1R1 but in contrast to IL-1 only contacts the Ig domains 1 and 2, but not domain 3. Therefore, IL-1ra does not trigger recruitment of IL-1RacP, and is unable to induce signaling (Schreuder et al., 1997). Knockout mice of IL-1ra spontaneously develop arthritis, showing the important function of IL-1ra to maintain homeostasis of the immune system (Horai et al., 2000). Besides, several new IL-1-related ligands have been recently identified, among which some also bind to IL-1R1 (Lin et al., 2001). Future studies will reveal if these proteins are also implicated in the control of IL-1 functioning.

On the other hand, IL-1 binds with high affinity to a decoy receptor IL-1R2, either to its membrane-bound form or its shed form (Colotta et al., 1993). IL-1R2 shows a high degree of similarity to IL-1R1 in its extracellular domain but has a very short intracellular domain of only 29 amino acids. Binding of IL-1 to IL-1R2 also triggers recruitment of IL-1RacP, but due to the absence of a TIR domain, no signaling molecules are recruited and no IL-1 response is generated (Lang et al., 1998). This means that IL-1R2 limits IL-1 activity in two ways, viz. by trapping IL-1 in an inactive receptor complex and by limiting the availability of IL-1RacP for IL-1R1. Absence of IL1R2 expression leads to uncontrolled IL-1 activity and results in severe inflammatory disorders, such as in human arthritis-affected cartilage (Attur et al., 2002).

Also intracellularly, TNF- and IL-1-initiated signal transduction pathways are negatively regulated by different mechanisms of which some involve depletion, translocation and secondary modifications of signaling proteins. For instance, the activity of IKKβ is tightly controlled by its phosphorylation state. Upon triggering by TNF or IL-1, initial phosphorylation of two serines in the activation loop of IKKβ leads to its activation. However, once activated, hyperphosphorylation of IKKβ by autophosphorylation at its C-terminal serine cluster results in decreased IKK activity, thereby preventing prolonged activation of the inflammatory response (Delhase et al., 1999).

In addition, triggering via TNF-R2 was suggested to lead to a redistribution of TRAF2 from a cytoplasmic fraction into a subcellular membrane fraction (Arch et al., 2000) or to ubiquitination and proteasome-dependent degradation of TRAF2 (Duckett & Thompson, 1997; Li et al., 2002b). Both phenomena probably result in a depletion of cytoplasmic TRAF2 available for recruitment to the TNF-R complex, finally leading to downregulation of TRAF2-dependent signal transduction. In the case of IL-1 signaling, phosphorylation, followed by proteasome-dependent degradation of IRAK, was observed, indicative of a similar regulation mechanism (Yamin & Miller, 1997). Alternatively, specific interference with TNF or IL-1-mediated signaling can also be achieved by expression of splice variants of intermediate signaling molecules. The TRAF2 splice variant TRAF2A contains an extra 21 base pairs, encoding a 7-amino acid insert within the effector RING finger domain. Overexpression studies indicate that TRAF2A interferes with TNF-R2-mediated NF-κB activation, whereas its effect on TNF-R1-mediated NF-κB activation has not been analyzed. TRAF2A/TRAF2 shows the highest expression levels in spleen, adipose tissue and thymus, suggesting that TRAF2A might interfere in a tissue- or cell-specific manner with TNF-induced signaling (Brink & Lodish, 1998). However, the precise role of TRAF2A remains to be identified, but indicates the importance of an intact RING finger domain. Alternatively, the IL-1-signaling pathway is negatively regulated by a splice variant of MyD88, viz. $MyD88_S$, which lacks the ID (Janssens et al., 2002). Although $MyD88_S$ can still interact with the IL-1R complex and IRAK, it is defective to induce NF-κB activation, probably by its inability to induce IRAK phosphorylation. $MyD88_S$ is predominantly expressed in spleen, indicative of a tissue-specific regulation of IL-1 activity by this splice variant.

Moreover, negative feedback regulation to terminate activation of NF-κB is obtained by NF-κB-dependent upregulation of cellular proteins. The best identified among these proteins is the NF-κB-interacting protein IκB. IκB not only retains NF-κB in its inactive state in the cytoplasm, but has also the capacity to interfere with nuclear DNA-bound NF-κB, resulting in a transient activation of NF-κB. This transient activation profile also requires the zinc finger protein A20: disruption of A20 leads to a prolonged activation of NF-κB due to a constitutively active IKK complex, indicating that A20 interferes with the signal transduction pathway upstream of this complex. Furthermore, these gene disruption studies indicate that A20 specifically interferes with TNF-mediated signal transduction, whereas IL-1 signaling is unaffected (Lee et al., 2000). The underlying mechanism for the inhibitory effect of A20 is still unknown, but may involve additional A20-binding proteins (ABINs) (Heyninck et al., 1999; Beyaert et al., 2000; Van Huffel et al., 2001).

Another NF-κB-regulated inhibitor protein is TRAF1, the only TRAF family member that lacks an N-terminal RING finger. TRAF1 interacts with TRAF2 and is recruited together with TRAF2 in a TNF-dependent way to TNF-R2 (Rothe et al., 1994). Overexpression of TRAF1 inhibits specifically TNF-mediated NF-κB activation (Carpentier & Beyaert, 1999), and disruption of the TRAF1 gene results in enhanced activation of NF-κB by triggering of TNF-R2 (Tsitsikov et al., 2001). The presence of multiple mechanisms for downregulating NF-κB-dependent responses indicates the importance of this regulation, finally leading to a well-controlled, transient activation profile of NF-κB.

6. FUTURE DIRECTIONS

The knowledge of the TNF- and IL-1-initiated signal transduction pathways leading to activation of the IKK complex is increasing daily. The quest for the real IKK kinase(s) is still going on, since genetic disruption studies have excluded numerous candidates. The only two plausible IKK kinases are TAK1 and MEKK3 for IL-1 and TNF, respectively (Wang et al., 2001; Yang et al., 2001). It seems, however, more likely that redundancy with other as yet unidentified MAP3K might exist. It should also be noted that the induced proximity model, in which IKKα and IKKβ activate each other via trans-autophosphorylation when they are in close proximity, remains an attractive alternative (Poyet et al., 2000).

Apart from the cytoplasmic activation of NF-κB by removal of its inhibitor IκB, mechanisms governing NF-κB activity in the nucleus have been poorly documented for a long time. The importance of transactivation has been underscored until recent knockout studies of potential IKK kinases unexpectedly showed an effect on gene expression but not on nuclear translocation of NF-κB (Bonnard et al., 2000; Hoeflich et al., 2000; Leitges et al., 2001; Yin et al., 2001). For most of these kinases, it remains unclear how they interfere with NF-κB transactivation. In the future, people probably will have to look more and more into transcription and chromatin remodeling machinery to find out.

7. ACKNOWLEDGMENTS

Research was supported by the Instituut voor de Aanmoediging van Innovatie door Wetenschap en Technologie in Vlaanderen, the Fonds voor Wetenschappelijk Onderzoek—Vlaanderen, the Interuniversitaire Attractiepolen, the Belgische Federatie tegen Kanker and the Stichting Emmanuel van der Schueren. We apologize to those who made also important contributions to understand the TNF- and IL-1-induced signal transduction pathways to NF-κB, but could not be cited due to space limitations.

8. REFERENCES

Adachi O, Kawai T, Takeda K, Matsumoto M, Tsutsui H, Sakagami M, Nakanishi K & Akira S (1998) Targeted disruption of the MyD88 gene results in loss of IL-1- and IL- 18-mediated function. Immunity 9: 143-150.

Aderka D, Engelmann H, Maor Y, Brakebusch C & Wallach D (1992) Stabilization of the bioactivity of tumor necrosis factor by its soluble receptors. J. Exp. Med. 175: 323-329.

Aggarwal BB, Samanta A & Feldmann M (2000) TNFα, in Cytokine Reference (Oppenheim JJ & Feldmann M, eds), Vol. I, Academic Press, San Diego, pp. 413-434.

Arch RH, Gedrich RW & Thompson CB (2000) Translocation of TRAF proteins regulates apoptotic threshold of cells. Biochem. Biophys. Res. Commun. 272: 936-945.

Attur MG, Dave MN, Leung MY, Cipolletta C, Meseck M, Woo SL & Amin AR (2002) Functional genomic analysis of type II IL-1β decoy receptor: potential for gene therapy in human arthritis and inflammation. J. Immunol. 168: 2001-2010.

Banner DW, D'Arcy A, Janes W, Gentz R, Schoenfeld HJ, Broger C, Loetscher H & Lesslauer W (1993) Crystal structure of the soluble human 55 kd TNF receptor-human TNF β complex: implications for TNF receptor activation. Cell 73: 431-445.

Baud V, Liu ZG, Bennett B, Suzuki N, Xia Y & Karin M (1999) Signaling by proinflammatory cytokines: oligomerization of TRAF2 and TRAF6 is sufficient for JNK and IKK activation and target gene induction via an amino-terminal effector domain. Genes Dev. 13: 1297-1308.

Bergmann M, Hart L, Lindsay M, Barnes PJ & Newton R (1998) IκBα degradation and nuclear factor-κB DNA binding are insufficient for interleukin-1β and tumor necrosis factor-α-induced κB-dependent transcription. Requirement for an additional activation pathway. J. Biol. Chem. 273: 6607-6610.

Beyaert R, Cuenda A, Vanden Berghe W, Plaisance S, Lee JC, Haegeman G, Cohen P & Fiers W (1996) The p38/RK mitogen-activated protein kinase pathway regulates interleukin-6 synthesis response to tumor necrosis factor. EMBO J. 15: 1914-1923.

Beyaert R, Heyninck K & Van Huffel S (2000) A20 and A20-binding proteins as cellular inhibitors of nuclear factor-κB-dependent gene expression and apoptosis. Biochem. Pharmacol. 60: 1143-1151.

Bonnard M, Mirtsos C, Suzuki S, Graham K, Huang J, Ng M, Itie A, Wakeham A, Shahinian A, Henzel WJ, Elia AJ, Shillinglaw W, Mak TW, Cao Z & Yeh WC (2000) Deficiency of T2K leads to apoptotic liver degeneration and impaired NF-κB-dependent gene transcription. EMBO J. 19: 4976-4985.

Brink R & Lodish HF (1998) Tumor necrosis factor receptor (TNFR)-associated factor 2A (TRAF2A), a TRAF2 splice variant with an extended RING finger domain that inhibits TNFR2-mediated NF-κB activation. J. Biol. Chem. 273: 4129-4134.

Burns K, Clatworthy J, Martin L, Martinon F, Plumpton C, Maschera B, Lewis A, Ray K, Tschopp J & Volpe F (2000) Tollip, a new component of the IL-1RI pathway, links IRAK to the IL-1 receptor. Nat. Cell Biol. 2: 346-351.

Cao Z, Henzel WJ & Gao X (1996a) IRAK: a kinase associated with the interleukin-1 receptor. Science 271: 1128-1131.

Cao Z, Xiong J, Takeuchi M, Kurama T & Goeddel DV (1996b) TRAF6 is a signal transducer for interleukin-1. Nature 383: 443-446.

Carpentier I & Beyaert R (1999) TRAF1 is a TNF inducible regulator of NF-κB activation. FEBS Lett. 460: 246-250.

Carter AB, Knudtson KL, Monick MM & Hunninghake GW (1999) The p38 mitogen-activated protein kinase is required for NF-κB-dependent gene expression. The role of TATA-binding protein (TBP). J. Biol. Chem. 274: 30858-30863.

Chan FK, Chun HJ, Zheng L, Siegel RM, Bui KL & Lenardo MJ (2000) A domain in TNF receptors that mediates ligand-independent receptor assembly and signaling. Science 288: 2351-2354.

Chung JY, Park YC, Ye H & Wu H (2002) All TRAFs are not created equal: common and distinct molecular mechanisms of TRAF-mediated signal transduction. J. Cell Sci. 115: 679-688.

Colotta F, Re F, Muzio M, Bertini R, Polentarutti N, Sironi M, Giri JG, Dower SK, Sims JE &

Mantovani A (1993) Interleukin-1 type II receptor: a decoy target for IL-1 that is regulated by IL-4. Science 261: 472-475.

Cooke EL, Uings IJ, Xia CL, Woo P & Ray KP (2001) Functional analysis of the interleukin-1-receptor-associated kinase (IRAK-1) in interleukin-1 β-stimulated nuclear factor κB (NF-κB) pathway activation: IRAK-1 associates with the NF-κB essential modulator (NEMO) upon receptor stimulation. Biochem. J. 359: 403-410.

Croston GE, Cao Z & Goeddel DV (1995) NF-κB activation by interleukin-1 (IL-1) requires an IL-1 receptor-associated protein kinase activity. J. Biol. Chem. 270: 16514-16517.

Deak M, Clifton AD, Lucocq LM & Alessi DR (1998) Mitogen- and stress-activated protein kinase-1 (MSK1) is directly activated by MAPK and SAPK2/p38, and may mediate activation of CREB. EMBO J. 17: 4426-4441.

Declercq W, Denecker G, Fiers W & Vandenabeele P (1998) Cooperation of both TNF receptors in inducing apoptosis: involvement of the TNF receptor-associated factor binding domain of the TNF receptor 75. J. Immunol. 161: 390-399.

Delhase M, Hayakawa M, Chen Y & Karin M (1999) Positive and negative regulation of IκB kinase activity through IKKβ subunit phosphorylation. Science 284: 309-313.

Deng L, Wang C, Spencer E, Yang L, Braun A, You J, Slaughter C, Pickart C & Chen ZJ (2000) Activation of the IκB kinase complex by TRAF6 requires a dimeric ubiquitin-conjugating enzyme complex and a unique polyubiquitin chain. Cell 103: 351-361.

Devin A, Cook A, Lin Y, Rodriguez Y, Kelliher M & Liu Z (2000) The distinct roles of TRAF2 and RIP in IKK activation by TNF-R1: TRAF2 recruits IKK to TNF-R1 while RIP mediates IKK activation. Immunity 12: 419-429.

Dinarello CA (2000a) IL1α, in Cytokine Reference (Oppenheim JJ & Feldmann M, eds), Vol. I, Academic Press, San Diego, pp. 307-318.

Dinarello CA (2000b) IL1β, in Cytokine Reference (Oppenheim JJ & Feldmann M, eds), Vol. I, Academic Press, San Diego, pp. 351-374.

Duckett CS & Thompson CB (1997) CD30-dependent degradation of TRAF2: implications for negative regulation of TRAF signaling and the control of cell survival. Genes Dev. 11: 2810-2821.

Feinstein E, Kimchi A, Wallach D, Boldin M & Varfolomeev E (1995) The death domain: a module shared by proteins with diverse cellular functions. Trends Biochem. Sci. 20: 342-344.

Fitzgerald KA, Palsson-McDermott EM, Bowie AG, Jefferies CA, Mansell AS, Brady G, Brint E, Dunne A, Gray P, Harte MT, McMurray D, Smith DE, Sims JE, Bird TA & O'Neill LA (2001) Mal (MyD88-adapter-like) is required for Toll-like receptor-4 signal transduction. Nature 413: 78-83.

Gerritsen ME, Williams AJ, Neish AS, Moore S, Shi Y & Collins T (1997) CREB-binding protein/p300 are transcriptional coactivators of p65. Proc. Natl. Acad. Sci. USA 94: 2927-2932.

Heyninck K, De Valck D, Vanden Berghe W, Van Criekinge W, Contreras R, Fiers W, Haegeman G & Beyaert R (1999) The zinc finger protein A20 inhibits TNF-induced NF-κB-dependent gene expression by interfering with an RIP- or TRAF2-mediated transactivation signal and directly binds to a novel NF-κB-inhibiting protein ABIN. J. Cell Biol. 145: 1471-1482.

Hoeflich KP, Luo J, Rubie EA, Tsao MS, Jin O & Woodgett JR (2000) Requirement for glycogen synthase kinase-3β in cell survival and NF-κB activation. Nature 406: 86-90.

Horai R, Saijo S, Tanioka H, Nakae S, Sudo K, Okahara A, Ikuse T, Asano M & Iwakura Y (2000) Development of chronic inflammatory arthropathy resembling rheumatoid arthritis in interleukin 1 receptor antagonist-deficient mice. J. Exp. Med. 191: 313-320.

Hsu H, Huang J, Shu HB, Baichwal V & Goeddel DV (1996a) TNF-dependent recruitment of the protein kinase RIP to the TNF receptor-1 signaling complex. Immunity 4: 387-396.

Hsu H, Shu HB, Pan MG & Goeddel DV (1996b) TRADD-TRAF2 and TRADD-FADD interactions define two distinct TNF receptor 1 signal transduction pathways. Cell 84: 299-308.

Janssens S, Burns K, Tschopp J & Beyaert R (2002) Regulation of interleukin-1- and lipopolysaccharide-induced NF-κB activation by alternative splicing of MyD88. Curr. Biol. 12: 467–471.

Jiang Y, Woronicz JD, Liu W & Goeddel DV (1999) Prevention of constitutive TNF receptor 1 signaling by silencer of death domains. Science 283: 543-546.

Jones SJ, Ledgerwood EC, Prins JB, Galbraith J, Johnson DR, Pober JS & Bradley JR (1999) TNF recruits TRADD to the plasma membrane but not the trans-Golgi network, the principal subcellular location of TNF-R1. J. Immunol. 162: 1042-1048.
Kanakaraj P, Schafer PH, Cavender DE, Wu Y, Ngo K, Grealish PF, Wadsworth SA, Peterson PA, Siekierka JJ, Harris CA & Fung-Leung WP (1998) Interleukin (IL)-1 receptor-associated kinase (IRAK) requirement for optimal induction of multiple IL-1 signaling pathways and IL-6 production. J. Exp. Med. 187: 2073-2079.
Kelliher MA, Grimm S, Ishida Y, Kuo F, Stanger BZ & Leder P (1998) The death domain kinase RIP mediates the TNF-induced NF-κB signal. Immunity 8: 297-303.
Kobayashi K, Hernandez LD, Galan JE, Janeway CA, Jr., Medzhitov R & Flavell RA (2002a) IRAK-M is a negative regulator of Toll-like receptor signaling. Cell 110: 191-202.
Kobayashi K, Inohara N, Hernandez LD, Galan JE, Nunez G, Janeway CA, Medzhitov R & Flavell RA (2002b) RICK/Rip2/CARDIAK mediates signalling for receptors of the innate and adaptive immune systems. Nature 416: 194-199.
Kobayashi N, Kadono Y, Naito A, Matsumoto K, Yamamoto T, Tanaka S & Inoue J (2001) Segregation of TRAF6-mediated signaling pathways clarifies its role in osteoclastogenesis. EMBO J. 20: 1271-1280.
Kopp E, Medzhitov R, Carothers J, Xiao C, Douglas I, Janeway CA & Ghosh S (1999) ECSIT is an evolutionarily conserved intermediate in the Toll/IL-1 signal transduction pathway. Genes Dev. 13: 2059-2071.
Kriegler M, Perez C, DeFay K, Albert I & Lu SD (1988) A novel form of TNF/cachectin is a cell surface cytotoxic transmembrane protein: ramifications for the complex physiology of TNF. Cell 53: 45-53.
Kyriakis JM & Avruch J (2001) Mammalian mitogen-activated protein kinase signal transduction pathways activated by stress and inflammation. Physiol. Rev. 81: 807-869.
Labow M, Shuster D, Zetterstrom M, Nunes P, Terry R, Cullinan EB, Bartfai T, Solorzano C, Moldawer LL, Chizzonite R & McIntyre KW (1997) Absence of IL-1 signaling and reduced inflammatory response in IL-1 type I receptor-deficient mice. J. Immunol. 159: 2452-2461.
Lang D, Knop J, Wesche H, Raffetseder U, Kurrle R, Boraschi D & Martin MU (1998) The type II IL-1 receptor interacts with the IL-1 receptor accessory protein: a novel mechanism of regulation of IL-1 responsiveness. J. Immunol. 161: 6871-6877.
Lee EG, Boone DL, Chai S, Libby SL, Chien M, Lodolce JP & Ma A (2000) Failure to regulate TNF-induced NF-κB and cell death responses in A20-deficient mice. Science 289: 2350-2354.
Leitges M, Sanz L, Martin P, Duran A, Braun U, Garcia JF, Camacho F, Diaz-Meco MT, Rennert PD & Moscat J (2001) Targeted disruption of the ζPKC gene results in the impairment of the NF-κB pathway. Mol. Cell 8: 771-780.
Li S, Strelow A, Fontana EJ & Wesche H (2002a) IRAK-4: a novel member of the IRAK family with the properties of an IRAK-kinase. Proc. Natl. Acad. Sci. USA 99: 5567-5572.
Li X, Commane M, Burns C, Vithalani K, Cao Z & Stark GR (1999) Mutant cells that do not respond to interleukin-1 (IL-1) reveal a novel role for IL-1 receptor-associated kinase. Mol. Cell. Biol. 19: 4643-4652.
Li X, Yang Y & Ashwell JD (2002b) TNF-RII and c-IAP1 mediate ubiquitination and degradation of TRAF2. Nature 416: 345-347.
Lin H, Ho AS, Haley-Vicente D, Zhang J, Bernal-Fussell J, Pace AM, Hansen D, Schweighofer K, Mize NK & Ford JE (2001) Cloning and characterization of IL-1HY2, a novel interleukin-1 family member. J. Biol. Chem. 276: 20597-20602.
Lomaga MA, Yeh WC, Sarosi I, Duncan GS, Furlonger C, Ho A, Morony S, Capparelli C, Van G, Kaufman S, van der Heiden A, Itie A, Wakeham A, Khoo W, Sasaki T, Cao Z, Penninger JM, Paige CJ, Lacey DL, Dunstan CR, Boyle WJ, Goeddel DV & Mak TW (1999) TRAF6 deficiency results in osteopetrosis and defective interleukin-1, CD40, and LPS signaling. Genes Dev. 13: 1015-1024.
Lord KA, Hoffman-Liebermann B & Liebermann DA (1990) Nucleotide sequence and expression of a cDNA encoding MyD88, a novel myeloid differentiation primary response gene induced by IL6.

Oncogene 5: 1095-1097.

Lorick KL, Jensen JP, Fang S, Ong AM, Hatakeyama S & Weissman AM (1999) RING fingers mediate ubiquitin-conjugating enzyme (E2)-dependent ubiquitination. Proc. Natl. Acad. Sci. USA 96: 11364-11369.

Mercurio F, Zhu H, Murray BW, Shevchenko A, Bennett BL, Li J, Young DB, Barbosa M, Mann M, Manning A & Rao A (1997) IKK-1 and IKK-2: cytokine-activated IκB kinases essential for NF-κB activation. Science 278: 860-866.

Muzio M, Ni J, Feng P & Dixit VM (1997) IRAK (Pelle) family member IRAK-2 and MyD88 as proximal mediators of IL-1 signaling. Science 278: 1612-1615.

Nakano H, Shindo M, Sakon S, Nishinaka S, Mihara M, Yagita H & Okumura K (1998) Differential regulation of IκB kinase α and β by two upstream kinases, NF-κB-inducing kinase and mitogen-activated protein kinase/ERK kinase kinase-1. Proc. Natl. Acad. Sci. USA 95: 3537-3542.

Ninomiya-Tsuji J, Kishimoto K, Hiyama A, Inoue J, Cao Z & Matsumoto K (1999) The kinase TAK1 can activate the NIK-I κB as well as the MAP kinase cascade in the IL-1 signalling pathway. Nature 398: 252-256.

O'Neill L (2000) The Toll/interleukin-1 receptor domain: a molecular switch for inflammation and host defence. Biochem. Soc. Trans. 28: 557-563.

Park YC, Burkitt V, Villa AR, Tong L & Wu H (1999) Structural basis for self-association and receptor recognition of human TRAF2. Nature 398: 533-538.

Park YC, Ye H, Hsia C, Segal D, Rich RL, Liou HC, Myszka DG & Wu H (2000) A novel mechanism of TRAF signaling revealed by structural and functional analyses of the TRADD-TRAF2 interaction. Cell 101: 777-787.

Pimentel-Muinos FX & Seed B (1999) Regulated commitment of TNF receptor signaling: a molecular switch for death or activation. Immunity 11: 783-793.

Pinckard JK, Sheehan KC, Arthur CD & Schreiber RD (1997) Constitutive shedding of both p55 and p75 murine TNF receptors in vivo. J. Immunol. 158: 3869-3873.

Poyet JL, Srinivasula SM, Lin JH, Fernandes-Alnemri T, Yamaoka S, Tsichlis PN & Alnemri ES (2000) Activation of the IκB kinases by RIP via IKKγ/NEMO-mediated oligomerization. J. Biol. Chem. 275: 37966-37977.

Qian Y, Commane M, Ninomiya-Tsuji J, Matsumoto K & Li X (2001) IRAK-mediated translocation of TRAF6 and TAB2 in the interleukin-1-induced activation of NFκB. J. Biol. Chem. 276: 41661-41667.

Rothe M, Wong SC, Henzel WJ & Goeddel DV (1994) A novel family of putative signal transducers associated with the cytoplasmic domain of the 75 kDa tumor necrosis factor receptor. Cell 78: 681-692.

Rothe M, Sarma V, Dixit VM & Goeddel DV (1995) TRAF2-mediated activation of NF-κB by TNF receptor 2 and CD40. Science 269: 1424-1427.

Saccani S, Pantano S & Natoli G (2002) p38-Dependent marking of inflammatory genes for increased NF-κB recruitment. Nat. Immunol. 3: 69-75.

Sakurai H, Chiba H, Miyoshi H, Sugita T & Toriumi W (1999) IκB kinases phosphorylate NF-κB p65 subunit on serine 536 in the transactivation domain. J. Biol. Chem. 274: 30353-30356.

Sanz L, Sanchez P, Lallena MJ, Diaz-Meco MT & Moscat J (1999) The interaction of p62 with RIP links the atypical PKCs to NF-κB activation. EMBO J. 18: 3044-3053.

Sanz L, Diaz-Meco MT, Nakano H & Moscat J (2000) The atypical PKC-interacting protein p62 channels NF-κB activation by the IL-1-TRAF6 pathway. EMBO J. 19: 1576-1586.

Schreuder H, Tardif C, Trump-Kallmeyer S, Soffientini A, Sarubbi E, Akeson A, Bowlin T, Yanofsky S & Barrett RW (1997) A new cytokine-receptor binding mode revealed by the crystal structure of the IL-1 receptor with an antagonist. Nature 386: 194-200.

Senftleben U, Cao Y, Xiao G, Greten FR, Krahn G, Bonizzi G, Chen Y, Hu Y, Fong A, Sun SC & Karin M (2001) Activation by IKKα of a second, evolutionary conserved, NF-κB signaling pathway. Science 293: 1495-1499.

Sheppard KA, Rose DW, Haque ZK, Kurokawa R, McInerney E, Westin S, Thanos D, Rosenfeld MG, Glass CK & Collins T (1999) Transcriptional activation by NF-κB requires multiple coactivators. Mol. Cell. Biol. 19: 6367-6378.

Sims JE (2002) IL-1 and IL-18 receptors, and their extended family. Curr. Opin. Immunol. 14: 117-122.

Sizemore N, Leung S & Stark GR (1999) Activation of phosphatidylinositol 3-kinase in response to interleukin-1 leads to phosphorylation and activation of the NF-κB p65/RelA subunit. Mol. Cell. Biol. 19: 4798-4805.

Smith CA, Farrah T & Goodwin RG (1994) The TNF receptor superfamily of cellular and viral proteins: activation, costimulation, and death. Cell 76: 959-962.

Stanger BZ, Leder P, Lee TH, Kim E & Seed B (1995) RIP: a novel protein containing a death domain that interacts with Fas/APO-1 (CD95) in yeast and causes cell death. Cell 81: 513-523.

Suzuki N, Suzuki S, Duncan GS, Millar DG, Wada T, Mirtsos C, Takada H, Wakeham A, Itie A, Li S, Penninger JM, Wesche H, Ohashi PS, Mak TW & Yeh WC (2002) Severe impairment of interleukin-1 and Toll-like receptor signalling in mice lacking IRAK-4. Nature 416: 750-756.

Tada K, Okazaki T, Sakon S, Kobarai T, Kurosawa K, Yamaoka S, Hashimoto H, Mak TW, Yagita H, Okumura K, Yeh WC & Nakano H (2001) Critical roles of TRAF2 and TRAF5 in tumor necrosis factor-induced NF-κB activation and protection from cell death. J. Biol. Chem. 276: 36530-36534.

Takaesu G, Kishida S, Hiyama A, Yamaguchi K, Shibuya H, Irie K, Ninomiya-Tsuji J & Matsumoto K (2000) TAB2, a novel adaptor protein, mediates activation of TAK1 MAPKKK by linking TAK1 to TRAF6 in the IL-1 signal transduction pathway. Mol. Cell 5: 649-658.

Tartaglia LA, Weber RF, Figari IS, Reynolds C, Palladino MA, Jr. & Goeddel DV (1991) The two different receptors for tumor necrosis factor mediate distinct cellular responses. Proc. Natl. Acad. Sci. USA 88: 9292-9296.

Tartaglia LA & Goeddel DV (1992) Two TNF receptors. Immunol. Today 13: 151-153.

Tartaglia LA, Ayres TM, Wong GH & Goeddel DV (1993) A novel domain within the 55 kd TNF receptor signals cell death. Cell 74: 845-853.

Thome M, Hofmann K, Burns K, Martinon F, Bodmer JL, Mattmann C & Tschopp J (1998) Identification of CARDIAK, a RIP-like kinase that associates with caspase-1. Curr. Biol. 8: 885-888.

Ting AT, Pimentel-Muinos FX & Seed B (1996) RIP mediates tumor necrosis factor receptor 1 activation of NF-κB but not Fas/APO-1-initiated apoptosis. EMBO J. 15: 6189-6196.

Tsitsikov EN, Laouini D, Dunn IF, Sannikova TY, Davidson L, Alt FW & Geha RS (2001) TRAF1 is a negative regulator of TNF signaling. enhanced TNF signaling in TRAF1-deficient mice. Immunity 15: 647-657.

Van Huffel S, Delaei F, Heyninck K, De Valck D & Beyaert R (2001) Identification of a novel A20-binding inhibitor of nuclear factor-κB activation termed ABIN-2. J. Biol. Chem. 276: 30216-30223.

Vanden Berghe W, Plaisance S, Boone E, De Bosscher K, Schmitz ML, Fiers W & Haegeman G (1998) p38 and extracellular signal-regulated kinase mitogen-activated protein kinase pathways are required for nuclear factor-κB p65 transactivation mediated by tumor necrosis factor. J. Biol. Chem. 273: 3285-3290.

Vandenabeele P, Declercq W, Vercammen D, Van de Craen M, Grooten J, Loetscher H, Brockhaus M, Lesslauer W & Fiers W (1992) Functional characterization of the human tumor necrosis factor receptor p75 in a transfected rat/mouse T cell hybridoma. J. Exp. Med. 176: 1015-1024.

Vandenabeele P, Declercq W, Beyaert R & Fiers W (1995) Two tumour necrosis factor receptors: structure and function. Trends Cell Biol. 5: 392-399.

Vigers GP, Anderson LJ, Caffes P & Brandhuber BJ (1997) Crystal structure of the type-I interleukin-1 receptor complexed with interleukin-1β. Nature 386: 190-194.

Wang C, Deng L, Hong M, Akkaraju GR, Inoue J & Chen ZJ (2001) TAK1 is a ubiquitin-dependent kinase of MKK and IKK. Nature 412: 346-351.

Wang D, Westerheide SD, Hanson JL & Baldwin AS, Jr. (2000) Tumor necrosis factor α-induced phosphorylation of RelA/p65 on Ser529 is controlled by casein kinase II. J. Biol. Chem. 275: 32592-32597.

Wesche H, Henzel WJ, Shillinglaw W, Li S & Cao Z (1997a) MyD88: an adapter that recruits IRAK to the IL-1 receptor complex. Immunity 7: 837-847.

Wesche H, Korherr C, Kracht M, Falk W, Resch K & Martin MU (1997b) The interleukin-1 receptor accessory protein (IL-1RAcP) is essential for IL-1-induced activation of interleukin-1 receptor-associated kinase (IRAK) and stress-activated protein kinases (SAP kinases). J. Biol. Chem. 272: 7727-7731.

Wesche H, Gao X, Li X, Kirschning CJ, Stark GR & Cao Z (1999) IRAK-M is a novel member of the Pelle/interleukin-1 receptor-associated kinase (IRAK) family. J. Biol. Chem. 274: 19403-19410.

Woronicz JD, Gao X, Cao Z, Rothe M & Goeddel DV (1997) IκB kinase-β: NF-κB activation and complex formation with IκB kinase-α and NIK. Science 278: 866-869.

Yamin TT & Miller DK (1997) The interleukin-1 receptor-associated kinase is degraded by proteasomes following its phosphorylation. J. Biol. Chem. 272: 21540-21547.

Yang J, Lin Y, Guo Z, Cheng J, Huang J, Deng L, Liao W, Chen Z, Liu Z & Su B (2001) The essential role of MEKK3 in TNF-induced NF-κB activation. Nat. Immunol. 2: 620-624.

Ye H, Park YC, Kreishman M, Kieff E & Wu H (1999) The structural basis for the recognition of diverse receptor sequences by TRAF2. Mol. Cell 4: 321-330.

Yeh WC, Shahinian A, Speiser D, Kraunus J, Billia F, Wakeham A, de la Pompa JL, Ferrick D, Hum B, Iscove N, Ohashi P, Rothe M, Goeddel DV & Mak TW (1997) Early lethality, functional NF-κB activation, and increased sensitivity to TNF-induced cell death in TRAF2-deficient mice. Immunity 7: 715-725.

Yin L, Wu L, Wesche H, Arthur CD, White JM, Goeddel DV & Schreiber RD (2001) Defective lymphotoxin-β receptor-induced NF-κB transcriptional activity in NIK-deficient mice. Science 291: 2162-2165.

Ythier A, Gascon MP, Juillard P, Vesin C, Wallach D & Grau GE (1993) Protective effect of natural TNF-binding protein on human TNF-induced toxicity in mice. Cytokine 5: 459-462.

Yujiri T, Ware M, Widmann C, Oyer R, Russell D, Chan E, Zaitsu Y, Clarke P, Tyler K, Oka Y, Fanger GR, Henson P & Johnson GL (2000) MEK kinase 1 gene disruption alters cell migration and c-Jun NH_2-terminal kinase regulation but does not cause a measurable defect in NF-κB activation. Proc. Natl. Acad. Sci. USA 97: 7272-7277.

Zhang SQ, Kovalenko A, Cantarella G & Wallach D (2000) Recruitment of the IKK signalosome to the p55 TNF receptor: RIP and A20 bind to NEMO (IKKγ) upon receptor stimulation. Immunity 12: 301-311.

Zhao Q & Lee FS (1999) Mitogen-activated protein kinase/ERK kinase kinases 2 and 3 activate nuclear factor-κB through IκB kinase-α and IκB kinase-β. J. Biol. Chem. 274: 8355-8358.

9. AFFILIATION

Ghent University and Flanders Interuniversity Institute for Biotechnology, Department for Molecular Biomedical Research, Unit for Molecular Signal Transduction in Inflammation, K.L. Ledeganckstraat 35, 9000 Gent, Belgium.

J.-L. POYET & E.S. ALNEMRI

NF-κB ACTIVATION BY CARD PROTEINS

1. INTRODUCTION

Apoptosis (programmed cell death) has a central role in the normal development and homeostasis of all multicellular organisms (Vaux & Strasser, 1996). Deregulation of this process, resulting in too much or too little cell death, can cause both developmental defects and a wide variety of diseases such as cancer and neurodegenerative disorders (Mullauer et al., 2001). At the heart of this complex machinery is a conserved family of cysteine proteases, the caspases, which are produced in cells as catalytically inactive zymogenes and require proteolytic activation during apoptosis (Cohen, 1997). Caspases are divided into two classes depending on their place in the apoptotic caspase cascade. The initiator caspase zymogens can autoactivate in response to apoptotic triggers and then activate the downstream effector caspases, producing a signaling cascade which results in the cleavage of substrates characteristic of the final stages of apoptosis. Autoactivation of the initiator caspases is mediated by adaptor molecules, which contain small protein-protein interaction domains that act as molecular Velcro, allowing them to bind similar domains present in the N-termini of the initiator caspase zymogens and also other signaling molecules in the apoptotic pathway. By sequence similarities, four families of homophilic interaction domains have been identified in proteins participating in caspase activation. These domains include the death domain (DD), the death effector domain (DED), the caspase-associated recruitment domain (CARD) and the pyrin domain (DAPIN/PAAD) (Hofmann et al., 1997; Aravind et al., 1999; Bertin & DiStefano, 2000; Pawlowski et al., 2001). Structural characterization of several of these domains revealed that the DD, DED and CARD families possess a highly similar structure comprising an antiparallel six-helical bundle in the Greek key topology, thereby suggesting a common evolutionary lineage (Fesik, 2000; Weber & Vincenz, 2001; Shi, 2002). Each of these families interacts with other binding proteins by homotypic interactions in which DD-DD, DED-DED and CARD-CARD contacts are formed exclusively (Weber & Vincenz, 2001).

The CARD was originally described as a specific motif of about 90 amino acids in the prodomains of some initiator procaspases, such as caspase-2 and -9 and their adaptor molecules, that mediates the association of procaspases to their adaptor proteins by heterodimerization of their respective CARDs. A well-

R. Beyaert (ed.), Nuclear Factor κB. Regulation and Role in Disease, 69-88.

characterized paradigm involves procaspase-9 activation by the adaptor protein Apaf-1. During apoptosis, cytochrome c is released from the mitochondria into the cytoplasm. Binding of cytochrome c to the WD-40 repeats in Apaf-1 leads to formation of an oligomeric complex named the apoptosome, which then recruits procaspase-9 by a CARD-CARD interaction between procaspase-9 and Apaf-1 and facilitates its autoactivation by an induced proximity mechanism (Salvesen & Dixit, 1999; Adrain & Martin, 2001; Shi, 2002). Although the CARD was initially identified in some caspases and their adaptors, this domain does not appear to be restricted to caspases and their adaptor proteins. Indeed, a number of CARD-containing proteins that do not promote caspase activation or apoptosis have been recently identified. Rather than participating in pathways leading to cell death, these proteins mediate the assembly of multi-protein complexes involved in the activation of the anti-apoptotic transcription factor NF-κB. The CARD thus appears now as a major regulator of life and death decisions, playing a pivotal role in signaling events that will decide the ultimate outcome of a cell's fate. We here provide an overview of the CARD-containing molecules that are involved in the activation of NF-κB as well as the mechanism by which they do so.

2. STRUCTURE OF THE CARD-CONTAINING PROTEINS INVOLVED IN NF-κB ACTIVATION

The CARD-containing proteins constitute a growing family, with new members being discovered regularly. We shall focus here only on the CARD-containing proteins that have been demonstrated to activate NF-κB. NF-κB is a collective term referring to dimeric transcription factors that belong to the Rel family (Ghosh et al., 1998). The antiapoptotic activity of NF-κB depends on gene induction. A comprehensive discussion of the NF-κB-responsive genes whose products can inhibit apoptosis is beyond the scope of this chapter. The reader is therefore referred to several other chapters in this volume for details on this particular topic. In most resting cells, NF-κB is sequestered in the cytoplasm by interaction with specific inhibitors, the IκBs (Baldwin, 1996). Exposure of cells to a wide variety of stimuli leads to the rapid phosphorylation, ubiquitination and ultimately proteolytic degradation of the IκBs (Brown et al., 1995; Chen et al., 1995; Scherer et al., 1995; Burke et al., 1998). This will allow the activated NF-κB to translocate to the nucleus and activate the transcription of the target genes (Wirth & Baltimore, 1988). The kinase activity responsible for phosphorylation of IκBs is present in a large (700-900 kDa) cytoplasmic complex composed of two catalytic subunits, IKKα and IKKβ, which can directly phosphorylate IκBs (DiDonato et al., 1997; Mercurio et al., 1997; Regnier et al., 1997; Woronicz et al., 1997; Zandi et al., 1997), and a regulatory subunit IKKγ (also known as NEMO), whose integrity is required for the activation of the complex (Rothwarf et al., 1998; Yamaoka et al., 1998; Li et al., 1999; Mercurio et al., 1999). So far, ten CARD-containing proteins involved in NF-κB activation have been identified. For clarity, we have subdivided the CARD-containing proteins involved in activating NF-κB into four subfamilies based on their domain structures (Fig. 1).

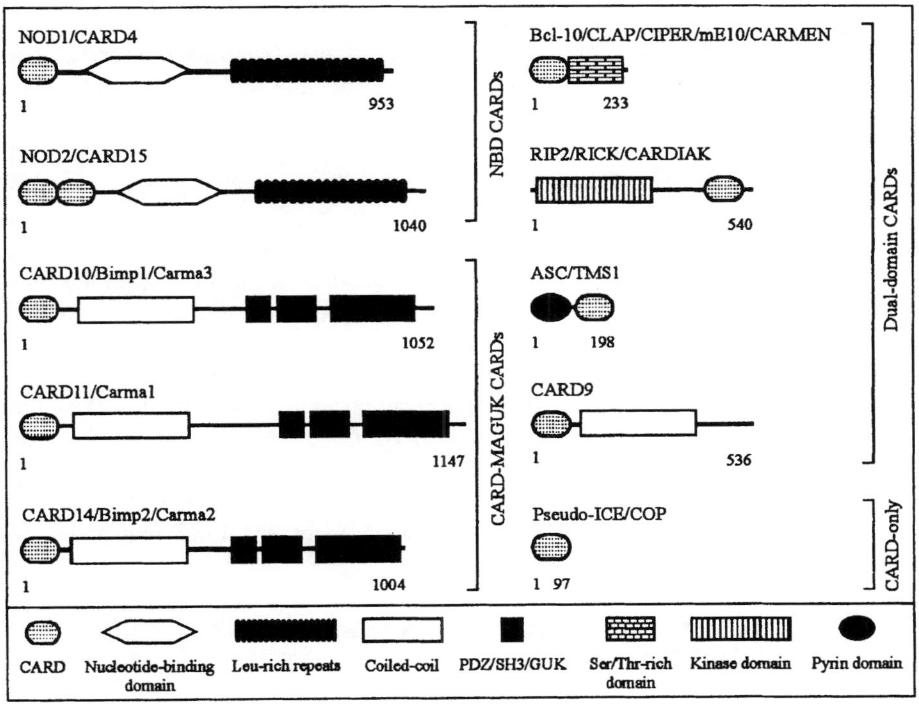

(1) The nucleotide-binding CARD proteins. This subfamily contains two members, viz. NOD1/CARD4 (Bertin et al., 1999; Inohara et al., 1999) and NOD2/CARD15 (Ogura et al., 2001b), and is implicated in NF-κB activation. These proteins contain a nucleotide-binding domain (NBD) in addition to one or two N-terminal CARDs as well as a C-terminal leucine-rich repeat (LRR) domain. The centrally located nucleotide-binding domain most likely functions as a self-association and oligomerization domain by analogy with the caspase-9-activating adaptor Apaf-1. The C-terminal LRR domain probably serves as a sensory domain that regulates oligomerization of the nucleotide-binding domain upon binding to an intracellular ligand, which might be lipopolysaccharide (LPS).

(2) The CARD-MAGUK proteins. Three members of this subfamily, viz. CARD10/Bimp1/Carma3 (Gaide et al., 2001; McAllister-Lucas et al., 2001; Wang et al., 2001), CARD11/Carma1 (Bertin et al., 2001; Gaide et al., 2001) and CARD14/Bimp2/Carma2 (Bertin et al., 2001; Gaide et al., 2001; McAllister-Lucas et al., 2001), have been implicated in NF-κB signaling in the Bcl-10-signaling pathway. Following the N-terminal CARD these proteins contain a coiled-coil (CC) domain and a C-terminal tripartite structure composed of PDZ,

Src homology 3 (SH3) and GUK domains with homology to guanylate kinase and thus belong to the family of the membrane-associated guanylate kinase (MAGUK) proteins. These proteins interact with Bcl-10 and could function as adaptor proteins for recruitment of Bcl-10 to large signaling complexes at the plasma membrane.

(3) The dual-domain CARD proteins. These are bipartite molecules that contain, in addition to the CARD, another protein module such as a pyrin domain (ASC/TMS1) (Masumoto et al., 1999; Conway et al., 2000), a kinase domain (RIP2/RICK/CARDIAK) (Inohara et al., 1998; McCarthy et al., 1998; Thome et al., 1998), a CC domain (CARD9) (Bertin et al., 2000) or an undefined domain (Bcl-10/CLAP/CIPER/mE10/CARMEN/cE10) (Costanzo et al., 1999; Koseki et al., 1999; Srinivasula et al., 1999; Thome et al., 1999; Willis et al., 1999; Yan et al., 1999). These CARD proteins also function as adaptor proteins for propagation of the NF-κB activation signals from upstream signaling molecules (e.g. NOD1, NOD2, CARD11) to the IκB kinases.

(4) The CARD-only proteins. Pseudo-ICE/COP is the only known member of this subfamily, which can induce NF-κB activation upon overexpression (Druilhe et al., 2001; Lee et al., 2001). This protein is composed of a CARD domain only and may also function as a decoy molecule in the caspase-1 activation pathway.

The CARD-containing proteins are able to interact not only with members of their family but also with a wide array of proteins involved in multiple functions (Table 1).

3. ROLE OF THE CARD-CONTAINING PROTEINS IN THE DIFFERENT NF-κB-SIGNALING PATHWAYS

3.1. The nucleotide-binding CARD proteins

The two members of this subfamily, NOD1 and NOD2, share significant structural homology with the caspase-9 activator Apaf-1 (Bertin et al., 1999; Inohara et al., 1999; Ogura et al., 2001b). Like Apaf-1, NOD1 and NOD2 have one (NOD1) or two (NOD2) amino-terminal CARDs, a centrally located nucleotide-binding domain (NBD) and a C-terminal-regulatory domain (Fig. 1). However, whereas Apaf-1 has a C-terminal WD-40 repeat domain, NOD1 and NOD2 contain LRR domains in their C-termini (Fig. 1). The structural homology between these proteins suggests that, like Apaf-1, the NOD proteins promote the assembly of large protein complexes upon stimulation by upstream signaling molecules. However, unlike Apaf-1, the NOD proteins seem to engage IκB kinases, rather than caspases. Indeed, transient overexpression of NOD1 or NOD2 in cells induced NF-κB activation via IKK activation (Bertin et al., 1999; Inohara et al., 1999; Ogura et al., 2001b). Overexpression of the NOD proteins did not induce apoptosis, but enhanced apoptosis induced by caspase-9 expression (Bertin et al., 1999; Inohara et al., 1999; Ogura et al., 2001b). However, the physiological relevance of this observation remains to be demonstrated.

Table 1. NF-κB-activating CARDs and their binding partners.

Subfamily type	*Protein name*	*Binding partners*	*References*
Nucleotide-binding	NOD1/CARD4 NOD2/CARD15	RIP2, caspase-9, LPS RIP2	Bertin et al., 1999 Inohara et al., 1999 Inohara et al., 2000 Girardin et al., 2001 Inohara et al., 2001
CARD-MAGUK	CARD10/Bimp1/ Carma3	Bcl-10	Bertin et al., 2001a McAllister-Lucas et al., 2001
	CARD11/Carma1	Bcl-10	Gaide et al., 2001
	CARD14/Bimp2/ Carma2	Bcl-10	Wang et al., 2001
Dual-domain	Bcl-10/CLAP/ CIPER/cE10/ mE10/CARMEN	DR4, TNF-R1, TRADD, α-actinin, TRAF2, cIAP1, MALT1, CARD9, CARD10, CARD11, CARD14	Costanzo et al., 1999 Koseki et al., 1999 Guiet et al., 2000 Uren et al., 2000 Yoneda et al., 2000 Lucas et al., 2001
	RIP2/RICK/ CARDIAK	CLARP, cIAP1, TRAF1, TRAF2, TRAF5, TRAF6, caspase-1, NOD1, NOD2, IKKγ, pseudo-ICE, p75NTR, TLR	Inohara et al., 1998 McCarthy et al., 1998 Thome et al., 1998 Inohara et al., 2000 Druilhe et al., 2001 Khursigara et al., 2001 Lee et al., 2001
	ASC	Pyrin, caspase-1, IPAF, NALP1, cryopyrin, PYPAF1, PYPAF7	Geddes et al., 2001 Martinon et al., 2001 Masumoto et al., 2001 Richards et al., 2001 Manji et al., 2002 Martino et al., 2002 Srinivasula et al., 2002 Wang et al., 2002
	CARD9	Bcl-10	Bertin et al., 2000
CARD-only	Pseudo-ICE/COP	RIP2, caspase-1	Druilhe et al., 2001 Lee et al., 2001

How do the NOD proteins activate the IKKs? NOD1 and NOD2 associate, by CARD-CARD interaction, with RIP2, a bipartite CARD-containing kinase that could activate NF-κB on its own upon overexpression in mammalian cells (Inohara et al., 1998; McCarthy et al., 1998; Thome et al., 1998; Bertin et al., 1999; Inohara et al., 1999; Ogura et al., 2001b). RIP2 contains an intermediate region between the kinase domain and the CARD domain that interacts with the essential regulatory subunit of the IKK complex, known as IKKγ (Inohara et al., 2000). Like Apaf-1, which provides a platform for procaspases-9 to assemble into an active oligomeric complex (Saleh et al., 1999; Zou et al., 1999), the NOD proteins are thought to provide a similar platform for RIP2 and the IKKs (Fig. 2). Indeed, it has been demonstrated that the IKKs are activated by an oligomerization-induced proximity mechanism (Inohara et al., 2000; Poyet et al., 2000). IKKγ-mediated oligomerization of the IKK complex is sufficient to activate NF-κB, probably by an induced autophosphorylation/activation of the IKK kinases (Poyet et al., 2000). In response to inflammatory signals, the NOD proteins could self-associate via their nucleotide-binding domain and are therefore likely to activate, through RIP2, the IKK complex by acting as a scaffold for aggregation of the IKK subunits. This adaptor-induced multi-molecular aggregation could enforce a locally high concentration of the IKK components, allowing the IKK kinases to self-activate, due to their low intrinsic enzymatic activity, by autophosphorylation of two serines located in their T-loop (Poyet et al., 2000). Consistent with this model, enforced oligomerization of the CARDs of NOD1 and NOD2 was able to induce NF-κB activation (Inohara et al., 2000). Overexpression of a variant of NOD1 containing a point mutation (K208R) of the NBD domain, which has reduced ability to self-associate, was, in contrast to the wild-type protein, unable to induce NF-κB activation (Inohara et al., 2000).

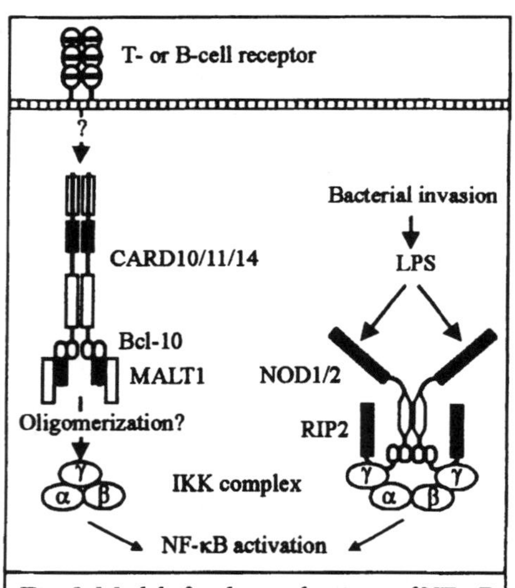

Fig. 2. Models for the mechanisms of NF-κB activation by the CARD-containing proteins.

The next important question is what regulates the oligomerization/activation of the NOD proteins. Interestingly, the NOD proteins have an overall domain structure which is strikingly similar to the cytoplasmic plant disease resistance (R) gene products, called R proteins (Dangl & Jones, 2001). These receptor-like proteins mediate plant resistance to pathogen invasion. The LRRs of R proteins are highly diverse and are involved in the recognition of a wide array of bacterially derived components, whereas their N-terminal domains appear to mediate a signaling cascade that regulates gene expression (Parniske et al., 1997; Dixon et al., 2000). The structural similarity between NODs and plant R proteins suggests that, like plant R proteins, NOD1 and NOD2 might also be involved in the detection of intracellular pathogen components. It has been suspected long ago that an LPS detection system, different from the one involving the Toll receptors (TLRs), was likely to exist in mammalian cells in order to respond to intracellular LPS (Philpott et al., 2000). Interestingly, expression of NOD1 and NOD2 conferred responsiveness, through NF-κB activation, to LPS in normally LPS-insensitive HEK293T cells (Inohara et al., 2001). Moreover, LPS was found to bind directly to NOD1, probably through the LRRs (Inohara et al., 2001). NOD1 and NOD2 mutants lacking the LRRs were unable to confer LPS responsiveness to HEK293T cells, though they were still able to activate NF-κB when overexpressed (Inohara et al., 2001). Overexpression of the LRRs of NOD1 alone was able to inhibit NF-κB activation after infection with invasive Shigella flexneri and microinjection of LPS (Philpott et al., 2000). Thus the NOD proteins appear to act as intracellular receptors for LPS. Binding of LPS to the LRRs of the NOD proteins probably induces a conformational change, which unmasks their CARDs and allows them to bind and activate the downstream component RIP2 (Fig. 2). Consistent with this model, it was shown that infection with invasive Shigella flexneri promotes NOD1 self-association and formation of a transient complex containing NOD1, RIP2 and components of the IKK complex (Girardin et al., 2001). The physiological importance of NOD2 as a sensor of bacterial components was confirmed. NOD2 was mapped to chromosome 16q12 at a site overlapping IBD1, a locus associated with susceptibility to Crohn's disease (CD), a chronic inflammatory bowel disorder (IBD) of the gastrointestinal tract (Hugot et al., 2001; Ogura et al., 2001a). An abnormal inflammatory response directed against enteric microflora by genetically predisposed hosts was suspected, though the etiology of IBD is unknown (Fiocchi, 1998).

Much evidence supports bacteria-induced NF-κB dysregulation in CD (Ogura et al., 2001a). A frameshift and two missense mutations in the NOD2 gene were found to be associated with susceptibility to CD. The mutations result in a truncated protein lacking part of the last LRRs. The LRR-truncated NOD2 associated with CD was found to be incapable of mediating LPS-induced NF-κB activation (Ogura et al., 2001a). Even though the mechanisms leading to IBD remain unclear, NOD2 might be involved in a signaling pathway which prevents invasion from enteric bacteria. Surprisingly, point mutations in the NOD2 gene have also been reported to be associated with the susceptibility to a second granulomatous disorder, viz. the Blau syndrome (Miceli-Richard et al., 2001). The Blau syndrome is a rare autosomal dominant disorder in which patients are affected with arthritis, uveitis and skin rash. The mutations associated with the susceptibility to Blau syndrome are located in the central nucleotide-binding domain of NOD2. Although the physiological consequences of these mutations remain unclear, alterations of this domain have been demonstrated to alter the ability of the NODs to oligomerize and activate NF-κB (Inohara et al., 1999; Ogura et al., 2001b). In conclusion, it is remarkable that mutations in the NOD2 gene have been demonstrated to be associated with two distinct inflammatory diseases, namely CD and the Blau syndrome. This clearly underscores the importance of the NOD proteins in the inflammatory signaling pathway.

3.2. The CARD-MAGUK proteins

Members of this subfamily, like the NOD proteins, are multi-domain CARD proteins with structural similarity to Apaf-1 (Fig. 1). However, these proteins contain a central CC domain instead of a nucleotide-binding domain, which probably is involved in protein oligomerization. In addition to the central CC domain, these proteins are characterized by the presence at their C-termini of a tripartite structure composed of a PDZ, a Src homology 3 (SH3) and a GUK domain with homology to guanylate kinase. Because of the high similarity between these proteins, they are likely to participate in similar signaling pathways.

So far, three members of this subfamily have been described: CARD10, also called Bimp1/Carma3 (Gaide et al., 2001; McAllister-Lucas et al., 2001; Wang et al., 2001); CARD11, also called Carma1 (Bertin et al., 2001; Gaide et al., 2001); CARD14, also called Bimp2/Carma2 (Bertin et al., 2001; Gaide et al., 2001; McAllister-Lucas et al., 2001). CARD10, CARD11 and CARD14 have distinct tissue distribution patterns. Whereas CARD10 is expressed in non-lymphoid tissues, CARD11 showed abundant expression in lymphoid tissues, including thymus, spleen and peripheral blood leukocytes; CARD14 transcript showed expression only in placenta (Bertin et al., 2001; Gaide et al., 2001; McAllister-Lucas et al., 2001; Wang et al., 2001). It is therefore possible that the CARD paralogues mediate analogous signals in different cell types. Interestingly, all of these proteins are able to activate NF-κB by recruitment, via CARD-CARD interactions, of the CARD-containing protein Bcl-10 (Willis et al., 1999; Bertin et al., 2001; Gaide et al., 2001; McAllister-Lucas et al., 2001; Wang et al., 2001). Indeed, expression of CARD10 and CARD14 in Bcl-10$^{-/-}$ MEF cells failed to induce NF-κB activation, whereas expression of these proteins induced NF-κB activation in wild-type MEFs (McAllister-Lucas et al., 2001). Thus, the CARD-MAGUK proteins function as upstream activators of Bcl-10 in pathways leading to NF-κB activation. Bcl-10 appears to activate the IKK complex by mediating MALT1 oligomerization (Uren et al., 2000; Lucas et al., 2001). Interestingly, it has been demonstrated that CARD10, Bcl-10 and MALT1 can form a ternary protein complex in which Bcl-10 acts as a molecular bridge between CARD10 and MALT1 (Lucas et al., 2001). Therefore, it is likely that CARD10, and probably the other CARD-MAGUK proteins, cooperate with Bcl-10 and MALT1 in NF-κB activation. Because enforced oligomerization of the C-terminal part of Bcl-10 induces NF-κB activation (Srinivasula et al., 1999), it is tempting to say that CARD-MAGUK proteins function by recruiting Bcl-10 through its N-terminal CARD and transmitting the oligomerization signal to the C-terminal effector part of Bcl-10 (Fig. 2).

The expression of CARD11 and CARD14 has been shown to induce, in addition to NF-κB activation, phosphorylation of the C-terminal part of Bcl-10 (Bertin et al., 2001; Gaide et al., 2001). Though the functional significance of this modification is unclear, it is possible that it could be involved in the regulation of Bcl-10-mediated NF-κB activation.

What are the stimuli that induce activation of the CARD-MAGUK proteins? The answer could lie in the modular structure of the CARD-MAGUK proteins. Bcl-10 is now known to function as a positive regulator of lymphocyte proliferation that specifically connects antigen receptor signaling in B- and T-cells to NF-κB activation (Ruland et al., 2001). CARD10, CARD11 and CARD14 have a C-terminal tripartite PDZ/SH3/MAGUK structure, which identifies these proteins as members of the MAGUK family. Interestingly, MAGUK proteins are signaling proteins that recruit and localize other proteins at the plasma membrane by interaction with the cytoplasmic tail of surface receptors though their PDZ domain (Dimitratos et al., 1999; Fanning & Anderson, 1999). The members of this family act as molecular scaffolds for signaling pathways components at the plasma membrane of animal cells. It is therefore possible that CARD-MAGUK proteins function as scaffolding proteins to recruit Bcl-10 in a multi-protein complex that signals the activation of NF-κB. In response to an upstream signal, CARD-MAGUK proteins could be recruited at the plasma membrane by a receptor through their PDZ/SH3/MAGUK domain. These proteins could then self-associate through their CC domains, resulting in the engagement, aggregation and activation of Bcl-10. An important class of signaling molecules, the family of protein kinase C (PKC) enzymes, is known to mediate signaling of both T-cell and B-cell receptors (Dienz et al., 2000; Lin et al., 2000). Interestingly, a mutant of CARD10 lacking the CARD has been shown to be able to block NF-κB activation triggered by CD3 ligation as well as PMA/Ca^{2+} inophore-induced NF-κB in T-cells (McAllister-Lucas et al., 2001). Moreover, the same CARD10 mutant was also able to block both PKCα-induced and PKCϵ-induced NF-κB activation (McAllister-Lucas et al., 2001). These observations, combined with the fact that Bcl-10-deficient lymphoid cells failed to activate NF-κB in response to PMA/ionomycin treatment (Ruland et al., 2001), suggest that CARD10 and Bcl-10 operate downstream of PKC in a pathway leading to NF-κB activation. Thus, CARD10, as well as the other CARD-MAGUK members, is likely to function as a linker between T-cell receptor stimulation and PKC activation to Bcl-10-mediated NF-κB activation. Very recently, two groups reported independently that CARD11 is a lipid raft-associated molecule required for TCR-induced NF-κB activation (Gaide et al., 2002; Wang et al., 2002a). Deficiency in CARD11 or expression of a CARD11 mutant that cannot bind to Bcl-10 resulted in a selective impairment of activation of NF-κB following TCR stimulation. Moreover, CARD11 and Bcl-10 were able to co-localize with the TCR following anti-CD3 activation (Gaide et al., 2002; Wang et al., 2002a). Therefore, CARD11 plays a critical role in T-cell signaling. However, the intervening steps between CARD11 and the TCR are still unknown.

Embryos from mice lacking Bcl-10, like those lacking both IKKα and IKKβ (Li et al., 2000), exhibit defects in neuronal tube closure associated with excessive apoptosis in the neural epithelium (Ruland et al., 2001), thus indicating a major role for Bcl-10 in the development of the central nervous system. It will be important to determine whether the CARD-MAGUK proteins also participate, through Bcl-10, in neuronal survival. Moreover, whereas a role in T-cell signaling has now been demonstrated for CARD11, it remains to be determined if the CARD-MAGUK proteins are also involved in B-cell signaling. Future studies, including the use of mice deficient in these proteins, should clarify the physiological role of the CARD-MAGUK proteins in vivo.

3.3. The dual-domain CARD proteins

Members of this subfamily have a dual domain structure in which the CARD domain is present either at the N- or C-termini of these proteins. Their main function is to link upstream signaling molecules such as the NOD and CARD-MAGUK proteins with the downstream IKK complex.

3.3.1. Bcl-10

One of the first CARD proteins shown to be able to promote NF-κB activation was the Bcl-10 protein. Bcl-10 is the cellular homologue of the equine herpesvirus-2 E10 protein (vE10). Like its cellular counterpart, vE10 is also able to activate NF-κB when overexpressed in mammalian cells. Bcl-10, also called CLAP/CIPER/mE10/CARMEN/cE10 (Costanzo et al., 1999; Koseki et al., 1999; Srinivasula et al., 1999; Thome et al., 1999; Willis et al., 1999; Yan et al., 1999), is a protein of 233 amino acids containing an N-terminal CARD and a C-terminal domain that is rich in serine-threonine residues and which does not show significant homology to any other known protein (Fig. 1). Bcl-10 is able to interact with a wide array of proteins implicated in various functions, such as signal transduction, like TRAF2 or TRADD (Costanzo et al., 1999; Yoneda et al., 2000) or proteins implicated in the cell structure like α-actinin (Guiet & Vito, 2000) (Table 1).

The Bcl-10 gene has been shown to be involved in the chromosomal translocation t(1;14)(p22;q32), which is observed in some mucosa-associated lymphoid tissue (MALT) lymphomas (Willis et al., 1999; Zhang et al., 1999). Translocation t(1;14)(p22;q32) in MALT lymphoma results in overexpression of Bcl-10 and is associated with frameshift mutations that cause C-terminal truncations distal to the CARD (Willis et al., 1999; Zhang et al., 1999; Du et al., 2000). Bcl-10 mutations have also been described in cases of follicular lymphoma and diffuse large B-cell lymphoma (Willis et al., 1999; Du et al., 2000). However, it is still not clear whether these truncation mutants are actually expressed in cancer cells.

Based on overexpression studies, a proapoptotic activity as well as an NF-κB-inducing activity were attributed to Bcl-10 (Koseki et al., 1999; Srinivasula et al., 1999; Yan et al., 1999). However, targeted inactivation of the Bcl-10 gene indicated that Bcl-10 was not essential for the execution of apoptosis as the susceptibility of embryonic fibroblasts or thymocytes from Bcl-10$^{-/-}$ mice to a wide array of apoptotic stimuli was similar to that of Bcl-10$^{+/+}$ mice (Ruland et al., 2001). Instead, Bcl-10 appeared to function as a crucial positive regulator of antigen receptor-induced NF-κB activation in B- and T-cells. Lymphocyte activation and proliferation following antigenic challenge were severely impaired in Bcl-10$^{-/-}$ mice. Moreover, Bcl-10-deficient T-cells were unable to activate NF-κB in response to CD3/CD28 (Ruland et al., 2001). Thus, Bcl-10 appears to connect the antigen receptor signaling to NF-κB activation in B- and T-cells. The intervening steps between the B- and T-cell receptor complexes and Bcl-10 are not characterized, but increasing evidence suggests that the CARD-MAGUK proteins are involved in this process. The observed phenotype of Bcl-10$^{-/-}$ mice and the role of Bcl-10 as a mediator of NF-κB activation suggest a role for Bcl-10 in malignancy. The outgrowth of a pathological lymphoid population that progressively replaces the normal lymphoid population, giving rise to a MALT lymphoma, is probably driven by persistent antigenic stimulation. It is therefore possible that upregulation of Bcl-10 could, in the absence of stimulation, mimic antigen receptor signaling, resulting in a constitutive activation of NF-κB. This constitutive activation would promote antigen-independent clonal expansion of the tumor clones and therefore results in lymphoma progression. Bcl-10$^{-/-}$ embryos exhibit defects in neuronal tube closure, as well as enhanced apoptosis in the hindbrain neuroepithelium, a phenotype resembling that of the IKKα/IKKβ$^{-/-}$ embryos. These observations indicate that Bcl-10-mediated NF-κB activation may also act in a survival pathway to protect developing neurons.

Although the physiological function of Bcl-10 is now known, the mechanism by which Bcl-10 activates NF-κB is not fully understood. It has been demonstrated that NF-κB activation by Bcl-10 is mediated by the IKK complex, as no Bcl-10-mediated NF-κB activation is observed in cells lacking IKKγ, the essential regulatory subunit of the IKK complex (Lucas et al., 2001; Poyet et al., 2001). How is Bcl-10 connected to the IKK complex? Very interestingly, it was demonstrated that Bcl-10, through its C-terminal domain, interacts with MALT1, the other major target of chromosomal translocation in MALT lymphomas (Uren et al., 2000; Lucas et al., 2001). MALT1 is a recently discovered protein composed of a death domain, two immunoglobulin-like domains involved in the Bcl-10/MALT1 interaction and a C-terminal caspase-like domain. Although unable to activate NF-κB on its own when overexpressed, MALT1 was able to dramatically enhance Bcl-10-induced NF-κB activation when the two proteins were coexpressed (Lucas et al., 2001). Because enforced oligomerization of the C-terminal domain of Bcl-10 (Srinivasula et al., 1999) and of the caspase-like domain of MALT1 (Lucas et al., 2001) induce NF-κB activation, it is possible that Bcl-10 mediates the recruitment/oligomerization of the IKK complex via MALT1 (Fig. 2). Therefore, MALT1 could function as a downstream effector of Bcl-10, acting as a bridge between Bcl-10 and the IKK complex and by transmitting the oligomerization/activation signal to the IKK kinases. How MALT1 interacts with the IKK complex remains to be determined. Moreover, it is not known if MALT1 is the only downstream effector of Bcl-10 or whether other molecules also function downstream of Bcl-10 to activate NF-κB.

3.3.2. RIP2

RIP2, also called RICK/CARDIAK (Inohara et al., 1998; McCarthy et al., 1998; Thome et al., 1998), is a serine/threonine kinase related to the TNF receptor-interacting protein RIP. RIP2 is a component of both the TNF-R1- and the CD40-signaling complexes and could activate NF-κB, JNK and cell death when ectopically expressed in mammalian cells. However, unlike RIP, RIP2 appears also to be engaged in a receptor-independent NF-κB activation via interaction with the NOD proteins NOD1 and NOD2 (Fig. 2) (Inohara et al., 2000). RIP2 contains an N-terminal serine/threonine kinase domain and a C-terminal CARD. In vitro studies have shown that RIP2 can associate with a variety of other signaling molecules involved in both apoptosis and inflammation (Table 1).

Overexpression experiments suggested a proapoptotic function for RIP2 (McCarthy et al., 1998; Thome et al., 1998). However, it has been shown that susceptibility of embryonic fibroblasts or thymocytes from $RIP2^{-/-}$ mice to a wide array of apoptotic stimuli was similar to that of $RIP2^{+/+}$ mice (Kobayashi et al., 2002). Thus the role of RIP2 in the process of apoptosis remains unclear. In vitro studies have also suggested that RIP2 associates, by CARD-CARD interactions, and activates procaspase-1, leading to generation of IL-1β (Thome et al., 1998). However, macrophages of $RIP2^{-/-}$ mice showed no reduction in IL-1β production when cells were stimulated with LPS, CpG DNA and peptidoglycan, suggesting that RIP2 is dispensable for activation of caspase-1 (Kobayashi et al., 2002).

RIP2 has been shown to be required for signaling through TLR protein family members. Cytokine production after stimulation of TLRs with LPS, peptidoglycan and ds-RNA was found to be reduced in RIP2-deficient mice (Chin et al., 2002; Kobayashi et al., 2002). RIP2 was also found to be recruited by the receptor TLR2 after stimulation with peptidoglycan (Kobayashi et al., 2002). $RIP2^{-/-}$ mice were also more resistant to LPS than wild-type mice, and LPS-induced NF-κB activation was reduced in macrophages from RIP2-deficient mice, indicating a role for this molecule in the LPS-mediated response in vivo (Chin et al., 2002; Kobayashi et al., 2002). RIP2 is also implicated in the activation of NF-κB by the CARD proteins NOD1 and NOD2 after presentation of pathogens in the cytosol (Kobayashi et al., 2002). Finally, RIP2 is also required for appropriate TCR signaling and T-cell differentiation. Hence, RIP2 is a crucial adaptor involved in multiple pathways regulating the response of both the innate (TLR and NOD proteins) and the adaptive (TCR) immune systems.

The mechanism by which RIP2 activates NF-κB is very similar to that of RIP (Inohara et al., 2000; Poyet et al., 2000). Indeed, both RIP2 and RIP have been shown to associate directly with the regulatory subunit of the IKK complex (IKKγ) via their intermediate regions (Inohara et al., 2000; Poyet et al., 2000). In agreement with this observation, RIP2 was unable to induce NF-κB activation in IKKγ-deficient cells, thus confirming that interaction of RIP2 with IKKγ is essential for NF-κB activation (Inohara et al., 2000). RIP2 appears to induce the activation of the IKK complex by the induced proximity mechanism (Inohara et al., 2000). In this model, the CARD of RIP2 functions as a receptor docking domain and an oligomerizer of the intermediate domain, which in turn could induce the clustering-dependent activation of the associated IKK complex (Fig. 2). Indeed, artificially induced oligomerization of the intermediate region of RIP2 was sufficient for induction of NF-κB activation independently of any proinflammatory stimulation (Inohara et al., 2000). Although not essential, the kinase activity of RIP2 seems to be required for optimal NF-κB activation. Indeed, variants of RIP2 carrying critical point mutations of the kinase domain exhibited reduced NF-κB activation property (Inohara et al., 2000). At the present time, the target(s) of the RIP2 kinase are unknown.

What cellular events trigger the activation of RIP2? It is likely that RIP2 acts, at least, as a downstream effector of the NOD proteins in pathways leading to NF-κB activation, probably following exposure of the cell to cytoplasmic LPS (Fig. 2). In this model, following self-oligomerization upon LPS stimulation, the NOD proteins associate with the IKK complex through RIP2, which could relay the oligomerization/activation signal to the IKK kinases. Indeed, NF-κB activation by NOD1/NOD2 expression was completely abolished in embryonic fibroblasts of $RIP2^{-/-}$ mice, whereas complementation of $RIP2^{-/-}$ fibroblasts with a RIP2 expression vector restored the responsiveness to NOD1 and NOD2 (Kobayashi et al., 2002). These results demonstrate that RIP2 is essential for NF-κB activation mediated by both NOD1 and NOD2. Targeted disruption of the RIP2 gene has shown that RIP2 is also critically implicated in signaling from various immune receptors (both TLRs and TCR) (Inohara et al., 2000; Chin et al., 2002). Clearly, more work is needed in order to characterize the multiple pathways requiring the recruitment of this multifaceted kinase.

3.3.3. ASC

ASC (also called TMS1) is a bipartite molecule composed of a C-terminal CARD and an N-terminal pyrin (PYD) domain (Fig. 1) (Masumoto et al., 1999; Conway et al., 2000). ASC was first identified as a protein that aggregates into large hollow-centered structures called specks during apoptosis of HL-60 cells (Masumoto et al., 1999). Because the number of specks directly correlates with the number of apoptotic cells, it was concluded that the formation of specks is associated with the apoptotic process. Moreover, because the speck formation occurs upstream of caspase-9 activation, it is possible that ASC is a proapoptotic protein, although this remains to be demonstrated.

Recent observations suggest that ASC could be involved in caspase-1 activation and pro-IL-1β cytokine processing (Srinivasula et al., 2002). ASC was able to directly interact with procaspase-1, by CARD-CARD interaction, and induce its processing. Artificial oligomerization of the CARD of ASC was sufficient for caspase-1 activation (Srinivasula et al., 2002). Therefore, it is possible that the CARD is the effector domain of ASC and that the role of the pyrin domain is to regulate the oligomerization status of the CARD by upstream signaling molecules in this particular caspase-1 activation pathway. Indeed, the pyrin domain of ASC has been shown to be a homophilic protein-protein interaction domain that plays a role in oligomerization of ASC (Masumoto et al., 2001).

New observations suggest that ASC could also play another role in inflammatory signaling. Indeed, ASC was found to activate NF-κB when ectopically expressed in 293T cells (Manji et al., 2002). The mechanisms by which ASC activates NF-κB are currently unknown, but this requires a functional IKK complex as ASC is unable to elicit NF-κB activation when expressed in IKKγ-deficient cells (Manji et al., 2002). Because ASC is able to oligomerize, it is tempting to speculate that ASC induces activation of the IKK kinases via the induced-proximity mechanism. Whether ASC can bind directly to the IKK complex or whether an intermediary molecule is needed, is currently unknown.

Very recently, ASC was found to interact with three newly identified pyrin-containing proteins: pyrin, a protein mutated in familial mediterranean fever (FMV); PYPAF1, also called cryopyrin, and PYPAF7, two proteins related to the NOD proteins and expressed mainly in immune cells (Richards et al., 2001; Manji et al., 2002; Wang et al., 2002b). Although not able to activate NF-κB by themselves, PYPAF1 and PYPAF7 can synergistically activate ASC-induced NF-κB activity (Manji et al., 2002; Wang et al., 2002b). PYPAF7 was also demonstrated to potentiate ASC-induced caspase-1 activation (Wang et al., 2002b). Consequently, PYPAF1 and PYPAF7 may be upstream activators of ASC activity in NF-κB signaling and pro-IL-1β cytokine processing. Activation of ASC by PYPAF1 or PYPAF7 may occur by an induced-proximity mechanism analogous to the activation of caspase-9 by Apaf-1. Mutations within the PYPAF1 gene were reported to be associated with the Muckle-Wells syndrome (MWS) and the familial cold auto-inflammatory syndrome (FCAS), two genetically determined fever disorders (Hoffman et al., 2001). This underscores the role of the PYPAF proteins, and perhaps ASC, in inflammatory signaling. It remains to be determined whether pyrin is also able to modulate ASC-induced NF-κB activation.

3.3.4. CARD9

CARD9 is a bipartite molecule that comprises an N-terminal CARD and a C-terminal CC domain (Fig. 1). CARD9, like CARD10, CARD11 and CARD14, is able to activate NF-κB by engagement, via CARD-CARD interaction, of Bcl-10 (Bertin et al., 2000). Although the identity of the upstream activators of CARD9 remains to be determined, it is likely that CARD9 recruits and activates Bcl-10, perhaps by a mechanism similar to that of the CARD-MAGUK proteins.

3.4. The CARD-only proteins

The only member of this family, pseudo-ICE (also called COP), is a CARD-only protein with high similarity with the CARD of procaspase-1 (Druilhe et al., 2001; Lee et al., 2001). Pseudo-ICE is able to bind to both procaspase-1 and RIP2, and has been shown to prevent RIP2-induced caspase-1 activation (Druilhe et al., 2001; Lee et al., 2001). Moreover, pseudo-ICE diminishes IL-1β secretion induced by LPS stimulation of TLR4, indicating that pseudo-ICE can regulate at least one pathway known to be physiologically important in pathogen responses and could act as a modulator of IL-1β activation. Pseudo-ICE has also been shown to induce NF-κB activation when overexpressed (Druilhe et al., 2001). This activity requires the IKK complex, as a kinase-inactive mutant of the kinase IKKα inhibits pseudo-ICE-induced NF-κB activation (Druilhe et al., 2001). Interestingly, pseudo-ICE is able to bind to the CARD-containing kinase RIP2. Moreover, pseudo-ICE is able to interact with itself, thus to oligomerize. It is therefore likely that pseudo-ICE activates NF-κB by recruitment/oligomerization of RIP2. RIP2 in turn could transmit the oligomerization signal to the IKK complex, which could result in its autoactivation by the induced-proximity mechanism.

4. CONCLUDING REMARKS

Since the identification of the first caspases, the membership of the CARD-containing protein family has steadily expanded. The family growth has dramatically accelerated, as the use of the distinctive CARD pattern facilitated the cloning in silico from expressed sequence tag (EST) databases. CARD-containing proteins have been demonstrated to play a crucial role in the apoptosis network. Initially, all the characterized proteins found to contain a CARD were reported to act in apoptotic signaling. However, investigations over the past two years have allowed the discovery of a whole raft of new CARD-containing molecules that do not interact with caspases; rather, many of these proteins have been implicated in the activation of the anti-apoptotic transcription factor NF-κB. Among the twenty-six potentially CARD-containing proteins found in the human genome (Gozani et al., 2002), it is likely that many of them could be involved in NF-κB signaling. Thus, the CARD domain is a very versatile signaling component, with fundamental importance in the cellular pathways that ultimately decide whether a cell should survive or succumb to death in response to a wide array of stimuli. Therefore, the identification of new NF-κB activating CARD proteins will allow a much better understanding of the cellular pathways controlling this vital process.

Interestingly, the CARD-containing proteins seem to activate their downstream effector proteins, such as caspases or the IKK complex, by inducing their oligomerization according to the induced-proximity model. It is quite fascinating that two opposing pathways, apoptosis and survival, use the same interaction motif and the same mechanism to transduce opposite signals. Moreover, several CARD-containing proteins, such as RIP2 or NOD1 and NOD2, have been shown to be able to promote both caspase and NF-κB activation. Thus the regulation of the CARD-CARD interaction appears to be more complicated than initially thought.

The physiological importance of the NF-κB-activating CARD-containing proteins in inflammation has been underscored by recent observations. Indeed, mutations of NOD2 have been associated with two distinct chronic inflammatory diseases. A frameshift mutation in NOD2, resulting in a protein that lacks part of the LRRs, has been associated with susceptibility to Crohn's disease, whereas point mutations in the NBD of NOD2 were found to be associated with susceptibility to the Blau syndrome (Hugot et al., 2001; Miceli-Richard et al., 2001; Ogura et al., 2001a). At this time, the mechanisms by which genetic variations in NOD2 increase the susceptibility to Crohn's disease or the Blau syndrome, are largely unknown. Moreover, it will be extremely interesting to determine if mutations of NOD2, or any other gene encoding a CARD-containing protein, are involved in other clinical disorders. Clearly, more work, including the generation of mice deficient in CARD-containing proteins, needs to be done in order to understand the detailed mechanisms by which the CARD-containing proteins are acting in the cell.

5. ACKNOWLEDGMENTS

Research in the laboratory of E.S. Alnemri is supported by a grant CA85421 from the National Institutes of Health.

6. REFERENCES

Adrain C & Martin SJ (2001) The mitochondrial apoptosome: a killer unleashed by the cytochrome seas. Trends Biochem. Sci. 26: 390-397.

Aravind L, Dixit VM & Koonin EV (1999) The domains of death: evolution of the apoptosis machinery. Trends Biochem. Sci. 24: 47-53.

Baldwin AS (1996) The NF-κB and IκB proteins: new discoveries and insights. Annu. Rev. Immunol. 14: 649-683.

Bertin J & DiStefano PS (2000) The PYRIN domain: a novel motif found in apoptosis and inflammation proteins. Cell Death Differ. 7: 1273-1274.

Bertin J, Guo Y, Wang L, Srinivasula SM, Jacobson MD, Poyet JL, Merriam S, Du MQ, Dyer MJ, Robison KE, DiStefano PS & Alnemri ES (2000) CARD9 is a novel caspase recruitment domain-containing protein that interacts with BCL10/CLAP and activates NF-κB. J. Biol. Chem. 275: 41082-41086.

Bertin J, Nir WJ, Fischer CM, Tayber OV, Errada PR, Grant JR, Keilty JJ, Gosselin ML, Robison KE, Wong GH, Glucksmann MA & DiStefano PS (1999) Human CARD4 protein is a novel CED-4/Apaf-1 cell death family member that activates NF-κB. J. Biol. Chem. 274: 12955-12958.

Bertin J, Wang L, Guo Y, Jacobson MD, Poyet JL, Srinivasula SM, Merriam S, DiStefano PS & Alnemri ES (2001) CARD11 and CARD14 are novel caspase recruitment domain (CARD)/membrane-associated guanylate kinase (MAGUK) family members that interact with BCL10 and activate NF-κB. J. Biol. Chem. 276: 11877-11882.

Brown K, Gerstberger S, Carlson L, Franzoso G & Siebenlist U (1995) Control of IκB-α proteolysis by site-specific, signal-induced phosphorylation. Science 267: 1485-1488.

Burke JR, Miller KR, Wood MK & Meyers CA (1998) The multisubunit IκB kinase complex shows random sequential kinetics and is activated by the C-terminal domain of IκB α. J. Biol. Chem. 273: 12041-12046.

Chen Z, Hagler J, Palombella VJ, Melandri F, Scherer D, Ballard D & Maniatis T (1995) Signal-induced site-specific phosphorylation targets IκBα to the ubiquitin-proteasome pathway. Genes Dev. 9: 1586-1597.

Chin AI, Dempsey PW, Bruhn K, Miller JF, Xu Y & Cheng G (2002) Involvement of receptor-interacting protein 2 in innate and adaptive immune responses. Nature 416: 190-194.

Cohen GM (1997) Caspases: the executioners of apoptosis. Biochem. J. 326: 1-16.

Conway KE, McConnell BB, Bowring CE, Donald CD, Warren ST & Vertino PM (2000) TMS1, a novel proapoptotic caspase recruitment domain protein, is a target of methylation-induced gene silencing in human breast cancers. Cancer Res. 60: 6236-6242.

Costanzo A, Guiet C & Vito P (1999) c-E10 is a caspase-recruiting domain-containing protein that interacts with components of death receptors signaling pathway and activates nuclear factor-κB. J. Biol. Chem. 274: 20127-20132.

Dangl JL & Jones JD (2001) Plant pathogens and integrated defence responses to infection. Nature 411: 826-833.

DiDonato JA, Hayakawa M, Rothwarf DM, Zandi E & Karin M (1997) A cytokine-responsive IκB kinase that activates the transcription factor NF-κB. Nature 388: 548-554.

Dienz O, Hehner SP, Droge W & Schmitz ML (2000) Synergistic activation of NF-κB by functional cooperation between vav and PKCθ in T lymphocytes. J. Biol. Chem. 275: 24547-24551.

Dimitratos SD, Woods DF, Stathakis DG & Bryant PJ (1999) Signaling pathways are focused at specialized regions of the plasma membrane by scaffolding proteins of the MAGUK family. Bioessays 21: 912-921.

Dixon MS, Golstein C, Thomas CM, van Der Biezen EA & Jones JD (2000) Genetic complexity of pathogen perception by plants: the example of Rcr3, a tomato gene required specifically by Cf-2. Proc. Natl. Acad. Sci. USA 97: 8807-8814.

Druilhe A, Srinivasula SM, Razmara M, Ahmad M & Alnemri ES (2001) Regulation of IL-1β generation by pseudo-ICE and ICEBERG, two dominant negative caspase recruitment domain proteins. Cell Death Differ. 8: 649-657.

Du MQ, Peng H, Liu H, Hamoudi RA, Diss TC, Willis TG, Ye H, Dogan A, Wotherspoon AC, Dyer MJ & Isaacson PG (2000) BCL10 gene mutation in lymphoma. Blood 95: 3885-3890.

Fanning AS & Anderson JM (1999) Protein modules as organizers of membrane structure. Curr. Opin. Cell Biol. 11: 432-439.

Fesik SW (2000) Insights into programmed cell death through structural biology. Cell 103: 273-282.

Fiocchi C (1998) Inflammatory bowel disease: etiology and pathogenesis. Gastroenterology 115: 182-205.

Gaide O, Favier B, Legler DF, Bonnet D, Brissoni B, Valitutti S, Bron C, Tschopp J & Thome M (2002) CARMA1 is a critical lipid raft#150;associated regulator of TCR-induced NF-κB activation. Nat. Immunol. 3: 836-843.

Gaide O, Martinon F, Micheau O, Bonnet D, Thome M & Tschopp J (2001) Carma1, a CARD-containing binding partner of Bcl10, induces Bcl10 phosphorylation and NF-κB activation. FEBS Lett. 496: 121-127.

Ghosh S, May MJ & Kopp EB (1998) NF-κB and Rel proteins: evolutionarily conserved mediators of immune responses. Annu. Rev. Immunol. 16: 225-260.

Girardin SE, Tournebize R, Mavris M, Page AL, Li X, Stark GR, Bertin J, DiStefano PS, Yaniv M, Sansonetti PJ & Philpott DJ (2001) CARD4/Nod1 mediates NF-κB and JNK activation by invasive Shigella flexneri. EMBO Rep. 2: 736-742.

Gozani O, Boyce M, Yoo L, Karuman P & Yuan J (2002) Life and death in paradise. Nat. Cell Biol. 4: E159-162.

Guiet C & Vito P (2000) Caspase recruitment domain (CARD)-dependent cytoplasmic filaments mediate bcl10-induced NF-κB activation. J. Cell Biol. 148: 1131-1140.

Hoffman HM, Mueller JL, Broide DH, Wanderer AA & Kolodner RD (2001) Mutation of a new gene encoding a putative pyrin-like protein causes familial cold autoinflammatory syndrome and Muckle-Wells syndrome. Nat. Genet. 29: 301-305.

Hofmann K, Bucher P & Tschopp J (1997) The CARD domain: a new apoptotic signalling motif. Trends Biochem. Sci. 22: 155-156.

Hugot JP, Chamaillard M, Zouali H, Lesage S, Cezard JP, Belaiche J, Almer S, Tysk C, O'Morain CA, Gassull M, Binder V, Finkel Y, Cortot A, Modigliani R, Laurent-Puig P, Gower-Rousseau C, Macry J, Colombel JF, Sahbatou M & Thomas G (2001) Association of NOD2 leucine-rich repeat variants with susceptibility to Crohn's disease. Nature 411: 599-603.

Humke EW, Shriver SK, Starovasnik MA, Fairbrother WJ & Dixit VM (2000) ICEBERG: a novel inhibitor of interleukin-1β generation. Cell 103: 99-111.

Inohara N, del Peso L, Koseki T, Chen S & Nunez G (1998) RICK, a novel protein kinase containing a caspase recruitment domain, interacts with CLARP and regulates CD95-mediated apoptosis. J.

Biol. Chem. 273: 12296-12300.

Inohara N, Koseki T, del Peso L, Hu Y, Yee C, Chen S, Carrio R, Merino J, Liu D, Ni J & Nunez G (1999) Nod1, an Apaf-1-like activator of caspase-9 and nuclear factor-κB. J. Biol. Chem. 274: 14560-14567.

Inohara N, Koseki T, Lin J, del Peso L, Lucas PC, Chen FF, Ogura Y & Nunez G (2000) An induced proximity model for NF-κB activation in the Nod1/RICK and RIP signaling pathways. J. Biol. Chem. 275: 27823-27831.

Inohara N, Ogura Y, Chen FF, Muto A & Nunez G (2001) Human Nod1 confers responsiveness to bacterial lipopolysaccharides. J. Biol. Chem. 276: 2551-2554.

Kobayashi K, Inohara N, Hernandez LD, Galan JE, Nunez G, Janeway CA, Medzhitov R & Flavell RA (2002) RICK/Rip2/CARDIAK mediates signalling for receptors of the innate and adaptive immune systems. Nature 416: 194-199.

Koseki T, Inohara N, Chen S, Carrio R, Merino J, Hottiger MO, Nabel GJ & Nunez G (1999) CIPER, a novel NFκB-activating protein containing a caspase recruitment domain with homology to herpesvirus-2 protein E10. J. Biol. Chem. 274: 9955-9961.

Lee SH, Stehlik C & Reed JC (2001) Cop, a caspase recruitment domain-containing protein and inhibitor of caspase-1 activation processing. J. Biol. Chem. 276: 34495-34500.

Li Q, Estepa G, Memet S, Israel A & Verma IM (2000) Complete lack of NF-κB activity in IKK1 and IKK2 double-deficient mice: additional defect in neurulation. Genes Dev. 14: 1729-1733.

Li Y, Kang J, Friedman J, Tarassishin L, Ye J, Kovalenko A, Wallach D & Horwitz MS (1999) Identification of a cell protein (FIP-3) as a modulator of NF-κB activity and as a target of an adenovirus inhibitor of tumor necrosis factor α-induced apoptosis. Proc. Natl. Acad. Sci. USA 96: 1042-1047.

Lin X, O'Mahony A, Mu Y, Geleziunas R & Greene WC (2000) Protein kinase C-θ participates in NF-κB activation induced by CD3-CD28 costimulation through selective activation of IκB kinase β. Mol. Cell. Biol. 20: 2933-2940.

Lucas PC, Yonezumi M, Inohara N, McAllister-Lucas LM, Abazeed ME, Chen FF, Yamaoka S, Seto M & Nunez G (2001) Bcl10 and MALT1, independent targets of chromosomal translocation in malt lymphoma, cooperate in a novel NF-κB signaling pathway. J. Biol. Chem. 276: 19012-19019.

Manji GA, Wang L, Geddes BJ, Brown M, Merriam S, Al-Garawi A, Mak S, Lora JM, Briskin M, Jurman M, Cao J, DiStefano PS & Bertin J (2002) PYPAF1, a PYRIN-containing Apaf1-like protein that assembles with ASC and regulates activation of NF-κB. J. Biol. Chem. 277: 11570-11575.

Martinon F, Burns K & Tschopp J (2002) The inflammasome. A molecular platform triggering activation of inflammatory caspases and processing of proIL-β. Mol. Cell 10: 417.

Masumoto J, Taniguchi S, Ayukawa K, Sarvotham H, Kishino T, Niikawa N, Hidaka E, Katsuyama T, Higuchi T & Sagara J (1999) ASC, a novel 22-kDa protein, aggregates during apoptosis of human promyelocytic leukemia HL-60 cells. J. Biol. Chem. 274: 33835-33838.

Masumoto J, Taniguchi S & Sagara J (2001) Pyrin N-terminal homology domain- and caspase recruitment domain-dependent oligomerization of ASC. Biochem. Biophys. Res. Commun. 280: 652-655.

McAllister-Lucas LM, Inohara N, Lucas PC, Ruland J, Benito A, Li Q, Chen S, Chen FF, Yamaoka S, Verma IM, Mak TW & Nunez G (2001) Bimp1, a MAGUK family member linking protein kinase C activation to Bcl10-mediated NF-κB induction. J. Biol. Chem. 276: 30589-30597.

McCarthy JV, Ni J & Dixit VM (1998) RIP2 is a novel NF-κB-activating and cell death-inducing kinase. J. Biol. Chem. 273: 16968-16975.

Mercurio F, Murray BW, Shevchenko A, Bennett BL, Young DB, Li JW, Pascual G, Motiwala A, Zhu H, Mann M & Manning AM (1999) IκB kinase (IKK)-associated protein 1, a common component of the heterogeneous IKK complex. Mol. Cell. Biol. 19: 1526-1538.

Mercurio F, Zhu H, Murray BW, Shevchenko A, Bennett BL, Li J, Young DB, Barbosa M, Mann M, Manning A & Rao A (1997) IKK-1 and IKK-2: cytokine-activated IκB kinases essential for NF-κB activation. Science 278: 860-866.

Miceli-Richard C, Lesage S, Rybojad M, Prieur AM, Manouvrier-Hanu S, Hafner R, Chamaillard M, Zouali H, Thomas G & Hugot JP (2001) CARD15 mutations in Blau syndrome. Nat. Genet.

29: 19-20.

Mullauer L, Gruber P, Sebinger D, Buch J, Wohlfart S & Chott A (2001) Mutations in apoptosis genes: a pathogenetic factor for human disease. Mutat. Res. 488: 211-231.

Ogura Y, Bonen DK, Inohara N, Nicolae DL, Chen FF, Ramos R, Britton H, Moran T, Karaliuskas R, Duerr RH, Achkar JP, Brant SR, Bayless TM, Kirschner BS, Hanauer SB, Nunez G & Cho JH (2001a) A frameshift mutation in NOD2 associated with susceptibility to Crohn's disease. Nature 411: 603-606.

Ogura Y, Inohara N, Benito A, Chen FF, Yamaoka S & Nunez G (2001b) Nod2, a Nod1/Apaf-1 family member that is restricted to monocytes and activates NF-κB. J. Biol. Chem. 276: 4812-4818.

Parniske M, Hammond-Kosack KE, Golstein C, Thomas CM, Jones DA, Harrison K, Wulff BB & Jones JD (1997) Novel disease resistance specificities result from sequence exchange between tandemly repeated genes at the Cf-4/9 locus of tomato. Cell 91: 821-832.

Pawlowski K, Pio F, Chu Z, Reed JC & Godzik A (2001) PAAD: a new protein domain associated with apoptosis, cancer and autoimmune diseases. Trends Biochem. Sci. 26: 85-87.

Philpott DJ, Yamaoka S, Israel A & Sansonetti PJ (2000) Invasive Shigella flexneri activates NF-κB through a lipopolysaccharide-dependent innate intracellular response and leads to IL-8 expression in epithelial cells. J. Immunol. 165: 903-914.

Poyet JL, Srinivasula SM & Alnemri ES (2001) vCLAP, a caspase-recruitment domain-containing protein of equine Herpesvirus-2, persistently activates the IκB kinases through oligomerization of IKKγ. J. Biol. Chem. 276: 3183-3187.

Poyet JL, Srinivasula SM, Lin JH, Fernandes-Alnemri T, Yamaoka S, Tsichlis PN & Alnemri ES (2000) Activation of the IκB kinases by RIP via IKKγ/NEMO-mediated oligomerization. J. Biol. Chem. 275: 37966-37977.

Regnier CH, Song HY, Gao X, Goeddel DV, Cao Z & Rothe M (1997) Identification and characterization of an IκB kinase. Cell 90: 373-383.

Richards N, Schaner P, Diaz A, Stuckey J, Shelden E, Wadhwa A & Gumucio DL (2001) Interaction between pyrin and the apoptotic speck protein (ASC) modulates ASC-induced apoptosis. J. Biol. Chem. 276: 39320-39329.

Rothwarf DM, Zandi E, Natoli G & Karin M (1998) IKK-γ is an essential regulatory subunit of the IκB kinase complex. Nature 395: 297-300.

Ruland J, Duncan GS, Elia A, del Barco Barrantes I, Nguyen L, Plyte S, Millar DG, Bouchard D, Wakeham A, Ohashi PS & Mak TW (2001) Bcl10 is a positive regulator of antigen receptor-induced activation of NF-κB and neural tube closure. Cell 104: 33-42.

Saleh A, Srinivasula SM, Acharya S, Fishel R & Alnemri ES (1999) Cytochrome c and dATP-mediated oligomerization of Apaf-1 is a prerequisite for procaspase-9 activation. J. Biol. Chem. 274: 17941-17945.

Salvesen GS & Dixit VM (1999) Caspase activation: the induced-proximity model. Proc. Natl. Acad. Sci. USA 96: 10964-10967.

Scherer DC, Brockman JA, Chen Z, Maniatis T & Ballard DW (1995) Signal-induced degradation of IκBα requires site-specific ubiquitination. Proc. Natl. Acad. Sci. USA 92: 11259-11263.

Shi Y (2002) Mechanisms of caspase activation and inhibition during apoptosis. Mol. Cell 9: 459-470.

Srinivasula SM, Ahmad M, Lin JH, Poyet JL, Fernandes-Alnemri T, Tsichlis PN & Alnemri ES (1999) CLAP, a novel caspase recruitment domain-containing protein in the tumor necrosis factor receptor pathway, regulates NF-κB activation and apoptosis. J. Biol. Chem. 274: 17946-17954.

Srinivasula SM, Poyet JL, Razmara M, Datta P, Zhang Z & Alnemri ES (2002) The PYRIN-CARD protein ASC is an activating adaptor for caspase-1. J. Biol. Chem. 277: 21119-21122.

Thome M, Hofmann K, Burns K, Martinon F, Bodmer JL, Mattmann C & Tschopp J (1998) Identification of CARDIAK, a RIP-like kinase that associates with caspase-1. Curr. Biol. 8: 885-888.

Thome M, Martinon F, Hofmann K, Rubio V, Steiner V, Schneider P, Mattmann C & Tschopp J (1999) Equine herpesvirus-2 E10 gene product, but not its cellular homologue, activates NF-κB transcription factor and c-Jun N-terminal kinase. J. Biol. Chem. 274: 9962-9968.

Uren AG, O'Rourke K, Aravind LA, Pisabarro MT, Seshagiri S, Koonin EV & Dixit VM (2000) Identification of paracaspases and metacaspases: two ancient families of caspase-like proteins, one of which plays a key role in MALT lymphoma. Mol. Cell 6: 961-967.

Vaux DL & Strasser A (1996) The molecular biology of apoptosis. Proc. Natl. Acad. Sci. USA 93: 2239-2244.

Wang D, You Y, Case SM, McAllister-Lucas LM, Wang L, DiStefano PS, Nunez G, Bertin J & Lin X (2002a) A requirement for CARMA1 in TCR-induced NF-κB activation. Nat. Immunol. 3: 830-835.

Wang L, Guo Y, Huang WJ, Ke X, Poyet JL, Manji GA, Merriam S, Glucksmann MA, DiStefano PS, Alnemri ES & Bertin J (2001) Card10 is a novel caspase recruitment domain/membrane-associated guanylate kinase family member that interacts with BCL10 and activates NF-κB. J. Biol. Chem. 276: 21405-21409.

Wang L, Manji G, Grenier JM, Al-Garawi A, Merriam S, Lora JM, Geddes BJ, Briskin M, DiStefano PS & Bertin J (2002b) PYPAF7, a novel PYRIN-containing Apaf1-like protein that regulates activation of NF-κB and caspase-1-dependent cytokine processing. J. Biol. Chem. 277: 29874-29880.

Weber CH & Vincenz C (2001) The death domain superfamily: a tale of two interfaces? Trends Biochem. Sci. 26: 475-481.

Willis TG, Jadayel DM, Du MQ, Peng H, Perry AR, Abdul-Rauf M, Price H, Karran L, Majekodunmi O, Wlodarska I, Pan L, Crook T, Hamoudi R, Isaacson PG & Dyer MJ (1999) Bcl10 is involved in t(1;14)(p22;q32) of MALT B cell lymphoma and mutated in multiple tumor types. Cell 96: 35-45.

Wirth T & Baltimore D (1988) Nuclear factor NF-κB can interact functionally with its cognate binding site to provide lymphoid-specific promoter function. EMBO J. 7: 3109-3113.

Woronicz JD, Gao X, Cao Z, Rothe M & Goeddel DV (1997) IκB kinase-β: NF-κB activation and complex formation with IκB kinase-α and NIK. Science 278: 866-869.

Yamaoka S, Courtois G, Bessia C, Whiteside ST, Weil R, Agou F, Kirk HE, Kay RJ & Israel A (1998) Complementation cloning of NEMO, a component of the IκB kinase complex essential for NF-κB activation. Cell 93: 1231-1240.

Yan M, Lee J, Schilbach S, Goddard A & Dixit V (1999) mE10, a novel caspase recruitment domain-containing proapoptotic molecule. J. Biol. Chem. 274: 10287-10292.

Yoneda T, Imaizumi K, Maeda M, Yui D, Manabe T, Katayama T, Sato N, Gomi F, Morihara T, Mori Y, Miyoshi K, Hitomi J, Ugawa S, Yamada S, Okabe M & Tohyama M (2000) Regulatory mechanisms of TRAF2-mediated signal transduction by Bcl10, a MALT lymphoma-associated protein. J. Biol. Chem. 275: 11114-11120.

Zandi E, Rothwarf DM, Delhase M, Hayakawa M & Karin M (1997) The IκB kinase complex (IKK) contains two kinase subunits, IKKα and IKKβ, necessary for IκB phosphorylation and NF-κB activation. Cell 91: 243-252.

Zhang Q, Siebert R, Yan M, Hinzmann B, Cui X, Xue L, Rakestraw KM, Naeve CW, Beckmann G, Weisenburger DD, Sanger WG, Nowotny H, Vesely M, Callet-Bauchu E, Salles G, Dixit VM, Rosenthal A, Schlegelberger B & Morris SW (1999) Inactivating mutations and overexpression of BCL10, a caspase recruitment domain-containing gene, in MALT lymphoma with t(1;14)(p22;q32). Nat. Genet. 22: 63-68.

Zou H, Li Y, Liu X & Wang X (1999) An APAF-1.cytochrome c multimeric complex is a functional apoptosome that activates procaspase-9. J. Biol. Chem. 274: 11549-11556.

7. AFFILIATION

Thomas Jefferson University, Kimmel Cancer Institute, Center for Apoptosis Research and Department of Microbiology and Immunology, 233 South 10th Street, Philadelphia, PA 19107, USA.

T.D. GILMORE & G. MOSIALOS

VIRUSES AS INTRUDERS IN THE Rel/NF-κB SIGNALING PATHWAY

1. THE NUTS AND BOLTS OF Rel/NF-κB SIGNALING

In vertebrates, the Rel/NF-κB family of transcription factors contains five members: c-Rel, RelA (p65), RelB, p50/p105 and p52/p100. These factors are related at the level of sequence and structure and in their regulation by subcellular localization. When bound to DNA, Rel/NF-κB proteins influence the expression of hundreds of cellular genes and, as discussed in this review, many viral genes. Many of the Rel/NF-κB cellular target genes control immune responses (e.g. cytokines, chemokines and their receptors), cell cycle and growth (e.g. cyclins, c-Myc), apoptosis (Bcl-2 family members, inhibitors of apoptosis [IAPs]), adhesion and angiogenesis (e.g. ICAM, VCAM, VEGF), and cellular stress (Pahl, 1999). Thus, viruses have targeted the Rel/NF-κB system to affect immune responses of their hosts, cell survival, cell growth, and to activate the expression of their own viral genes.

All Rel/NF-κB transcription factors affect specific gene transcription by binding to a collection of related 9-10 bp DNA sites, called κB sites, as dimers. These dimers can be either homodimers or heterodimers. The most common dimer is a p50-RelA heterodimer, often specifically called NF-κB. Dimerization, DNA binding and nuclear localization are mediated by sequences within the conserved N-terminal Rel homology domain, which defines this family of transcription factors. One subset of Rel/NF-κB protein, including c-Rel, RelA and RelB, has C-terminal sequences that contain transcriptional activation domains. In contrast, the p105 and p100 proteins have C-terminal ankyrin repeat domains that fold back to inhibit DNA binding by the Rel homology domain sequences; thus these C-terminal domains must be removed to create the active DNA-binding subunits p50 and p52, respectively. Thus, dimers that contain at least one subunit of the first class of protein (c-Rel, RelA or RelB) usually activate transcription, whereas p50-p50 or p52-p52 homodimers generally act as repressors of transcription.

Rel/NF-κB transcription factors are regulated by subcellular localization through interaction with a family of inhibitor proteins, called IκBs (Karin, 1999). That is, under most circumstances Rel/NF-κB dimers are inactive IκB-bound complexes in the cytoplasm. Upon activation, the Rel/NF-κB dimer enters the

R. Beyaert (ed.), Nuclear Factor κB. Regulation and Role in Disease, 89-113.

nucleus to affect transcription. Activation of the Rel/NF-κB complex is effected primarily by proteolysis of the IκB protein. The IκBs are either separate molecules (IκBα, IκBβ, IκBε, Bcl-3) or, as mentioned above, the C-terminal sequences of p105 and p100. All IκBs contain 6-8 copies of the ankyrin repeat, which is the protein-protein interaction domain that mediates interaction with the Rel/NF-κB dimer.

The Rel/NF-κB pathway can be activated by a large variety of extracellular signals and intracellular conditions (Pahl, 1999). For example, activation occurs in response to many cytokines (tumor necrosis factor, interleukins), growth factors (PDGF, TGFβ), stress (as induced by UV irradiation, the production of oxygen-free radicals, or the accumulation of protein in intracellular organelles [e.g. the ER]), and double-stranded RNA. In almost all cases, induction of Rel/NF-κB by these myriad signals converges at the activation of a multi-component kinase complex called IKK (IκB kinase) (Karin, 1999). In the best-studied case, activated IKK phosphorylates two closely spaced N-terminal serine residues on IκBα, which causes IκBα to undergo ubiquitination at nearby lysine residues, and then to be degraded by the proteasome. The NF-κB complex is then free to enter the nucleus and bind to DNA. In the case of p100 or p105 complexes, the C-terminal ankyrin repeat domain undergoes limited proteolysis to liberate the complex. The gene encoding IκBα contains upstream κB sites, and consequently shows increased expression upon activation of NF-κB; thus, the NF-κB response is turned off by newly synthesized IκBα, usually about one hour after induction. Thus, activation of NF-κB is normally a transient event, except in a few normal cell types (e.g. B-cells). However, in many pathogenic states, including chronic inflammatory diseases, some cancers, and in response to several viruses, the Rel/NF-κB pathway is constitutively activated.

The IKK complex contains two related catalytic kinase subunits, called IKKα and IKKβ, and a regulatory subunit called IKKγ. In most cases, IKKγ is required for activation of IKK, and likely serves as an activator platform for the kinase subunits. The kinase components are not completely redundant: for example, IKKβ appears to be the kinase that primarily signals the activation of IκBα-containing complexes, whereas IKKα signals the activation of p100-containing complexes. Moreover, IKKβ knockout mice die embryonically due to liver apoptosis, while IKKα mice have defects in skin development. Activation and inactivation of IKK occur via phosphorylation. That is, IKKγ undergoes signal-induced phosphorylation, and IKKα and IKKβ require specific phosphorylation first to activate them and then to inactivate them. As the central integrator of normal Rel/NF-κB signaling, the IKK complex is also the target of many viruses.

The understanding of the molecular events of Rel/NF-κB activation has allowed for the development of specific inhibitors of this pathway. For example, expression of nondegradable (superrepressor) forms of IκBα, which lack sequences required for its signal-induced degradation, dominant-negative forms of IKKα/β, and proteasome inhibitors can all block NF-κB activation. Similarly, viruses, in many cases, encode proteins that interact directly with components of the Rel/NF-κB signaling pathway to modulate its activity (Table 1).

Table 1. Viral proteins that modulate the Rel/NF-κB pathway.

Virus	*Viral protein*	*Protein activity*
—Activators of NF-κB		
Rev-T	v-Rel	Activated c-Rel
Epstein-Barr virus	LMP-1	Activated CD40-like
	EBNA-2	Transcription coactivator
Human cytomegalovirus	IE1	Nuclear protein
	US28	Constitutive transmembrane receptor
Mouse cytomegalovirus	M33	Constitutive transmembrane receptor
Hepatitis B virus	HBx	Adaptor
Hepatitis C virus	Core protein*	Unknown
Herpesvirus saimiri	Tip; StpC	Adaptor for LCK; interacts with TRAF2
Human herpesvirus-8	ORF74	G-protein-coupled chemokine receptor
	vFLIP	Associates with IKK
HIV-1	Vpr*	Virion protein
	Tat	Transactivator
	Nef	Multifunctional adaptor
	gp120	Engages CD4 receptor
Human T-cell leukemia virus-1	Tax	Adaptor for IKKγ
Human T-cell leukemia virus-2	Tax	Adaptor for IKKγ
Influenza A	HA, M, NP*	Hemagglutinin, matrix, nuclear protein
African swine fever virus	A224L	IAP-like activator of IKK
Adenovirus	E1A**	Adaptor
Rotavirus	VP4 capsid protein	Activates IKK
—Inhibitors of NF-κB		
African swine fever virus	A238L	Nondegradable IκB
Adenovirus	E1A**	Adaptor
	E1B	Adaptor
	E3-14.7K	Binds to IKKγ
	E3-10.4/14.5K	Regulator of Fas
Hepatitis C virus	NS5A	Adaptor
Influenza A	NS1	Binds double-stranded RNA
Human papillomavirus	E7	IKK inhibitor

* *Activation could be due to ER overload response (Pahl & Baeuerle, 1997).*
** *Adenovirus E1A has been reported to have both stimulatory and inhibitory effects on NF-κB activation.*

2. THE AVIAN Rev-T RETROVIRUS ENCODES THE HIGHLY EFFECTIVE ONCOPROTEIN v-Rel

The first indication that aberrant expression of Rel/NF-κB transcription factors could lead to disease came from the study of the avian Rev-T retrovirus. Rev-T has a single gene (v-rel), which encodes an extensively mutated version of the avian c-rel proto-oncogene (Gilmore, 1999). Rev-T was isolated from a turkey with a reticular malignancy in 1958. Prolonged in vitro and in vivo passage of Rev-T has generated a highly oncogenic virus that can induce a rapidly fatal lymphoma/leukemia in young birds and that can efficiently transform and immortalize chicken lymphoid cells in vitro. Retroviral vectors that direct the overexpression of chicken, mouse, or human c-Rel can also transform and immortalize chicken lymphoid cells in vitro, although these normal c-Rel proteins do so less efficiently than v-Rel (Gilmore, 1999; Gilmore et al., 2001). In addition, T-cell-specific expression of v-rel in transgenic mice results in the development of T-cell lymphomas (Carrasco et al., 1996). A variety of mutagenesis studies has shown that four factors are critical for the transforming activity of v-Rel: it must be expressed at a relatively high level, and it must form homodimers, bind to DNA and activate transcription (Gilmore, 1999).

v-Rel is missing two N-terminal and 118 C-terminal amino acids of c-Rel and has viral Envelope-derived amino acids in their place; thus, v-Rel is an Env-Rel-Env fusion protein. In addition, v-Rel has several internal mutations as compared to c-Rel. In many cases, it is known how these changes from c-Rel have altered the function of v-Rel and contribute to its enhanced oncogenicity. The mutation that confers most of the increased oncogenicity onto v-Rel is the deletion of the C-terminal c-Rel residues: that is, the deletion of C-terminal residues from chicken c-Rel increases its in vitro transforming activity by approximately 10-fold (Kamens et al., 1990; Hrdlicková et al., 1994), and c-Rel proteins with C-terminal deletions often arise from transformation assays conducted with the full-length chicken c-Rel protein (Hrdlicková et al., 1994; Gilmore et al., 1995). The C-terminal deletion in v-Rel removes a strong c-Rel transactivation domain; thus, v-Rel is a weaker activator of transcription than c-Rel in many assays. However, one cannot simply conclude that reduced transactivation by v-Rel causes its oncogenicity, in that other mutations cause v-Rel to be a more effective activator of transcription than c-Rel and contribute to the increased transforming activity of v-Rel. First, v-Rel is also less effectively inhibited by IκBα than is c-Rel, in part due to internal mutations in v-Rel that reduce its affinity for IκBα (Sachdev & Hannnik, 1998). Second, v-Rel has an additional transactivation domain as compared to c-Rel, due to the presence of virus-derived Envelope amino acids at its N-terminus (Epinat et al., 2000). Third, certain internal mutations in v-Rel affect its DNA-binding site specificity (Nehyba et al., 1997; Huang et al., 2001a) and may impair regulation of its activity by cellular mechanisms (Mosialos & Gilmore, 1993).

Perhaps not surprisingly, many of the target genes that are affected by v-Rel in transformed lymphoid cells control growth or apoptosis. These genes include ones encoding growth promoting transcription factors (c-Rel, c-Jun, STAT1 and IRF4), cytokine receptors (IL-2Rα) and antiapoptotic proteins (IAP1) (Gilmore, 1999). Thus, it appears that v-Rel impinges on various growth promoting and survival pathways as part of its transforming process. Although v-Rel can transform a variety of cell types in vitro, its most potent oncogenic activity appears to be directed towards cells in the B-cell lineage. As discussed elsewhere in this book, REL gene amplifications are quite common in human B-cell malignancies, especially in diffuse B-cell lymphomas and Hodgkin's lymphomas (Gilmore, 2003). Although not formally proven, it is hard to imagine that the mechanism by which v-Rel and overexpressed c-Rel proteins cause the malignant transformation of avian lymphoid cells, is not similar to the mechanism by which REL gene amplification contributes to human B-cell malignancies.

3. TRANSMEMBRANE ONCOPROTEIN LMP-1 OF EPSTEIN-BARR VIRUS ACTIVATES NF-κB BY MIMICKING A CONSTITUTIVELY ACTIVE TNF RECEPTOR

Epstein-Barr virus (EBV) is an oncogenic human herpesvirus with a tropism for epithelial cells and B-lymphocytes (Kieff & Rickinson, 2001). Latent EBV infection is closely associated with the development of several human malignancies, including immunoblastic lymphomas in immunocompromised patients, Burkitt's lymphoma, Hodgkin's disease, nasopharyngeal carcinoma and gastric carcinomas (Rickinson & Kieff, 2001). The most commonly used in vitro model for EBV's oncogenicity assesses the growth transformation of primary human B-lymphocytes into perpetually proliferating lymphoblastoid cell lines upon infection with EBV. Activation of NF-κB plays an essential role in the growth transformation of human B-cells generated by EBV infection in vitro (Cahir-McFarland et al., 2000; Feuillard et al., 2000; He et al., 2000). The principal activator of NF-κB by EBV is the latent membrane protein 1 (LMP1), which is also a major viral transforming factor.

3.1. Structural and functional features of LMP1

LMP-1 mimics constitutively activated members of the TNF receptor family, located primarily at the plasma membrane. LMP1 consists of three major domains (Fig. 1): a short 24 amino acid cytoplasmic N-terminus; six transmembrane domains connected by short turns; a 200 amino acid cytoplasmic C-terminus (referred to here as CCT). Deletion analyses have demonstrated that the six transmembrane domains and the CCT of LMP1 are essential for B-cell transformation, whereas the cytoplasmic N-terminus is largely dispensable for transformation. LMP1 has pleiotropic phenotypic effects on various cell lines of lymphoid and epithelial origin. It upregulates most of the cell surface activation markers that are induced by EBV in growth-transformed B-lymphocytes. LMP1 also upregulates the expression of genes encoding antiapoptotic molecules such as Bcl2, Bfl1, Mcl1 and A20, and prevents apoptosis of lymphoid and epithelial cells in response to serum withdrawal (Henderson et al., 1991; Laherty et al., 1992; Wang et al., 1996; D'Souza et al., 2000). When expressed in epithelial cells, LMP1 prevents differentiation and induces the expression of the epidermal growth factor (EGF) receptor (Dawson et al., 1990; Miller et al., 1995). Activation of the transcription factor NF-κB mediates many of the phenotypic effects of LMP1, including the upregulation of ICAM-1, LFA-3, A20, Bfl1, Bcl-2, CD40, IL-6, Fas, TRAF1, EBI3 and cyclooxygenase-2 (Eliopoulos et al., 1997; Devergne et al., 1998; Cahir-McFarland et al., 2000; Feuillard et al., 2000; Murono et al., 2001).

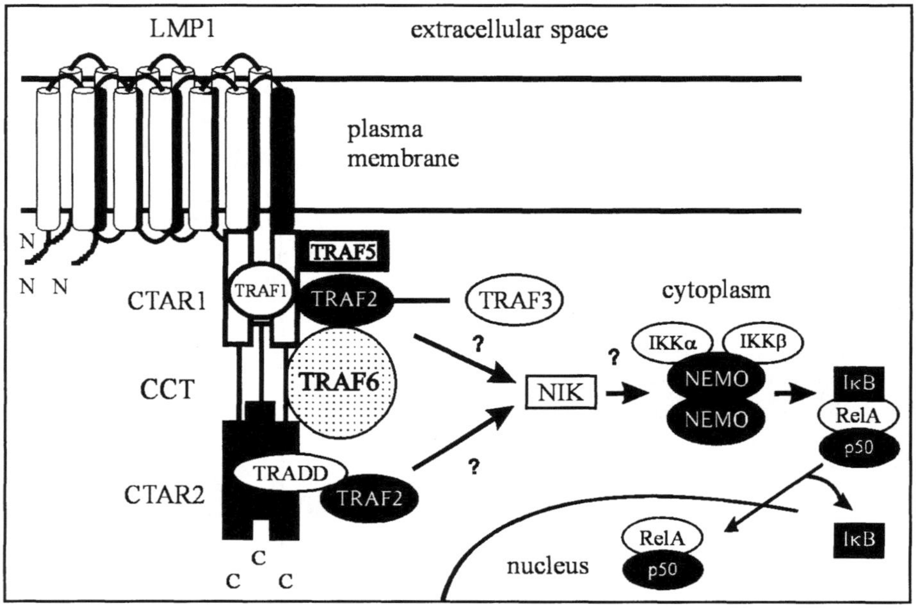

Fig. 1. Schematic representation of the mechanism of NF-κB activation by the Epstein-Barr virus oncoprotein LMP1. LMP1 is shown as a trimer at the plasma membrane to indicate the oligomeric structure of the protein; however, the exact stoichiometry of LMP1 oligomers at the plasma membrane is unknown. LMP1 consists of a short N-terminal region (N), six transmembrane domains (shown as cylinders) connected by short turns and a 200-amino acid cytoplasmic C-terminus (CCT). Two NF-κB-activating domains (CTAR1 and CTAR2) are present in the CCT. CTAR1 interacts with TRAF1, TRAF2, TRAF3 and TRAF5, whereas CTAR2 interacts with TRADD and possibly with TRAF2. TRAF6 is also implicated in NF-κB activation and interacts with an unknown domain of the CCT. After recruitment of TRAF molecules, downstream kinases of the MAP3K family (NIK or a related kinase) are activated and mediate the activation of the IκB kinase (IKK) complex. Activated IKK phosphorylates IκB to cause its degradation, which allows active dimers of NF-κB (p50-RelA) to enter the nucleus and activate transcription.

As with B-cell transformation, activation of NF-κB by LMP1 requires sequences from both the transmembrane domain and the CCT. The six transmembrane domains mediate the constitutive aggregation of LMP1 that is required for activation of NF-κB. In addition, two regions within the CCT contribute to activation of NF-κB. One region, called CTAR1, spans amino acids 185-231 and is located in the portion of CCT that is near the plasma membrane (Huen et al., 1995; Mitchell & Sugden, 1995). The second region (CTAR2) maps within the last 35 amino acids of CCT. Approximately 20-30% of the LMP1-induced NF-κB activation is mediated by CTAR1, whereas 70-80% is transduced by CTAR2. Two lines of evidence indicate that activation of NF-κB is required for transformation by EBV LMP1. First, CTAR1 and CTAR2 overlap with two domains (transformation effector sites [TES] 1 and 2, respectively) in the CCT that play a critical role in B-lymphocyte transformation (Izumi et al., 1997; Izumi & Kieff, 1997). Second, overexpression of the IκBα superrepressor of NF-κB causes apoptotic death in EBV-transformed B-cell cultures and inhibits transformation of rodent fibroblasts by LMP1 (Cahir-McFarland et al., 2000; Feuillard et al., 2000; He et al., 2000).

3.2. Mechanism of NF-κB activation by LMP1

NF-κB activation by LMP1 is mediated by the recruitment of cellular adaptor proteins, TRAFs and TRADD, to the LMP1 CCT (Mosialos et al., 1995; Devergne et al., 1996; Kaye et al., 1996; Izumi & Kieff, 1997). TRAFs constitute a family of proteins that are characterized by a highly homologous C-terminal region of approximately 200 amino acids known as the TRAF domain (Arch et al., 1998). The N-terminal halves of TRAFs are rich in cysteine and histidine residues that can form several zinc finger-like motifs and in some TRAFs a RING-finger motif. Six mammalian TRAF molecules have been identified. Several members of this family mediate signal transduction from the cytoplasmic tail of certain TNF receptors including TNF-R1, TNF-R2, lymphotoxin-β receptor (LTβ-R), CD40, CD30 and CD27. TRAF1, TRAF2, TRAF3 and TRAF5 associate with the LMP1 CTAR1 (Mosialos et al., 1995; Devergne et al., 1996; Devergne et al., 1998; Schultheiss et al., 2001). Site-directed mutagenesis of LMP1 and overexpression of wild-type and dominant-negative mutants of TRAFs indicate that TRAF1, TRAF2, and possibly TRAF5, mediate NF-κB activation by CTAR1, whereas TRAF3 is likely to have an inhibitory role in this process (Schultheiss et al., 2001; Hatzivassiliou & Mosialos, 2002). TRAF6 binding to LMP1 depends on intact CTAR1 and CTAR2. The principal mediator of NF-κB signaling from CTAR2 is TRADD. TRADD also interacts with the cytoplasmic domain of TNF-R1 and appears to play the role of an anchoring core element for the assembly of multi-subunit complexes that mediate proapoptotic and antiapoptotic effects of TNF. TRAF2 has been also implicated in NF-κB activation from CTAR2, even though it has not been possible to detect an interaction between CTAR2 and TRAF2. It is likely that TRADD mediates a weak or transient association of CTAR2 with TRAF2 since TRADD can bind TRAF2.

The signaling events that follow recruitment of TRAFs and TRADD to the LMP1 CCT are not precisely known. However, NF-κB activation by LMP1 or TNF-Rs depends on activation of IKK with the subsequent phosphorylation and degradation of IκB. The protein kinase NIK is a MAP kinase kinase kinase

(MAP3K) that was identified as a TRAF2-associated kinase mediating NF-κB activation by several TNF-Rs. As such, NIK or a related MAP3K may activate IKK as part of both TNF-R and LMP-1 signaling. Indeed, dominant-negative forms of NIK, IKKα and IKKβ all reduce activation of NF-κB by LMP1 (Sylla et al., 1998). Tpl2 is another MAP3K that is activated by LMP1 and has been implicated in NF-κB activation by LMP1 downstream of TRAF2 (Eliopoulos et al., 2002).

TRAFs are generally thought to activate downstream kinases by forcing their clustering at the plasma membrane. However, it has recently been demonstrated that TRAF2 and TRAF6 can also act as ubiquitin ligases and that their auto-ubiquitination can activate downstream MAP3Ks (Deng et al., 2000; Wang et al., 2001). Activation of the ubiquitin ligase activity of TRAF2 and TRAF6 is mediated by their oligomerization and depends on the integrity of their N-terminal RING finger motif, which is also essential for NF-κB activation. The mechanism of activation of MAP3K by polyubiquitinated TRAF2 or TRAF6 is unclear. The ultimate evaluation of the role of each signaling molecule implicated in NF-κB activation by LMP1 awaits the analysis of LMP1 signaling in cells from mice lacking the corresponding genes.

3.3. Activation of NF-κB by the EBV latent antigen EBNA2

NF-κB is also activated by the viral protein EBNA2, another latent antigen that is essential for B-lymphocyte transformation by EBV (Hammarskjold & Simurda, 1992). EBNA2 is a powerful transcription activator which targets viral and cellular genes via its interaction with the cellular transcription factor RBPJκ/CBF1 (Grossman et al., 1994; Henkel et al., 1994; Zimber-Strobl et al., 1994). EBNA2 is a weaker activator of NF-κB than LMP1 (Chen & Cooper, 1996). The mechanism of NF-κB activation by EBNA2 is likely to involve a direct effect at the level of DNA, possibly through RBPJκ/CBF1. Consistent with this possibility is the fact that the DNA-binding sites of NF-κB and RBPJκ/CBF1 frequently overlap (Shirakata et al., 1996; Lee et al., 2000).

4. THE Tax ONCOPROTEIN OF HUMAN T-CELL LEUKEMIA VIRUS TYPE 1 ACTIVATES MULTIPLE COMPONENTS OF THE Rel/NF-κB SIGNALING PATHWAY

4.1. Overview of Rel/NF-κB activation induced by human T-cell leukemia virus type 1 (HTLV-1)

HTLV-1 is a human retrovirus involved in the development of three distinct human pathologies: adult T-cell leukemia (ATL) of CD4$^+$ T-cells, as well as the neurological disorders HTLV-1-associated myelopathy (HAM) and tropical spastic paraparesis (TSP) (Yoshida, 2001). A highly related virus (HTLV-2) also causes T-cell leukemias, but it is less prevalent than HTLV-1 and has been less studied; nevertheless, in most cases, the conclusions drawn in this review for HTLV-1 replication and pathogenesis also apply to HTLV-2.

In addition to the usual retroviral replicative genes (gag, pol and env), HTLV-1 has a 3' gene (tax), which is essential for both viral replication and pathogenesis. The Tax protein is an approximately 40 kDa polypeptide that is found primarily in the nucleus, but also in the cytoplasm of HTLV-1-infected cells (Sun & Ballard, 1999). Tax lacks catalytic activity and appears to function as an adaptor protein. As such, Tax interacts with several cellular proteins and can activate various cellular signaling pathways, including the CREB/ATF and NF-κB pathways. Much early work focused on the role of Tax in viral replication, namely on the ability of Tax to transactivate the HTLV-1 promoter through a 21-bp element in the viral long-terminal repeat (LTR) (Yoshida, 2001). In that Tax cannot bind to the Tax-responsive element in the LTR, it was clear that Tax was indirectly affecting transcription. It is now known that Tax increases LTR-dependent transcription primarily by interaction of a Tax-CREB complex with the viral 21-bp Tax-responsive element (Yoshida, 2001).

Several lines of evidence indicate that Tax is also required for the oncogenic effects of HTLV-1. First, although leukemic cells from ATL patients frequently have deleted HTLV-1 proviral genomes, the tax region is selectively retained in these cells (Korber et al., 1991). Second, Tax can immortalize human T-cells in vitro (Grassmann et al., 1989; Grassmann et al., 1992). Third, Tax can transform rodent cells in vitro (Pozatti et al., 1990; Tanaka et al., 1990) and can cause tumors when expressed in transgenic mice (Nerenberg et al., 1987; Grossman et al., 1995).

Whereas the effects of Tax on viral replication are mediated through effects on CREB, the oncogenic activity of Tax is thought to be primarily a result of its effects on the NF-κB pathway. Mutants of Tax that can no longer activate NF-κB (but can still activate CREB) also do not immortalize human T-cells (Robek & Ratner, 1999). Moreover, a phenotypically normal revertant cell clone of a Tax-transformed rat cell line was found to have an inactivating mutation in IKKγ, such that NF-κB could no longer be activated in the revertant (Yamaoka et al., 1998). Agents that block the activity of NF-κB in Tax-induced mouse tumor cells, also block the proliferation of these cells and sensitize them to inducers of apoptosis (Portis et al., 2001). Taken together, these results suggest that the constitutive activation of the Rel/NF-κB pathway by Tax alters the expression of specific target genes that contributes to both the proliferation and survival of leukemic T-cells in ATL patients.

Tax induces the appearance of nuclear DNA-binding complexes that contain p50, p52, RelA and c-Rel (Arima et al., 1991). However, these complexes appear in two phases: an early phase wherein p50-RelA complexes are induced and a late phase wherein p52 and c-Rel complexes appear (Kanno et al., 1994). In part, this is likely attributable to the presence of κB sites in the promoters for nfkb2 and c-rel, such that expression of these genes is induced by the early phase activation of latent p50-RelA complexes. Tax affects Rel/NF-κB signaling in both the nucleus and the cytoplasm, and Tax has been demonstrated to interact directly with many components of the Rel/NF-κB pathway (Table 2).

Table 2. Rel/NF-κB-signaling proteins interacting with the Tax oncoprotein of HTLV-1.

Protein	*References*
p50	Suzuki et al., 1993
RelA	Suzuki et al., 1994
c-Rel	Suzuki et al., 1994
IκBα	Hirai et al., 1994; Petropoulos & Hiscott, 1998
IKKγ	Chu et al., 1999; Harhaj & Sun, 1999; Xiao et al., 2000
p100	Béraud et al., 1994; Lanoix et al., 1994; Murakami et al., 1995; Xiao et al., 2001
p105	Hirai et al., 1992

4.2. Cytoplasmic effects of Tax on Rel/NF-κB signaling

Through interactions in the cytoplasm, Tax promotes processing of IκBα, IκBβ and the NF-κB precursor p100, and possibly p105 (Sun & Ballard, 1999).

4.2.1. Tax-induced degradation of IκB proteins

Two genetic experiments indicated that the IKKγ noncatalytic subunit of IKK was one target for Tax-induced activation of NF-κB. First, as mentioned above, in an elegant somatic cell-genetic experiment, a flat revertant cell clone of a Tax-transformed cell line was found to be defective in both the ability to be transformed by Tax and for activation of NF-κB, due to the loss of IKKγ expression (Yamaoka et al., 1998). Second, antisense inhibition of IKKγ expression blocks the ability of Tax to activate NF-κB in human Jurkat T-cells (Chu et al., 1999). Further studies have shown that Tax can interact directly with IKKγ and this interaction requires leucine-rich sequences in both proteins (Chu et al., 1999; Harhaj & Sun, 1999; Xiao et al., 2000). Moreover, mutations in Tax that abolish its interaction with IKKγ also cripple its ability to activate NF-κB (Chu et al., 1999; Harhaj & Sun, 1999; Xiao et al., 2000). The Tax-IKKγ interaction leads to activation of the β-catalytic subunit of IKK (Chu et al., 1999; Harhaj & Sun, 1999), presumably because the Tax-IKKγ complex promotes clustering of the IKK components, and this induced proximity leads to persistent phosphorylation of the activation loop in IKKβ (Carter et al., 2001). Indeed, fusion of IKKγ, α or β to Tax is sufficient to activate IKK (Xiao & Sun, 2000), and Tax itself is known to have an efficient dimerization ability (Tie et al., 1996; Jin & Jeang, 1997). Tax-induced activation of IKK then targets IκBα and IκBβ for degradation, causing the chronic induction of NF-κB that has been observed in Tax-transformed cells.

Tax has also been reported to bind directly to IκBα (Hirai et al., 1994; Suzuki et al., 1995; Petropoulos & Hiscott, 1998), leading to the suggestion that Tax can directly destabilize NF-κB/IκB complexes to liberate NF-κB. However, the evidence for this pathway of Tax-mediated induction of Rel/NF-κB is less compelling than for the Tax → IKKγ pathway discussed above. For example, some mutants of Tax that cannot activate NF-κB can still bind to IκBα (McKinsey et al., 1996; Petropoulos et al., 1996), and proteasome-mediated degradation of IκBα is required for Tax-induced activation of NF-κB (Maggirwar et al., 1995). Nevertheless, given that the ikba gene is strongly induced by NF-κB complexes, that newly-synthesized IκBα usually attenuates the NF-κB response, and that most Tax protein is in the nucleus of HTLV-1-infected cells, it is possible that a nuclear Tax-IκBα interaction blocks the post-induction removal of Rel/NF-κB complexes from DNA, and in this manner Tax-IκBα contributes to constitutive activation of Rel/NF-κB complexes.

4.2.2. Tax-induced processing of p100 and possibly p105

Tax can also interact with the NF-κB p52 and p50 precursor molecules p100 and p105 (Hirai et al., 1992; Béraud et al., 1994; Lanoix et al., 1994; Murakami et al., 1995; Xiao et al., 2001). In the case of p100, Tax binds to sequences within the Rel homology domain of p100 to bridge an interaction to the IKK complex through IKKγ (Xiao et al., 2001). This interaction then activates the catalytic activity of IKKα, and leads to the phosphorylation, ubiquitination and proteasome-mediated processing of p100 to p52. This activity of Tax no doubt accounts for the increased levels of p52 that are observed in HTLV-1-infected human T-cells (Xiao et al., 2001). Of note, Tax-induced processing of p100 to p52 appears to be distinct from normal processing of p100, in that Tax does not need NIK, which is required for the processing of p100 in normal cells (Xiao et al., 2001).

The evidence that Tax can induce processing of p105 to p50 is less clear, and it is still unknown what role such processing might play in malignant transformation. Nevertheless, Tax has been shown to form ternary complexes with p105 and the HC9 subunit of the proteasome when these proteins are overexpressed in cells (Rousset et al., 1996). This suggests that Tax may direct p105 to the proteasome.

4.3. Nuclear effects of Tax on Rel/NF-κB activity

In the nucleus, Tax appears to stimulate the activity of Rel/NF-κB complexes on DNA. First, Tax co-localizes in discrete nuclear structures with p50 and RelA (Bex et al., 1997). Second, Tax can directly interact with p50, c-Rel and RelA on DNA (Suzuki et al., 1993; Suzuki et al., 1994). Third, Tax has transactivating activity when fused to a heterologous DNA-binding domain such as GAL4 (Fujii et al., 1991; Fujisawa et al., 1991), and Tax can interact with TATA box-binding protein (TBP) and the TBP-associated factor TAF28 (Caron et al., 1993; Caron et al., 1997). Thus, Tax may serve as a "coactivator" for NF-κB complexes by bridging an interaction between them and components of the general transcriptional machinery.

4.4. Target genes affected by Tax-mediated activation of Rel/NF-κB

ATL develops as a clonal expansion of leukemic $CD4^+$ T-cells in HTLV-1-infected individuals only after a long latency, indicating that cellular alterations besides Tax expression are required for full transformation of the T-lymphocytes. Moreover, ATL cells from patients or T-lymphocytes infected with HTLV-1 in vitro show chromosomal abnormalities (Jeang, 2001). However, the nature or identity of the types of secondary changes that contribute to HTLV-1-induced (or Tax-induced) T-cell malignancy are not really known. The progressive nature of the disease in vivo is perhaps mimicked in cell culture, wherein in vitro-infected primary human T-cells undergo an initial phase of proliferation that is dependent on IL-2, and a subsequent phase wherein their growth becomes independent of IL-2 and during which the cells acquire chromosomal alterations (Ressler et al., 1996). In the case of HTLV-2, the analysis of Tax mutants selectively defective in NF-κB/Rel activation vs. CREB/ATF activation suggests that the initial phase is dependent on NF-κB/Rel activation, while the second phase is dependent on CREB/ATF activation (Ross et al., 2000).

Interestingly, ATL cells express high levels of both IL-2 and the IL-2Rα chain (Sun & Ballard, 1999), both of which are induced by Tax and contain upstream κB sites (Leung & Nabel, 1988; Wano et al., 1988; Hoyos et al., 1989). Thus, one aspect of Tax-induced proliferation of T-cells appears to be the NF-κB-mediated upregulation of IL-2 and a component of its receptor. Furthermore, Tax activates the expression of a number of other NF-κB-dependent cytokine genes, including several interleukins and TNF (Sun & Ballard, 1999).

In addition to the autocrine stimulation of cell growth described above, Tax (via NF-κB) may also regulate genes directly contributing to the survival and proliferation of T-cells. So the levels of several NF-κB target genes/proteins, including Bcl-2, Bcl-X_L, A1 and IAP, that promote the survival of lymphoid cells, are increased in HTLV-1-transformed T-cells and in cells from ATL patients (Harhaj et al., 1999; Nicot et al., 2000). Moreover, some cell cycle genes and cellular oncogenes, such as cyclin D1, cyclin D2, c-myc and c-rel, show increased expression due to the induction of NF-κB by Tax (Duyao et al., 1992; Li et al., 1993; Harhaj et al., 1999; Huang et al., 2001b; Mori et al., 2002).

In summary, a number of genes affecting lymphoid cell growth and survival show the increased expression that one would predict from constitutive activation of Rel/NF-κB by Tax. Thus one would also predict that inhibitors of Rel/NF-κB would block the growth of ATL cells or sensitize them to chemotherapeutics that induce apoptosis, much as has been found in several other cancer cell types with chronic Rel/NF-κB activity (Gilmore, 2003). However, such experiments remain to be performed with Tax-transformed human cells. In some cases, Tax-mediated activation of NF-κB may be involved in early phases of HTLV-induced T-cell leukemia, which are not required later for the maintenance of the transformed state; as such, different ATL patients may show different responses to Rel/NF-κB inhibition.

5. HUMAN HEPATITIS VIRUSES MODULATE Rel/NF-κB SIGNALING

5.1. The hepatitis B virus (HBV) X protein (HBx) activates NF-κB

Hepatitis B virus is a double-stranded DNA virus, which causes chronic and acute hepatitis in humans and predisposes infected individuals to hepatocellular carcinoma. The 17-kDa X protein of HBV (HBx), which has been associated with the tumorigenic potential of HBV, is a potent activator of several transcription factors, including NF-κB, AP1, C/EBP and TBP (Diao et al., 2001; Hiscott et al., 2001).

Activation of NF-κB by HBx is mediated by degradation of IκBα and processing of p105. Thus, one obvious target for HBx would be the IKK complex, but it is unclear how HBx might activate IKK; however, HBx can activate Ras and protein kinase-C pathways, which have been associated under some conditions with activation of IKK.

In addition, HBx can interact with the NF-κB inhibitor IκBα, and HBx can be transported into the nucleus as a complex with IκBα (Weil et al., 1999). Thus, HBx may also cause sustained activation of NF-κB by preventing nuclear IκBα from binding and exporting NF-κB subunits. Activation of NF-κB has been associated with protecting liver cells from apoptosis: for example, RelA knockout mice die embryonically of extensive liver apoptosis (Beg & Baltimore, 1996). Therefore, HBx may contribute to tumor development by prolonging the survival of hepatocytes.

5.2. The hepatitis C virus (HCV) may have multiple effects on Rel/NF-κB signaling

HCV is a single-stranded RNA virus, which has emerged as another major causative agent of acute and chronic hepatitis. HCV expresses at least two proteins that have opposite effects on the NF-κB pathway. The HCV nonstructural protein 5A (NS5A) inhibits NF-κB activation in response to TNF by interacting with TRADD and presumably interfering with the interaction of TRADD with TRAF2 (Majumder et al., 2002; Park et al., 2002). Despite its inhibitory effect on NF-κB, NS5A blocks TNF-mediated apoptosis by disrupting the interaction between TRADD and FADD.

On the other hand, the HCV core protein potentiates NF-κB activation in response to TNF in a TRAF2-dependent manner (Chung et al., 2001; Yoshida et al., 2001). Consequently, the HCV core protein results in an increase in the resistance of cells to TNF-induced apoptosis (Tai et al., 2000). The HCV core protein can associate with the TNF-R1, TRADD and TRAF2, and may influence TNF signaling by stabilizing this NF-κB-promoting complex. The ability of NS5A as well as of the HCV core protein to block TNF-mediated apoptosis suggests that they may, like HBx, contribute to hepatogenesis by blocking liver cell death.

6. AFRICAN SWINE FEVER VIRUS ENCODES PROTEINS WITH OPPOSING EFFECTS ON Rel/NF-κB SIGNALING

African swine fever virus (ASFV) is a large double-stranded DNA virus that has over 150 genes, several of which encode proteins similar to cellular proteins (Yanez et al., 1995). ASFV induces a rapidly fatal hemorrhagic disease in pigs, and infects primarily macrophages and monocytes. Notably, ASFV infection is characterized by the absence of an acute inflammatory response and immune response. For example, many of the NF-κB-dependent cytokine genes are not induced upon infection of macrophages by ASFV (Powell et al., 1996).

Relevant to this review, two viral proteins, A224L and A238L, appear to have direct effects on the Rel/NF-κB pathway. IAPs are cellular proteins that block apoptosis by inhibiting caspase activity, and several IAPs can also activate NF-κB. It has recently been shown that the ASFV IAP-like protein A224L can activate IKK, and consequently NF-κB when stably expressed in Jurkat T-cells (Rodríguez et al., 2002). By contrast, the A238L protein has extensive sequence similarity to IκB proteins, and can bind to and inhibit NF-κB (Powell et al., 1996; Revilla et al., 1998). Interestingly, the A238L protein is missing the serine residues of IκBα that are required for signal-induced phosphorylation by IKK and for degradation. As such, A238L is resistant to signal-induced degradation (Tait et al., 2000). Thus, while ASFV infection initially induces NF-κB, the now degraded cellular IκBα is replaced by the viral IκB A238L, which now stably inhibits the NF-κB complex (Tait et al., 2000). Of note, A238L also inhibits calcineurin, a protein phosphatase that is required for the induction of the NFAT-dependent immune response genes (Miskin et al., 1998).

In short, it appears that ASFV has acquired and adapted cellular genes to modulate the host's immune, inflammatory and apoptotic responses. The viral IκB homologue A238L, which is expressed early in ASFV infection, likely directly inhibits NF-κB- and NFAT-dependent immune and inflammatory responses. In contrast, the IAP homologue A224L protein, which is expressed later in infection, may prevent apoptosis (and thus prolong the viral infection state) both by blocking caspase activity and by inducing NF-κB activity.

7. OTHER NOTABLE EFFECTS OF VIRUSES ON Rel/NF-κB SIGNALING

In addition to the better known viruses described in previous sections, several other viruses have been reported to affect Rel/NF-κB signaling, and there are no doubt more to be discovered. These lesser studied viruses are briefly summarized (Table 1).

7.1. Viral oncoproteins that affect Rel/NF-κB signaling

Besides v-Rel, LMP-1, Tax and the hepatitis virus proteins, several other viruses associated with oncogenesis may constitutively activate Rel/NF-κB. In particular, emerging evidence suggests that two herpesviruses, namely human herpesvirus 8 (HHV8) and Herpesvirus saimiri associated with lymphoid cell transformation, encode multiple proteins that cooperate to activate NF-κB.

HHV8 is associated with the development of both Kaposi's sarcoma in immune-compromised individuals and primary effusion lymphomas. Consistent with persistent activation of NF-κB being involved in HHV8-associated disease, inhibition of NF-κB induces apoptosis in primary effusion lymphoma cells (Keller et al., 2000). HHV8 ORF74 is a homologue of a G-protein-coupled chemokine receptor and can transform the morphology of human skin endothelial cells in culture (Pati et al., 2001). Expression of ORF74 can activate NF-κB and hence induce the expression of several NF-κB target genes, including ones encoding cytokines, angiogenesis factors and cell-surface adhesion molecules (Pati et al., 2001; Schwarz & Murphy, 2001). A second HHV8 activator of NF-κB is vFLIP, which is a viral homologue of FLICE inhibitory protein that prevents apoptotic cell death by inhibiting the activity of the protease FLICE (Liu et al., 2002). However, vFLIP also associates with and activates the IKK complex to cause persistent activation of NF-κB (Liu et al., 2002).

The transforming proteins Tip and StpC of Herpesvirus saimiri can induce fatal T-cell lymphoproliferation in primates and can transform human T-cell in vitro. These transforming proteins of Herpesvirus saimiri also cooperatively induce the activity of NF-κB (Merlo & Tsygankov, 2001); however, it is not yet clear whether NF-κB is involved in their oncogenic effects.

7.2. Viruses that activate Rel/NF-κB

A number of viruses have viral proteins that activate NF-κB as part of their life cycle (Table 3). Several other viruses, including herpes simplex virus-1, sendai virus, measles virus, respiratory syncytial virus, encephalomyocarditis virus and rhinovirus, also induce activation of NF-κB, but the mechanisms by which they do so are not known (Pahl, 1999; Hiscott et al., 2001). In some cases, viruses may "purposely" induce NF-κB because of its antiapoptotic activity to prolong cell survival, and thus maximize viral replication. In other cases, activation of NF-κB may be part of the host's response to viral infection, because of the key role that NF-κB plays in regulating genes of the innate immune response, which is mediated in part by NF-κB-activated cytokine genes. Two of the more general mechanisms for activation of NF-κB are mentioned here.

Double-stranded RNA, an intermediate in many viral infections, induces NF-κB through a mechanism that involves the double-stranded RNA-activated protein kinase PKR (Kaufman, 1999). Thus, all viruses that use a double-stranded RNA intermediate in their replication are likely to induce NF-κB, except for some that have evolved mechanisms to block activation of NF-κB. In addition, some viral products are likely to induce NF-κB by a mechanism called the ER overload response (Pahl & Baeuerle, 1997); that is, the nonspecific accumulation of proteins in the ER membrane, as may occur during many viral infections, induces a release of calcium ions, which activates NF-κB. For example, the E3/19K protein of adenovirus induces NF-κB by the ER overload response (Pahl et al., 1996).

7.3. Viruses that inhibit Rel/NF-κB

A smaller number of viruses has been shown to inhibit NF-κB. In many cases, they may do so to block the innate and adaptive host immune responses that are mediated by NF-κB. In one interesting case, human adenovirus contains an early-expressed region, termed E3, that encodes an activity blocking activation of NF-κB by agents such as TNFα or IL-1. This is thought to prolong adenoviral infection in that host-mediated clearance of adenovirus is mediated by TNFα. As the E3 region is not required for viral replication, it has been deleted from many adenovirus vectors designed for gene therapy, and thus these vectors are often rapidly cleared by infected hosts due to an inability to block the activity of TNFα. The adenovirus NF-κB inhibition activity was mapped to two proteins in the E3-10.4K/14.5K region (Friedman & Horwitz, 2002). It is not known how E3-10.4K/14.5K block activation of TNFα-mediated induction of NF-κB, but they appear to act downstream of TNF-R engagement and upstream of degradation of IκBα. Interestingly, another E3 protein (E3-14.7K) has been shown to interact with IKKγ (Li et al., 1999), but it is not clear whether E3-14.7K affects IKK activity (Friedman & Horwitz, 2002).

8. Rel/NF-κB REGULATES THE TRANSCRIPTION OF SEVERAL VIRAL PROMOTERS/ENHANCERS

8.1. Introduction to the regulation of viral transcription by NF-κB

Many viruses, including several human pathogens, have κB sites in their promoters/enhancers (Table 3). In most cases, inhibition of NF-κB has been shown to reduce expression from the viral promoter, but generally NF-κB is only one of the cellular factors required for full expression from the promoter. Viruses are no doubt molecular parasites of the Rel/NF-κB-signaling pathway, either because of the cell type that the virus targets (i.e. cells in which NF-κB is constitutively active) or because the inflammatory response that the virus engenders, promotes activation of NF-κB, which in turn activates virus transcription.

8.2. NF-κB regulation of transcription from the HIV-1 long-terminal repeat

The κB sites in the HIV-1 long-terminal repeat (LTR) promoter have undoubtedly received the most attention among the viral promoters; the role of these sites in the viral life cycle, however, continues to be the subject of some controversy (Hiscott et al., 2001). Namely, the κB sites appear to be required for HIV transcription and replication in some, but not all, cell types. For example, expression of the IκBα superrepressor reduces viral replication in T-cell cultures in vitro (Kwon et al., 1998; Quinto et al., 1999), but mutation of the κB sites does not absolutely block virus growth in vitro (Chen et al., 1997).

Table 3. Viral promoters with Rel/NF-κB-binding sites.

Virus	*Viral promoter*	*References*
Avian leukosis virus	LTR	Bowers et al., 1996
Bovine leukemia virus	LTR	Brooks et al., 1998
Simian virus 40	Control region	Kanno et al., 1989
Simian immunodeficiency virus	LTR	Bellas et al., 1993
Adenovirus	Early E3 promoter	Williams et al., 1990
Cytomegalovirus	Immediate early promoter	Prösch et al., 1995
Epstein-Barr virus	Wp	Sugano et al., 1997
Hepatitis B virus	Enhancer II/pregenomic	Kwon & Rho, 2002
Human immunodeficiency virus 1	LTR	Nabel & Baltimore, 1987
Human papillomavirus type 16	Long control region	Fontaine et al., 2000
Herpes simplex virus	ICP90, Vmw65	Rong et al., 1992
JC virus	Viral control region	Ranganathan & Khalili, 1993

In most commonly studied viral isolates, the HIV-1 promoter has two NF-κB-binding sites located approximately 100 bp from the start site of transcription (Hiscott et al., 2001). These sites act in synergy with nearby Sp1-binding sites. Interestingly, different HIV-1 isolates have different numbers of NF-κB-binding sites, and there is a general correlation between the number of κB sites and transcription from the viral promoter: that is, HIV-1 clade E has only one κB site and low transcriptional activity, clade B has two κB sites and moderate activity, and clade C has three κB sites and the most activity. Nevertheless, the number of κB sites does not necessarily correlate with the infectivity or replication efficiency of these viral isolates.

Activation of NF-κB can also increase transcription from the HIV-1 LTR. That is, unstimulated $CD4^+$ T-cells (one of the primary in vivo target cells for HIV-1 pathogenicity) have primarily p50-p50 DNA-binding activity, and activation of these cells leads to induction of p50-RelA complexes and enhanced expression from the HIV-1 LTR. Such experiments suggest that activation of T-cells, such as might occur due to infections by agents other than HIV-1, would lead to increased HIV-1 virus production. In addition, several HIV-1 gene products (Tat, Vpr and Nef) have been shown to activate NF-κB (Table 1; Hiscott et al., 2001), with the concomitant induction of NF-κB target genes encoding cytokines and growth factors, which themselves can further induce NF-κB. Thus, HIV-1 may both use NF-κB for its replication, and have evolved complex and synergistic mechanisms for inducing NF-κB.

9. CONCLUSIONS

As discussed in this review, viruses have devised many mechanisms to subvert the Rel/NF-κB signal transduction pathway for their replication or pathogenicity. That is, viruses may encode proteins that directly affect steps in this pathway (Tables 1-2) or may respond to the Rel/NF-κB transcription factors (Table 3). In most cases, viruses and their proteins act to activate or respond positively to Rel/NF-κB. It is likely that many other viruses will be found to affect Rel/NF-κB signaling. In particular, viruses that are associated with lymphoid cell malignancies, induce or suppress the host immune response, block or induce apoptosis, or use a double-stranded RNA intermediate in their replication, are likely to modulate NF-κB. As such, the interplay between viruses and NF-κB signaling is certain to maintain a prominent place in molecular virology, and inhibitors of Rel/NF-κB signaling, which are being developed to treat inflammation or cancer, may also have effects on certain viral pathogenicities in animals and humans. However, one caution is that NF-κB inhibitors, in addition to having effects on the viral life cycle or pathogenicity, may also have deleterious effects on the host's immune response to the virus.

10. NOTES

Research in the laboratory of T.D. Gilmore on Rel/NF-κB is supported by a grant from the National Institutes of Health (CA47763). Research in the laboratory of G. Mosialos is supported by an International Research Scholarship from the Howard Hughes Medical Institute, a Human Frontier Science Program grant and an EMBO Young Investigator Award. The reader is also referred to www.nf-kb.org for further information on Rel/NF-κB proteins and signaling.

11. REFERENCES

Arch RH, Gedrich RW & Thompson CB (1998) Tumor necrosis factor receptor-associated factors (TRAFs). A family of adapter proteins that regulates life and death. Genes Dev. 12: 2821-2830.

Arima N, Molitor JA, Smith MR, Kim JH, Daitoku Y & Greene WC (1991) Human T-cell leukemia virus type I induces expression of the Rel-related family of κB enhancer-binding proteins: evidence for a pretranslation component of regulation. J. Virol. 65: 6892-6899.

Beg AA & Baltimore D (1996) An essential role for NF-κB in preventing TNF-α-induced cell death. Science 274: 782-784.

Bellas RE, Hopkins N & Li Y (1993) The NF-κB binding site is necessary for efficient replication of simian immunodeficiency virus of macaques in primary macrophages but not in T cells in vivo. J. Virol. 67: 2908-2913.

Béraud C, Sun S-C, Ganchi P, Ballard DW & Greene WC (1994) Human T-cell leukemia virus type I Tax associates with and is negatively regulated by the NF-κB2 p100 gene product: implications for viral latency. Mol. Cell. Biol. 14: 1374-1382.

Bex F, McDowall A, Burny A & Gaynor R (1997) The human T-cell leukemia virus type 1 transactivator protein Tax colocalizes in unique nuclear structures with NF-κB proteins. J. Virol. 71: 3484-3497.

Bowers WJ, Baglia LA & Ruddel A (1996) Regulation of avian leukosis virus long terminal repeat-

enhanced transcription by C/EBP-Rel interactions. J. Virol. 70: 3051-3059.
Brooks PA, Cockerell GL & Nyborg JK (1998) Activation of BLV transcription by NF-κB and Tax. Virology 243: 94-98.
Cahir-McFarland ED, Davidson DM, Schauer SL, Duong J & Kieff E (2000) NF-κB inhibition causes spontaneous apoptosis in Epstein-Barr virus-transformed lymphoblastoid cells. Proc. Natl. Acad. Sci. USA 97: 6055-6060.
Caron C, Rousset R, Beraud C, Moncollin V, Egly JM & Jalinot P (1993) Functional and biochemical interaction of the HTLV-1 Tax1 transactivator with TBP. EMBO J. 12: 4269-4278.
Caron C, Mengus G, Dubrowskaya V, Roisin A, Davidson I & Jalinot P (1997) Human TAF(II)28 interacts with the human T-cell leukemia virus type I Tax transactivator and promotes its transcriptional activity. Proc. Natl. Acad. Sci. USA 94: 3662-3667.
Carrasco D, Rizzo CA, Dorfman K & Bravo R (1996) The v-rel oncogene promotes malignant T-cell leukemia/lymphoma in transgenic mice. EMBO J. 15: 3640-3650.
Carter RS, Geyer BC, Xie M, Acevedo-Suárez CA & Ballard DW (2001) Persistent activation of NF-κB by the Tax transforming protein involves chronic phosphorylation of IκB kinase subunits IKKβ and IKKγ. J. Biol. Chem. 276: 24445-24448.
Chen BJ, Feinberg MB & Baltimore D (1997) The κB sites in the human immunodeficiency virus type 1 long terminal repeat enhance virus replication yet are not absolutely required for virus growth. J. Virol. 73: 3893-3903.
Chen W & Cooper NR (1996) Epstein-Barr virus nuclear antigen 2 and latent membrane protein independently transactivate p53 through induction of NF-κB activity. J. Virol. 70: 4849-4853.
Chu Z-L, Shin Y-A, Yang J-M, DiDonato JA & Ballard DW (1999) IKKγ mediates the interaction of cellular IκB kinases with the Tax transforming protein of human T cell leukemia virus type 1. J. Biol. Chem. 274: 15297-15300.
Chung YM, Park KJ, Choi SY, Hwang SB & Lee SY (2001) Hepatitis C virus core protein potentiates TNF-α-induced NF-κB activation through TRAF2-IKKβ-dependent pathway. Biochem. Biophys. Res. Commun. 284: 15-19.
Dawson, CW, Rickinson AB & Young LS (1990) Epstein-Barr virus latent membrane protein inhibits human epithelial cell differentiation. Nature 344: 777-780.
Deng L, Wang C, Spencer E, Yang L, Braun A, You J, Slaughter C, Pickart C & Chen ZJ (2000) Activation of the IκB kinase complex by TRAF6 requires a dimeric ubiquitin-conjugating enzyme complex and a unique polyubiquitin chain. Cell 103: 351-361.
Devergne O, Hatzivassiliou E, Izumi KM, Kaye KM, Kleijnen MF, Kieff E & Mosialos G (1996) Association of TRAF1, TRAF2, and TRAF3 with an Epstein-Barr virus LMP1 domain important for B-lymphocyte transformation: role in NF-κB activation. Mol. Cell. Biol. 16: 7098-7108.
Devergne O, McFarland EC, Mosialos G, Izumi KM, Ware CF & Kieff E (1998) Role of the TRAF binding site and NF-κB activation in Epstein-Barr virus latent membrane protein 1-induced cell gene expression. J. Virol. 72: 7900-7908.
Diao J, Garces R & Richardson CD (2001) X protein of hepatitis B virus modulates cytokine and growth factor related signal transduction pathways during the course of viral infections and hepatocarcinogenesis. Cytokine Growth Factor Rev. 12: 189-205.
D'Souza B, Rowe M & Walls D (2000) The bfl-1 gene is transcriptionally upregulated by the Epstein-Barr virus LMP1, and its expression promotes the survival of a Burkitt's lymphoma cell line. J. Virol. 74: 6652-6658.
Duyao MP, Kessler DJ, Spicer DB, Bartholomew C, Cleveland JL, Siekevitz M & Sonenshein GE (1992) Transactivation of the c-myc promoter by human T cell leukemia virus type 1 tax is mediated by NFκB. J. Biol. Chem. 267: 16288-16291.
Eliopoulos AG, Davies C, Blake SS, Murray P, Najafipour S, Tsichlis PN & Young LS (2002) The oncogenic protein kinase Tpl-2/Cot contributes to Epstein-Barr virus-encoded latent infection membrane protein 1-induced NF-κB signaling downstream of TRAF2. J. Virol. 76: 4567-4579.
Eliopoulos AG, Stack M, Dawson CW, Kaye KM, Hodgkin L, Sihota S, Rowe M & Young LS (1997) Epstein-Barr virus-encoded LMP1 and CD40 mediate IL-6 production in epithelial cells via an NF-κB pathway involving TNF receptor-associated factors. Oncogene 14: 2899-2916.

Epinat J-C, Kazandjian D, Harkness DD, Petros S, Dave J, White DW & Gilmore TD (2000) Mutant Envelope residues confer a transactivation function onto N-terminal sequences of the v-Rel oncoprotein. Oncogene 19: 599-607.

Feuillard J, Schuhmacher M, Kohanna S, Asso-Bonnet M, Ledeur F, Joubert-Caron R, Bissieres P, Polack A, Bornkamm GW & Raphael M (2000) Inducible loss of NF-κB activity is associated with apoptosis and Bcl-2 down-regulation in Epstein-Barr virus-transformed B lymphocytes. Blood 95: 2068-2075.

Fontaine V, van der Meijeden E, de Graaf J, ter Schegget J & Struyk L (2000) A functional NF-κB binding site in the human papillomavirus type 16 long control region. Virology 272: 40-49.

Friedman JM & Horwitz MS (2002) Inhibition of tumor necrosis factor α-induced NF-κB activation by the adenovirus E3-10.4/14.5K complex. J. Virol. 76: 5515-5521.

Fujii M, Tsuchiya H & Seiki M (1991) HTLV-1 Tax has distinct but overlapping domains for transcriptional activation and for enhancer specificity. Oncogene 6: 2349-2352.

Fujisawa J, Toita M, Yoshimura T & Yoshida M (1991) The indirect association of human T-cell leukemia virus tax protein with DNA results in transcriptional activation. J. Virol. 65: 4525-4528.

Gilmore TD, White DW, Sarkar S & Sif S (1995) Malignant transformation of cells by the v-Rel oncoprotein, in: The DNA Provirus. Howard Temin's Scientific Legacy (Cooper GM, Greenberg TR & Sugden B, eds), ASM Press, Washington, p. 109-128.

Gilmore TD (1999) Multiple mutations contribute to the oncogenicity of the retroviral oncoprotein v-Rel. Oncogene 18: 6925-6937.

Gilmore TD, Cormier C, Jean-Jacques J & Gapuzan M-E (2001) Malignant transformation of primary chicken spleen cells by human transcription factor c-Rel. Oncogene 20: 7098-7103.

Gilmore TD (2003) The Rel/NF-κB signal transduction pathway and cancer, in Signal Transduction in Cancer (Frank DA, ed), in press.

Grassmann R, Berchtold S, Radant I, Alt M, Fleckenstein B, Sodroski JG, Haseltine WA & Ramstedt U (1992) Role of human T-cell leukemia virus type 1 X region proteins in immortalization of primary human lymphocytes in culture. J. Virol. 66: 4570-4575.

Grassmann R, Dengler C, Müller-Fleckenstein I, Flackenstein B, McGuire K, Dokhelar MC, Sodroski JG & Haseltine WA (1989) Transformation to continuous growth of primary human T lymphocytes by human T-cell leukemia virus type I X-region genes transduced by a Herpesvirus saimiri vector. Proc. Natl. Acad. Sci. USA 86: 3351-3355.

Grossman SR, Johannsen E, Tong X, Yalamanchili R & Kieff E (1994) The Epstein-Barr virus nuclear antigen 2 transactivator is directed to response elements by the Jκ recombination signal binding protein. Proc. Natl. Acad. Sci. USA 91: 7568-7572.

Grossman WJ, Kimata JT, Wong FH, Zutter M, Ley TJ & Ratner L (1995) Development of leukemia in mice transgenic for the tax gene of human T-cell leukemia virus type I. Proc. Natl. Acad. Sci. USA 92: 1057-1061.

Hammarskjold ML & Simurda MC (1992) Epstein-Barr virus latent membrane protein transactivates the human immunodeficiency virus type 1 long terminal repeat through induction of NF-κB activity. J. Virol. 66: 6496-6501.

Harhaj EW, Good L, Xiao G & Sun S-C (1999) Gene expression profiles in HTLV-1-immortalized T cells: deregulated expression of genes involved in apoptosis regulation. Oncogene 18: 1341-1349.

Harhaj EW & Sun S-C (1999) IKKγ serves as a docking subunit of the IκB kinase (IKK) and mediates interaction of IKK with the human T-cell leukemia virus Tax protein. J. Biol. Chem. 274: 22911-22914.

Hatzivassiliou E & Mosialos G (2002) Cellular signaling pathways engaged by the Epstein-Barr virus transforming protein LMP1. Front. Biosci. 7: d319-d329.

He Z, Xin B, Yang X, Chan C & Cao L (2000) Nuclear factor-κB activation is involved in LMP1-mediated transformation and tumorigenesis of rat-1 fibroblasts. Cancer Res. 60: 1845-1848.

Henderson S, Rowe M, Gregory C, Croom-Carter D, Wang F, Longnecker R, Kieff E & Rickinson A (1991) Induction of bcl-2 expression by Epstein-Barr virus latent membrane protein 1 protects infected B cells from programmed cell death. Cell 65: 1107-1115.

Henkel T, Ling PD, Hayward SD & Peterson MG (1994) Mediation of Epstein-Barr virus EBNA2 transactivation by recombination signal-binding protein Jκ. Science 265: 92-95.

Hirai H, Fujisawa J, Suzuki T, Ueda K, Muramatsu M, Tsuboi A, Arai N & Yoshida M (1992) Transcriptional activator Tax of HTLV-1 binds to the NF-κB precursor p105. Oncogene 7: 1737-1742.

Hirai H, Suzuki T, Fujisawa J, Inoue J & Yoshida M (1994) Tax protein of human T-cell leukemia virus type I binds to the ankyrin motifs of inhibitory factor κB and induces nuclear translocation of transcription factor NF-κB proteins for transcriptional activation. Proc. Natl. Acad. Sci. USA 91: 3584-3588.

Hiscott J, Kwon H & Génin P (2001) Hostile takeovers: viral appropriation of the NF-κB pathway. J. Clin. Invest. 107: 143-151.

Hoyos B, Ballard DW, Böhnlein E, Siekevitz M & Greene WC (1989) κB-specific DNA binding proteins: role in the regulation of human interleukin-2 gene expression. Science 244: 457-460.

Hrdlicková R, Nehyba J & Humphries EH (1994) In vivo evolution of c-rel oncogenic potential. J. Virol. 68: 2371-2382.

Huang DB, Chen YQ, Ruetsche M, Phelps CB & Ghosh G (2001a) X-ray crystal structure of proto-oncogene product c-Rel bound at the CD28 response element of IL-2. Structure 9: 669-678.

Huang Y, Obtani K, Iwanaga R, Matsumura Y & Nakamura M (2001b) Direct trans-activation of the human cyclin D2 gene by the oncogene product Tax of human T-cell leukmia virus type I. Oncogene 20: 1094-1102.

Huen DS, Henderson SA, Croom-Carter D & Rowe M (1995) The Epstein-Barr virus latent membrane protein-1 (LMP1) mediates activation of NF-κB and cell surface phenotype via two effector regions in its carboxy-terminal cytoplasmic domain. Oncogene 10: 549-560.

Izumi KM, Kaye KM & Kieff ED (1997) The Epstein-Barr virus LMP1 amino acid sequence that engages tumor necrosis factor receptor associated factors is critical for primary B lymphocyte growth transformation. Proc. Natl. Acad. Sci. USA 94: 1447-1452.

Izumi KM & Kieff ED (1997) The Epstein-Barr virus oncogene product latent membrane protein 1 engages the tumor necrosis factor receptor-associated death domain protein to mediate B lymphocyte growth transformation and activate NF-κB. Proc. Natl. Acad. Sci. USA 94: 12592-12597.

Jeang K-T (2001) Functional activities of the human T-cell leukemia virus type I Tax oncoprotein: cellular signaling through NF-κB. Cytokine Growth Factor Rev. 12: 207-217.

Jin D-Y & Jeang K-T (1997) HTLV-I Tax self-association in optimal trans-activation function. Nucleic Acids Res. 15: 379-387.

Kamens J, Richardson P, Mosialos G, Brent R & Gilmore TD (1990) Oncogenic transformation by v-Rel requires an amino-terminal activation domain. Mol. Cell. Biol. 10: 2840-2847.

Kanno M, Fromental C, Staub A, Ruffenach F, Davidson I & Chambon P (1989) The SV40 TC-II(κB) and the related H-2Kβ enhansons exhibit different cell type specific and inducible proto-enhancer activities, but the SV40 core sequence and the AP-2 binding site have no enhanson properties. EMBO J. 8: 4205-4214.

Kanno T, Brown K, Franzoso G & Siebenlist U (1994) Kinetic analysis of human T-cell leukemia virus type-I Tax-mediated activation of NF-κB. Mol. Cell. Biol. 14: 6443-6451.

Karin M (1999) How NF-κB is activated: the role of the IκB kinase (IKK) complex. Oncogene 18: 6867-6874.

Kaufman RJ (1999) Double-stranded RNA-activated protein kinase mediates virus induced apoptosis: a new role for an old actor. Proc. Natl. Acad. Sci. USA 96: 11693-11695.

Kaye KM, Devergne O, Harada JN, Izumi KM, Yalamanchili R, Kieff E & Mosialos G (1996) Tumor necrosis factor receptor associated factor 2 is a mediator of NF-κB activation by latent infection membrane protein 1, the Epstein-Barr virus transforming protein. Proc. Natl. Acad. Sci. USA 93: 11085-11090.

Keller SA, Schattner EJ & Cesarman E (2000) Inhibition of NF-κB induces apoptosis of KSHV-infected primary effusion lymphoma cells. Blood 96: 2537-2542.

Kieff E & Rickinson AB (2001) Epstein-Barr virus and its replication, in: Fields Virology, 4th ed (Knipe DM & Howley PM, eds), Lippincott, Williams & Wilkins, Philadelphia, p. 2511-2573.

Korber B, Okayama A, Donnelly T, Tachibana N & Essex M (1991) Polymerase chain reaction analysis of defective human T-cell leukemia virus type I proviral genomes in leukemic cells of patients with adult T-cell leukemia. J. Virol. 65: 5471-5476.

Kwon H, Pelletier N, DeLuca C, Genin P, Cisternas S, Lin R, Wainberg MA & Hiscott J (1998) Inducible expression of IκBα repressor mutants interferes with NF-κB activity and HIV-1 replication in Jurkat T cells. J. Biol. Chem. 273: 7431-7440.

Kwon JA & Rho HM (2002) Hepatitis B viral core protein activates the hepatitis B viral enhancer II/pregenomic promoter through the nuclear factor κB binding site. Biochem. Cell. Biol. 80: 445-455.

Laherty CD, Hu HM, Opipari AW, Wang F & Dixit VM (1992) The Epstein-Barr virus LMP1 gene product induces A20 zinc finger protein expression by activating nuclear factor κB. J. Biol. Chem. 267: 24157-24160.

Lanoix J, Lacoste J, Pepin N, Rice N & Hiscott J (1994) Overproduction of NFKB2 (lyt-10) and c-Rel: a mechanism for HTLV-I Tax-mediated trans-activation via the NF-κB signalling pathway. Oncogene 9: 841-852.

Lee SH, Wang X & DeJong J (2000) Functional interactions between an atypical NF-κB site from the rat CYP2B1 promoter and the transcriptional repressor RBP-Jκ/CBF1. Nucleic Acids Res. 28: 2091-2098.

Leung K & Nabel GJ (1988) HTLV-1 transactivator induces interleukin 2 receptor expression through an NF-κB-like factor. Nature 333: 776-778.

Li C-CH, Ruscetti FW, Rice NR, Chen E, Yang N-S, Mikovits J & Longo DL (1993) Differential expression of Rel family members in human T-cell leukemia virus type I-infected cells: transcriptional activation of c-rel by Tax protein. J. Virol. 67: 4205-4213.

Li Y, Kang J, Friedman J, Tarassishin L, Ye J, Kovalenko A, Wallach D & Horwitz MS (1999) Identification of a cell protein (FIP-3) as a modulator of NF-κB activity and as a target of an adenovirus inhibitor of tumor necrosis factor α-induced apoptosis. Proc. Natl. Acad. Sci. USA 96: 1042-1047.

Liu L, Eby MT, Rathore N, Sinha SK, Kumar A & Chaudhary PM (2002) The human herpes virus 8-encoded viral FLICE inhibitory protein physically associates with and persistently activates the IκB kinase complex. J. Biol. Chem. 277: 13745-13751.

Maggirwar SB, Harhaj E & Sun S-C (1995) Activation of NF-κB/rel by Tax involves degradation of IκBα and is blocked by a proteasome inhibitor. Oncogene 11: 993-998.

Majumder M, Ghosh AK, Steele R, Zhou XY, Phillips NJ, Ray R & Ray RB (2002) Hepatitis C virus NS5A protein impairs TNF-mediated hepatic apoptosis, but not by an anti-FAS antibody, in transgenic mice. Virology 294: 94-105.

McKinsey TA, Brockman JA, Scherer DC, Al-murrani SW, Green PL & Ballard DW (1996) Inactivation of IκBβ by the Tax protein of human T-cell leukemia virus type 1: a potential mechanism for constitutive induction of NF-κB. Mol. Cell. Biol. 16: 2083-2090.

Merlo JJ & Tsygankov AY (2001) Herpesvirus saimiri oncoproteins Tip and StpC synergistically stimulate NF-κB activity and interleukin-2 gene expression. Virology 279: 325-338.

Miller WE, Earp HS & Raab-Traub N (1995) The Epstein-Barr virus latent membrane protein 1 induces expression of the epidermal growth factor receptor. J. Virol. 69: 4390-4398.

Miskin JE, Abrams CC, Goatley LC & Dixon LK (1998) A viral mechanism for inhibition of the cellular phosphatase calcineurin. Science 281: 562-565.

Mitchell T & Sugden B (1995) Stimulation of NF-κB-mediated transcription by mutant derivatives of the latent membrane protein of Epstein-Barr virus. J. Virol. 69: 2968-2976.

Mori N, Fujii M, Hinz M, Nakayama K, Yamada Y, Ikeda S, Yamasaki Y, Kashanchi F, Tanaka Y, Tomonaga M & Yamamoto N (2002) Activation of cyclin D1 and D2 promoters by human T-cell leukemia virus type 1 Tax protein is associated with IL-2-independent growth of T cells. Int. J. Cancer 99: 378-385.

Mosialos G & Gilmore TD (1993) v-Rel and c-Rel are differentially affected by mutations at a consensus protein kinase recognition sequence. Oncogene 8: 721-730.

Mosialos G, Birkenbach M, Yalamanchili R, VanArsdale T, Ware C & Kieff E (1995) The Epstein-Barr virus transforming protein LMP1 engages signaling proteins for the tumor necrosis factor receptor family. Cell 80: 389-399.

Murakami T, Hirai H, Suzuki T, Fujisawa J-I & Yoshida M (1995) HTLV-1 Tax enhances NF-κB2 expression and binds to the products p52 and p100, but does not suppress the inhibitory function of p100. Virology 206: 1066-1074.

Murono S, Inoue H, Tanabe T, Joab I, Yoshizaki T, Furukawa M & Pagano JS (2001) Induction of cyclooxygenase-2 by Epstein-Barr virus latent membrane protein 1 is involved in vascular endothelial growth factor production in nasopharyngeal carcinoma cells. Proc. Natl. Acad. Sci. USA 98: 6905-6910.

Nabel G & Baltimore D (1987) An inducible transcription factor activates expression of human immunodeficiency virus in T cells. Nature 326: 711-713.

Nehyba J, Hrdlicková R & Bose HR (1997) Differences in κB DNA-binding properties of v-Rel and c-Rel are the result of oncogenic mutations in three distinct functional regions of the Rel protein. Oncogene 14: 2881-2897.

Nerenberg M, Hinrichs SH, Reynolds RK, Khoury G & Jay G (1987) The tat gene of human T lymphotropic virus type I induces mesenchymal tumors in transgenic mice. Science 237: 1324-1229.

Nicot C, Mahieux R, Takemoto S & Franchini G (2000) Bcl-X_L is up-regulated in HTLV-I and HTLV-II in vitro and in ex vivo ATLL samples. Blood 96: 275-281.

Pahl HL, Sester M, Burgert HG & Baeuerle PA (1996) Activation of transcription factor NF-κB by the adenovirus E3/19K protein requires its ER retention. J. Cell Biol. 132: 511-522.

Pahl HL & Baeuerle PA (1997) The ER-overload response: activation of NF-κB. Trends Biochem. Sci. 22: 63-67.

Pahl HL (1999) Activators and target genes of Rel/NF-κB transcription factors. Oncogene 18: 6853-6866.

Park KJ, Choi SH, Lee SY, Hwang SB & Lai MM (2002) Nonstructural 5A protein of hepatitis C virus modulates tumor necrosis factor α-stimulated nuclear factor κB activation. J. Biol. Chem. 277: 13122-13128.

Pati S, Cavrois M, Guo HG, Foulke JS, Kim J, Feldman RA & Reitz M (2001) Activation of NF-κB by the human herpesvirus 8 chemokine receptor ORF74: evidence for a paracrine model of Kaposi's sarcoma pathogenesis. J. Virol. 75: 8660-8673.

Petropoulos L, Lin R & Hiscott J (1996) Human T cell leukemia virus type 1 Tax protein increases NF-κB dimer formation and antagonizes the inhibitory activity of the IκBα regulatory protein. Virology 225: 52-64.

Petropoulos L & Hiscott J (1998) Association between HTLV-1 Tax and IκBα is dependent on the IκBα phosphorylation state. Virology 252: 189-199.

Portis T, Harding JC & Ratner L (2001) The contribution of NF-κB activity to spontaneous proliferation and resistance to apoptosis in human T-cell leukemia virus type 1 Tax-induced tumors. Blood 98: 1200-1208.

Powell PP, Dixon LK & Parknouse RM (1996) An IκB homolog encoded by African swine fever virus provides a novel mechanism for downregulation of proinflammatory cytokine responses in host macrophages. J. Virol. 70: 8527-8533.

Pozatti T, Vogel J & Jay G (1990) The human T-lymphotropic virus type I tax gene can cooperate with the ras oncogene to induce neoplastic transformation of cells. Mol. Cell. Biol. 10: 413-417.

Prösch S, Staak K, Stein J, Liebenthal C, Stamminger T, Volk HD & Krüger DH (1995) Stimulation of the human cytomegalovirus IE enhancer/promoter in HL-60 cells by TNFα is mediated via induction of NF-κB. Virology 208: 197-206.

Quinto I, Mallardo M, Baldassarre F, Scala G, Englund G & Jeang KT (1999) Potent and stable attenuation of live HIV-1 by gain of a proteolysis-resistant inhibitor of NF-κB (IκBα S32/36A) and the implications for vaccine development. J. Biol. Chem. 274: 17567-17572.

Ranganathan P & Khalili K (1993) The transcriptional enhancer element, κB, regulates promoter activity of the human neurotropic virus, JCV, in cells derived from the CNS. Nucleic Acids Res. 21: 1959-1964.

Ressler S, Connor LM & Marriott SJ (1996) Cellular transformation by human T-cell leukemia virus type I. FEMS Microbiol. Lett. 140: 99-109.

Revilla Y, Callejo M, Rodriguez JM, Culebras E, Nogal ML, Slaas ML, Viñuela E & Fresno M (1998) Inhibition of nuclear factor κB activation by a virus encoded IκB-like protein. J. Biol. Chem. 273: 5405-5411.

Rickinson AB & Kieff E (2001) Epstein-Barr virus, in: Fields Virology, 4th ed (Knipe DM & Howley PM, eds), Lippincott, Williams & Wilkins, Philadelphia, p. 2575-2628.

Robek M & Ratner L (1999) Immortalization of $CD4^+$ and $CD8^+$ T lymphocytes by human T-cell leukemia virus type I Tax mutants expressed in a functional molecular clone. J. Virol. 73: 4856-4865.

Rodríguez CI, Nogal ML, Carrascosa AL, Salax ML, Fresno M & Bevilla Y (2002) African swine fever virus IAP-like protein induces the activation of nuclear factor κB. J. Virol. 76: 3936-3942.

Rong BL, Libermann TA, Kogawa K, Ghosh S, Cao LX, Pavan-Langston D & Dunkel EC (1992) HSV-1-inducible proteins bind to NF-κB-like sites in the HSV-1 genome. Virology 189: 750-756.

Ross TM, Narayan M, Fang Z-Y, Minella AC & Green PL (2000) Human T-cell leukemia virus type 2 Tax mutants that selectively abrograte NFκB or CREB/ATF activation fail to trasnform primary T cells. J. Virol. 74: 2655-2662.

Rousset R, Desbois C, Bantignies F & Jalinot P (1996) Effects on NF-κB1/p105 processing of the interaction between the HTLV-1 transactivator Tax and the proteasome. Nature 381: 328-331.

Sachdev S & Hannink M (1998) Loss of IκBα-mediated control over nuclear import and DNA-binding enables oncogenic activation of c-Rel. Mol. Cell. Biol. 18: 5445-5456.

Schultheiss U, Puschner S, Kremmer E, Mak TW, Engelmann H, Hammerschmidt W & Kieser A (2001) TRAF6 is a critical mediator of signal transduction by the viral oncogene latent membrane protein 1. EMBO J. 20: 5678-5691.

Schwarz M & Murphy PM (2001) Kaposi's sarcoma-associated herpesvirus G protein-coupled receptor constitutively activates NF-κB and induces proinflammatory cytokine and chemokine production via a C-terminal signaling determinant. J. Immunol. 167: 505-513.

Shirakata Y, Shuman JD, & Coligan JE (1996) Purification of a novel MHC class I element binding activity from thymus nuclear extracts reveals that thymic RBP-Jκ/CBF1 binds to NF-κB-like elements. J. Immunol. 156: 4672-4679.

Sugano N, Chen W, Roberts ML & Cooper NR (1997) Epstein-Barr virus binding to CD21 activates the initial viral promoter via NF-κB induction. J. Exp. Med. 186: 731-737.

Sun S-C & Ballard DW (1999) Peristent activation of NF-κB by the Tax transforming protein of HTLV-1: hijacking cellular IκB kinases. Oncogene 18: 6948-6958.

Suzuki T, Hirai H, Fujisawa J, Fujita T & Yoshida M (1993) A trans-activator Tax of human T-cell leukemia virus type 1 binds to NF-κB p50 and serum response factor (SRF) and associates with enhancer DNAs of the NF-κB site and CArG box. Oncogene 8: 2391-2397.

Suzuki T, Hirai H & Yoshida M (1994) Tax protein of HTLV-1 interacts with the Rel homology domain of NF-κB p65 and c-Rel proteins bound to the NF-κB binding site and activates transcription. Oncogene 9: 3099-3105.

Suzuki T, Hirai H, Murakami T & Yoshida M (1995) Tax protein of HTLV-1 destabilizes the complexes of NF-κB and IκB-α and induces nuclear translocation of NF-κB for transcriptional activation. Oncogene 10: 1199-1207.

Sylla BS, Hung SC, Davidson DM, Hatzivassiliou E, Malinin NL, Wallach D, Gilmore TD, Kieff E & Mosialos G (1998) Epstein-Barr virus-transforming protein latent infection membrane protein 1 activates transcription factor NF-κB through a pathway that includes the NF-κB-inducing kinase and the IκB kinases IKKα and IKKβ. Proc. Natl. Acad. Sci. USA 95: 10106-10111.

Tai DI, Tsai SL, Chen YM, Chuang YL, Peng CY, Sheen IS, Yeh CT, Chang KS, Huang SN, Kuo GC & Liaw YF (2000) Activation of nuclear factor κB in hepatitis C virus infection: implications for pathogenesis and hepatocarcinogenesis. Hepatology 31: 656-664.

Tait SWG, Reid EB, Greaves DR, Wileman TE & Powell PP (2000) Mechanism of inactivation of NF-κB by a viral homologue of IκBα. J. Biol. Chem. 275: 34656-34664.

Tanaka A, Takahashi C, Tamaoka S, Nosaka T, Maki M & Hatanaka M (1990) Oncogenic transformation by the tax gene of human T-cell leukemia virus type I in vitro. Proc. Natl. Acad. Sci. USA

87: 1071-1075.

Tie T, Adya N, Greene WC & Giam C-Z (1996) Interaction of the human T-lymphotropic virus type 1 Tax dimer with CREB and the viral 21 base-pair repeat. J. Virol. 70: 8368-8374.

Wang C, Deng L, Hong M, Akkaraju GR, Inoue J & Chen ZJ (2001) TAK1 is a ubiquitin-dependent kinase of MKK and IKK. Nature 412: 346-351.

Wang S, Rowe M & Lundgren E (1996) Expression of the Epstein Barr virus transforming protein LMP1 causes a rapid and transient stimulation of the Bcl-2 homologue Mcl-1 levels in B-cell lines. Cancer Res. 56: 4610-4613.

Wano Y, Feinberg M, Hosking JB, Bogerd H & Greene WC (1988) Stable expression of the tax gene of type I human T-cell leukemia virus in human T cells activates specific genes involved in growth. Proc. Natl. Acad. Sci. USA 85: 9733-9737.

Weil R, Sirma H, Giannini C, Kremsdorf D, Bessia C, Dargemont C, Bréchot C & Israël A (1999) Direct association and nuclear import of the hepatitis B virus X protein with the NF-κB inhibitor IκBα. Mol. Cell. Biol. 19: 6345-6354.

Williams JL, Garcia J, Harrich D, Pearson L, Wu F & Gaynor R (1990) Lymphoid specific gene expression of adenovirus early region 3 promoter is mediated by NF-κB binding motifs. EMBO J. 9: 4435-4442.

Xiao G, Cvijic ME, Fong A, Warhaj EW, Uhik MT, Waterfield M & Sun S-C (2001) Retroviral oncoprotein Tax induces processing of NF-κB2/p100 in T cells: evidence for the involvement of IKKα. EMBO J. 20: 6805-6815.

Xiao G, Harhaj EW & Sun S-C (2000) Domain-specific interaction with IκB kinase (IKK) regulatory subunit IKKγ is an essential step in Tax-mediated activation of IKK. J. Biol. Chem. 275: 34060-34067.

Xiao G & Sun S-C (2000) Activation of IKKα and IKKβ through their fusion with HTLV-1 Tax protein. Oncogene 19: 5198-5203.

Yamaoka S, Courtois G, Bessia C, Whiteside ST, Weil R, Agou F, Kirk HE, Kay RJ & Israël A (1998) Complementation cloning of NEMO, a component of the IκB kinase complex essential for NF-κB activation. Cell 93: 1231-1240.

Yañex RJ, Rodríguez JM, Nogal ML, Yuste L, Enríquez C, Rodriguez JF & Viñuela E (1995) Analysis of the complete nucleotide sequence of African swine fever virus. Virology 208: 249-278.

Yoshida H, Kato N, Shiratori Y, Otsuka M, Maeda S, Kato J & Omata M (2001) Hepatitis C virus core protein activates nuclear factor κB-dependent signaling through tumor necrosis factor receptor-associated factor. J. Biol. Chem. 276: 16399-16405.

Yoshida M (2001) Multiple strategies of HTLV-1 for dysregulation of cell growth control. Annu. Rev. Immunol. 19: 475-496.

Zimber-Strobl, U, Strobl LJ, Meitinger C, Hinrichs R, Sakai T, Furukawa T, Honjo T & Bornkamm GW (1994) Epstein-Barr virus nuclear antigen 2 exerts its transactivating function through interaction with recombination signal binding protein RBP-Jκ, the homologue of Drosophila Suppressor of Hairless. EMBO J. 13: 4973-4982.

12. AFFILIATIONS

Boston University, Biology Department, 5 Cummington Street, Boston, MA 02215, USA (T.D. Gilmore). Biomedical Sciences Research Center — Alexander Fleming, Institute of Immunology, 14-16 Alexander Fleming Street, 16672 Vari, Greece (G. Mosialos).

U. SENFTLEBEN & M. KARIN

THE IκB KINASE (IKK) COMPLEX

1. INTRODUCTION

The IκB kinase (IKK) complex is a central element of NF-κB-related signaling. It is the kinase which, upon cellular stimulation by a variety of physical, chemical and biological activators, activates cytoplasmic NF-κB indirectly via phosphorylation of its inhibitor protein IκB. Phosphorylated IκBs rapidly undergo ubiquitin-dependent degradation and subsequently free NF-κB dimers enter the nucleus to regulate transcription of target genes. This inducible activation and nuclear entry of NF-κB depends on a complex interaction of constitutively active enzymes and nonenzymatic proteins (Karin & Ben-Neriah, 2000; Ghosh & Karin, 2002). Thus, the activation of the IKK complex is the critical step in NF-κB signaling and therefore the regulation and function of this pathway are tightly linked to the subunits of IKK.

As a complex consisting of three subunits, IKK is not only a kinase which determines the level of NF-κB activity, but also modulates the quality of NF-κB-activated transcription. Thus, there is evidence that distinct upstream signals are converted by corresponding IKK subunits to activate specific NF-κB hetero- or homodimers and to regulate the composition of NF-κB subunits. This leads to the transcription of different patterns of NF-κB target genes.

The functional studies of the IKK complex regarding the connection of upstream activators with specific NF-κB target genes are only at their beginning. With respect to specific pharmacological inhibition of IKK subunits this is a rapidly growing field of interest. Nevertheless, the purpose of this article is to summarize the knowledge about the structure, the biochemical and physiological function of the IKK complex. In particular, genetic approaches form the basis to study IKK function in vivo and will be the focus of this review. Few data about the role of IKK in diseases are available so far and are discussed at the end of this chapter.

2. STRUCTURE OF IKK

The IKK complex consists of two catalytic subunits, viz. IKKα and IKKβ (also referred to as IKK1 and IKK2, respectively) and the regulatory IKKγ (also known as NF-κB essential modulator or NEMO, IKK-associated protein 1 or IKKAP1, and 14.7-interacting protein or FIP-3) (DiDonato et al., 1997; Regnier

R. Beyaert (ed.), Nuclear Factor κB. Regulation and Role in Disease, 115-135.

et al., 1997; Zandi et al., 1997; Rothwarf et al., 1998; Yamaoka et al., 1998; Li et al., 1999b; Mercurio et al., 1999). IKKα and IKKβ are 85-kDa and 87-kDa proteins, respectively, whereas the γ-subunit is a 47-kDa protein. These α- and β-subunits are highly homologous proteins with 50% sequence identity (70% sequence similarity) and contain N-terminal protein kinase domains as well as leucine zipper (LZ) and helix-loop-helix (HLH) motifs. IKKγ is predominantly helical with stretches of coiled-coil structure and an LZ motif near the C-terminus (Rothwarf et al., 1998). There are several reports of proteins, for example IKAP (Cohen et al., 1998), NIK (Regnier et al., 1997; Cohen et al., 1998), MEKK1 (Rao et al., 1996) or members of the IκB and Rel family (Heilker et al., 1999), that are considered to be additional components of the 700-900 kDa IKK complex. These data could not be confirmed by others because these proteins do not co-elute with the IKKα-IKKβ-IKKγ complex under stringent conditions (Karin & Ben-Neriah, 2000).

The stoichiometry of the known IKK components has not yet been finally determined in mammalian cells. The main part of the 700-900 kDa complex appears to contain equivalent amounts of IKKα and IKKβ (Rothwarf et al., 1998). In mammalian cells, their activation depends on the ability to dimerize via the LZ motifs (Mercurio et al., 1997; Woronicz et al., 1997; Zandi et al., 1997) and on their association with the regulatory IKKγ (Rothwarf et al., 1998; Yamaoka et al., 1998; Makris et al., 2000). The dimerization of IKKα and IKKβ, mediated by the LZ regions (Woronicz et al., 1997; Zandi et al., 1997; Zandi et al., 1998), is a prerequisite for association with IKKγ via an interaction motif at the C-terminus of either catalytic subunit (May et al., 2000). The assembly of IKKα and IKKβ dimers with IKKγ, essential for IKK activation, forms the large 700-900 kDa complex usually composed of IKKα:IKKβ heterodimers, held together via dimeric interactions between two IKKγ molecules (dimer of dimers) (Rothwarf & Karin, 1999; Miller & Zandi, 2001). It is still unclear whether this is the sole composition characteristic for a ubiquitously expressed IKK complex or whether there are other variants of the assembly.

3. BIOCHEMICAL FUNCTION

In 1996 a cytokine-inducible protein kinase activity specific for the N-terminal regulatory serines Ser32 and Ser36 of IκBα and IκBβ was identified (DiDonato et al., 1997). Although numerous other enzymes were suggested to be responsible for IκB phosphorylation and NF-κB induction, only IKK matches the kinetic characteristics of signal-induced activation caused by simultaneous phosphorylation of Ser32/Ser36 and rapid IκBα degradation (DiDonato et al., 1997).

3.1. Activation of the IKK complex

The activation of IKK, for example by TNFα or IL-1, depends on phosphorylation at two conserved serines of either IKKα (Ser176 and Ser180) or IKKβ

(Ser177 and Ser181) located within their activation loops (Mercurio et al., 1997; Ling et al., 1998; Delhase et al., 1999). The phosphorylation of the activation loop is followed by a conformational change which releases the catalytic pocket and thus allows the interaction with substrates (Johnson et al., 1996). Such phosphorylation may either be due to action of an upstream kinase or is the result of autophosphorylation between two catalytic subunits within the same complex. In the case of IKK, autophosphorylation is essential for its activation (Zandi et al., 1998; Miller & Zandi, 2001), but the mechanisms leading to IKK activation in response to upstream signals are not fully understood. Replacement of the activation loop serines in IKKβ by alanines generates a mutant form of the β-subunit that, when integrated into the IKK complex, prevents its activation upon stimulation with TNFα, IL-1 or LPS (Delhase et al., 1999). Interestingly, a similar replacement of the IKKα activation loop serines by alanines prevents autophosphorylation, but has no effect on stimulation of IKK activity by TNFα or IL-1 (Delhase et al., 1999; Cao et al., 2001). Consistent with these findings is the observation that IKKβ, but not IKKα is essential for IKK and NF-κB activation by proinflammatory stimuli (Li et al., 1999b-c; Tanaka et al., 1999). However, more recent data support a specific role of IKKα in IKK and NF-κB activation upon stimulation with other than proinflammatory molecules. Mammary epithelial cells and B-lymphocytes that express the kinase-inactive IKKα^{AA} form exhibit defective IKK and NF-κB activation upon stimulation with the TNF cytokine family members called receptor activator of NF-κB (RANK) ligand and B-lymphocyte stimulator (BLyS), respectively (Cao et al., 2001). However, IKK and NF-κB activation in response to TNFα, IL-1 and LPS stimulation was normal in these mutant cells (Cao et al., 2001).

In this context it needs to be emphasized that Ikk$\alpha^{-/-}$ B-cells do not develop into mature cells (Kaisho et al., 2001; Senftleben et al., 2001a). Most intriguingly, a relatively normal NF-κB DNA-binding activity is induced in these mutant cells upon stimulation with LPS, but the cells exhibit defective NF-κB2/p100 processing to the mature p52 protein. A more detailed analysis revealed that the induction of NF-κB2/p100 processing by ectopic NIK (Xiao et al., 2001) can be blocked by the absence of IKKα (Senftleben et al., 2001a). Knockin mice expressing the kinase-inactive IKKα^{AA} mutant also exhibit defects of NF-κB2/p100 processing, B-cell maturation and lymphoid organ formation (Senftleben et al., 2001a). These results suggest that activated NIK targets IKKα to induce phosphorylation-dependent processing of NF-κB2/p100.

As IKKα was also described as NIK-interacting protein (Regnier et al., 1997), a different IKK complex containing NIK might actually exist. This is consistent with the observation that NIK-induced NF-κB2/p100 processing is not affected by the lack of IKKβ or IKKγ (Senftleben et al., 2001a). However, this NIK-IKKα pathway seems to be specifically activated via the lymphotoxin-β receptor (LTβ-R), a TNF receptor family member, as LTβ-R stimulation activates NF-κB in wild-type, but not in IKKα-deficient or NIK-mutant (aly) murine embryonic fibroblasts (MEFs) (Matsushima et al., 2001). Furthermore, Ltβr$^{-/-}$

(Futterer et al., 1998) as well as Nik$^{-/-}$ (Yin et al., 2001) and Nik$^{aly/aly}$ (Shinkura et al., 1999; Fagarasan et al., 2000) mice are characterized by defective lymphoid organ development. As LTβ-R is preferably expressed in stromal cells, most likely another receptor specific for B-cells accounts for activation of the putative NIK-IKKα pathway. Recent studies demonstrated that mice lacking BLyS (also known as BAFF) (Schiemann et al., 2001) or BAFF receptor (BAFF-R) (Thompson et al., 2001) exhibit defective B-cell maturation and lymphoid organ development similar to IKKα-deficient radiation chimeras.

The data discussed above suggest a functional separation of IKKα- and IKKβ-dependent NF-κB signaling. NF-κB activation via proinflammatory stimuli, such as TNFα, IL-1 or LPS, depend on IKKβ/γ, whereas other stimuli, possibly LTβ or BLyS, use the NIK-IKKα pathway. Consecutively, activated IKKβ induces IκB phosphorylation followed by nuclear translocation of liberated NF-κB heterodimers. On the other hand, IKKα-induced phosphorylation of NF-κB2/p100 processing also results in degradation of the inhibitory C-terminal IκB-like domain. Accordingly, increased levels of p52 containing NF-κB dimers enter the nucleus to regulate the transcription of a specific subset of p52-dependent target genes (Senftleben et al., 2001a). This pathway is not involved in proinflammatory effector mechanisms, but most likely plays a central role in lymphoid organ development and adaptive immunity (Fig. 1).

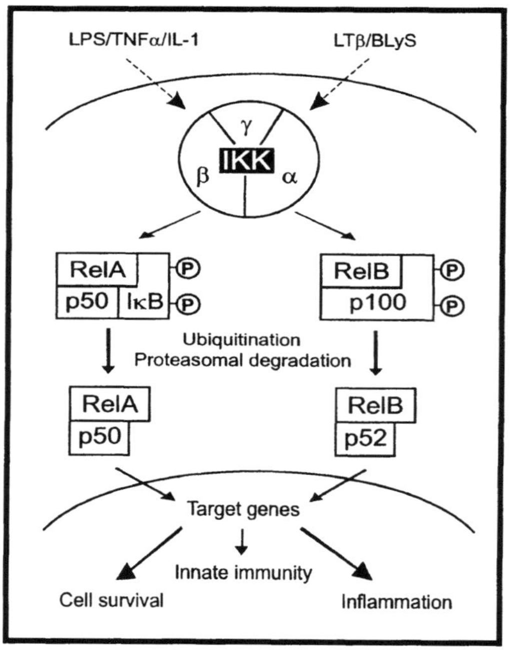

Fig. 1. Activation of NF-κB transcription factors by IKK via two distinct pathways. The canonical pathway triggered by TNFα, IL-1, or products of bacterial or viral origin (such as LPS and dsRNA) requires the IKKβ subunit and is accomplished by IκB phosphorylation and ubiquitin-dependent degradation. The α-specific pathway is triggered by members of the TNF family. After stimulation, for example with lymphotoxin-β (LTβ) or BLyS (also known as BAFF), the phosphorylation-dependent processing of p100 (which results in mature p52) depends on the IKKα subunit. Subsequently, distinct p52-containing NF-κB dimers, such as RelB:p52, are involved in the transcription of specific target genes.

So far, the exact role of IKKγ is not clear. In addition to its essential function in IKK activation via complex assembly IKKγ plays a role in signaling as well. Thus, its characterization as regulatory IKK subunit which recruits upstream signals to the complex (Rothwarf et al., 1998; Yamaoka et al., 1998) has led to the identification of a small number of potential IKKγ-binding proteins (Leonardi et al., 2000; Hong et al., 2001). However, the exact physiological function of any of these proteins is not clear, rendering the possibility that IKKγ serves as a target for regulated phosphorylation most likely (Carter et al., 2001). A more detailed analysis needs to determine the location of the phosphorylation sites and their regulatory function, which, as in the case of IKKβ, might be either inhibitory or activating (Delhase et al., 1999).

3.2. Induced proximity or upstream kinase: two potential mechanisms activating IKK

It is still a matter of question how any signaling pathway targets the IKK complex and finally leads to its activation. Two different mechanisms were suggested to integrate upstream signals via the IKK complex to induce NF-κB transcription (Ghosh & Karin, 2002). Thus, activation of IKK, which depends on activation loop phosphorylation of IKKα and IKKβ, may be achieved either through an induced proximity mechanism based on protein-protein interactions followed by transautophosphorylation within the complex or through the action of an upstream kinase activity (Ghosh & Karin, 2002). Specific components of the TNF-R1-signaling complex, TRAF2 and TRAF5 and the protein kinase RIP1, have been described to be involved in activation of IKK (Kelliher et al., 1998; Devin et al., 2000; Tada et al., 2001). It has been suggested that TRAF2 or TRAF5 recruit the IKK complex to the receptor, whereas RIP1 activates IKK (Devin et al., 2000). Whether the recruitment is mediated via IKKγ which binds to RIP1 (Zhang et al., 2000) or via the catalytic subunits (Devin et al., 2000) is controversially discussed. Though the more apparent recruitment works via IKKγ, no direct interactions between this IKK subunit and components of the TNFα and IL-1 receptors have been detected so far. Interestingly, the kinase activity of RIP1 is dispensible for IKK activation (Ghosh & Karin, 2002). This suggests that RIP1 acts via protein-protein interaction with IKK or another unknown kinase. In the context of IL-1 receptor signaling it is noteworthy that the kinase activity of IL-1 receptor-associated kinase 1 (IRAK1) is not necessary for IKK activation either (Li et al., 1999a). PKR is another protein kinase associated with IKK activation which is required for the response to dsRNA and viral infection. As a catalytically inactive mutant of PKR is able to activate IKK to the same extent as the wild-type form, it probably acts via protein-protein interactions as well (Chu et al., 1999; Bonnet et al., 2000). In this context it is noteworthy that peripheral T-lymphocytes of PKCθ-deficient mice cannot be activated due to defective NF-κB activation (Sun et al., 2000). As PKCθ is not a direct IKK activator, its action may involve additional proteins such as Bcl-10 and MALT1 (Gaide et al., 2001; Lucas et al., 2001). Unlike other protein kinases, PKCθ-associated activation of NF-κB depends on its kinase activity (Khoshnan et al., 2000). However, the exact substrate for PKCθ remains to be identified.

Recently, another mechanism involving ubiquitination and the MAP3 kinase TAK1 in the activation of IKK has been described (Deng et al., 2000; Wang et al., 2001). In an in vitro system, ubiquitination of TRAF6 was shown to result in its activation. Subsequently, ubiquitinated TRAF6 activates TAK1, which in turn activates MKK6 and IKK, thereby turning on both the JNK and the NF-κB pathways. However, all of the evidence presented so far is based on the analysis of in vitro and overexpression systems, which requires to be confirmed genetically.

In conclusion, proteins such as PKR, IRAK1 and RIP1 may activate IKK directly via an induced proximity mechanism based on protein-protein interactions which increase the proximity of the catalytic subunits within the IKK complex and facilitate mutual autophosphorylation. In this scenario, an IKK kinase is dispensable. This idea is supported by genetic disruptions of putative IKK kinases. Mice deficient in NIK (Yin et al., 2001), NAK/T2K/TBK (Bonnard et al., 2000), MEKK1 (Xia et al., 2000; Yujiri et al., 2000), Cot/TPL2 (Dumitru et al., 2000) do not present defects in IKK activation or induction of NF-κB DNA-binding activity. Therefore, based on experimental data, the existence of an upstream IKK kinase activity cannot be completely ruled out, but presently the hypothesis of induced proximity leading to IKK activation seems to be more plausible.

3.3. Inactivation of the IKK complex

IKK activity has to be transient because persistant activation of NF-κB might result in deleterious conditions, such as uncontrolled inflammation or even septic shock. Following stimulation with TNFα, IKK activity peaks within 5-15 min in most cell types and then decreases to about 25% of its maximum value (DiDonato et al., 1997; Zandi et al., 1997; Delhase et al., 1999). These kinetics are consistent with complete IκBα degradation within 15 min, followed by resynthesized cytoplasmic IκBα pools within 60 min after TNFα treatment. Downregulation of IKK activity seems to be the result of extensive autophosphorylation of IKKα and IKKβ at their C-terminal regions following TNFα stimulation (Delhase et al., 1999). Replacement of ten autophosphorylated serines by alanines in the C-terminal portion of IKKβ significantly prolongs IKK activation, whereas exchange of these serines by the phosphoserine mimic glutamic acid results in diminished TNFα-induced IKK activity (Delhase et al., 1999). Consequently, hyperphosphorylation may result in a conformational change of the C-terminus, which contains the HLH motif leading to decreased IKK activity. Another timing mechanism to limit a strong and permanent IKK activity is its downregulation by a phosphatase, which is recruited to the complex and decreases the efficiency of activation loop autophosphorylation. The existence of a constitutive phosphatase has been suggested based on the observation that the phosphatase inhibitor okadaic acid is capable of activating IKK in mammalian cells (DiDonato et al., 1997). Future research will have to prove whether an inducible phosphatase could also play a role in IKK inactivation. Furthermore, a combination of both mechanisms, viz. downregulation via phosphatases and via autophosphorylation of the C-terminal region of IKKα and IKKβ, may be necessary to terminate the activity of IKK.

4. IKK IN DROSOPHILA

Two distinct pathways for activation of the Drosophila immune response have been identified, viz. the Toll/antifungal and the antibacterial pathway (Silverman & Maniatis, 2001). Three Drosophila Rel proteins are differentially required for those two immune response pathways. The antifungal response results in degradation of the Drosophila IκB homologue Cactus in a manner thought to resemble IκBα degradation followed by nuclear translocation of two Drosophila Rel proteins (Dif and Dorsal) (Reach et al., 1996; Wu & Anderson, 1998; Meng et al., 1999; Rutschmann et al., 2000a). Activation of the antibacterial pathway leads to cleavage of another Drosophila Rel protein (Relish), a p100/p105 homologue (Hedengren et al., 1999; Rutschmann et al., 2000b).

By analogy with mammalian systems, signal-dependent serine phosphorylation is regarded to be a key step of NF-κB activation in Drosophila as well. However, a kinase activity responsible for Cactus degradation has not been found so far (Karin & Ben-Neriah, 2000; Silverman & Maniatis, 2001). There is one IKK-related kinase (DmIKKβ), encoded in the Drosophila genome, that is required for the antibacterial signaling pathway (Kim et al., 2000; Silverman et al., 2000). This protein has little homology to the human IKKα or IKKβ in its C-terminus, but its kinase domain is highly similar to that of the human enzymes (Silverman & Maniatis, 2001). A high-molecular weight Drosophila IKK complex was shown to contain DmIKKβ and DmIKKγ, a homologue of mammalian IKKγ/NEMO (Silverman et al., 2000). Activated by LPS, this DmIKK complex is responsible for Relish phosphorylation, a prerequisite for Relish cleavage and activation, respectively, leading to antibacterial peptide gene transcription (Silverman et al., 2000). It was also shown that DmIKKβ phosphorylates Cactus at the N-terminal regulatory domain (Kim et al., 2000), but the significance of this activity is not fully understood. Interestingly, mutations in the DmIKKβ and DmIKKγ genes cause defects in antibacterial immunity, but do not affect the antifungal response (Rutschmann et al., 2000b; Lu et al., 2001). A comparison of mammalian and Drosophila IKK-related activation of NF-κB has revealed several similarities, but also some differences. Most interestingly, the Drosophila IKK complex, which contains only one catalytic subunit similar to both IKKα and IKKβ, does not phosphorylate the Drosophila IκBα homologue Cactus. Instead, it leads to activation of antibacterial genes through phosphorylation-induced processing of the Drosophila NFκB1/2 homologue Relish (Silverman et al., 2000; Lu et al., 2001). Whereas this pathway is required for innate immune responses in insects, in mammals it is important for adaptive immunity (Senftleben et al., 2001a). Consequently, both the duplication and functional divergence of IKK catalytic subunits correlate with the evolution of the immune system.

5. PHYSIOLOGICAL ROLE OF THE IKK COMPLEX

5.1. IKK function evaluated by genetic approaches

Although the molecular and biochemical aspects of the IKK/NF-κB-signaling pathway have been characterized in detail (Rothwarf & Karin, 1999; Karin & Ben-Neriah, 2000), evaluation of the role of the IKK complex in vivo is only at its beginning. Surprisingly, gene targeting experiments revealed that the catalytic IKKα and IKKβ subunits and the regulatory IKKγ play crucial, but entirely different roles in embryonic development and survival (Hu et al., 1999; Li et al., 1999a,b,c; Takeda et al., 1999; Tanaka et al., 1999). It was also learned from these experiments that, depending on the cell type, the IKK subunits have specific biological functions.

5.1.1. IKKα

IKKα-deficient mice show impaired limb and skeletal patterning associated with altered skin proliferation and differentiation (Hu et al., 1999; Li et al., 1999a; Takeda et al., 1999). These mice die within 1 h after birth, possibly due to impaired ventilation mechanics. IKKα has a unique function in skin and skeletal development which cannot be replaced by IKKβ because neither Ikk$\beta^{-/-}$ embryos (Li et al., 1999a,c; Tanaka et al., 1999) nor rescued Tnfr1$^{-/-}$Ikk$\beta^{-/-}$ double-mutant mice (Senftleben et al., 2001b) present any skin phenotype. Also none of the Rel/NF-κB single-mutant and double-mutant mice exhibit a skin phenotype comparable to IKKα-deficient newborn mice. Additionally, normal IKK activation and IκBα degradation were detectable in Ikk$\alpha^{-/-}$ mice and MEFs (Hu et al., 1999). Thus, IKKα is required for normal skin development via a mechanism that does not involve the canonical NF-κB pathway. Recently, it was found that during keratinocyte differentiation IKKα acts independently of NF-κB via a differentiation factor named KDIF (Hu et al., 2001).

Furthermore, it was shown that the IKKα subunit plays a key role during B-cell maturation. In radiation chimeras, IKKα is essential for the regular maturation of virgin IgM^{hi} to mature IgM^{lo}IgD^{hi} B-cells (Kaisho et al., 2001; Senftleben et al., 2001a). In contrast to the canonical NF-κB activity, which was unchanged in isolated Ikk$\alpha^{-/-}$ B-cells following stimulation with LPS, supershift analysis revealed a decreased level of p52-containing NF-κB complexes (Senftleben et al., 2001a). This was associated with a reduced transcription of NF-κB target genes, such as MIP1α or inducible nitric oxide synthetase (iNOS), whereas other target genes, such as TNFα or cyclin D2, were not affected (Senftleben et al., 2001a). Obviously, virgin B-cells depend on the protein kinase activity of IKKα to produce significant amounts of p52 for the transcription of specific NF-κB target genes (Senftleben et al., 2001a), which are responsible for B-cell maturation. A more detailed investigation showed that IKKα promotes the processing of the precursor NF-κB protein p100 to the mature p52 subunit (Senftleben et al., 2001a).

New data also revealed a role of IKKα during the proliferation of mammary epithelial cells (Cao et al., 2001). Ikk^{AA} knockin mice, which express catalytically inactive, but structurally intact IKKα, displayed a severe lactation defect. This phenotype is due to reduced NF-κB activity in response to RANK signaling and impaired cyclin D1 induction (Cao et al., 2001). These studies are good examples of the complex role of IKK, which is not only responsible for activation of NF-κB, but also determines which type of NF-κB complex is activated.

5.1.2. IKKβ

By contrast to IKKα, the IKKβ subunit plays an essential and dominant role in NF-κB activation upon stimulation by proinflammatory stimuli. Similar to $Rela^{-/-}$ mice (Beg et al., 1995; Doi et al., 1999; Alcamo et al., 2001), $Ikk\beta^{-/-}$ mice die at E12.5 from massive liver apoptosis, but survive after the deletion of TNF-R1 (Li et al., 1999b-c; Tanaka et al., 1999; Senftleben et al., 2001b). These experiments indicate that up to midgestation the most critical function of IKKβ is the activation of RelA-containing NF-κB complexes in response to TNFα in fetal hepatocytes. Biochemical analysis of $Ikk\beta^{-/-}$ MEFs reveals a dramatic defect in IKK and NF-κB activation in response to proinflammatory stimuli (Li et al., 1999c). Again IKKβ radiation chimeras revealed the function of IKKβ in hematopoietic cells (Senftleben et al., 2001b). Thus, erythrocytes and monocytes/granulocytes were generated, whereas a complete absence of B- and T-cells, as well as of apoptotic thymic rudiments, was observed (Senftleben et al., 2001b). Interestingly, B- and T-lymphopoiesis could be completely restored after reconstitution of lethally irradiated hosts with $Tnfr1^{-/-}Ikk\beta^{-/-}$ double-mutant hematopoietic stem cells. Also, normal thymic tissue was formed in these mice (Senftleben et al., 2001b). In conclusion, these results separate an antiapoptotic function of IKKβ in early precursors of lymphocytes compared to IKKα, which is an essential kinase for the late maturation steps in B-cells.

5.1.3. IKKγ

Similar to IKKβ-deficient mice an even earlier lethality due to liver degeneration was observed in embryos that lack the regulatory subunit IKKγ/NEMO (Makris et al., 2000; Rudolph et al., 2000), which could be prevented in $Tnfr1^{-/-}Ikk\gamma^{-/-}$ embryos (C. Makris, personal communication). Thus, the phenotypes of these mice indicate related biochemical and biological functions in vivo. Indeed, the β- and γ-subunits work together to phosphorylate IκBα in response to various stimuli. The block of this pathway in $Ikk\gamma^{-/-}$ as well as in $Ikk\beta^{-/-}$ MEFs results in defective IKK and NF-κB activation (Li et al., 1999c; Makris et al., 2000; Rudolph et al., 2000). Consequently, such cells display defective induction of various cytokine and chemokine genes (Chu et al., 1999). $Ikk\gamma^{-/-}$ radiation chimeras were not described so far. However, chimeric mice generated by injection of ES cells carrying a deleted Ikkγ/Nemo allele into blastocysts did not produce IKKγ-deficient B- or T-cells (Schmidt-Supprian et al., 2000), a similar defect to what was detected in IKKβ- and RelA-deficient radiation chimeras

(Horwitz et al., 1997; Senftleben et al., 2001b). Thus, like IKKβ, IKKγ is required for cell survival.

As IKKγ is an X-linked gene, hemizygote IKKγ/NEMO female mice develop a dermatopathy similar to the human disorder incontinentia pigmenti (Makris et al., 2000; Schmidt-Supprian et al., 2000), a genodermatosis characterized by a sequence of skin inflammation, hyperkeratosis and increased apoptosis (Makris et al., 2000; Schmidt-Supprian et al., 2000). Interestingly, this transient phenotype appears in previously normal newborn mice in about 4 days after birth and thus might be the effect of additional environmental factors like pathogens or stress factors on mutant X-chromosome expressing IKKγ-deficient cells. Another explanation is an expansion of X-inactivated mutant cells due to the absence of proliferation-inhibiting NF-κB activity (Seitz et al., 1998) followed by inflammation and, finally, apoptosis of the mutant cells (Makris et al., 2000).

In conclusion, none of the phenotypes of $Ikk\beta^{-/-}$, $Ikk\gamma^{-/-}$ or $Rela^{-/-}$ mice and radiation chimeras is comparable to IKKα-deficient mice and $Ikka^{-/-}$ radiation chimeras or to Ikk^{AA} knockin mice. However, certain defects detected in $Ikka^{-/-}$ radiation chimeras (Senftleben et al., 2001a), such as impaired formation of secondary lymphoid organs or disturbed B-cell maturation, were also seen in NF-κB2-deficient mice (Caamano et al., 1998; Franzoso et al., 1998). A similar defective lymphoid organ development was also seen in $Nik^{-/-}$ (Yin et al., 2001), $Nik^{aly/aly}$ (Shinkura et al., 1999; Fagarasan et al., 2000) and $Lt\beta r^{-/-}$ (Futterer et al., 1998) mice, whereas additional impaired B-cell maturation could be detected in BLyS (BAFF)-deficient mice (Schiemann et al., 2001) and BLyS receptor (BLyS-R/BAFF-R)-deficient mice (Thompson et al., 2001). Thus, the separation of the canonical IKKβ/NF-κB pathway and the α-specific pathway described above seems to be plausible, not only at the biochemical, but also at the physiological level. Although not formally proven, it appears that the canonical NF-κB pathway, which is based on IKKβ/γ-dependent IκB degradation, is essential for innate immunity, whereas the IKKα-specific pathway, which is based on NF-κB2 processing, is mostly involved in lymphoid organ development and adaptive immunity.

5.2. Comparison of IKK and Rel knockouts

Interestingly, several Rel knockout models reveal similarities to mice or radiation chimeras deficient in components of the IKK complex. Most often, proliferation or functional defects of the lymphopoietic system was described (Gerondakis et al., 1999). They also exhibited disturbances of the innate immune system. Thus, granulocytosis was detected in mice deficient for RelB (Weih et al., 1995), $p50^{+}RelB^{-}$ (Weih et al., 1997), $RelA^{+}c\text{-}Rel$ (Grossmann et al., 1999) and $p50^{+}p52^{-}$ (Franzoso et al., 1997), as well as in radiation chimeras generated by transfer of $p50^{+}p65^{-}$ (Horwitz et al., 1997) and IKKβ-deficient (Senftleben et al., 2001b) stem cells. In addition, increased granulopoiesis was also found in mice lacking IκBα (Klement et al., 1996). Those mice also exhibit widespread inflammatory dermatitis and are characterized by increased activation of NF-κB associated with elevated mRNA levels of target genes in splenocytes (Klement et al., 1996). Other rel-deficient mice fail to resist the challenge with pathogens such as Listeria monocytogenes or Streptococcus pneumoniae (Attar et al., 1997). Reduced clearance of Listeria monocytogenes was also found in mice deficient for RelB (Weih et al., 1995). Interestingly, these mice also exhibit multi-organ inflammation which is T-cell-dependent. $Rel\beta^{-/-}$ mice insufficiently eliminate autoreactive thymocytes and their splenocytes generate autoreactive responses (Weih et al., 1995; Laufer et al., 1996). These results suggest that autoreactive RelB-deficient T-cells promote inflammatory infiltrates. The severity of this phenotype was increased by additional disruption of NF-κB1 (Weih et al., 1997).

These defects usually are interpreted as directly caused by intrinsic dysregulation or malfunction of the IKK/NF-κB pathway. However, presently it cannot be ruled out that these phenotypes evolve secondarily in response to exogenous factors, such as pathogens or physico-chemical stress. Indeed, granulocytosis or dermatitis can be the consequence of persistent latent infection with opportunistic organisms on the background of an impaired innate or adaptive immune system, as it was proposed for IKK/NF-κB deficiency (Kopp & Ghosh, 1995; Ghosh et al., 1998; Sha, 1998).

In this context it is also interesting that $Tnfr1^{-/-}Ikk\beta^{-/-}$ double-mutant mice, which cannot activate NF-κB, exhibit multi-organ inflammation and die within 14 days after birth, possibly due to infection (U. Senftleben, unpublished results), whereas adoptive transfer of $Ikk\beta^{-/-}$ or $Tnfr1^{-/-}Ikk\beta^{-/-}$ hematopoietic stem cells into lethally irradiated hosts results in apparently healthy mice (Senftleben et al., 2001b) similar to $Tnfr1^{-/-}Rela^{-/-}$ radiation chimeras (Alcamo et al., 2001). It has to be emphasized that LPS-treated TNF-R1/RelA double-mutant young mice show a reduced migration of neutrophils into lung tissue compared with wild-type or $Tnfr1^{-/-}$ mice, whereas $Tnfr1^{-/-}Rela^{-/-}$ radiation chimeras exhibit a normal neutrophilic recruitment (Alcamo et al., 2001). Interestingly, reconstitution of $Tnfr1^{-/-}Rela^{-/-}$ mice with wild-type myeloid and lymphoid cells did not reduce the susceptibility of the host to infections (Alcamo et al., 2001). Collectively, these results suggest that IKKβ- and p65/RelA-containing NF-κB dimers are necessary for innate immune responses. It can be assumed that in epithelial surfaces, the first lines of defence, an intact NF-κB pathway is crucial to protect the organism from invading pathogens. Consecutively, damaged barriers require a regularly working NF-κB pathway in immunocompetent cells in order to control pathogenic invasions and to prevent infections. Nevertheless, the role of IKK/NF-κB in adaptive immune responses seems to be of a more complex nature. There is growing evidence that Rel proteins are involved in antibody production and antibody class switching, in differentiation, activation and proliferation of B- and T-cells, as well as in organization and formation of secondary lymphoid organs (Sha, 1998). However, tissue-specific functions of the IKK subunits as well as expression patterns of distinct Rel proteins, such as p52 and RelB in antigen-presenting cells (Sha, 1998), furthermore suggest an important role of the

IKK/NF-κB pathway to integrate information of both innate and adaptive immunity. Consequently, the importance of IKK/NF-κB as a central regulator of the immune system is evident; however, its exact role, especially during diseases, remains to be elucidated.

5.3. IKK and the cellular stress response

IKK/NF-κB is not only involved in the immediate immune response, but also controls the transcription of many genes which are necessary for the cell to respond to stress. The IKK complex is considered to be the key enzyme of NF-κB activation. In addition, a main feature of IKK/NF-κB activation is the prompt induction of this system, which makes this pathway an ideal mechanism to quickly respond to a variety of harmful stress factors, infections or injuries (Karin, 1999). Thus, there are many inducers of the NF-κB pathway via activation of IKK, which are not of viral or bacterial origin, such as tissue ischemia-reperfusion, hyperosmolarity, ionizing radiation and oxidative stress (Pahl, 1999). Conclusively, following tissue damage, inflammatory defence mechanisms immediately respond to induce tissue repair and to eliminate invading pathogens. The IKK/NF-κB pathway is likely to play a central role in this scenario (Barnes & Karin, 1997; Blackwell & Christman, 1997) via transcription of stress response genes such as iNOS (Geller et al., 1993), cyclooxygenase-2 (COX-2) (Yamamoto et al., 1995) or phospholipase A2 (PLA2) (Morri et al., 1994). The dynamic behavior of this pathway strongly suggests a central coordinating role for IKK in the response to acute insults (Senftleben & Karin, 2002). However, data about the importance of the IKK complex and its subunits with regard to cellular or tissue stress are scarce. The rapidly increasing IKKβ activity in the ischemic heart in vivo (Li et al., 2001) has to be considered suggestive for the roles of IKK subunits during the stress response.

5.4. IKK and programmed cell death

A long known, but increasingly important cellular mechanism associated with NF-κB signaling is the inhibition of apoptosis. Although there is limited evidence for a cell type-specific proapoptotic function for NF-κB (Hettmann et al., 1999), it is generally accepted that the IKK/NF-κB pathway is mostly acting to prevent the induction of programmed cell death by a variety of proapoptotic stimuli (Sonenshein, 1997; Hu et al., 2001). In fact, the antiapoptotic activity of NF-κB is tightly linked to its essential proinflammatory activity and depends on gene induction (Liu et al., 1996; Karin & Lin, 2002). The inhibition of TNFα-induced programmed cell death is well studied and is likely to be a major consequence of NF-κB activation (Beg & Baltimore, 1996; Wang et al., 1996; Wu et al., 1996; Alcamo et al., 2001). Through NF-κB, the IKK complex plays a major role in preventing TNFα-induced apoptosis. Thus, IKKβ- as well as IKKγ-deficient embryos die of liver apoptosis, which can be inhibited by deletion of the Tnfr1 gene (Senftleben et al., 2001b; C. Makris, personal communication). There is evidence that hepatic apoptosis is due to the intense production of TNFα by hematopoietic progenitor cells that populate the fetal liver (Rosenfeld et al., 2000). Adoptive transfer experiments of Ikk$\beta^{-/-}$ hematopoietic stem cells revealed a similar antiapoptotic function for the β-subunit of the IKK complex in lymphocytes (Senftleben et al., 2001b). IKKβ-deficient thymic rudiments were full of apoptotic cells, whereas Tnfr1$^{-/-}$Ikk$\beta^{-/-}$ hematopoietic stem cells give rise to normal thymocytes (Senftleben et al., 2001b). These experiments indicate an essential antiapoptotic function of IKKβ to prevent TNFα-induced apoptosis in T-cells (Senftleben et al., 2001b). Similar apoptotic rates could be detected in thymus and spleen of female mice heterozygous for the X-linked IKKγ deficiency (Makris et al., 2000). Thus, the antiapoptotic function of the IKKβ and IKKγ subunits might be an essential mechanism in lymphocytes to maintain immune homeostasis.

Taken into consideration that the majority of the experimental data regarding the antiapoptotic function of the NF-κB pathway are derived from RelA- or IKKβ/γ-deficient mice and cells, or involve the inhibition of NF-κB activity (Wu et al., 1996; Sonenshein, 1997), the role of IKKα with respect to programmed cell death is not clear. In IKKα-deficient mice increased apoptosis could not be detected in any of the investigated tissues (Hu et al., 1999). Also, T-lymphopoiesis of Ikk$\alpha^{-/-}$ radiation chimeras did not show a substantial proliferation or differentiation defect in vivo, whereas isolated resting IKKα-deficient B-cells derived from these mice exhibited increased apoptotic rates in vitro (Senftleben et al., 2001b). However, transgenic expression of bcl-2 could only partially rescue the maturation defect of B-cells in Ikk$\alpha^{-/-}$ radiation chimeras (Kaisho et al., 2001). Thus, it remains to be elucidated whether programmed cell death can be prevented via the IKKα-specific pathway as well, and if so, whether p52-containing NF-κB dimers activate the same antiapoptotic genes as those induced by the p50-RelA dimers.

6. THE IKK COMPLEX IN HUMAN DISEASES

Inappropriate regulation of NF-κB is considered to be directly involved in a wide range of human diseases, including a variety of cancers (Karin et al., 2002), neuro-degenerative diseases (Grilli & Memo, 1999), ataxia telangiectasia (Lee et al., 1998), arthritis (Foxwell et al., 1998), asthma (Barnes & Adcock, 1998), inflammatory bowel disease (Neurath et al., 1998), sepsis/systemic inflammatory response syndrome (SIRS) (Senftleben & Karin, 2002) and numerous other inflammatory or degenerative diseases. However, up to now, the knowledge about the role of the IKK complex in human diseases is limited. As key element of NF-κB activation, IKK most likely is also involved in diseases with significant NF-κB activation. To avoid duplication of remarks on NF-κB-associated diseases elsewhere in this book, in this section diseases will be discussed which, based on experimental data, clearly depend on activated or dysregulated IKK.

The human genodermatosis incontinentia pigmenti (IP) is a prime example for the involvement of a defective IKK complex in diseases. It was observed in parallel by three groups that deletions or mutations in the X-linked Ikkγ gene, which result in impaired IKK and NF-κB activation, are associated with the histological and clinical pattern of IP in heterozygous female mice (Makris et al., 2000; Schmidt-Supprian et al., 2000) and humans (Smahi et al., 2000). Female heterozygous IP patients show a succession of erythema, grouped vesicles, hyperkeratotic papules leading to a bizarre hyperpigmented striped skin pattern. Male embryos die in utero. Both, in mice and man, IP is a self-limiting disease. In particular no signs of immunodeficiency could be detected. This temporary disease pattern might result from a biological disadvantage of cells which express the mutant Ikkγ allele and undergo apoptosis due to defective NF-κB activation. As soon as all those cells are eliminated, the disease is resolved. There are reports of sporadic surviving males in IP families (Aradhya et al., 2001b; Mansour et al., 2001). Interestingly, survival of male IP patients is associated with specific mutations leading to diminished, but not totally defective function of IKKγ and NF-κB, respectively (Aradhya et al., 2001b; Mansour et al., 2001). Recently, a detailed genetic analysis of IKKγ/NEMO mutations causing IP in humans has been published (Aradhya et al., 2001a-c).

In this context, another X-linked disease that shows impaired, but not completely defective NF-κB activation that is linked to IKKγ, provides additional evidence about the role of IKK in diseases. It is called hypo- or anhidrotic ectodermal dysplasia with immunodeficiency (EDA-ID). The association with a mutation of the Ikkγ gene was described by several groups (Zonana et al., 2000; Doffinger et al., 2001; Kosaki et al., 2001). EDA is a developmental syndrome characterized by absence of sweat glands, spars scalp hair and rare conical teeth (Pinheiro & Freire-Maia, 1994). EDA-ID types are known to represent a separate genetic entity caused by mutations of the Ikkγ gene leading to defects of NF-κB activation (Zonana et al., 2000; Doffinger et al., 2001). Genetic analysis revealed a subtype of EDA-ID, clinically identified by additional osteopetrosis and lymphedema (OL-EDA-ID), that is caused by stop codon mutations in the Ikkγ gene (Doffinger et al., 2001). Interestingly, cellular responses upon stimulation with NF-κB activators were studied in one OL-EDA-ID patient. Thus, stimulation of fibroblasts and peripheral mononuclear blood cells (PMBC) with either TNFα or with LPS, IL-1β + IL-12 and IL-18 + IL-12, respectively, revealed reduced NF-κB DNA-binding activity in fibroblasts, and impaired TNFα and IFNγ production in PMBCs (Doffinger et al., 2001). Furthermore, all studied patients were highly susceptible to infectious diseases with the highest grade presented by the two OL-EDA-ID patients (Doffinger et al., 2001). Those often lethal infections were mainly caused by gram-positive and gram-negative bacteria and mycobacteria, whereas viruses, fungi and protozoa were of minor importance (Doffinger et al., 2001).

It is most interesting to compare NF-κB activation in cells of IP and EDA-ID patients. Obviously, a basal level of inducible NF-κB is necessary to protect cells from apoptosis, whereas full NF-κB activation is required for sufficient immune responses in vivo. Thus, distinct mutations of the Ikkγ gene, resulting in varying degrees of defective NF-κB activation, determine phenotypes and the extent of immune incompetence. These studies provide significant evidence for a nonredundant role of IKK/NF-κB signaling during human immune responses. Especially the ability of mammalian organisms to fight bacterial infections seems to crucially depend on intact IKK-dependent NF-κB activation. Both this fact and the anti-apoptotic function of NF-κB need to be considered when inhibitory agents of the IKK/NF-κB pathway are developed for therapeutic purposes.

7. REFERENCES

Alcamo E, Mizgerd JP, Horwitz BH, Bronson R, Beg AA, Scott M, Doerschuk CM, Hynes RO & Baltimore D (2001) Targeted mutation of TNF receptor I rescues the RelA-deficient mouse and reveals a critical role for NF-κB in leukocyte recruitment. J. Immunol. 167: 1592-1600.

Aradhya S, Bardaro T, Galgoczy P, Yamagata T, Esposito T, Patlan H, Ciccodicola A, Munnich A, Kenwrick S, Platzer M, D'Urso M & Nelson DL (2001a) Multiple pathogenic and benign genomic rearrangements occur at a 35 kb duplication involving the NEMO and LAGE2 genes. Hum. Mol. Genet. 10: 2557-2567.

Aradhya S, Courtois G, Rajkovic A, Lewis RA, Levy M, Israel A & Nelson DL (2001b) Atypical forms of incontinentia pigmenti in male individuals result from mutations of a cytosine tract in exon 10 of NEMO (IKK-γ) Am. J. Hum. Genet. 68: 765-771.

Aradhya S, Woffendin H, Jakins T, Bardaro T, Esposito T, Smahi A, Shaw C, Levy M, Munnich A, D'Urso M, Lewis RA, Kenwrick S & Nelson DL (2001c) A recurrent deletion in the ubiquitously expressed NEMO (IKK-γ) gene accounts for the vast majority of incontinentia pigmenti mutations. Hum. Mol. Genet. 10: 2171-2179.

Attar RM, Caamano J, Carrasco D, Iotsova V, Ishikawa H, Ryseck RP, Weih F & Bravo R (1997) Genetic approaches to study Rel/NF-κB/IκB function in mice. Semin. Cancer Biol. 8: 93-101.

Barnes PJ & Adcock IM (1998) Transcription factors and asthma. Eur. Respir. J. 12: 221-234.

Barnes PJ & Karin M (1997) Nuclear factor-κB: a pivotal transcription factor in chronic inflammatory diseases. N. Engl. J. Med. 336: 1066-1071.

Beg A & Baltimore D (1996) An essential role for NF-κB in preventing TNF-α-induced cell death. Science 274: 782-784.

Beg A, Sha WC, Bronson RT, Ghosh S & Baltimore D (1995) Embryonic lethality and liver degeneration in mice lacking the RelA component of NF-κB. Nature 376: 167-170.

Blackwell TS & Christman JW (1997) The role of nuclear factor-κB in cytokine gene regulation. Am. J. Respir. Cell Mol. Biol. 17: 3-9.

Bonnard M, Mirtsos C, Suzuki S, Graham K, Huang J, Ng M, Itie A, Wakeham A, Shahinian A, Henzel WJ, Elia AJ, Shillinglaw W, Mak TW, Cao Z & Yeh WC (2000) Deficiency of T2K leads to apoptotic liver degeneration and impaired NF-κB-dependent gene transcription. EMBO J. 19: 4976-4985.

Bonnet MC, Weil R, Dam E, Hovanessian AG & Meurs EF (2000) PKR stimulates NF-κB irrespective of its kinase function by interacting with the IκB kinase complex. Mol. Cell Biol. 20: 4532-4542.

Caamano JH, Rizzo CA, Durham SK, Barton DS, Raventos-Suarez C, Snapper CM & Bravo R (1998) Nuclear factor (NF)-κB2 (p100/p52) is required for normal splenic microarchitecture and B cell-mediated immune responses. J. Exp. Med. 187: 185-196.

Cao Y, Bonizzi G, Seagroves TN, Greten FR, Johnson R, Schmidt EV & Karin M (2001) IKKα provides an essential link between RANK signaling and cyclin D1 expression during mammary gland development. Cell 107: 763-775.

Carter RS, Geyer BC, Xie M, Acevedo-Suarez CA & Ballard DW (2001) Persistent activation of NF-

κB by the tax transforming protein involves chronic phosphorylation of IκB kinase subunits IKKβ and IKKγ. J. Biol. Chem. 276: 24445-24448.

Chu WM, Ostertag D, Li ZW, Chang LF, Chen Y, Hu YL, Williams B, Perrault J & Karin M (1999) JNK2 and IKKβ are required for activating the innate response to viral infection. Immunity 11: 721-731.

Cohen L, Henzel WJ & Baeuerle PA (1998) IKAP is a scaffold protein of the IκB kinase complex. Nature 395: 292-296.

Delhase M, Hayakawa M, Chen Y & Karin M (1999) Positive and negative regulation of IκB kinase activity through IKKβ subunit phosphorylation. Science 284: 309-313.

Deng L, Wang C, Spencer E, Yang L, Braun A, You J, Slaughter C, Pickart C & Chen ZJ (2000) Activation of the IκB kinase complex by TRAF6 requires a dimeric ubiquitin-conjugating enzyme complex and a unique polyubiquitin chain. Cell 103: 351-361.

Devin A, Cook A, Lin Y, Rodriguez Y, Kelliher M & Liu Z (2000) The distinct roles of TRAF2 and RIP in IKK activation by TNF-R1: TRAF2 recruits IKK to TNF-R1 while RIP mediates IKK activation. Immunity 12: 419-429.

DiDonato J, Hayakawa M, Rothwarf DM, Zandi E & Karin M (1997) A cytokine-responsive IκB kinase that activates the transkription factor NFκB. Nature 388: 548-554.

Doffinger R, Smahi A, Bessia C, Geissmann F, Feinberg J, Durandy A, Bodemer C, Kenwrick S, Dupuis-Girod S, Blanche S, Wood P, Rabia SH, Headon DJ, Overbeek PA, Le Deist F, Holland SM, Belani K, Kumararatne DS, Fischer A, Shapiro R, Conley ME, Reimund E, Kalhoff H, Abinun M, Munnich A, Israel A, Courtois G & Casanova JL (2001) X-linked anhidrotic ectodermal dysplasia with immunodeficiency is caused by impaired NF-κB signaling. Nat. Genet. 27: 277-285.

Doi TS, Marino MW, Takahashi T, Yoshida T, Sakakura T, Old LJ & Obata Y (1999) Absence of tumor necrosis factor rescues RelA-deficient mice from embryonic lethality. Proc. Natl. Acad. Sci. USA 96: 2994-2999.

Dumitru CD, Ceci JD, Tsatsanis C, Kontoyiannis D, Stamatakis K, Lin JH, Patriotis C, Jenkins NA, Copeland NG, Kollias G & Tsichlis PN (2000) TNF-α induction by LPS is regulated posttranscriptionally via a Tpl2/ERK-dependent pathway. Cell 103: 1071-1083.

Fagarasan S, Shinkura R, Kamata T, Nogaki F, Ikuta K, Tashiro K & Honjo T (2000) Alymphoplasia (aly)-type nuclear factor κB-inducing kinase (NIK) causes defects in secondary lymphoid tissue chemokine receptor signaling and homing of peritoneal cells to the gut-associated lymphatic tissue system. J. Exp. Med. 191: 1477-1486.

Foxwell B, Browne K, Bondeson J, Clarke C, de Martin R, Brennan F & Feldmann M (1998) Efficient adenoviral infection with IκBα reveals that macrophage tumor necrosis factor α production in rheumatoid arthritis is NF-κB dependent. Proc. Natl. Acad. Sci. USA 95: 8211-8215.

Franzoso G, Carlson L, Poljak L, Shores EW, Epstein S, Leonardi A, Grinberg A, Tran T, Scharton-Kersten T, Anver M, Love P, Brown K & Siebenlist U (1998) Mice deficient in nuclear factor (NF)-κB/p52 present with defects in humoral responses, germinal center reactions, and splenic microarchitecture. J. Exp. Med. 187: 147-159.

Franzoso G, Carlson L, Xing LP, Poljak L, Shores EW, Brown KD, Leonardi A, Tran T, Boyce BF & Siebenlist U (1997) Requirement for NF-κB in osteoclast and B-cell development. Genes Dev. 11: 3482-3496.

Futterer A, Mink K, Luz A, Kosco-Vilbois MH & Pfeffer K (1998) The lymphotoxin β receptor controls organogenesis and affinity maturation in peripheral lymphoid tissues. Immunity 9: 59-70.

Gaide O, Martinon F, Micheau O, Bonnet D, Thome M & Tschopp J (2001) Carma1, a CARD-containing binding partner of Bcl10, induces Bcl10 phosphorylation and NF-κB activation. FEBS Lett. 496: 121-127.

Geller DA, Nussler AK, Di Silvio M, Lowenstein CJ, Shapiro RA, Wang SC, Simmons RL & Billiar TR (1993) Cytokines, endotoxin, and glucocorticoids regulate the expression of inducible nitric oxide synthase in hepatocytes. Proc. Natl. Acad. Sci. USA 90: 522-526.

Gerondakis S, Grossmann M, Nakamura Y, Pohl T & Grumont R (1999) Genetic approaches in mice to understand Rel/NF-κB and IκB function: transgenics and knockouts. Oncogene 18: 6888-6895.

Ghosh S & Karin M (2002) Missing pieces in the NF-κB puzzle. Cell 109: S81-S96.

Ghosh S, May MJ & Kopp EB (1998) NF-κB and rel proteins: evolutionarily conserved mediators of immune responses. Annu. Rev. Immunol. 16: 225-260.
Grilli M & Memo M (1999) Nuclear factor-κB/Rel proteins: a point of convergence of signalling pathways relevant in neuronal function and dysfunction. Biochem. Pharmacol. 57: 1-7.
Grossmann M, Metcalf D, Merryfull J, Beg A, Baltimore D & Gerondakis S (1999) The combined absence of the transcription factors Rel and RelA leads to multiple hemopoietic cell defects. Proc. Natl. Acad. Sci. USA 96: 11848-11853.
Hedengren M, Asling B, Dushay MS, Ando I, Ekengren S, Wihlborg M & Hultmark D (1999) Relish, a central factor in the control of humoral but not cellular immunity in Drosophila. Mol. Cell 4: 827-837.
Heilker R, Freuler F, Pulfer R, Di Padova F & Eder J (1999) All three IκB isoforms and most Rel family members are stably associated with the IκB kinase 1/2 complex. Eur. J. Biochem. 259: 253-261.
Hettmann T, DiDonato J, Karin M & Leiden JM (1999) An essential role for nuclear factor κB in promoting double positive thymocyte apoptosis. J. Exp. Med. 189: 145-157.
Hong X, Xu L, Li X, Zhai Z & Shu H (2001) CSN3 interacts with IKKγ and inhibits TNF- but not IL-1-induced NF-κB activation. FEBS Lett. 499: 133-136.
Horwitz BH, Scott ML, Cherry SR, Bronson RT & Baltimore D (1997) Failure of lymphopoiesis after adoptive transfer of NF-κB-deficient fetal liver cells. Immunity 6: 765-772.
Hu Y, Baud V, Oga T, Kim KI, Yoshida K & Karin M (2001) IKKα controls formation of the epidermis independently of NF-κB. Nature 410: 710-714.
Hu YL, Baud V, Delhase M, Zhang PL, Deerinck T, Ellisman M, Johnson R & Karin M (1999) Abnormal morphogenesis but intact IKK activation in mice lacking the IKKα subunit of IκB kinase. Science 284: 316-320.
Johnson LN, Noble ME & Owen DJ (1996) Active and inactive protein kinases: structural basis for regulation. Cell 85: 149-158.
Kaisho T, Takeda K, Tsujimura T, Kawai T, Nomura F, Terada N & Akira S (2001) IκB kinase α is essential for mature B cell development and function. J. Exp. Med. 193: 417-426.
Karin M (1999) The beginning of the end: IκB kinase (IKK) and NF-κB activation. J. Biol. Chem. 274: 27339-27342.
Karin M & Ben-Neriah Y (2000) Phosphorylation meets ubiquitination: the control of NF-κB activity. Annu. Rev. Immunol. 18: 621-663.
Karin M, Cao Y, Greten FR & Li ZW (2002) NF-κB in cancer: from innocent bystander to major culprit. Nat. Rev. Cancer 2: 301-310.
Karin M & Lin A (2002) NF-κB at the crossroads of life and death. Nat. Immunol. 3: 221-227.
Kelliher MA, Grimm S, Ishida Y, Kuo F, Stanger BZ & Leder P (1998) The death domain kinase RIP mediates the TNF-induced NF-κB signal. Immunity 8: 297-303.
Khoshnan A, Bae D, Tindell CA & Nel AE (2000) The physical association of protein kinase Cθ with a lipid raft-associated inhibitor of κB factor kinase (IKK) complex plays a role in the activation of the NF-κB cascade by TCR and CD28. J. Immunol. 165: 6933-6940.
Kim YS, Han SJ, Ryu JH, Choi KH, Hong YS, Chung YH, Perrot S, Raibaud A, Brey PT & Lee WJ (2000) Lipopolysaccharide-activated kinase, an essential component for the induction of the antimicrobial peptide genes in Drosophila melanogaster cells. J. Biol. Chem. 275: 2071-2079.
Klement JF, Rice NR, Car BD, Abbondanzo SJ, Powers GD, Bhatt H, Chen CH, Rosen CA & Stewart CL (1996) IκBα deficiency results in a sustained NF-κB response and severe widespread dermatitis in mice. Mol. Cell. Biol. 16: 2341-2349.
Kopp EB & Ghosh S (1995) NF-κB and rel proteins in innate immunity. Adv. Immunol. 58: 1-27.
Kosaki K, Shimasaki N, Fukushima H, Hara M, Ogata T & Matsuo N (2001) Female patient showing hypohidrotic ectodermal dysplasia and immunodeficiency (HED-ID) Am. J. Hum. Genet. 69: 664-666.
Laufer TM, DeKoning J, Markowitz JS, Lo D & Glimcher LH (1996) Unopposed positive selection and autoreactivity in mice expressing class II MHC only on thymic cortex. Nature 383: 81-85.
Lee SJ, Dimtchev A, Lavin MF, Dritschilo A & Jung M (1998) A novel ionizing radiation-induced

signaling pathway that activates the transcription factor NF-κB. Oncogene 17: 1821-1826.

Leonardi A, Chariot A, Claudio E, Cunningham K & Siebenlist U (2000) CIKS, a connection to IκB kinase and stress-activated protein kinase. Proc. Natl. Acad. Sci. USA 97: 10494-10499.

Li C, Kao RL, Ha T, Kelley J, Browder IW & Williams DL (2001) Early activation of IKKβ during in vivo myocardial ischemia. Am. J. Physiol. Heart Circ. Physiol. 280: H1264-H1271.

Li QT, Lu QX, Hwang JY, Buscher D, Lee KF, Izpisua-Belmonte JC & Verma IM (1999a) IKK1-deficient mice exhibit abnormal development of skin and skeleton. Genes Dev. 13: 1322-1328.

Li QT, Van Antwerp D, Mercurio F, Lee KF & Verma IM (1999b) Severe liver degeneration in mice lacking the IκB kinase 2 gene. Science 284: 321-325.

Li ZW, Chu WM, Hu YL, Delhase M, Deerinck T, Ellisman M, Johnson R & Karin M (1999c) The IKKβ subunit of IκB kinase (IKK) is essential for nuclear factor κB activation and prevention of apoptosis. J. Exp. Med. 189: 1839-1845.

Li X, Commane M, Burns C, Vithalani K, Cao Z & Stark GR (1999d) Mutant cells that do not respond to interleukin-1 (IL-1) reveal a novel role for IL-1 receptor-associated kinase. Mol. Cell. Biol. 19: 4643-4652.

Li Y, Kang J, Friedman J, Tarassishin L, Ye J, Kovalenko A, Wallach D & Horwitz MS (1999e) Identification of a cell protein (FIP-3) as a modulator of NF-κB activity and as a target of an adenovirus inhibitor of tumor necrosis factor α-induced apoptosis. Proc. Natl. Acad. Sci. USA 96: 1042-1047.

Ling L, Cao Z & Goeddel DV (1998) NF-κB-inducing kinase activates IKK-α by phosphorylation of Ser-176. Proc. Natl. Acad. Sci. USA 95: 3792-3797.

Liu ZG, Hsu H, Goeddel DV & Karin M (1996) Dissection of TNF receptor 1 effector functions: JNK activation is not linked to apoptosis while NF-κB activation prevents cell death. Cell 87: 565-576.

Lu Y, Wu LP & Anderson KV (2001) The antibacterial arm of the drosophila innate immune response requires an IκB kinase. Genes Dev. 15: 104-110.

Lucas PC, Yonezumi M, Inohara N, McAllister-Lucas LM, Abazeed ME, Chen FF, Yamaoka S, Seto M & Nunez G (2001) Bcl10 and MALT1, independent targets of chromosomal translocation in malt lymphoma, cooperate in a novel NF-κB signaling pathway. J. Biol. Chem. 276: 19012-19019.

Makris C, Godfrey VL, Kraehn-Senftleben G, Takahashi T, Roberts JL, Schwarz T, Feng L, Johnson RS & Karin M (2000) Female mice heterozygous for IKKγ/NEMO deficiencies develop a dermatopathy similar to the human X-linked disorder Incontinentia Pigmenti. Mol. Cell 5: 969-979.

Mansour S, Woffendin H, Mitton S, Jeffery I, Jakins T, Kenwrick S & Murday VA (2001) Incontinentia pigmenti in a surviving male is accompanied by hypohidrotic ectodermal dysplasia and recurrent infection. Am. J. Med. Genet. 99: 172-177.

Matsushima A, Kaisho T, Rennert PD, Nakano H, Kurosawa K, Uchida D, Takeda K, Akira S & Matsumoto M (2001) Essential role of nuclear factor (NF)-κB-inducing kinase and inhibitor of κB (IκB) kinase α in NF-κB activation through lymphotoxin β receptor, but not through tumor necrosis factor receptor I. J. Exp. Med. 193: 631-636.

May MJ, D'Acquisto F, Madge LA, Glockner J, Pober JS & Ghosh S (2000) Selective inhibition of NF-κB activation by a peptide that blocks the interaction of NEMO with the IκB kinase complex. Science 289: 1550-1554.

Meng X, Khanuja BS & Ip YT (1999) Toll receptor-mediated Drosophila immune response requires Dif, an NF-κB factor. Genes Dev. 13: 792-797.

Mercurio F, Murray BW, Shevchenko A, Bennett BL, Young DB, Li JW, Pascual G, Motiwala A, Zhu H, Mann M & Manning AM (1999) IκB kinase (IKK)-associated protein 1, a common component of the heterogeneous IKK complex. Mol. Cell. Biol. 19: 1526-1538.

Mercurio F, Zhu HY, Murray BW, Shevchenko A, Bennett BL, Li JW, Young DB, Barbosa M & Mann M (1997) IKK-1 and IKK-2: cytokine-activated IκB kinases essential for NF-κB activation. Science 278: 860-866.

Miller BS & Zandi E (2001) Complete reconstitution of human IκB kinase (IKK) complex in yeast. J. Biol. Chem. 276: 36320-36326.

Morri H, Ozaki M & Watanabe Y (1994) 5'-flanking region surrounding a human cytosolic phospho-

lipase A2 gene. Biochem. Biophys. Res. Commun. 205: 6-11.
Neurath MF, Becker C & Barbulescu K (1998) Role of NF-κB in immune and inflammatory responses in the gut. Gut 43: 856-860.
Pahl HL (1999) Activators and target genes of rel/NF-κB transcription factors. Oncogene 18: 6853-6866.
Pinheiro M & Freire-Maia N (1994) Ectodermal dysplasias: a clinical classification and a causal review. Am. J. Med. Genet. 53: 153-162.
Rao GN, Glasgow WC, Eling TE & Runge MS (1996) Role of hydroperoxyeicosatetraenoic acids in oxidative stress-induced activating protein 1 (AP-1) activity. J. Biol. Chem. 271: 27760-27764.
Reach M, Galindo RL, Towb P, Allen JL, Karin M & Wasserman SA (1996) A gradient of cactus protein degradation establishes dorsoventral polarity in the Drosophila embryo. Dev. Biol. 180: 353-364.
Regnier CH, Song HY, Gao X, Goeddel DV, Cao ZD & Rothe M (1997) Identification and characterization of an IκB kinase. Cell 90: 373-383.
Rosenfeld ME, Prichard L, Shiojiri N & Fausto N (2000) Prevention of hepatic apoptosis and embryonic lethality in RelA/TNFR-1 double knockout mice. Am. J. Pathol. 156: 997-1007.
Rothwarf DM & Karin M (1999) The NF-κB pathway: a paradigm in information transfer from membrane to nucleus. Science's STKE: www.stke.org/cgi/content/full/OC_sigtrans;1999/5/rel.
Rothwarf DM, Zandi E, Natoli G & Karin M (1998) IKK-γ is an essential regulatory subunit of the IκB kinase complex. Nature 395: 297-300.
Rudolph D, Yeh WC, Wakeham A, Rudolph B, Nallainathan D, Potter J, Elia AJ & Mak TW (2000) Severe liver degeneration and lack of NF-κB activation in NEMO/IKKγ-deficient mice. Genes Dev. 14: 854-862.
Rutschmann S, Jung AC, Hetru C, Reichhart JM, Hoffmann JA & Ferrandon D (2000a) The Rel protein DIF mediates the antifungal but not the antibacterial host defense in Drosophila. Immunity 12: 569-580.
Rutschmann S, Jung AC, Zhou R, Silverman N, Hoffmann JA & Ferrandon D (2000b) Role of Drosophila IKKγ in a toll-independent antibacterial immune response. Nat. Immunol. 1: 342-347.
Schiemann B, Gommerman JL, Vora K, Cachero TG, Shulga-Morskaya S, Dobles M, Frew E & Scott ML (2001) An essential role for BAFF in the normal development of B cells through a BCMA-independent pathway. Science 293: 2111-2114.
Schmidt-Supprian M, Bloch W, Courtois G, Addicks K, Israel A, Rajewsky K & Pasparakis M (2000) NEMO/IKKγ-deficient mice model incontinentia pigmenti. Mol. Cell 5: 981-992.
Seitz CS, Lin Q, Deng H & Khavari PA (1998) Alterations in NF-κB function in transgenic epithelial tissue demonstrate a growth inhibitory role for NF-κB. Proc. Natl. Acad. Sci. USA 95: 2307-2312.
Senftleben U, Cao Y, Xiao G, Greten FR, Krahn G, Bonizzi G, Chen Y, Hu Y, Fong A, Sun SC & Karin M (2001a) Activation by IKKα of a second, evolutionary conserved, NF-κB signaling pathway. Science 293: 1495-1499.
Senftleben U, Li ZW, Baud V & Karin M (2001b) IKKβ is essential for protecting T cells from TNFα-induced apoptosis. Immunity 14: 217-230.
Senftleben U & Karin M (2002) The IKK/NF-κB pathway. Crit. Care Med. 30: S18-S26.
Sha WC (1998) Regulation of immune responses by NF-κB/Rel transcription factors. J. Exp. Med. 187: 143-146.
Shinkura R, Kitada K, Matsuda F, Tashiro K, Ikuta K, Suzuki M, Kogishi K, Serikawa T & Honjo T (1999) Alymphoplasia is caused by a point mutation in the mouse gene encoding NF-κB-inducing kinase. Nat. Genet. 22: 74-77.
Silverman N & Maniatis T (2001) NF-κB signaling pathways in mammalian and insect innate immunity. Genes Dev. 15: 2321-2342.
Silverman N, Zhou R, Stoven S, Pandey N, Hultmark D & Maniatis T (2000) A Drosophila IκB kinase complex required for Relish cleavage and antibacterial immunity. Genes Dev. 14: 2461-2471.
Smahi A, Courtois G, Vabres P, Yamaoka S, Heuertz S, Munnich A, Israel A, Heiss NS, Klauck SM, Kioschis P, Wiemann S, Poustka A, Esposito T, Bardaro T, Gianfrancesco F, Ciccodicola A,

D'Urso M, Woffendin H, Jakins T, Donnai D, Stewart H, Kenwrick SJ, Aradhya S, Yamagata T, Levy M, Lewis RA & Nelson DL (2000) Genomic rearrangement in NEMO impairs NF-κB activation and is a cause of incontinentia pigmenti. The International Incontinentia Pigmenti (IP) Consortium. Nature 405: 466-472.

Sonenshein GE (1997) Rel/NF-κB transcription factors and the control of apoptosis. Semin. Cancer Biol. 8: 113-119.

Sun Z, Arendt CW, Ellmeier W, Schaeffer EM, Sunshine MJ, Gandhi L, Annes J, Petrzilka D, Kupfer A, Schwartzberg PL & Littman DR (2000) PKC-θ is required for TCR-induced NF-κB activation in mature but not immature T lymphocytes. Nature 404: 402-407.

Tada K, Okazaki T, Sakon S, Kobarai T, Kurosawa K, Yamaoka S, Hashimoto H, Mak TW, Yagita H, Okumura K, Yeh WC & Nakano H (2001) Critical roles of TRAF2 and TRAF5 in tumor necrosis factor-induced NF-κB activation and protection from cell death. J. Biol. Chem. 276: 36530-36534.

Takeda K, Takeuchi O, Tsujimura T, Itami S, Adachi O, Kawai T, Sanjo H, Yoshikawa K, Terada N & Akira S (1999) Limb and skin abnormalities in mice lacking IKKα. Science 284: 313-316.

Tanaka M, Fuentes ME, Yamaguchi K, Durnin MH, Dalrymple SA, Hardy KL & Goeddel DV (1999) Embryonic lethality, liver degeneration, and impaired NF-κB activation in IKK-β-deficient mice. Immunity 10: 421-429.

Thompson JS, Bixler SA, Qian F, Vora K, Scott ML, Cachero TG, Hession C, Schneider P, Sizing ID, Mullen C, Strauch K, Zafari M, Benjamin CD, Tschopp J, Browning JL & Ambrose C (2001) BAFF-R, a newly identified TNF receptor that specifically interacts with BAFF. Science 293: 2108-2111.

Wang C, Deng L, Hong M, Akkaraju GR, Inoue J & Chen ZJ (2001) TAK1 is a ubiquitin-dependent kinase of MKK and IKK. Nature 412: 346-351.

Wang CY, Mayo MW & Baldwin AS (1996) TNF- and cancer therapy-induced apoptosis: potentiation by inhibition of NF-κB. Science 274: 784-787.

Weih F, Carrasco D, Durham SK, Barton DS, Rizzo CA, Ryseck RP, Lira SA & Bravo R (1995) Multiorgan inflammation and hematopoietic abnormalities in mice with a targeted disruption of RelB, a member of the NF-κB/Rel family. Cell 80: 331-340.

Weih F, Durham SK, Barton DS, Sha WC, Baltimore D & Bravo R (1997) p50-NF-κB complexes partially compensate for the absence of relB: severely increased pathology in $p50^{-/-}relB^{-/-}$ double-knockout mice. J. Exp. Med. 185: 1359-1370.

Woronicz JD, Gao X, Cao Z, Rothe M & Goeddel DV (1997) IκB kinase-β: NF-κB activation and complex formation with IκB kinase-α and NIK. Science 278: 866-869.

Wu LP & Anderson KV (1998) Regulated nuclear import of Rel proteins in the Drosophila immune response. Nature 392: 93-97.

Wu M, Lee HY, Bellas RE, Schauer SL, Arsura M, Katz D, FitzGerald MJ, Rothstein TL, Sherr DH & Sonenshein GE (1996) Inhibition of NF-κB/Rel induces apoptosis of murine B cells. EMBO J. 15: 4682-4690.

Xia Y, Makris C, Su B, Li E, Yang J, Nemerow GR & Karin M (2000) MEK kinase 1 is critically required for c-Jun N-terminal kinase activation by proinflammatory stimuli and growth factor-induced cell migration. Proc. Natl. Acad. Sci. USA 97: 5243-5248.

Xiao G, Harhaj EW & Sun SC (2001) NF-κB-inducing kinase regulates the processing of NF-κB2 p100. Mol. Cell 7: 401-409.

Yamamoto K, Arakawa T, Ueda N & Yamamoto S (1995) Transcriptional roles of nuclear factor κB and nuclear factor-interleukin-6 in the tumor necrosis factor α-dependent induction of cyclooxygenase-2 in MC3T3-E1 cells. J. Biol. Chem. 270: 31315-31320.

Yamaoka S, Courtois G, Bessia C, Whiteside ST, Weil R, Agou F, Kirk HE, Kay RJ & Israel A (1998) Complementation cloning of NEMO, a component of the IκB kinase complex essential for NF-κB activation. Cell 93: 1231-1240.

Yin L, Wu L, Wesche H, Arthur CD, White JM, Goeddel DV & Schreiber RD (2001) Defective lymphotoxin-β receptor-induced NF-κB transcriptional activity in NIK-deficient mice. Science 291: 2162-2165.

Yujiri T, Ware M, Widmann C, Oyer R, Russell D, Chan E, Zaitsu Y, Clarke P, Tyler K, Oka Y, Fanger GR, Henson P & Johnson GL (2000) MEK kinase 1 gene disruption alters cell migration and c-Jun NH_2-terminal kinase regulation but does not cause a measurable defect in NF-κB activation. Proc. Natl. Acad. Sci. USA 97: 7272-7277.

Zandi E, Chen Y & Karin M (1998) Direct phosphorylation of IκB by IKKα and IKKβ: discrimination between free and NF-κB-bound substrate. Science 281: 1360-1363.

Zandi E, Rothwarf DM, Delhase M, Hayakawa M & Karin M (1997) The IκB kinase complex (IKK) contains two kinase subunits, IKKα and IKKβ, necessary for IκB phosphorylation and NFκB activation. Cell 91: 243-252.

Zhang SQ, Kovalenko A, Cantarella G & Wallach D (2000) Recruitment of the IKK signalosome to the p55 TNF receptor: RIP and A20 bind to NEMO (IKKγ) upon receptor stimulation. Immunity 12: 301-311.

Zonana J, Elder ME, Schneider LC, Orlow SJ, Moss C, Golabi M, Shapira SK, Farndon PA, Wara DW, Emmal SA & Ferguson BM (2000) A novel X-linked disorder of immune deficiency and hypohidrotic ectodermal dysplasia is allelic to incontinentia pigmenti and due to mutations in IKK-γ (NEMO). Am. J. Hum. Genet. 67: 1555-1562.

8. AFFILIATIONS

University of Ulm, Clinic for Anesthesiology, Steinhövelstraße 9, 89075 Ulm, Germany (U. Senftleben). University of California, San Diego, Department of Pharmacology, Laboratory of Gene Regulation and Signal Transduction, 9500 Gilman Drive, La Jolla, CA 92093, USA (M. Karin).

L. DENG & Z.J. CHEN

THE ROLE OF UBIQUITIN IN NF-κB SIGNALING

1. INTRODUCTION

Ubiquitin is a 76-amino acid polypeptide that is ubiquitously expressed and highly conserved in all eukaryotes. The best characterized function of ubiquitin is to mark a target protein for degradation by the proteasome (Hershko & Ciechanover, 1998). This is achieved through covalent conjugation via an isopeptide bond between the C-terminus of ubiquitin and the ε-amino group of a lysine residue on the target protein. This conjugation, usually referred to as ubiquitination, is a three-step enzymatic process (Hershko & Ciechanover, 1992). In the first step, ubiquitin is activated by a ubiquitin-activating enzyme (E1) in an ATP-dependent reaction to form a high-energy thioester bond between the C-terminal glycine of ubiquitin and the catalytic cysteine of E1. In the second step, the activated ubiquitin is transferred to a ubiquitin-conjugating enzyme (E2). Finally, in the presence of a ubiquitin-protein ligase (E3), ubiquitin is conjugated to a lysine residue of the protein substrate through an isopeptide linkage. Ubiquitin itself has seven lysines. One of these lysines, typically Lys48, can be further conjugated by another ubiquitin to form a polyubiquitin chain through the processive action of the ubiquitination machinery (Chau et al., 1989). It is this polyubiquitin chain that is preferentially recognized by the proteasome, allowing the conjugated proteins to be degraded. Polyubiquitin chains linked through other lysines of ubiquitin also exist in cells, but the functions of these chains are largely unknown (Pickart, 2000). One exception is the K63-linked polyubiquitin chain, which has recently been shown to be involved in DNA repair (Spence et al., 1995; Hofmann & Pickart, 1999), stress response (Arnason & Ellison, 1994) and the activation of protein kinases, such as IκB kinase (IKK) (Deng et al., 2000). These polyubiquitin chains can be disassembled by a group of enzymes known as isopeptidases or deubiquitination enzymes (Dubs) (Wilkinson, 2000). Thus, ubiquitination/deubiquitination is a rapidly reversible form of covalent modification similar to phosphorylation/dephosphorylation or acetylation/deacetylation.

The genome sequences of several eukaryotes, including those of man, have revealed that there are one or two E1s, dozens of E2s (also known as Ubcs) and isopeptidases (or Dubs), and hundreds of E3s (von Arnim, 2001). Such a diversity of E2s and E3s dictates the specificity of protein ubiquitination and determines the vastly different stability of intracellular proteins. While E1, E2s and Dubs can be recognized by highly conserved domains that define these enzyme

R. Beyaert (ed.), Nuclear Factor κB. Regulation and Role in Disease, 137-158.

classes, E3s are structurally more diverse. Most of the E3s known so far contain either the HECT domain or the RING domain (Pickart, 2001). The HECT domain defines a family of proteins that are homologous to the C-terminus of E6AP, an E3 that binds to the human papilloma virus (HPV) E6 protein and mediates the ubiquitination of p53 (Scheffner et al., 1993; Huibregtse et al., 1995). Within the HECT domain there is a conserved cysteine residue that is capable of forming a thioester with ubiquitin and transferring ubiquitin to a target protein. Thus, the HECT E3s not only bind E2 and substrates, but also function catalytically in relaying ubiquitin to protein substrates (Scheffner et al., 1995). In contrast, none of the RING E3s are known to be capable of forming a thioester with ubiquitin. The RING domain is a cysteine- and histidine-rich motif present in a large number of proteins. Some of the RING E3s, such as Rbx1/Roc1, are the core components of multiprotein E3 complexes known as the SCF complexes, which consist of Skp1, Cul1 and an F-box protein that recognizes a specific protein substrate (Deshaies, 1999). For example, one of the first identified F-box proteins (Cdc4) binds specifically to the phosphorylated form of Sic1, an inhibitor of G1 cyclin-dependent kinase (CDK) (Feldman et al., 1997; Skowyra et al., 1997). Similarly, another F-box protein (Slimb/βTrCP) binds to the phosphorylated form of IκBα, thus targeting IκBα for ubiquitination and subsequent degradation (Yaron et al., 1998; Spencer et al., 1999; Winston et al., 1999). Although the SCF complexes are the largest family of E3s that target protein ubiquitination, the majority of RING domain-containing proteins are not associated with SCF complexes. These RING proteins, many of which are capable of promoting polyubiquitination in vitro, are involved in various biological processes, such as apoptosis (Mdm2 and XIAP) (Honda et al., 1997; Fang et al., 2000; Yang et al., 2000), DNA repair (BRCA1) (Lorick et al., 1999; Ruffner et al., 2001), development (Siah-1) (Hu & Fearon, 1999; Lorick et al., 1999), cell division (Apc11) (Gmachl et al., 2000; Tang et al., 2001) and signal transduction (Cbl and TRAFs) (Joazeiro et al., 1999; Deng et al., 2000).

The 26S proteasome is composed of a 20S catalytic core and two 19S (also known as PA700) regulatory particles. A total of 28 subunits form the 20S complex, which consists of 4-stacked rings, each containing 7 subunits. Thus, the 20S complex is a barrel-shaped structure with catalytic sites located within the interior of the barrel (Baumeister et al., 1998). Proteins that are destined for degradation must gain access to these sequestered proteolytic sites through an axial channel, which is gated by the 19S complex. This regulatory particle can be physically stripped into two subassemblies: a base containing six ATPases of the AAA family (Rpt) plus three non-ATPase subunits (Rpn), and a lid containing eight Rpn subunits (Glickman et al., 1998). Interestingly, the lid is homologous to the COP9 signalosome, a 500-kDa particle that was first identified as a repressor of light responses in Arabidopsis (Chamovitz et al., 1996). The COP9 signalosome likely has many functions, one of which is to function as an isopeptidase to remove the ubiquitin-like protein Nedd8 from Cul1 (Lyapina et al., 2001). The catalytic subunit harboring this isopeptidase activity has recently been mapped to

JAB1/Csn5, which contains a novel metalloprotease motif termed JAMM (Cope et al., 2002). Significantly, such a JAMM motif was also found in Rpn11, a subunit located in the lid of the proteasome (Verma et al., 2002; Yao & Cohen, 2002). Mutation of the catalytic histidine residues of Rpn11 abolishes its ability to function as an isopeptidase to disassemble polyubiquitin chains. Moreover, such a mutation abolishes the ability of proteasome to degrade polyubiquitinated substrates, suggesting that proteasomal degradation and disassembly of polyubiquitin chains are tightly coupled. Another subunit in the lid (S5a/Rpn10) was found to be capable of binding to polyubiquitin chains (Deveraux et al., 1994). However, mutational analysis of Rpn10 in yeast did not support its role as the major receptor for ubiquitin chains (Fu et al., 1998). Recent studies indicate that an ATPase subunit located in the base (S6'/Rpt5) is the predominant subunit that is cross-linked to polyubiquitin chains (Lam et al., 2002). Presumably, binding of polyubiquitin chains by the 19S complex recruits the conjugated proteins to the proteasome. The ATPase activity of the base then unfolds the ubiquitinated polypeptide and facilitates the translocation of substrates into the catalytic chamber of the 20S complex through a narrow channel opened by Rpt2 ATPase (Braun et al., 1999; Kohler et al., 2001; Finley, 2002). Through its intrinsic chaperone activity, the base also promotes transcription in a manner that is independent of proteasomal degradation (Swaffield et al., 1995; Ferdous et al., 2001).

2. UBIQUITINATION AND PROCESSING OF NF-κB PRECURSORS

2.1. Ubiquitination and processing of NF-κB1/p105

The role of ubiquitin in the NF-κB pathway was first discovered in the course of studying NF-κB1/p105 processing. Pulse-chase experiments showed that the mature NF-κB subunit p50 was derived from p105 by posttranslational processing (Fan & Maniatis, 1991). Furthermore, in vitro reconstitution experiments revealed that p105 processing was mediated by an ATP-dependent process (Fan & Maniatis, 1991). Subsequent experiments showed that p105 processing was blocked by proteasome inhibitors or by mutation of proteasomal subunits in yeast. In addition, a ubiquitin mutant defective in forming K48-linked polyubiquitin chains inhibited p105 processing in vitro (Palombella et al., 1994; Sears et al., 1998). These results not only introduce the ubiquitin-proteasome pathway into the NF-κB arena, but also reveal a novel function of the proteasome in processing (limited degradation) rather than complete degradation of a protein target. More recently, several other proteins have been shown to be processed by the proteasome in a similar manner. For example, the Drosophila protein Cubitus interruptus (Ci) is proteolytically processed to generate a 75-kDa N-terminal fragment, which functions as a transcriptional repressor (Aza-Blanc et al., 1997; Chen et al., 1999). In the presence of Hedgehog, however, Ci phosphorylation and subsequent degradation by the proteasome is suppressed, resulting in the accumulation of the full-length Ci protein, which functions as a transcriptional activator. Other proteins that are processed by the proteasome include NF-κB2/p100 and two distant homologues of NF-κB in the budding yeast Saccharomyces cerevisiae, viz. Spt23p and Mga2p (Rape & Jentsch, 2002).

The proteolytic processing of p105 is mainly a constitutive event, as both mature p50 and p105 can be readily detected in unstimulated cells. However, certain stimuli, such as phorbol ester, have been reported to facilitate the processing or even complete degradation of p105 (Mercurio et al., 1993; MacKichan et al., 1996; Belich et al., 1999; Beinke et al., 2002). It is possible that both processing and degradation of p105 by the proteasome are utilized by certain stimuli as a posttranslational mechanism to activate NF-κB. In the case of constitutive processing of p105 to p50, a cotranslational proteasomal processing mechanism was proposed (Lin et al., 1998). This mechanism was based on pulse-chase experiments in CHO-CD14 cells, which showed that both p50 and p105 were generated simultaneously. An N-terminal fragment (1-497) of p105, when engineered with a stop codon, was not processed to produce p50, neither in vivo nor in vitro. However, when the stop codon was eliminated, this fragment was processed to generate p50 in vitro. These and other experiments imply that a nascent but unfolded p105 is sufficient for p50 production. Furthermore, in synchronized in vitro translation experiments, p50 is produced well before completion of translation of p105, suggesting that p50 is generated via a cotranslational mechanism. The Rel homology domain in p50 was shown to undergo cotranslational dimerization, which is required for efficient production of p50 (Lin et al., 2000). However, this dimerization-guided cotranslational processing generates only p105-p50 heterodimer. Thus, additional posttranslational steps are required to generate a p50 homodimer and to regulate the cellular ratio of p105 and p50. For example, Bcl3 might displace p105 to form a p50-Bcl3 heterodimer, which functions as a transcriptional activator (Watanabe et al., 1997).

The mechanism of p105 processing is a subject of intense interest. A glycine-rich region (GRR) at the C-terminal domain of p50 was shown to function as a processing signal (Lin & Ghosh, 1996). Moreover, amino acid residues 441-454 are important. Lys441 and Lys442 are putative ubiquitination sites, whereas residues 446-454 may be important for ligase recognition (Ciechanover et al., 2001). It is thought that ubiquitination may recruit the proteasome to initiate an endoproteolytic cut at a site downstream of (C-terminal to) GRR, which is likely an unstructured segment requiring little unfolding to gain entry into the proteasome. The unfolded polypeptide may form a hairpin loop to enter the axial channel leading to the proteolytic chamber of the proteasome, which then unravels and degrades polypeptides in both N- and C-terminal directions (Lee et al., 2001; Rape & Jentsch, 2002). The C-terminal degradation proceeds to the end, thus eliminating the C-terminus of p105. On the other hand, the degradation towards the N-terminal p50 domain comes to a halt when it encounters the dimerized Rel-homology domain, which is a tightly packed structure. This model not only provides a plausible explanation for how certain proteins are processed, but may also provide a general mechanism for complete degradation of ubiquitinated proteins. In most cases, proteins are completely degraded, simply because they do not possess tightly packed structures resistant to proteasome degradation.

2.2. Ubiquitination and processing of NF-κB2/p100

The paralogue of NF-κB1/p105, viz. NFκB2/p100, is also processed to the mature p52 subunit by the proteasome (Heusch et al., 1999). This processing is essential for the development of B-cells (Caamano et al., 1998; Franzoso et al., 1998; Poljak et al., 1999). Unlike p105, whose processing is mostly constitutive, the processing of p100 is tightly regulated. A death domain located near the C-terminus of p100 negatively regulates p100 processing, but this inhibition can be relieved by phosphorylation at Ser866 and Ser870 (Xiao et al., 2001).

The kinase cascade responsible for phosphorylation of p100 has been elucidated (Senftleben et al., 2001; Xiao et al., 2001). The first clue for the potential involvement of NIK in p100 processing was provided by the observation that a NIK mutant (aly) mouse exhibits B-cell developmental defects similar to those of NF-κB2 knockout mice (Shinkura et al., 1999). Later experiments showed that B-cells derived from the NIK mutant (aly) mice are defective in p100 processing (Xiao et al., 2001). Moreover, ectopic expression of NIK, but not of NIK mutants lacking catalytic activity or carrying the aly mutation, induced p100 ubiquitination and processing. A NIK-containing complex, immuno-purified from transfected cells, preferentially phosphorylates p100 at Ser866 and Ser870. The double mutant S866A/S870A functions as dominant negative inhibitor of p100 processing in vivo. These experiments provide strong support for an essential role of NIK in the induced phosphorylation and processing of p100. However, subsequent experiments showed that purified NIK does not phosphorylate p100 directly. Instead, NIK phosphorylates IKKα, which in turn phosphorylates p100 (Senftleben et al., 2001). Indeed, an IKKα dominant-negative mutant blocks NIK-induced p100 processing. Moreover, IKKα-deficient B-cells exhibit defects in p100 processing.

Following phosphorylation by IKKα, p100 is polyubiquitinated and then processed to mature p52 by the proteasome. Recently, the F-box protein βTrCP was found to bind specifically to the phosphorylated form of p100 (Fong & Sun, 2002). Moreover, suppression of βTrCP expression by dsRNA-mediated silencing blocked the NIK-induced processing of p100. Thus it is quite likely that the βTrCP-containing complex is the E3 responsible for ubiquitination of phosphorylated p100, although it remains to be seen whether the βTrCP complex can ubiquitinate phosphorylated p100 directly in vitro.

2.3. Ubiquitination and processing of the yeast NF-κB homologues Spt23p and Mga2p

Spt23p and Mga2p are two transcription factors with overlapping physiological functions controlling the synthesis of unsaturated fatty acid in yeast cells (Zhang et al., 1999). They regulate the expression of OLE1, which encodes Δ9 fatty-acid desaturase. Spt23p and Mga2p are synthesized as membrane-bound precursors, which are processed to generate mature transcription factors in response to the level of fatty-acid pool (Hoppe et al., 2000). Both Spt23p and Mga2p are structurally related to NF-κB precursor p105, containing N-terminal immunoglobulin-like putative DNA-binding domains and C-terminal ankyrin repeats.

Genetic experiments indicate that processing of the two proteins requires both the HECT-domain ubiquitin ligase RSP5 and the proteasome. Consistently, Spt23p directly interacts with RSP5. Like p105 and p100, Spt23p dimerizes through the immunoglobulin-like domain, and this dimerization is required for the processing to occur. After processing by the proteasome, the mature form of Spt23p remains ubiquitinated and bound to the unprocessed precursor. The ubiquitin on the processed Spt23p then recruits the $CDC48^{UFD1/NPL4}$ chaperone complex, which releases the mature Spt23p from the precursor, thus facilitating its translocation to the nucleus (Rape et al., 2001; Rape & Jentsch, 2002).

3. UBIQUITINATION AND DEGRADATION OF IκB

IκB comprises a family of ankyrin-repeats containing proteins such as IκBα, IκBβ, IκBε and Bcl3 (Ghosh et al., 1998). Except for Bcl3, which cooperates with the p52 subunit of NF-κB to activate transcription, other IκB proteins bind to and sequester NF-κB in the cytoplasm. To activate NF-κB, IκB must be removed or eliminated. Indeed, degradation of IκB is the principal mechanism for NF-κB activation by most, if not all, stimuli (Silverman & Maniatis, 2001). Proteasome inhibitors were previously shown to block signal-induced degradation, but not phosphorylation of IκBα; phosphorylated IκBα remained bound to NF-κB, thereby preventing NF-κB activation (Palombella et al., 1994).

Subsequently, direct evidence was provided that IκBα is ubiquitinated in a phosphorylation-dependent manner and that ubiquitinated IκBα remains bound to NF-κB, but is selectively degraded by the 26S proteasome (Chen et al., 1995). These studies set the stage for the isolation and characterization of the IκB ubiquitination machinery. An E2 capable of ubiquitinating IκBα was found to belong to the Ubc4/Ubc5 family (Chen et al., 1996), and the IκB E3 turned out to be a multi-protein complex consisting of Skp1, Cul1, Roc1/Rbx1 and the F-box protein Slimb/βTrCP (Deshaies, 1999; Maniatis, 1999).

Slimb was first discovered in a genetic screen for Drosophila mutants that display ectopic wing signaling (Jiang & Struhl, 1998). The slimb mutation led to the stabilization of Cubitus interruptus of the Hedgehog pathway and Armadillo/β-catenin, a transcriptional coactivator of the Wnt/Wingless pathway. Furthermore, in slimb mutant embryos, expression of Dorsal/NF-κB-dependent genes was severely impaired, a phenotype that is also found in Cactus/IκB gain of function mutants. The mammalian homologue of Slimb (βTrCP) was found to mediate the ubiquitination of IκBα, IκBβ, p105, p100 and β-catenin (Yaron et al., 1998; Latres et al., 1999; Shirane et al., 1999; Spencer et al., 1999; Winston et al., 1999; Wu & Ghosh, 1999; Orian et al., 2000; Fong & Sun, 2002). In each case, βTrCP binds specifically to the phosphorylated, but not to the unphosphorylated form of the substrates, which provides a molecular basis for phosphorylation-dependent ubiquitination of these proteins. βTrCP is also involved in the ubiquitination and degradation of CD4, a T-cell surface receptor that is targeted for degradation by the HIV Vpu protein, which binds to βTrCP (Margottin et al., 1998). A consensus sequence in these βTrCP substrates consists of DpSGXXpS, where pS represents phosphorylated serine.

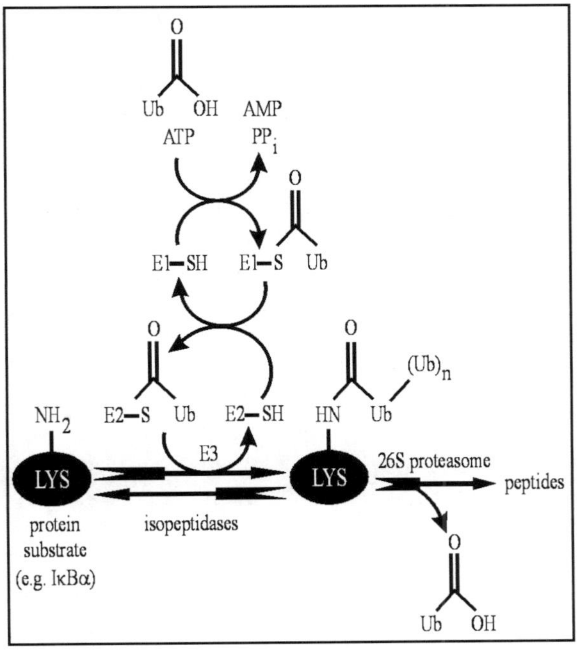

Fig. 1. The ubiquitin/proteasome pathway. Ubiquitination of a protein substrate (e.g. IκBα) is carried out by an enzymatic cascade consisting of E1, E2 and E3. The conjugated ubiquitin can be further ubiquitinated in a highly processive manner to form a polyubiquitin chain, which is preferentially recognized by the 26S proteasome. This allows the conjugated protein to translocate into the proteolytic chamber, where it is degraded. The polyubiquitin chains are then disassembled to regenerate monoubiquitin by the action of isopeptidases. This deubiquitination process may be coupled to degradation by the isopeptidase activity of the proteasome.

Slimb/βTrCP contains seven WD40 repeats, which are responsible for recognition of phosphorylated substrates such as IκBα (Fig. 1). The WD40 repeats, also known as β-transducin repeats, are present in a large number of proteins, including the G-protein Giβ1, whose crystal structure has been solved (Wall et al., 1995). In Giβ1, WD40 repeats form a β-propeller structure that stacks against Giγ1. Molecular modeling and mutagenesis studies of the F-box/WD40 protein Cdc4 reveal clusters of positive charge residues within the β-propeller that might be in direct contact with phosphorylated Sic1, an S-phase Cdk inhibitor ubiquitinated by Cdc4 (Nash et al., 2001).

It remains to be determined what residues within the WD40 repeats of βTrCP are involved in binding to phosphorylated IκB and β-catenin. This is an important issue for considering the development of anti-inflammatory agents that block βTrCP:phosphorylated IκB binding. Ideally, these inhibitors should not block degradation of β-catenin, which is an oncogene when its level accumulates. This is only possible if binding to IκB and β-catenin involves distinct regions of βTrCP.

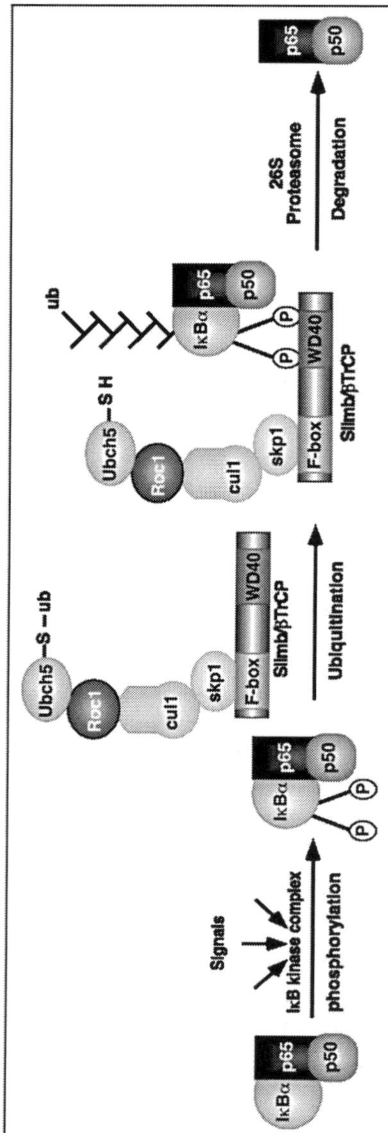

Fig. 2. Ubiquitination of IκBα.
Following signal-induced phosphorylation of IκBα at specific serine residues, phosphorylated IκBα is recruited to the WD40 repeats of βTrCP. Through its F-box, βTrCP binds to Skp1, which in turn binds to Cul1. Cul1 is complexed with the RING domain protein Rbx1/Roc1, which recruits the E2 Ubch5 to transfer ubiquitin to phosphorylated IκBα juxtaposed nearby. Ubiquitinated IκBα remains associated with NF-κB (p50/p65 heterodimer), but is selectively degraded by the 26S proteasome.

Slimb/βTrCP also contains an F-box that binds to Skp1, which in turn binds to Cul1. Cul1 then binds to the small RING domain protein Roc1/Rbx1, which binds to Ubch5 and promotes polyubiquitination by Ubch5 (Fig. 2). Thus a major function of the SCF/βTrCP complex is to function as a scaffold to bring phosphorylated IκBα into close proximity to Ubch5, thereby allowing ubiquitin conjugation to take place. This model is supported by the crystal structure of another SCF complex containing Skp2 as the F-box protein that recognizes the CDK inhibitor p27 (Zheng et al., 2002). A striking feature of this structure is that Cul1 forms a long stalk that connects Skp1 and the F-box protein Skp2 at one end, and Rbx1 at the other. Hence, Cul1 serves as a rigid scaffold that positions the substrate (through Skp2) and E2 (through Rbx1) on the same surface with proper orientation and spacing. The distance between E2 and the substrate-binding surface of Skp2 is approximately 50 Å, which could accommodate substrates of varying sizes, and perhaps additional proteins. None of the subunits of the SCF complex, including the RING protein Rbx1, appears to directly participate in acid/base catalysis. Instead, this E3 complex might facilitate the transfer of ubiquitin from the E2-Ub thioester to the substrates by positioning certain lysine side chains of the substrates in the immediate vicinity of the E2 active site. In addition, Rbx1 facilitates the elongation of the polyubiquitin chains by E2s through an unknown mechanism (Ohta et al., 1999; Seol et al., 1999; Tan et al., 1999; Wu et al., 2000; Tang et al., 2001; Furukawa et al., 2002). The binding of Rbx1 to E2 likely involves a hydrophobic groove that is also conserved in the c-Cbl RING domain (Zheng et al., 2000). Rbx1 binds to Cul1 through an intermolecular β-sheet, which greatly stabilizes this two-subunit catalytic core. The structure of this SCF^{Skp2} complex may also help explain why modification of Cul1 by the ubiquitin-like protein Nedd8 enhances the activity of SCF complexes towards some substrates such as IκB and p27. Cul1 neddylation occurs at Lys720, which is located in the interface between Cul1 and Rbx1. Thus, Nedd8 modification might optimize the orientation or positioning of the E2 or E2-Ub thioester.

One issue that was not resolved by the structural studies is the selection of specific E2s by SCF complexes for ubiquitination of specific substrates. For instance, ubiquitination of IκB, $p21^{Sic1}$ and $p27^{Kip1}$ requires the same core components of the SCF complex, including Skp1, Cul1 and Rbx1. However, IκB is ubiquitinated by Ubc4/5, whereas $p21^{Sic1}$ and $p27^{Kip1}$ are ubiquitinated by Ubc3 (Cdc34 in yeast). Thus, it appears that substrates might play some roles in the recruitment of specific E2s to the SCF complex. Alternatively, various E2s may come on and off the SCF complexes in a dynamic flux; only when the right E2 is positioned to its cognate substrate will ubiquitin transfer take place.

Two forms of βTrCP, viz. βTrCP1 and βTrCP2, were found in the human genome (Suzuki et al., 1999). These genes have been mapped to different chromosomes and encode proteins that are 78% identical in amino acid sequences. Despite this high degree of sequence similarity, βTrCP1 is predominantly localized to the nucleus, whereas βTrCP2 is found in the cytoplasm (Davis et al., 2002). βTrCP1 is sequestered in the nucleus by binding to hnRNP-U, an abundant nuclear protein that carries a nuclear localization sequence (NLS). Through its acidic N-terminus, hnRNP-U binds to the WD40 repeats of βTrCP1 in a manner competing with phosphorylated IκBα. It was proposed that hnRNP-U functions as a pseudosubstrate to sequester βTrCP1 in the nucleus until IκBα is phosphorylated. Implicit in this model is that phosphorylated IκBα then enters the nucleus for ubiquitination by βTrCP1, or a fraction of βTrCP1 somehow shuttles to the cytoplasm to ubiquitinate IκBα.

It should be noted that cytoplasmic extracts efficiently ubiquitinate IκBα in a phosphorylation-dependent manner, indicating the presence of the complete IκB ubiquitination machinery in the cytosol, including IκB ligase (Chen et al., 1995; Yaron et al., 1998). Given that β-TrCP2 is present in the cytosol, one simple possibility is that βTrCP2, but not βTrCP1, is the dominant ubiquitin ligase responsible for ubiquitination of IκBα in the cytosol. Despite the uncertainty about the location of IκB ubiquitination and the specific isoform of ligase involved, it is clear that βTrCP is the ubiquitin ligase for IκB and p100, as a small RNA duplex matching a common sequence of both βTrCP1 and βTrCP2 blocks signal-induced degradation of IκBα and processing of p100 (Fong & Sun, 2002).

4. UBIQUITIN-MEDIATED ACTIVATION OF THE IκB KINASE COMPLEX BY TRAF PROTEINS

Signal-induced phosphorylation targets IκB for ubiquitination and subsequent degradation by the proteasome (Palombella et al., 1994; Alkalay et al., 1995; Chen et al., 1995). Consequently, phosphorylation of IκB is the most important regulatory step in NF-κB pathways. Indeed, virtually all NF-κB stimuli induce phosphorylation of IκBα at Ser32 and Ser36, and mutation of one or both of these serines converts IκBα into a strong dominant-negative mutant that blocks NF-κB activation (Brockman et al., 1995; Brown et al., 1995; Traenckner et al., 1995). A protein kinase that phosphorylates IκBα at these specific serine residues, was first identified by biochemical fractionation of HeLa cell cytoplasmic extracts (Chen et al., 1996). Surprisingly, this kinase is a large (~700 kDa) complex whose activity is activated by polyubiquitination through a novel mechanism that does not involve proteasomal degradation. Further studies showed that the same kinase complex could be activated by MEKK1, a kinase of the JNK pathway (Lee et al., 1997). Hence it appears that the IκB kinase complex can be activated independently by polyubiquitination and phosphorylation. While kinase activation by phosphorylation cascades is well established, activation of protein kinase by polyubiquitination is unprecedented and poorly understood.

Two catalytic subunits of the IκB kinase complex, designated as IKKα and IKKβ, have been cloned (DiDonato et al., 1997; Mercurio et al., 1997; Regnier et al., 1997; Woronicz et al., 1997; Zandi et al., 1997). The regulatory subunit of the IKK complex, termed NEMO or IKKγ, has also been identified (Rothwarf et al., 1998; Yamaoka et al., 1998). Thus the core IKK complex consists of IKKα, IKKβ and NEMO, but the exact stoichiometry of these subunits in a complex has not yet been determined. A recent report suggests that the IKK complex also contains the chaperones Hsp90 and Cdc37 (Chen et al., 2002). Genetic studies have demonstrated that IKKβ and NEMO are essential for NF-κB activation by most (if not all) stimuli tested so far (Ghosh & Karin, 2002). On the other hand, IKKα has a minor role in NF-κB activation by proinflammatory cytokines, but is essential for B-cell development through its role in the phosphorylation of NF-κB2/p100 (Karin & Delhase, 2000; Senftleben et al., 2001). Furthermore, IKKα is involved in skin development through an unknown mechanism that does not involve NF-κB activation (Hu et al., 2001).

The IKK complex is activated by a large variety of NF-κB stimuli that turn on distinct signaling pathways (Silverman & Maniatis, 2001; Ben-Neriah, 2002). How such diverse pathways converge on the IKK complex to regulate its activity is a subject of intense interest and significant importance. Some of the most widely studied pathways that activate NF-κB, are those emanating from proinflammatory cytokines such as TNFα and IL-1β. An important mediator of these signaling pathways is a member of the TRAF family (Inoue et al., 2000; Bradley & Pober, 2001). The first TRAFs (TRAF1 and TRAF2) were identified as proteins associated with the TNF-R2 complex (Rothe et al., 1994). Later, four other

TRAF proteins were discovered, mostly by yeast two-hybrid screening (Wajant et al., 1999). A common feature present in all TRAF proteins is the C-terminal TRAF domain, consisting of a coiled-coil domain followed by a TRAF-C domain. This TRAF domain is responsible for self-association and interaction with receptors and other upstream signaling proteins such as TRADD (Hsu et al., 1995). The crystal structure of the TRAF domain of TRAF2 reveals a mushroom-like structure comprising a stalk of trimeric coiled-coil domain and a cap of trimeric TRAF-C domain, which forms an antiparallel-sandwich fold (McWhirter et al., 1999; Ye et al., 1999; Chung et al., 2002). Except for TRAF1, the other five TRAFs also contain an N-terminal RING domain and several zinc finger motifs important for downstream signaling (Baud et al., 1999).

Three TRAF proteins (TRAF2, TRAF5 and TRAF6) have been shown to be important for activation of NF-κB and AP1 pathways. Specifically, TRAF2 and TRAF5 have partially redundant function in mediating NF-κB activation in response to TNFα (Wajant & Scheurich, 2001), whereas TRAF6 is essential for signaling by IL-1R, IL-18R, CD40, RANK/TRANCE-R, and Toll-like receptors (Cao et al., 1996; Ishida et al., 1996; Lomaga et al., 1999; Naito et al., 1999; Schwandner et al., 2000; Naito et al., 2002). In an effort to understand how TRAF6 mediates activation of IKK, a cell-free system was set up that activates IKK in response to recombinant TRAF6 protein (Deng et al., 2000). Fractionation of the cell extracts led to the discovery of two factors, termed TRIKA1 (TRAF6-regulated IKK activator-1) and TRIKA2, which are required for IKK activation by TRAF6 in vitro. TRIKA1 turns out to be a ubiquitin-conjugating enzyme complex consisting of Ubc13 and a Ubc-like protein Uev1A (also known as Mms2). The yeast orthologue of this E2 complex has a unique specificity of synthesizing Lys63-linked polyubiquitin chains, which has been implicated in stress response and DNA repair through unknown mechanisms (Hofmann & Pickart, 1999). The crystal structure of the yeast Ubc13/Mms2 complex reveals two channels that could dock two ubiquitin molecules in such an orientation that only Lys63 of the acceptor ubiquitin is optimally positioned for conjugation by the Gly76 of the donor ubiquitin (Moraes et al., 2001; VanDemark et al., 2001). The discovery of the role of Ubc13/Uev1A in TRAF6 signaling immediately suggests a potential mechanism by which TRAF6 leads to the activation of IKK: TRAF6 may function as a ubiquitin ligase through its N-terminal RING domain. Indeed, in conjunction with Ubc13/Uev1A, TRAF6 greatly facilitates the synthesis of the K-63 polyubiquitin chains (Deng et al., 2000). The formation of these polyubiquitin chains is both necessary and sufficient to mediate IKK activation in vitro. Other TRAF proteins, such as TRAF2, may also mediate downstream signaling by functioning as ubiquitin ligases (L. Deng & Z. Chen, unpublished data). These findings provide an explanation for the well-known signaling function of the RING domain in TRAF proteins (Baud et al., 1999). They also place the earlier puzzling observation of the ubiquitin activation of the IκB kinase complex (Chen et al., 1996) in the context of TRAF signaling.

TRIKA2 was identified as a protein kinase complex comprising TAK1, TAB1 and TAB2 (Wang et al., 2001). TAK1 was initially discovered as a TGFβ-activated kinase (Yamaguchi et al., 1995), but was subsequently shown to be also involved in signaling both by IL-1 and Wnt (Ishitani et al., 1999; Ninomiya-Tsuji et al., 1999). In response to IL-1 stimulation, TAB2 translocates from a membranous location to the cytoplasm, where it functions as an adaptor to bring together TRAF6 and the TAK1/TAB1 complex (Takaesu et al., 2000). However, recruitment of TRAF6 to the TAK1/TAB1/TAB2 complex is not sufficient to activate TAK1. Instead, TAK1 activation requires TRAF6- and Ubc13/Uev1A-mediated polyubiquitination (Wang et al., 2001). One ubiquitination target is TRAF6 itself, whose polyubiquitination is induced by IL-1 or by forced oligomerization of TRAF6. It is not at all clear how TRAF6 polyubiquitination leads to activation of TAK1. In any case, once TAK1 is activated, it phosphorylates IKKβ in the activation loop and activates the IKK complex. In this regard, TAK1 is the IKK kinase (IKKK) of the TRAF6 pathways. TAK1 also phosphorylates and activates MKKs such as MKK6, which in turn activates the JNK and p38 kinase pathways (Wang et al., 2001). Thus, TRAF6-mediated polyubiquitination appears to be the initiating step that activates both IKK and JNK/p38 kinase cascades (Fig. 3).

In comparison to TRAF6 signaling, much less is known about the potential role of ubiquitination in the TRAF2 pathway that mediates IKK activation by TNFα. It has been shown that the dominant-negative mutant of Ubc13 blocks NF-κB activation by TNFα and TRAF2 (Deng et al., 2000). In addition, TRAF2 can also catalyze K63-linked polyubiquitination in the presence of Ubc13/Uev1A (L. Deng & Z. Chen, unpublished data). Furthermore, RIP, a TRAF2 partner that is essential for NF-κB activation by TNFα, has been shown to be polyubiquitinated, but not degraded in response to TNFα treatment (Zhang et al., 2000; Chen et al., 2002).

Other studies have suggested that IKK might be activated through its recruitment to a submembranous receptor complex consisting of TNF-R1, TRADD, TRAF2 and RIP (Devin et al., 2000; Zhang et al., 2000; Devin et al., 2001). It is not clear how IKK is activated after it is recruited to the membrane, but one possibility is that such a recruitment process might increase the local concentration of the IKK complex, leading to its aggregation and activation. According to this model, no upstream "IKKK" is required to mediate IKK activation by TRAF2 and RIP. However, recent genetic studies have shown that MEKK3 is essential for IKK activation by TNFα (Yang et al., 2001). It will be interesting to see whether MEKK3 is activated by a ubiquitin-mediated mechanism similar to that of TAK1 activation.

Whereas the RING domain of TRAF2 appears to be essential for its ability to activate NF-κB, removal of the RING domain of TRAF6 only partially compromises its NF-κB-stimulatory activity (Kobayashi et al., 2001). Thus, it is possible that TRAF6 has another partially redundant signaling domain that activates downstream pathways by a ubiquitination-independent mechanism. Consistent with this possibility, RNAi-mediated silencing of Ubc13 and TAK1 blocked IKK and JNK activation by the RING domain of TRAF6, but only partially blocked IL-1 signaling (M. Hong & Z.J. Chen, unpublished data).

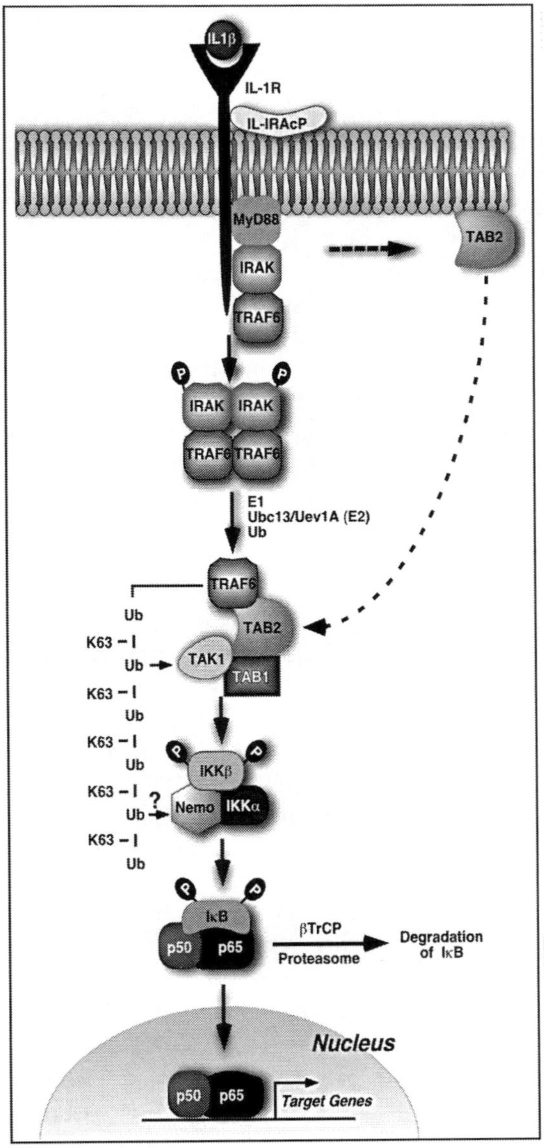

Fig. 3. Proposed signaling mechanism of NF-κB activation by TRAF6. Stimulation of cells with IL-1β leads to association of IL-1R and IL-1R accessory protein (IL-1RAcP), as well as to the recruitment of adaptor proteins, including MyD88, IRAK and TRAF6. IRAK is then phosphorylated and released to the cytoplasm as a complex with TRAF6. Recruitment of TRAF6 results in transient oligomerization, presumably bringing together the N-terminal RING domains and activating the ubiquitin ligase activity. Together with E2 Ubc13/Uev1A, TRAF6 facilitates the conjugation of K63-linked polyubiquitin chains to some substrates, including TRAF6. Ubiquitinated TRAF6 binds to TAB2 after IL-1β-induced translocation of TAB2 from a membranous location to the cytoplasm. TAB2 binds to TAK1 and TAB1, allowing ubiquitinated TRAF6 to activate TAK1 by an unknown mechanism. Activated TAK1 then phosphorylates IKKβ in the activation loop, thereby activating the IKK complex to phosphorylate IκB at specific N-terminal serine residues. Phosphorylated IκB is recruited by βTrCP for ubiquitination and degradation by the proteasome, permitting NF-κB to enter the nucleus and to activate a large variety of target genes. Such a putative signaling mechanism may be applicable to a number of other pathways that involve TRAF6, including those of Toll-like receptors (TLRs), CD40 and RANK.

Several other RING domain proteins have recently been shown to activate NF-κB and AP1. For example, XIAP and cIAP1 are inhibitors of apoptosis that contain a RING domain in their N-termini, and have been shown to possess ubiquitin ligase activity (Yang et al., 2000). Overexpression of these proteins leads to potent activation of IKK and JNK (Yamaguchi et al., 1999; Hofer-Warbinek et al., 2000). Furthermore, it has been shown that XIAP binds to and activates TAK1. Consequently, both XIAP and TRAF6 may share some common ubiquitination-dependent mechanism to activate downstream kinase cascades. TRAF2 has recently been shown to be ubiquitinated in response to stimulation of TNF-R2. This ubiquitination was mediated by cIAP1 and results in the degradation of TRAF2, which has an antiapoptotic role (Li et al., 2002). It was proposed that this might be the mechanism by which TNF-R2 stimulation by TNF potentiates the proapoptotic effect of TNF-R1 signaling. Since there is no evidence of TRAF2 polyubiquitination following TNFα-stimulation of TNF-R1, which also recruits TRAF2 and cIAP1 to the receptor complex, it is not clear why TRAF2 ubiquitination and degradation are only restricted to TNF-R2 signaling. TRAF6 is also ubiquitinated and then slowly degraded following stimulation of osteoclasts with RANK ligand (Takayanagi et al., 2000). It is possible that ubiquitination of TRAF2 and TRAF6 initially leads to the activation of downstream kinase cascades, and subsequent proteasomal degradation helps to terminate the signals. Such a built-in mechanism to ensure transient activation has been shown for some transcriptional activators that require ubiquitination for transcriptional activation and for subsequent degradation by the proteasome (Salghetti et al., 2001).

5. CONCLUSIONS AND PERSPECTIVES

Ubiquitination is closely intertwined with phosphorylation. This is best illustrated in the NF-κB pathway. It is nowadays well established that phosphorylation targets IκB for ubiquitination and subsequent degradation by the proteasome. This signal-dependent ubiquitination is accomplished by a specialized SCF ubiquitin ligase complex containing βTrCP, which binds specifically to phosphorylated IκB. Phosphorylation also targets the NF-κB precursor NF-κB2/p100, and to a less extent NF-κB1/p105, for ubiquitination and subsequent processing by the proteasome. The processing rather than complete degradation of the NF-κB precursor is likely due to the presence of the dimerized Rel-homology domain, which is a tightly packed structure that may be resistant to unfolding by the proteasome. One surprise that emerges from studying NF-κB signaling, is the finding that ubiquitination can also regulate phosphorylation through direct activation of protein kinases. The IκB kinase complex was initially identified as a ubiquitination-dependent kinase activity, but it was not until recently that this finding began to fit into a well-known signaling pathway, namely the TRAF6 pathway. Future studies should determine whether this novel mechanism of protein kinase activation by ubiquitin is applicable to other pathways involving TRAFs as well as other signaling proteins. Of immediate interest is to determine whether IKK and JNK activation by TRAF2 in the TNF pathway also involves TRAF2-mediated ubiquitination, and whether a kinase other than TAK1 might be responsible for activation of IKK and JNK in the TNF pathway. In the TRAF6 pathway, it is also crucial to understand in details the biochemical mechanisms underlying ubiquitin-mediated activation of TAK1 and IKK. Finally, given the important role of Ubc13/Mms2 in DNA repair (Hofmann & Pickart, 1999), and the fact that DNA damage leads to activation of several protein kinases including IKK (Huang et al., 2000), it is curious whether ubiquitin might play a role in signaling cascades emanating from DNA damage and other stress conditions.

6. NOTES

Research is supported by grants from NIH (GM59203 and GM63692), the Robert A. Welch foundation and the American Cancer Society. Z.J. Chen is a Lymphoma and Leukemia Society scholar, as well as a Burroughs Wellcome Fund investigator in the pathogenesis of infectious diseases.

7. REFERENCES

Alkalay I, Yaron A, Hatzubai A, Orian A, Ciechanover A & Ben-Neriah Y (1995) Stimulation-dependent IκBα phosphorylation marks the NF-κB inhibitor for degradation via the ubiquitin-proteasome pathway. Proc. Natl. Acad. Sci. USA 92: 10599-10603.

Arnason T & Ellison MJ (1994) Stress resistance in Saccharomyces cerevisiae is strongly correlated with assembly of a novel type of multiubiquitin chain. Mol. Cell. Biol. 14: 7876-7883.

Aza-Blanc P, Ramirez-Weber FA, Laget MP, Schwartz C & Kornberg TB (1997) Proteolysis that is inhibited by hedgehog targets Cubitus interruptus protein to the nucleus and converts it to a repressor. Cell 89: 1043-1053.

Baud V, Liu ZG, Bennett B, Suzuki N, Xia Y & Karin M (1999) Signaling by proinflammatory cytokines: oligomerization of TRAF2 and TRAF6 is sufficient for JNK and IKK activation and target gene induction via an amino-terminal effector domain. Genes Dev. 13: 1297-1308.

Baumeister W, Walz J, Zuhl F & Seemuller E (1998) The proteasome: paradigm of a self-compartmentalizing protease. Cell 92: 367-380.

Beinke S, Belich MP & Ley SC (2002) The death domain of NF-κB1 p105 is essential for signal-induced p105 proteolysis. J. Biol. Chem. 277: 24162-24168.

Belich MP, Salmeron A, Johnston LH & Ley SC (1999) TPL-2 kinase regulates the proteolysis of the NF-κB-inhibitory protein NF-κB1 p105. Nature 397: 363-368.

Ben-Neriah Y (2002) Regulatory functions of ubiquitination in the immune system. Nat. Immunol. 3: 20-26.

Bradley JR & Pober JS (2001) Tumor necrosis factor receptor-associated factors (TRAFs). Oncogene 20: 6482-6491.

Braun BC, Glickman M, Kraft R, Dahlmann B, Kloetzel PM, Finley D & Schmidt M (1999) The base of the proteasome regulatory particle exhibits chaperone-like activity. Nat. Cell Biol. 1: 221-226.

Brockman JA, Scherer DC, McKinsey TA, Hall SM, Qi X, Lee WY & Ballard DW (1995) Coupling of a signal response domain in IκBα to multiple pathways for NF-κB activation. Mol. Cell. Biol. 15: 2809-2818.

Brown K, Gerstberger S, Carlson L, Franzoso G & Siebenlist U (1995) Control of IκB-α proteolysis by site-specific, signal-induced phosphorylation. Science 267: 1485-1488.

Caamano JH, Rizzo CA, Durham SK, Barton DS, Raventos-Suarez C, Snapper CM & Bravo R (1998) Nuclear factor (NF)-κB2 (p100/p52) is required for normal splenic microarchitecture and B cell-mediated immune responses. J. Exp. Med. 187: 185-196.

Cao Z, Xiong J, Takeuchi M, Kurama T & Goeddel DV (1996) TRAF6 is a signal transducer for interleukin-1. Nature 383: 443-446.

Chamovitz DA, Wei N, Osterlund MT, von Arnim AG, Staub JM, Matsui M & Deng XW (1996)

The COP9 complex, a novel multisubunit nuclear regulator involved in light control of a plant developmental switch. Cell 86: 115-121.

Chau V, Tobias JW, Bachmair A, Marriott D, Ecker DJ, Gonda DK & Varshavsky A (1989) A multiubiquitin chain is confined to specific lysine in a targeted short-lived protein. Science 243: 1576-1583.

Chen CH, von Kessler DP, Park W, Wang B, Ma Y & Beachy PA (1999) Nuclear trafficking of Cubitus interruptus in the transcriptional regulation of Hedgehog target gene expression. Cell 98: 305-316.

Chen G, Cao P & Goeddel DV (2002) TNF-induced recruitment and activation of the IKK complex require Cdc37 and Hsp90. Mol. Cell 9: 401-410.

Chen Z, Hagler J, Palombella VJ, Melandri F, Scherer D, Ballard D & Maniatis T (1995) Signal-induced site-specific phosphorylation targets IκBα to the ubiquitin-proteasome pathway. Genes Dev. 9: 1586-1597.

Chen ZJ, Parent L & Maniatis T (1996) Site-specific phosphorylation of IκBα by a novel ubiquitination-dependent protein kinase activity. Cell 84: 853-862.

Chung JY, Park YC, Ye H & Wu H (2002) All TRAFs are not created equal: common and distinct molecular mechanisms of TRAF-mediated signal transduction. J. Cell Sci. 115: 679-688.

Ciechanover A, Gonen H, Bercovich B, Cohen S, Fajerman I, Israel A, Mercurio F, Kahana C, Schwartz AL, Iwai K & Orian A (2001) Mechanisms of ubiquitin-mediated, limited processing of the NF-κB1 precursor protein p105. Biochimie 83: 341-349.

Cope GA, Suh GS, Aravind L, Schwarz SE, Zipursky SL, Koonin EV & Deshaies RJ (2002) Role of predicted metalloprotease motif of Jab1/Csn5 in cleavage of Nedd8 from Cul1. Science 298: 608-611.

Davis M, Hatzubai A, Andersen JS, Ben-Shushan E, Fisher GZ, Yaron A, Bauskin A, Mercurio F, Mann M & Ben-Neriah Y (2002) Pseudosubstrate regulation of the SCF(β-TrCP) ubiquitin ligase by hnRNP-U. Genes Dev. 16: 439-451.

Deng L, Wang C, Spencer E, Yang L, Braun A, You J, Slaughter C, Pickart C & Chen ZJ (2000) Activation of the IκB kinase complex by TRAF6 requires a dimeric ubiquitin-conjugating enzyme complex and a unique polyubiquitin chain. Cell 103: 351-361.

Deshaies RJ (1999) SCF and Cullin/Ring H2-based ubiquitin ligases. Annu. Rev. Cell Dev. Biol. 15: 435-467.

Deveraux Q, Ustrell V, Pickart C & Rechsteiner M (1994) A 26 S protease subunit that binds ubiquitin conjugates. J. Biol. Chem. 269: 7059-7061.

Devin A, Cook A, Lin Y, Rodriguez Y, Kelliher M & Liu Z (2000) The distinct roles of TRAF2 and RIP in IKK activation by TNF-R1: TRAF2 recruits IKK to TNF-R1 while RIP mediates IKK activation. Immunity 12: 419-429.

Devin A, Lin Y, Yamaoka S, Li Z, Karin M & Liu Z (2001) The α and β subunits of IκB kinase (IKK) mediate TRAF2-dependent IKK recruitment to tumor necrosis factor (TNF) receptor 1 in response to TNF. Mol. Cell. Biol. 21: 3986-3994.

DiDonato JA, Hayakawa M, Rothwarf DM, Zandi E & Karin M (1997) A cytokine-responsive IκB kinase that activates the transcription factor NF-κB. Nature 388: 548-554.

Fan CM & Maniatis T (1991) Generation of p50 subunit of NF-κB by processing of p105 through an ATP-dependent pathway. Nature 354: 395-398.

Fang S, Jensen JP, Ludwig RL, Vousden KH & Weissman AM (2000) Mdm2 is a RING finger-dependent ubiquitin protein ligase for itself and p53. J. Biol. Chem. 275: 8945-8951.

Feldman RM, Correll CC, Kaplan KB & Deshaies RJ (1997) A complex of Cdc4p, Skp1p, and Cdc53p/cullin catalyzes ubiquitination of the phosphorylated CDK inhibitor Sic1p. Cell 91: 221-230.

Ferdous A, Gonzalez F, Sun L, Kodadek T & Johnston SA (2001) The 19S regulatory particle of the proteasome is required for efficient transcription elongation by RNA polymerase II. Mol. Cell 7: 981-991.

Finley D (2002) Ubiquitin chained and crosslinked. Nat. Cell Biol. 4: E121-123.

Fong A & Sun SC (2002) Genetic evidence for the essential role of β-transducin repeat-containing protein in the inducible processing of NF-κB2/p100. J. Biol. Chem. 277: 22111-22114.

Franzoso G, Carlson L, Poljak L, Shores EW, Epstein S, Leonardi A, Grinberg A, Tran T, Scharton-Kersten T, Anver M, Love P, Brown K & Siebenlist U (1998) Mice deficient in nuclear factor (NF)-κB/p52 present with defects in humoral responses, germinal center reactions, and splenic microarchitecture. J. Exp. Med. 187: 147-159.

Fu H, Sadis S, Rubin DM, Glickman M, van Nocker S, Finley D & Vierstra RD (1998) Multiubiquitin chain binding and protein degradation are mediated by distinct domains within the 26 S proteasome subunit Mcb1. J. Biol. Chem. 273: 1970-1981.

Furukawa M, Ohta T & Xiong Y (2002) Activation of UBC5 ubiquitin-conjugating enzyme by the RING finger of ROC1 and assembly of active ubiquitin ligases by all cullins. J. Biol. Chem. 277: 15758-15765.

Ghosh S, May MJ & Kopp EB (1998) NF-κB and Rel proteins: evolutionarily conserved mediators of immune responses. Annu. Rev. Immunol. 16: 225-260.

Ghosh S & Karin M (2002) Missing pieces in the NF-κB puzzle. Cell 109: S81-96.

Glickman MH, Rubin DM, Fried VA & Finley D (1998) The regulatory particle of the Saccharomyces cerevisiae proteasome. Mol. Cell. Biol. 18: 3149-3162.

Gmachl M, Gieffers C, Podtelejnikov AV, Mann M & Peters JM (2000) The RING-H2 finger protein APC11 and the E2 enzyme UBC4 are sufficient to ubiquitinate substrates of the anaphase-promoting complex. Proc. Natl. Acad. Sci. USA 97: 8973-8978.

Hershko A & Ciechanover A (1992) The ubiquitin system for protein degradation. Annu. Rev. Biochem. 61: 761-807.

Hershko A & Ciechanover A (1998) The ubiquitin system. Annu. Rev. Biochem. 67: 425-479.

Heusch M, Lin L, Geleziunas R & Greene WC (1999) The generation of nfkb2 p52: mechanism and efficiency. Oncogene 18: 6201-6208.

Hofer-Warbinek R, Schmid JA, Stehlik C, Binder BR, Lipp J & de Martin R (2000) Activation of NF-κB by XIAP, the X chromosome-linked inhibitor of apoptosis, in endothelial cells involves TAK1. J. Biol. Chem. 275: 22064-22068.

Hofmann RM & Pickart CM (1999) Noncanonical MMS2-encoded ubiquitin-conjugating enzyme functions in assembly of novel polyubiquitin chains for DNA repair. Cell 96: 645-653.

Honda R, Tanaka H & Yasuda H (1997) Oncoprotein MDM2 is a ubiquitin ligase E3 for tumor suppressor p53. FEBS Lett. 420: 25-27.

Hoppe T, Matuschewski K, Rape M, Schlenker S, Ulrich HD & Jentsch S (2000) Activation of a membrane-bound transcription factor by regulated ubiquitin/proteasome-dependent processing. Cell 102: 577-586.

Hsu H, Xiong J & Goeddel DV (1995) The TNF receptor 1-associated protein TRADD signals cell death and NF-κB activation. Cell 81: 495-504.

Hu G & Fearon ER (1999) Siah-1 N-terminal RING domain is required for proteolysis function, and C-terminal sequences regulate oligomerization and binding to target proteins. Mol. Cell. Biol. 19: 724-732.

Hu Y, Baud V, Oga T, Kim KI, Yoshida K & Karin M (2001) IKKα controls formation of the epidermis independently of NF-κB. Nature 410: 710-714.

Huang TT, Wuerzberger-Davis SM, Seufzer BJ, Shumway SD, Kurama T, Boothman DA & Miyamoto S (2000) NF-κB activation by camptothecin. A linkage between nuclear DNA damage and cytoplasmic signaling events. J. Biol. Chem. 275: 9501-9509.

Huibregtse JM, Scheffner M, Beaudenon S & Howley PM (1995) A family of proteins structurally and functionally related to the E6-AP ubiquitin-protein ligase. Proc. Natl. Acad. Sci. USA 92: 2563-2567.

Inoue J, Ishida T, Tsukamoto N, Kobayashi N, Naito A, Azuma S & Yamamoto T (2000) Tumor necrosis factor receptor-associated factor (TRAF) family: adapter proteins that mediate cytokine signaling. Exp. Cell. Res. 254: 14-24.

Ishida T, Mizushima S, Azuma S, Kobayashi N, Tojo T, Suzuki K, Aizawa S, Watanabe T, Mosialos G, Kieff E, Yamamoto T & Inoue J (1996) Identification of TRAF6, a novel tumor necrosis factor receptor-associated factor protein that mediates signaling from an amino-terminal domain of the

CD40 cytoplasmic region. J. Biol. Chem. 271: 28745-28748.

Ishitani T, Ninomiya-Tsuji J, Nagai S, Nishita M, Meneghini M, Barker N, Waterman M, Bowerman B, Clevers H, Shibuya H & Matsumoto K (1999) The TAK1-NLK-MAPK-related pathway antagonizes signalling between β-catenin and transcription factor TCF. Nature 399: 798-802.

Jiang J & Struhl G (1998) Regulation of the Hedgehog and Wingless signalling pathways by the F-box/WD40-repeat protein Slimb. Nature 391: 493-496.

Joazeiro CA, Wing SS, Huang H, Leverson JD, Hunter T & Liu YC (1999) The tyrosine kinase negative regulator c-Cbl as a RING-type, E2-dependent ubiquitin-protein ligase. Science 286: 309-312.

Karin M & Delhase M (2000) The IκB kinase (IKK) and NF-κB: key elements of proinflammatory signalling. Semin. Immunol. 12: 85-98.

Kobayashi N, Kadono Y, Naito A, Matsumoto K, Yamamoto T, Tanaka S & Inoue J (2001) Segregation of TRAF6-mediated signaling pathways clarifies its role in osteoclastogenesis. EMBO J. 20: 1271-1280.

Kohler A, Cascio P, Leggett DS, Woo KM, Goldberg AL & Finley D (2001) The axial channel of the proteasome core particle is gated by the Rpt2 ATPase and controls both substrate entry and product release. Mol. Cell 7: 1143-1152.

Lam YA, Lawson TG, Velayutham M, Zweier JL & Pickart CM (2002) A proteasomal ATPase subunit recognizes the polyubiquitin degradation signal. Nature 416: 763-767.

Latres E, Chiaur DS & Pagano M (1999) The human F box protein β-Trcp associates with the Cul1/Skp1 complex and regulates the stability of β-catenin. Oncogene 18: 849-854.

Lee C, Schwartz MP, Prakash S, Iwakura M & Matouschek A (2001) ATP-dependent proteases degrade their substrates by processively unraveling them from the degradation signal. Mol. Cell 7: 627-637.

Lee FS, Hagler J, Chen ZJ & Maniatis T (1997) Activation of the IκBα kinase complex by MEKK1, a kinase of the JNK pathway. Cell 88: 213-222.

Li X, Yang Y & Ashwell JD (2002) TNF-RII and c-IAP1 mediate ubiquitination and degradation of TRAF2. Nature 416: 345-347.

Lin L & Ghosh S (1996) A glycine-rich region in NF-κB p105 functions as a processing signal for the generation of the p50 subunit. Mol. Cell. Biol. 16: 2248-2254.

Lin L, DeMartino GN & Greene WC (1998) Cotranslational biogenesis of NF-κB p50 by the 26S proteasome. Cell 92: 819-828.

Lin L, DeMartino GN & Greene WC (2000) Cotranslational dimerization of the Rel homology domain of NF-κB1 generates p50-p105 heterodimers and is required for effective p50 production. EMBO J. 19: 4712-4722.

Lomaga MA, Yeh WC, Sarosi I, Duncan GS, Furlonger C, Ho A, Morony S, Capparelli C, Van G, Kaufman S, van der Heiden A, Itie A, Wakeham A, Khoo W, Sasaki T, Cao Z, Penninger JM, Paige CJ, Lacey DL, Dunstan CR, Boyle WJ, Goeddel DV & Mak TW (1999) TRAF6 deficiency results in osteopetrosis and defective interleukin-1, CD40, and LPS signaling. Genes Dev. 13: 1015-1024.

Lorick KL, Jensen JP, Fang S, Ong AM, Hatakeyama S & Weissman AM (1999) RING fingers mediate ubiquitin-conjugating enzyme (E2)-dependent ubiquitination. Proc. Natl. Acad. Sci. USA 96: 11364-11369.

Lyapina S, Cope G, Shevchenko A, Serino G, Tsuge T, Zhou C, Wolf DA, Wei N & Deshaies RJ (2001) Promotion of NEDD-CUL1 conjugate cleavage by COP9 signalosome. Science 292: 1382-1385.

MacKichan ML, Logeat F & Israel A (1996) Phosphorylation of p105 PEST sequence via a redox-insensitive pathway up-regulates processing of p50 NF-κB. J. Biol. Chem. 271: 6084-6091.

Maniatis T (1999) A ubiquitin ligase complex essential for the NF-κB, Wnt/Wingless, and Hedgehog signaling pathways. Genes Dev. 13: 505-510.

Margottin F, Bour SP, Durand H, Selig L, Benichou S, Richard V, Thomas D, Strebel K & Benarous R (1998) A novel human WD protein, h-β TrCp, that interacts with HIV-1 Vpu connects CD4 to the ER degradation pathway through an F-box motif. Mol. Cell 1: 565-574.

McWhirter SM, Pullen SS, Holton JM, Crute JJ, Kehry MR & Alber T (1999) Crystallographic analysis of CD40 recognition and signaling by human TRAF2. Proc. Natl. Acad. Sci. USA 96: 8408-8413.

Mercurio F, DiDonato JA, Rosette C & Karin M (1993) p105 and p98 precursor proteins play an active role in NF-κB-mediated signal transduction. Genes Dev. 7: 705-718.

Mercurio F, Zhu H, Murray BW, Shevchenko A, Bennett BL, Li J, Young DB, Barbosa M, Mann M, Manning A & Rao A (1997) IKK-1 and IKK-2: cytokine-activated IκB kinases essential for NF-κB activation. Science 278: 860-866.

Moraes TF, Edwards RA, McKenna S, Pastushok L, Xiao W, Glover JN & Ellison MJ (2001) Crystal structure of the human ubiquitin conjugating enzyme complex, hMms2-hUbc13. Nat. Struct. Biol. 8: 669-673.

Naito A, Azuma S, Tanaka S, Miyazaki T, Takaki S, Takatsu K, Nakao K, Nakamura K, Katsuki M, Yamamoto T & Inoue J (1999) Severe osteopetrosis, defective interleukin-1 signalling and lymph node organogenesis in TRAF6-deficient mice. Genes Cells 4: 353-362.

Naito A, Yoshida H, Nishioka E, Satoh M, Azuma S, Yamamoto T, Nishikawa S & Inoue J (2002) TRAF6-deficient mice display hypohidrotic ectodermal dysplasia. Proc. Natl. Acad. Sci. USA 99: 8766-8771.

Nash P, Tang X, Orlicky S, Chen Q, Gertler FB, Mendenhall MD, Sicheri F, Pawson T & Tyers M (2001) Multisite phosphorylation of a CDK inhibitor sets a threshold for the onset of DNA replication. Nature 414: 514-521.

Ninomiya-Tsuji J, Kishimoto K, Hiyama A, Inoue J, Cao Z & Matsumoto K (1999) The kinase TAK1 can activate the NIK-IκB as well as the MAP kinase cascade in the IL-1 signalling pathway. Nature 398: 252-256.

Ohta T, Michel JJ, Schottelius AJ & Xiong Y (1999) ROC1, a homolog of APC11, represents a family of cullin partners with an associated ubiquitin ligase activity. Mol. Cell 3: 535-541.

Orian A, Gonen H, Bercovich B, Fajerman I, Eytan E, Israel A, Mercurio F, Iwai K, Schwartz AL & Ciechanover A (2000) SCF(β)(-TrCP) ubiquitin ligase-mediated processing of NF-κB p105 requires phosphorylation of its C-terminus by IκB kinase. EMBO J. 19: 2580-2591.

Palombella VJ, Rando OJ, Goldberg AL & Maniatis T (1994) The ubiquitin-proteasome pathway is required for processing the NF-κB1 precursor protein and the activation of NF-κB. Cell 78: 773-785.

Pickart CM (2000) Ubiquitin in chains. Trends Biochem. Sci. 25: 544-548.

Pickart CM (2001) Mechanisms underlying ubiquitination. Annu. Rev. Biochem. 70: 503-533.

Poljak L, Carlson L, Cunningham K, Kosco-Vilbois MH & Siebenlist U (1999) Distinct activities of p52/NF-κB required for proper secondary lymphoid organ microarchitecture: functions enhanced by Bcl-3. J. Immunol. 163: 6581-6588.

Rape M, Hoppe T, Gorr I, Kalocay M, Richly H & Jentsch S (2001) Mobilization of processed, membrane-tethered SPT23 transcription factor by CDC48(UFD1/NPL4), a ubiquitin-selective chaperone. Cell 107: 667-677.

Rape M & Jentsch S (2002) Taking a bite: proteasomal protein processing. Nat. Cell Biol. 4: E113-116.

Regnier CH, Song HY, Gao X, Goeddel DV, Cao Z & Rothe M (1997) Identification and characterization of an IκB kinase. Cell 90: 373-383.

Rothe M, Wong SC, Henzel WJ & Goeddel DV (1994) A novel family of putative signal transducers associated with the cytoplasmic domain of the 75 kDa tumor necrosis factor receptor. Cell 78: 681-692.

Rothwarf DM, Zandi E, Natoli G & Karin M (1998) IKK-γ is an essential regulatory subunit of the IκB kinase complex. Nature 395: 297-300.

Ruffner H, Joazeiro CA, Hemmati D, Hunter T & Verma IM (2001) Cancer-predisposing mutations within the RING domain of BRCA1: loss of ubiquitin protein ligase activity and protection from radiation hypersensitivity. Proc. Natl. Acad. Sci. USA 98: 5134-5139.

Salghetti SE, Caudy AA, Chenoweth JG & Tansey WP (2001) Regulation of transcriptional activation domain function by ubiquitin. Science 293: 1651-1653.

Scheffner M, Huibregtse JM, Vierstra RD & Howley PM (1993) The HPV-16 E6 and E6-AP complex functions as a ubiquitin-protein ligase in the ubiquitination of p53. Cell 75: 495-505.

Scheffner M, Nuber U & Huibregtse JM (1995) Protein ubiquitination involving an E1-E2-E3 enzyme ubiquitin thioester cascade. Nature 373: 81-83.

Schwandner R, Yamaguchi K & Cao Z (2000) Requirement of tumor necrosis factor receptor-associated factor (TRAF)6 in interleukin 17 signal transduction. J. Exp. Med. 191: 1233-1240.

Sears C, Olesen J, Rubin D, Finley D & Maniatis T (1998) NF-κB p105 processing via the ubiquitin-proteasome pathway. J. Biol. Chem. 273: 1409-1419.

Senftleben U, Cao Y, Xiao G, Greten FR, Krahn G, Bonizzi G, Chen Y, Hu Y, Fong A, Sun SC & Karin M (2001) Activation by IKKα of a second, evolutionary conserved, NF-κB signaling pathway. Science 293: 1495-1499.

Seol JH, Feldman RM, Zachariae W, Shevchenko A, Correll CC, Lyapina S, Chi Y, Galova M, Claypool J, Sandmeyer S, Nasmyth K & Deshaies RJ (1999) Cdc53/cullin and the essential Hrt1 RING-H2 subunit of SCF define a ubiquitin ligase module that activates the E2 enzyme Cdc34. Genes Dev. 13: 1614-1626.

Shinkura R, Kitada K, Matsuda F, Tashiro K, Ikuta K, Suzuki M, Kogishi K, Serikawa T & Honjo T (1999) Alymphoplasia is caused by a point mutation in the mouse gene encoding Nf-κ b-inducing kinase. Nat. Genet. 22: 74-77.

Shirane M, Hatakeyama S, Hattori K & Nakayama K (1999) Common pathway for the ubiquitination of IκBα, IκBβ, and IκBε mediated by the F-box protein FWD1. J. Biol. Chem. 274: 28169-28174.

Silverman N & Maniatis T (2001) NF-κB signaling pathways in mammalian and insect innate immunity. Genes Dev. 15: 2321-2342.

Skowyra D, Craig KL, Tyers M, Elledge SJ & Harper JW (1997) F-box proteins are receptors that recruit phosphorylated substrates to the SCF ubiquitin-ligase complex. Cell 91: 209-219.

Spence J, Sadis S, Haas AL & Finley D (1995) A ubiquitin mutant with specific defects in DNA repair and multiubiquitination. Mol. Cell. Biol. 15: 1265-1273.

Spencer E, Jiang J & Chen ZJ (1999) Signal-induced ubiquitination of IκBα by the F-box protein Slimb/β-TrCP. Genes Dev. 13: 284-294.

Suzuki H, Chiba T, Kobayashi M, Takeuchi M, Suzuki T, Ichiyama A, Ikenoue T, Omata M, Furuichi K & Tanaka K (1999) IκBα ubiquitination is catalyzed by an SCF-like complex containing Skp1, cullin-1, and two F-box/WD40-repeat proteins, βTrCP1 and βTrCP2. Biochem. Biophys. Res. Commun. 256: 127-132.

Swaffield JC, Melcher K & Johnston SA (1995) A highly conserved ATPase protein as a mediator between acidic activation domains and the TATA-binding protein. Nature 374: 88-91.

Takaesu G, Kishida S, Hiyama A, Yamaguchi K, Shibuya H, Irie K, Ninomiya-Tsuji J & Matsumoto K (2000) TAB2, a novel adaptor protein, mediates activation of TAK1 MAPKKK by linking TAK1 to TRAF6 in the IL-1 signal transduction pathway. Mol. Cell 5: 649-658.

Takayanagi H, Ogasawara K, Hida S, Chiba T, Murata S, Sato K, Takaoka A, Yokochi T, Oda H, Tanaka K, Nakamura K & Taniguchi T (2000) T-cell-mediated regulation of osteoclastogenesis by signalling cross-talk between RANKL and IFN-γ. Nature 408: 600-605.

Tan P, Fuchs SY, Chen A, Wu K, Gomez C, Ronai Z & Pan ZQ (1999) Recruitment of a ROC1-CUL1 ubiquitin ligase by Skp1 and HOS to catalyze the ubiquitination of IκBα. Mol. Cell 3: 527-533.

Tang Z, Li B, Bharadwaj R, Zhu H, Ozkan E, Hakala K, Deisenhofer J & Yu H (2001) APC2 Cullin protein and APC11 RING protein comprise the minimal ubiquitin ligase module of the anaphase-promoting complex. Mol. Biol. Cell 12: 3839-3851.

Traenckner EB, Pahl HL, Henkel T, Schmidt KN, Wilk S & Baeuerle PA (1995) Phosphorylation of human IκB-α on serines 32 and 36 controls IκB-α proteolysis and NF-κB activation in response to diverse stimuli. EMBO J. 14: 2876-2883.

VanDemark AP, Hofmann RM, Tsui C, Pickart CM & Wolberger C (2001) Molecular insights into polyubiquitin chain assembly: crystal structure of the Mms2/Ubc13 heterodimer. Cell 105: 711-720.

Verma R, Aravind L, Oania R, McDonald WH, Yates JR, Koonin EV & Deshaies RJ (2002) Role of Rpn11 metalloprotease in deubiquitination and degradation by the 26S proteasome. Science 298:

611-615.
von Arnim AG (2001) A hitchhiker's guide to the proteasome. Sci. STKE 2001: PE2.
Wajant H, Grell M & Scheurich P (1999) TNF receptor associated factors in cytokine signaling. Cytokine Growth Factor Rev. 10: 15-26.
Wajant H & Scheurich P (2001) Tumor necrosis factor receptor-associated factor (TRAF) 2 and its role in TNF signaling. Int. J. Biochem. Cell Biol. 33: 19-32.
Wall MA, Coleman DE, Lee E, Iniguez-Lluhi JA, Posner BA, Gilman AG & Sprang SR (1995) The structure of the G protein heterotrimer Giα1β1γ2. Cell 83: 1047-1058.
Wang C, Deng L, Hong M, Akkaraju GR, Inoue J & Chen ZJ (2001) TAK1 is a ubiquitin-dependent kinase of MKK and IKK. Nature 412: 346-351.
Watanabe N, Iwamura T, Shinoda T & Fujita T (1997) Regulation of NFKB1 proteins by the candidate oncoprotein BCL-3: generation of NF-κB homodimers from the cytoplasmic pool of p50-p105 and nuclear translocation. EMBO J. 16: 3609-3620.
Wilkinson KD (2000) Ubiquitination and deubiquitination: targeting of proteins for degradation by the proteasome. Semin. Cell Dev. Biol. 11: 141-148.
Winston JT, Strack P, Beer-Romero P, Chu CY, Elledge SJ & Harper JW (1999) The SCFβ-TRCP-ubiquitin ligase complex associates specifically with phosphorylated destruction motifs in IκBα and β-catenin and stimulates IκBα ubiquitination in vitro. Genes Dev. 13: 270-283.
Woronicz JD, Gao X, Cao Z, Rothe M & Goeddel DV (1997) IκB kinase-β: NF-κB activation and complex formation with IκB kinase-α and NIK. Science 278: 866-869.
Wu C & Ghosh S (1999) β-TrCP mediates the signal-induced ubiquitination of IκBβ. J. Biol. Chem. 274: 29591-29594.
Wu K, Fuchs SY, Chen A, Tan P, Gomez C, Ronai Z & Pan ZQ (2000) The SCF(HOS/β-TRCP)-ROC1 E3 ubiquitin ligase utilizes two distinct domains within CUL1 for substrate targeting and ubiquitin ligation. Mol. Cell. Biol. 20: 1382-1393.
Xiao G, Harhaj EW & Sun SC (2001) NF-κB-inducing kinase regulates the processing of NF-κB2 p100. Mol. Cell 7: 401-409.
Yamaguchi K, Shirakabe K, Shibuya H, Irie K, Oishi I, Ueno N, Taniguchi T, Nishida E & Matsumoto K (1995) Identification of a member of the MAPKKK family as a potential mediator of TGF-β signal transduction. Science 270: 2008-2011.
Yamaguchi K, Nagai S, Ninomiya-Tsuji J, Nishita M, Tamai K, Irie K, Ueno N, Nishida E, Shibuya H & Matsumoto K (1999) XIAP, a cellular member of the inhibitor of apoptosis protein family, links the receptors to TAB1-TAK1 in the BMP signaling pathway. EMBO J. 18: 179-187.
Yamaoka S, Courtois G, Bessia C, Whiteside ST, Weil R, Agou F, Kirk HE, Kay RJ & Israel A (1998) Complementation cloning of NEMO, a component of the IκB kinase complex essential for NF-κB activation. Cell 93: 1231-1240.
Yang J, Lin Y, Guo Z, Cheng J, Huang J, Deng L, Liao W, Chen Z, Liu Z & Su B (2001) The essential role of MEKK3 in TNF-induced NF-κB activation. Nat. Immunol. 2: 620-624.
Yang Y, Fang S, Jensen JP, Weissman AM & Ashwell JD (2000) Ubiquitin protein ligase activity of IAPs and their degradation in proteasomes in response to apoptotic stimuli. Science 288: 874-877.
Yao T & Cohen RE (2002) A cryptic protease couples deubiquitination and degradation by the proteasome. Nature 419: 403-407.
Yaron A, Hatzubai A, Davis M, Lavon I, Amit S, Manning AM, Andersen JS, Mann M, Mercurio F & Ben-Neriah Y (1998) Identification of the receptor component of the IκBα-ubiquitin ligase. Nature 396: 590-594.
Ye H, Park YC, Kreishman M, Kieff E & Wu H (1999) The structural basis for the recognition of diverse receptor sequences by TRAF2. Mol. Cell 4: 321-330.
Zandi E, Rothwarf DM, Delhase M, Hayakawa M & Karin M (1997) The IκB kinase complex (IKK) contains two kinase subunits, IKKα and IKKβ, necessary for IκB phosphorylation and NF-κB activation. Cell 91: 243-252.
Zhang S, Skalsky Y & Garfinkel DJ (1999) MGA2 or SPT23 is required for transcription of the δ9 fatty acid desaturase gene, OLE1, and nuclear membrane integrity in Saccharomyces cerevisiae. Genetics 151: 473-483.

Zhang SQ, Kovalenko A, Cantarella G & Wallach D (2000) Recruitment of the IKK signalosome to the p55 TNF receptor: RIP and A20 bind to NEMO (IKKγ) upon receptor stimulation. Immunity 12: 301-311.

Zheng N, Wang P, Jeffrey PD & Pavletich NP (2000) Structure of a c-Cbl-UbcH7 complex: RING domain function in ubiquitin-protein ligases. Cell 102: 533-539.

Zheng N, Schulman BA, Song L, Miller JJ, Jeffrey PD, Wang P, Chu C, Koepp DM, Elledge SJ, Pagano M, Conaway RC, Conaway JW, Harper JW & Pavletich NP (2002) Structure of the Cul1-Rbx1-Skp1-F boxSkp2 SCF ubiquitin ligase complex. Nature 416: 703-709.

8. AFFILIATION

University of Texas Southwestern Medical Center, Department of Molecular Biology, 5323 Harry Hines Boulevard, Dallas, TX 75390, USA.

A. O'MAHONY, L.F. CHEN & W.C. GREENE

NEW INSIGHTS INTO THE REGULATION OF NF-κB

ROLES FOR PHOSPHORYLATION AND ACETYLATION OF NF-κB SUBUNITS

1. INTRODUCTION

Long considered to be a pivotal transcription factor regulating both the immune and inflammatory responses, NF-κB is now known to regulate such diverse processes as cell proliferation, differentiation and apoptotic cell death. NF-κB belongs to a family of transcription factors, all sharing an N-terminal Rel homology domain. NF-κB is tightly regulated within the cell at multiple levels, both during its activation and mobilization from the cytoplasm and before or during its transcriptional activation in the nucleus. Considering the wide variety of signals activating this transcription factor and the relative transcriptional strength of the RelA subunit, it is not surprising that NF-κB is regulated at multiple levels. The classical pathway leading to NF-κB activation proceeds via a kinase cascade, culminating in phosphorylation and degradation of IκBα, one member of a family of inhibitors controlling NF-κB action (Senftleben & Karin, 2002). NF-κB activation can be viewed as two discrete phases. The first comprises the proximal events leading to degradation of IκBα and rapid translocation of the liberated NF-κB complex into the nucleus. The second phase involves posttranslational modification of the NF-κB subunits themselves by both phosphorylation and acetylation, which are required for full activity of the induced nuclear NF-κB complex. While acetylation appears to occur in the nucleus, certain phosphorylation events in the second phase are likely executed in the cytoplasm before nuclear entry of NF-κB. We focus on the intracellular events that comprise this second phase of NF-κB activation, emphasizing how these changes in phosphorylation and acetylation both shape the DNA-binding and/or transactivation properties of NF-κB, thereby controlling the strength and duration of the nuclear NF-κB response.

2. PHOSPHORYLATION OF RelA

The first clue that NF-κB as well as IκB might be subject to phosphorylation stems from the observation that phorbol ester-induced PKC activation leads to changes not only in the subcellular localization of NF-κB but also in its DNA-

R. Beyaert (ed.), Nuclear Factor κB. Regulation and Role in Disease, 159-176.

binding activity (Baeuerle et al., 1988; Baeuerle & Baltimore, 1996). Accordingly, the role of Rel protein phosphorylation has recently come under greater scrutiny. Of note, the Rel proteins are phosphorylated under basal conditions in Jurkat cells. However, PMA and PHA stimulation of these cells leads to a significant increase in [^{32}P]-orthophosphate incorporation into both p50 and RelA, the two subunits of the prototypical NF-κB complex (Li et al., 1994b). RelA contains two C-terminal transactivation domains designated TA_1 and TA_2. The transactivation potential of TA_2 is significantly enhanced after stimulation with PMA, and this change in activity correlated with increased phosphorylation of this region (Schmitz et al., 1995). The enhanced phosphorylation of NF-κB is also associated with increased DNA binding of the complex measured either in vitro or in vivo. Consistent with this finding, dephosphorylation of the activated nuclear extract yields NF-κB complexes displaying diminished DNA-binding activity (Li et al., 1994a). Similarly, when TNFα-stimulated HeLa cells are radiolabeled in vivo with [^{32}P]-orthophosphate, RelA, as well as p100 and p105, are phosphorylated, while c-Rel is not. Signal-coupled phosphorylation of RelA occurs in a transient manner as determined by comparing the ratio of [^{32}P]:[^{35}S] radiolabeled RelA immunoprecipitated from TNFα-stimulated cells. While this ratio increases to 3:1 after 5 min of TNFα stimulation, it falls rapidly thereafter (Diehl et al., 1995). In the case of TNFα stimulation of HeLa cells, RelA is phosphorylated mainly on Ser529, which is located within the TA_1 domain. While there was no significant difference in the abilities of wild-type RelA and the S529A mutant of RelA to translocate to the nucleus and bind DNA, the S529A mutant failed to activate κB reporter activity in RelA$^{-/-}$ cells. In contrast to the inducible nature of phosphorylation observed in T cells, RelA is constitutively phosphorylated in B cells. The effect of phosphorylation on the p50 subunit of NF-κB is less clear. Some reports show that p50 is hyperphosphorylated in PMA/PHA-stimulated Jurkat cells. However, phosphorylated p50 is not detectable in TNFα-stimulated HeLa cells (Li et al., 1994a; Naumann & Scheidereit, 1994).

The prevailing model for the first phase of NF-κB activation states that NF-κB is sequestered by IκB in the cytoplasm until the cell is stimulated. However, more recent studies show that the latent IκB/NF-κB complex shuttles between the cytoplasm and the nucleus. The complex exhibits a significantly longer cytoplasmic dwell time, underlying the chiefly cytoplasmic pattern of expression observed in static fractionation experiments (Johnson et al., 1999; Birbach et al., 2002). The nuclear NF-κB/IκB complex does not bind DNA or exert functional effects due to the presence of the IκB component. While it is clear that signal-coupled phosphorylation of IκBα occurs in the cytoplasm, it remains unclear where the additional posttranslational modifications of the Rel proteins themselves take place. RelA is hyperphosphorylated in response to TNFα signaling, and one report demonstrates that phospho-RelA can be detected even in the presence of the proteosome inhibitor N-acetyl-leucyl-leucyl-norleucinal (ALLN), which blocks phospho-IκB degradation and RelA nuclear translocation (Sakurai et al., 1999). These data support the notion that RelA is phosphorylated, at least in part, in the cytoplasm of stimulated cells. However, it is also possible that certain phosphorylation events can occur in the nucleus. Considerable attention has focused on identifying the kinases that mediate Rel phosphorylation and several candidate kinases have been proposed (Fig. 1). The high number of different kinases implicated in Rel phosphorylation likely reflects the panoply of agonists involved in NF-κB activation.

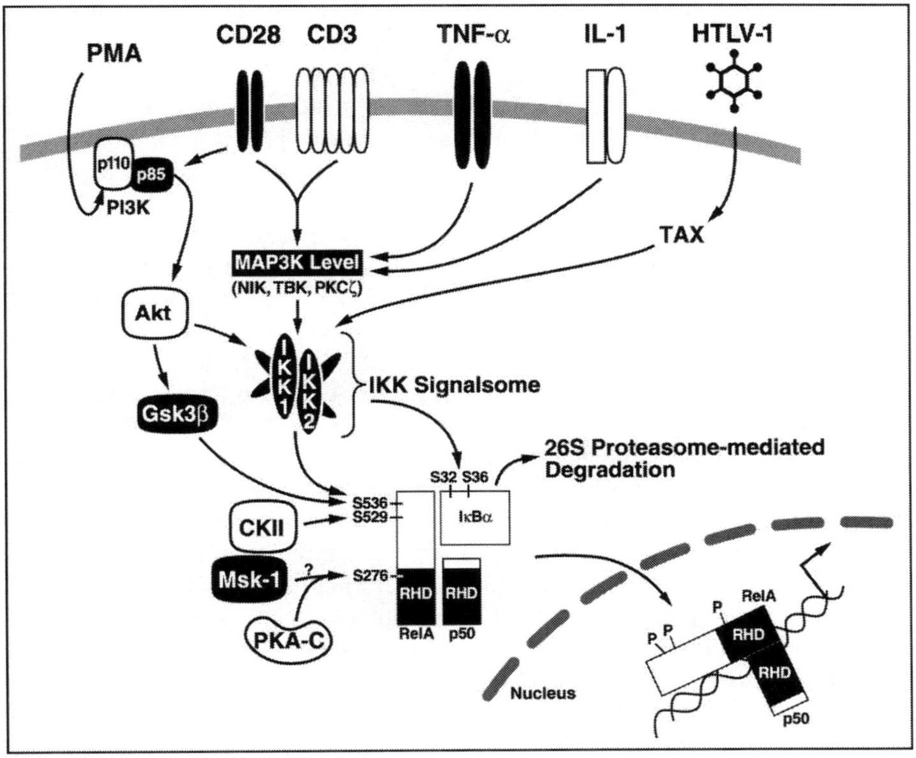

Fig. 1. Model depicting signaling pathways leading to RelA phosphorylation. In addition to the classical pathway culminating in IκBα phosphorylation, the RelA subunit is subject to modification by various cellular kinases. While RelA may be phosphorylated under basal conditions, this subunit is hyperphosphorylated upon signal-induced activation of NF-κB. Such phosphorylation appears to be critical for full target gene transactivation by the NF-κB complex.

2.1. Protein kinase A

In addition to the in vivo phosphorylation induced by TNFα or combinations of PMA and PHA, RelA, p100 and p105, but not p50 or p52, are phosphorylated in vitro by both PKA and PKC (Naumann & Scheidereit, 1994). Based on these results, several studies have focused on the cAMP-dependent protein kinase (PKA) as a potential mediator of Rel phosphorylation. In most Rel family members, a phospho-acceptor site for PKA (Arg-Arg-X-Ser) is present within the conserved Rel homology domain (Mosialos et al., 1991; Mosialos & Gilmore,

1993). The Drosophila Rel homologue (Dorsal), whose phosphorylation state is regulated by Toll receptor signaling, also exhibits enhanced nuclear translocation and transactivation activity when PKA is coexpressed (Gillespie & Wasserman, 1994). However, complicating the notion that PKA mediates Rel phosphorylation in a physiological context is the mechanism that underlies activation of this kinase in vivo. Normally present as an inactive holoenzyme in the cytoplasm (PKA-R_2C_2), elevated cAMP levels promote release of the catalytic subunits (PKA-C) by cAMP binding to the regulatory subunits (PKA-R) (Van Patten et al., 1986; Van Patten et al., 1988). However, many NF-κB-inducing stimuli like IL-1, LPS and PMA do not induce marked increases in intracellular cAMP levels. One model potentially providing a cAMP-independent mechanism for PKA regulation stems from the observation that PKA-C physically associates with the latent IκB/NF-κB complex. In this complex, IκB negatively regulates the constitutive activity of PKA-C. Following stimulation with LPS, which leads to the phosphorylation and degradation of IκB, the PKA-C kinase activity is no longer constrained. The RelA protein is then phosphorylated by the associated PKA-C at the PKA phospho-acceptor site (Ser276) located within the Rel homology domain. Phosphorylation at this site has little effect on DNA binding, but it enhances the transcriptional activity of RelA considerably (Zhong et al., 1997). Phosphorylation at Ser276 further promotes the association of RelA with CBP/p300, a transcriptional coactivator protein. These transcriptional coactivators form a key bridge between the transcription factors bound to their cognate DNA sequences with the components of the cellular transcriptional machinery. PKA phosphorylation of RelA is proposed to change the conformation of RelA, leading to loss of N- and C-terminal interactions, which in turn allows the efficient binding of p300 to a C-terminal domain of RelA (Zhong et al., 1998). Consistent with these findings, synergistic activation is obtained with combinations of RelA and CBP, however, these effects are lost after mutation of Ser276 in RelA or when a catalytically defective version of PKA-C (K73M) is expressed. It is important to note that mutation of Ser276 to alanine also can markedly impair homodimerization of RelA (Ganchi et al., 1993). Whether this defect in dimer assembly plays a role in the effects observed with Ser276, requires further clarification.

2.2. *Casein kinase II*

A second kinase reported to phosphorylate RelA is casein kinase II (CKII). A kinase capable of phosphorylating both α- and β-casein can be immunoprecipitated with activated NF-κB in IL-1-treated cells (Bird et al., 1997). The activity of this kinase is completely inhibited by heparin (20 μg/ml), a potent inhibitor of CKII. Purified CKII phosphorylates full-length, but not C-terminally truncated, forms of RelA. These findings suggest that the target site for CKII phosphorylation resides in the C-terminal portion of the RelA protein. Phospho-peptide mapping experiments show that Ser529 forms the major site of CKII phosphorylation. TNFα-induced phosphorylation of Ser529 in RelA by CKII also enhances the transcriptional activity of NF-κB (Wang & Baldwin, 1998). However, deletion or mutation of Ser529 results in only a partial loss of TNFα-induced RelA phosphorylation, raising the possibility that this agonist activates one or more additional kinases targeting RelA (Wang & Baldwin, 1998). CKII also constitutively phosphorylates the PEST sequence within the C-terminus of IκBα; however, the precise function of this modification in the induction of NF-κB remains poorly understood (Barroga et al., 1995; Krappmann et al., 1996; Lin et al., 1996). Although TNFα only weakly stimulates CKII activity, the degradation of IκBα in response to TNFα may facilitate CKII-mediated phosphorylation of RelA (Wang et al., 2000).

2.3. *p38 MAP kinase*

TNFα also activates the p38 MAP kinase pathway. This pathway appears to be required for optimal NF-κB transactivation, yet is dispensable for nuclear translocation and DNA binding by NF-κB (Beyaert et al., 1996). The p38 MAPK inhibitor SB203580 impairs TNFα-stimulated transactivation of κB-dependent reporter genes without affecting the DNA-binding activity of NF-κB. Further, the p38 MAPK inhibitors SB203580 and PD0980589 display direct repressive effects on RelA-mediated transactivation (Vanden Berghe et al., 1998). Of note, MSK1 is positioned downstream of p38 in the stress-response pathway. Intriguingly, MSK1 has been suggested to phosphorylate Ser276, the site also targeted by PKA (Vermeulen et al., 2003). Thus, more than one kinase may modify this apparently pivotal phospho-acceptor site in RelA.

2.4. *Phosphatidylinositol-3 kinase*

PI3K, a downstream effector in the IL-1-signaling pathway has also been implicated in the activation of NF-κB (Reddy et al., 1997). Upon IL-1 binding, the p110 catalytic subunit of PI3K is recruited by the p85 subunit to the cell membrane and mediates the phosphorylation of lipid substrates. In turn, these phospho-substrates activate secondary kinases, including Akt (PKB) and certain forms of PKC (Fruman & Cantley, 2002). Support for the involvement of PI3K in the NF-κB pathway derives from the observation that IL-1-stimulated cells expressing a mutated form of the IL-1R, which blocks interaction with, and activation of, the PI3 kinase, fail to mount a full NF-κB response (Marmiroli et al., 1998). The exact mechanism by which PI3K contributes to NF-κB function remains poorly defined with conflicting reports implicating this kinase at different steps in the NF-κB pathway. Some studies suggest that PI3K and Akt regulate the transactivation function of RelA by altering its state of phosphorylation. In cells labeled in vivo with [^{32}P]-orthophosphate, IL-1 induces robust phosphorylation of RelA, p50 and IκBα. However, pretreatment of these cells with the PI3K inhibitor LY294002 results in a 90% reduction in RelA phosphorylation and a marked inhibition of induced IL-8 gene expression in the apparent absence of changes in IκB degradation or nuclear translocation and DNA-binding activity of NF-κB. Overexpression of either the p110 catalytic subunit of PI3K or a constitutively active form of Akt similarly induces RelA-mediated transactivation (Sizemore et al., 1999). Further supporting a role for P13K, prostate cancer cells reconstituted with a natural PI3K inhibitor (PTEN) display normal TNFα-induced IκB degradation, RelA nuclear translocation and DNA-binding activity, yet fail to transactivate a κB-dependent reporter gene (Mayo et al., 2002). However, in a glioma cell line stably expressing PTEN, IL-1-induced DNA-binding activity and NF-κB-mediated transactivation are both inhibited without detectable changes in IκBα degradation or RelA/p50 translocation into the nucleus. In that study, p50 phosphorylation is inhibited, likely explaining the observed reduction in DNA-binding activity of the NF-κB complex. Conversely, expression of a dominant-negative version of the P13K regulatory subunit blocks transactivation of a κB reporter construct without affecting DNA binding (Reddy et al., 1997). One report suggests that PTEN may act by inhibiting the IL-1-induced physical interaction of the upstream activating Akt kinase and the IKK signalsome (Koul et al., 2001). In contrast, PTEN has been demonstrated to act by inhibiting TNFα-induced activation of the IKK complex through impairment of Akt catalytic activity (Gustin et al., 2001). A mutant Akt has also been reported to affect IκBα degradation (Kane et al., 1999). These varying findings emphasize that, while PI3K and the downstream Akt enzyme appear to play important roles in modulating NF-κB activity, the precise basis for their action remains controversial.

2.5. *Glycogen synthase kinase-3*

IL-1 also stimulates the GSK3 pathway through the action of PI3K and Akt. This event leads to phosphorylation of Ser21 on GSK3, a modification blocked by LY294002. Unlike most protein kinases, GSK3 is active under basal conditions and is inhibited after phosphorylation (Woodgett, 2001). In addition to being inhibited by PI3K and Akt, GSK3 is also phosphorylated in response to elevated levels of cyclic AMP (cAMP). The most convincing argument supporting a role for GSK3 in NF-κB activation stems from recent in vivo studies, where disruption of the gene encoding GSK3β in mice leads to embryonic lethality secondary to severe liver degeneration (Hoeflich et al., 2000). This phenotype is remarkably similar to that observed in mice where the RelA or IKK2 genes have been disrupted. Death is caused by accelerated TNFα-induced hepatocyte apoptosis due to the lack of the antiapoptotic effects of NF-κB. Consistent with this notion, $GSK3\beta^{-/-}$ embryos can be rescued from premature death by the administration of anti-TNF-α antibodies. Moreover, fibroblasts derived from $GSK3\beta^{-/-}$ embryos are hypersensitive to TNFα-induced death and display reduced NF-κB-mediated transactivation of a κB-reporter gene. Intriguingly, these $GSK3^{-/-}$ murine embryo fibroblasts display no defect in IκB degradation or NF-κB nuclear translocation. These findings support the concept that this kinase plays a role in the second phase of NF-κB activation, perhaps by directly or indirectly modifying RelA.

2.6. IκB kinases

While much of the focus on phosphorylation events occurring in the NF-κB pathway has centered on the IκB kinases (IKKs) and their direct phosphorylation of IκB, recent studies suggest that these kinases also target RelA. Recombinant IKK1 and IKK2 complexes were previously shown to phosphorylate not only an IκB substrate, but also RelA; however, the IKKs did not phosphorylate cRel or p52 (Mercurio et al., 1999). TNFα-activated IKKs were further shown to target the RelA component of NF-κB for phosphorylation after TNFα stimulation (Sakurai et al., 1999). Using GST fusion proteins comprising amino acids 1-305 or 354-551 of RelA, these investigators demonstrated that IKKs phosphorylate the C-terminal segment of RelA. In TNFα-stimulated HeLa cells, one IKK phospho-acceptor site corresponds to Ser536, a highly conserved residue within the Rel family of proteins. Transfection of IKK1 and IKK2 also results in robust phosphorylation of RelA (354-551), with IKK2 displaying constitutive activity. IKK1 phosphorylation of this site requires coexpression of TAK1 and TAB, which represent upstream activators of this kinase. Our own studies have shown that recombinant baculovirus-expressed IKK2 robustly phosphorylates the C-terminal portion of RelA. In contrast, IKK1-mediated phosphorylation of RelA was only observed in the presence of recombinant NIK, which potently activates IKK1. NIK alone fails to induce RelA phosphorylation (A. O'Mahony, M. Montano, K. Van Beneden & W.C. Greene, unpublished data). A role for NIK-induced IKK activation leading to p65 phosphorylation in vivo is supported by the analysis of NIK$^{-/-}$ mice. These mice display normal NF-κB DNA-binding activity after treatment with various cytokines; however, NF-κB-mediated gene transcription induced by stimulation with LTβ is markedly impaired (Yin et al., 2001). In terms of the IKK phospho-acceptor site in RelA, we observe diminished, though not absent, phosphorylation of the RelA-S536A mutant compared to wild-type RelA. This residual phosphorylation raises the possibility that the IKKs may also phosphorylate one or more sites in RelA (A. O'Mahony, M. Montano, K. Van Beneden & W.C. Greene, unpublished data). A role for the IKKs in regulating RelA transactivation has been confirmed using mouse embryonic fibroblast cultures from mice containing disruptions in the IKK1 or IKK2 genes (Sizemore et al., 2002). These in vivo studies revealed that IKK2 is the principal kinase governing the first phase of NF-κB activation occurring in response to cytokine stimulation, i.e. IκB phosphorylation, degradation and NF-κB nuclear translocation. The role played by IKK1 in the first phase appears more subtle. The role of IKK1 in these IKK1-deficient animals may be confounded by the genetic compensation provided by the formation of functional IKK2 homodimers. IKK1-deficient signalsomes were reported to be capable of phosphorylating IκBα and IκBβ in vitro; however, a defect in the expression of many NF-κB-regulated genes (including IκBα, M-CSF and IL-6) is evident in IKK1-deficient murine embryo fibroblasts (MEFs) (Li et al., 1999). Hence it was concluded that full NF-κB activation mediated by TNFα and IL-1 is diminished in IKK1-deficient cells and that these two homologous kinases mediate distinct functions with respect to NF-κB activation (Li et al., 2000). Both IKK1- and IKK2-deficient MEFs display diminished levels of the NF-κB-regulated genes IL-6 and IP10 in response to IL-1 or TNFα stimulation relative to wild-type MEFs. Conversely, IKK1- and IKK2-deficient MEFs exhibit intact ERK, Akt, JNK, $p90^{rsk}$ and p38-MAPK activation after TNFα or IL-1 stimulation, suggesting that the defect is related to an impairment in the NF-κB pathway. As expected from in vivo gene disruption studies, IKK1-deficient MEFs display normal IκBα degradation and NF-κB translocation. In contrast, these first-phase inductive events are completely defective in IKK2-deficient MEFs. While NF-κB translocates normally to the nucleus in IKK1-deficient MEFs, the NF-κB complex in these cells fails to transactivate target genes (A. O'Mahony, M. Montano, K. Van Beneden & W.C. Greene, unpublished data). This defect likely reflects a direct or indirect role for IKK1 in the phosphorylation of RelA. In this regard, immunoprecipitation of signalsomes from TNFα- or IL-1-stimulated IKK1- or IKK2-deficient MEFs demonstrates diminished phosphorylation of RelA (Sizemore et al., 2002).

Consistent with the notion that IKK1 plays a role in modifying the NF-κB complex, we find that the HTLV-1 Tax oncoprotein, a potent inducer of NF-κB, displays an absolute requirement for IKK1 to induce a full NF-κB transcriptional response (A. O'Mahony, M. Montano, K. Van Beneden & W.C. Greene, unpublished data). In $IKK1^{-/-}$ MEFs, Tax induces a normal first phase of NF-κB activation, with phosphorylation and degradation of IκBα and nuclear translocation of RelA proceeding normally. However, the nuclear NF-κB induced by Tax in these $IKK1^{-/-}$ cells is functionally impaired, as evidenced by its failure to activate several different κB enhancer-dependent gene reporters. This defect is selective, as Tax normally activates the CREB/ATF pathway in these $IKK1^{-/-}$ cells, leading to full stimulation of an HTLV-1 LTR. This functional defect also cannot be attributed to the embryonic stage of development of these $IKK1^{-/-}$cells, since Tax normally activates each of the κB reporters in wild-type MEFs. Further, reconstitution of $IKK1^{-/-}$ cells with wild-type IKK1, but not kinase-deficient IKK1, restores the Tax response. Finally, this defect cannot be explained by the failure of $IKK1^{-/-}$ MEFs to mediate the processing of p100 to p52 (Xiao et al., 2001), as expression of p52 by transfection does not repair the transactivation defect in these cells. The findings are consistent with distinct roles for the IKK1 and IKK2 kinases in the activation of NF-κB by Tax. IKK1 appears to play an important role in enhancing the transactivation function of NF-κB, while IKK2 plays a dominant role in signaling for IκBα degradation. Such specialization of function for IKK1 and IKK2 highlights an intriguing genetic economy within the signalsome complex.

2.7. Other kinases

Other kinases that act at the level of RelA phosphorylation and have been implicated in NF-κB regulation include PKCζ and NAK/TBK/T2K. Mice in which

PKCζ has been disrupted by homologous recombination display phenotypic alterations in secondary lymphoid organs. These PKCζ-deficient MEFs also display significantly diminished transactivation of a κB reporter following TNFα stimulation. Moreover, TNFα-induced phosphorylation of p65 is impaired in these PKCζ $^{-/-}$ cells (Leitges et al., 2001). Likewise, in another study in which the gene encoding the NAK/TBK/T2K kinase was disrupted in mice, the effects were largely confined to the second phase of activation, leading to a compromise in NF-κB transcriptional activity. Although T2K$^{-/-}$ animals die at approximately E14.5 of massive liver cell degeneration and apoptosis, MEFs exhibit normal degradation of IκBα- and DNA-binding activity in response to either TNFα or IL-1. NF-κB-mediated transactivation is dramatically decreased in these cells (Bonnard et al., 2000).

3. ACETYLATION OF RelA

In addition to the recruitment and binding of NF-κB to its cognate DNA enhancer, transcriptional regulation of NF-κB involves the actions of various coactivator and corepressor proteins. For example, transactivation mediated by NF-κB is greatly increased by coexpression of p300/CBP coactivators (Gerritsen et al., 1997; Sheppard et al., 1999; Vanden Berghe et al., 1999). Moreover, NF-κB has been shown to interact physically with both CBP and p300 (Gerritsen et al., 1997; Perkins et al., 1997; Sheppard et al., 1999). Each of these enzymes contains intrinsic histone acetyltransferase (HAT) activity (Chrivia et al., 1993; Eckner et al., 1994). The interplay of RelA and p300 involves both the N-terminal Rel homology domain and the C-terminal transcriptional activation domain of RelA interacting with the amino-terminal CH1 domain of p300 (Gerritsen et al., 1997; Sheppard et al., 1999). SRC-1/N-CoA-1, a nuclear receptor coactivator within the p160 family, also participates as a coactivator with NF-κB (Na et al., 1998; Sheppard et al., 1999). SRC-1 interacts with the p50 subunit of NF-κB and acts synergistically with p300 to enhance NF-κB-mediated transactivation (Na et al., 1998). This synergy likely derives from the cooperative recruitment of two different coactivator molecules (e.g. SRC-1 and CBP/p300) by the NF-κB p50 and RelA subunits, respectively. Other studies suggest that a second p160 family member (SRC-3) may also exert coactivating effects with NF-κB and serve as a target for phosphorylation by the IKKs (Werbajh et al., 2000; Wu et al., 2002). In contrast to p300 and CBP, the PCAF acetyltransferase has not been shown to interact directly with either RelA or p50. However, intracellular microinjection of anti-PCAF antibodies compromises NF-κB action, while antibodies specific for CBP are curiously without effect (Sheppard et al., 1999).

The HAT activity of p300/CBP regulates gene expression in part through acetylation of the N-terminal tails of histones (Berger, 1999). Acetylated histones are associated with transcriptionally active regions in the genome, while deacetylated histones accumulate in transcriptionally repressed regions (Imhof et al., 1997; Kuo et al., 1998; Ng et al., 2000). However, in addition to the acetylation of histones, p300/CBP also directly acetylates a variety of transcription factors, including p53, GATA-1, MyoD, TFIIEβ, and E2F (Gu & Roeder, 1997; Imhof et al., 1997; Boyes et al., 1998; Sartorelli et al., 1999; Martinez-Balbas et al., 2000). Acetylation of these transcription factors leads to alterations in various functions, such as DNA binding, transcriptional activity, protein-protein interactions and protein stability (Martinez-Balbas et al., 2000; Sterner & Berger, 2000; Chen et al., 2001). Recent studies discussed in greater detail below indicate that RelA is also subject to acetylation and that this modification controls many of its key biological properties, specifically DNA binding, transactivation and assembly with IκBα.

In addition to binding to various coactivators, NF-κB also interacts with a family of corepressors including specific histone deacetylases (HDACs) (Ashburner et al., 2001; Chen et al., 2001; Zhong et al., 2002). These HDACs inhibit NF-κB action through a variety of mechanisms, such as direct modification of the Rel proteins themselves and targeting of the histone tails surrounding various NF-κB-regulated genes. The N-terminal region of the RelA subunit of NF-κB selectively interacts with HDAC3. Specific association of these two proteins leads to the deacetylation of RelA, which in turn promotes IκBα assembly and rapid IκBα-dependent export of RelA out of the nucleus via a CRM-1-dependent pathway (Chen et al., 2001). As such, the deacetylation of RelA limits the duration of nuclear NF-κB action. The association of NF-κB with the HDAC1 and HDAC2 corepressor proteins represses expression of NF-κB-regulated genes, including the IL-8 gene (Ashburner et al., 2001). In contrast to the mechanism underlying HDAC3 action, HDAC1 and HDAC2 appear to operate at the level of chromatin structure. The N-terminal region of p50 appears to mediate the interaction of HDAC1 with the NF-κB complex (Zhong et al., 2002). As noted earlier, changes in the phosphorylation status of nuclear NF-κB regulate its association with CBP/p300 and HDAC-1, and as such these events control the activity of NF-κB.

3.1. RelA is acetylated by p300/CBP and deacetylated by HDAC3

Recent studies have highlighted the role of acetylation and deacetylation in the regulation of NF-κB action both in vitro and in vivo (Chen et al., 2001; Furia et al., 2002). [^{3}H]-acetate radiolabeling indicates that when overexpressed, RelA is acetylated in vivo by an endogenous HAT. More importantly, endogenous RelA is similarly acetylated following TNFα stimulation, indicating that acetylation occurs under conditions of physiological expression and in a signal-coupled manner. In vivo acetylation of RelA is detectable using antiacetylated lysine antibodies, provided cells are cotransfected with expression vectors encoding a relevant HAT. Both p300 and CBP, but not PCAF, acetylate RelA under these conditions. Not surprisingly, mutation of the acetyltransferase domain within p300 prevents this enzyme from acetylating RelA (Chen et al., 2002). Further, coexpression of either a HAT-deficient mutant of p300 or the E1A gene product of adenovirus, an inhibitor of endogenous p300/CBP activity (Chakravarti et al., 1999; Hamamori et al., 1999), blocks the acetylation of RelA induced by endogenous HAT(s). These findings suggest that p300/CBP plays a key role in the regulation of RelA acetylation in vivo.

RelA acetylation occurs at three major sites Lys218, Lys221 and Lys310. Substitution of all three of these lysines with arginine, which preserves overall charge but blocks acetylation, markedly inhibits the acetylation of RelA in vivo. Lys218 and Lys221 are highly conserved within all members of the Rel family, while K310 uniquely appears in RelA. Not unexpectedly, RelA is also subject to deacetylation. This reaction involves a selective action of HDAC3 in vivo. Overexpression of HDAC3 not only leads to deacetylation of RelA, but also markedly impairs transactivation of κB enhancer-dependent gene expression. This inhibitory effect of HDAC3 is blocked by addition of trichostatin A (TSA), a general HDAC inhibitor, indicating that the deacetylase function of HDAC3 is required for these effects (Chen et al., 2001).

The p50 subunit of NF-κB is also subject to acetylation by p300/CBP in vitro (Furia et al., 2002). However, this acetylation of p50 was reported to depend on the presence of the HIV Tat protein, raising questions concerning the generality of this posttranslational modification. Conversely, our recent studies suggest that p50 is acetylated by p300 and CBP, independently of the presence of Tat. Lys431, Lys440 and Lys441, located within the C-terminal portion of p50, appear to correspond to the major sites of acetylation. Sequence similarities have been noted for the lysine acetylation sites and flanking sequences in p50, p53, Tat and ACTR (Furia et al., 2002).

3.2. Regulation of RelA transcriptional activity by acetylation

A potential role for acetylation in the regulation of NF-κB-mediated transactivation emerged with the finding that TSA enhances κB-luciferase reporter gene expression in the presence of TNFα (Chen et al., 2001). TSA also potentiates TNFα-mediated activation of the IL-6 promoter (Vanden Berghe et al., 1999). NF-κB activation is also increased following coexpression of wild-type p300/CBP; however, these effects are not observed when a HAT-deficient mutant of p300 (HAT^-) is coexpressed (Gerritsen et al., 1997; Sheppard et al., 1999; Chen et al., 2002). Conversely, expression of HDAC3, but not of HDAC1, 2, 4, 5 and 6, inhibits TNFα-mediated activation of NF-κB. In the presence of TSA, these inhibitory effects of HDAC3 are not observed (Chen et al., 2001). Analysis of arginine substitution mutants in RelA, corresponding to the sites of acetylation, indicated that acetylation of Lys221 and Lys310 is important in regulating the overall transcriptional activity of NF-κB (Chen et al., 2002). As discussed below, acetylation of Lys221 also is centrally involved in regulating DNA binding and IκBα assembly. Changes in these activities likely contribute to the observed decrease in RelA transcriptional activity. In contrast, mutation of Lys310 to arginine does not alter DNA binding or IκBα assembly, yet does markedly inhibit transactivation. We speculate that Lys310 forms a platform for recruitment of an unidentified factor or factors contributing to the RelA transcriptional response. Acetylation of the p50 subunit of NF-κB also has been shown to enhance transcription on the HIV long-terminal repeat (Furia et al., 2002).

3.3. Acetylation and binding to the κB enhancer

The DNA-binding activity of various transcription factors is known to be altered by acetylation (Bannister & Miska, 2000; Sterner & Berger, 2000). For example, acetylation of both p53 and GATA-1 augments their intrinsic DNA-binding activity measured either in vitro or in vivo (Gu & Roeder, 1997; Boyes et al., 1998). Acetylation also modulates the DNA-binding affinity of NF-κB. Blockade of HDAC activity by addition of TSA leads to the enhancement of TNFα-induced NF-κB DNA binding (Chen et al., 2001). With regard to RelA homodimers, mutation of Lys221 produces a sharp decline in overall DNA-binding activity. In contrast, mutation of the two other major acetylation sites in RelA, Lys218 and Lys310, does not alter DNA-binding activity of these RelA homodimers. When these RelA mutants are tested in the context of p50/RelA NF-κB complexes, steady-state DNA-binding activity measured in electrophoretic mobility shift assays (EMSAs) is not markedly changed. However, when unlabeled κB enhancer DNA is added as a competitor, the off-rate of DNA binding to the κB enhancer is accelerated when acetylation is blocked, either by introduction of the arginine for lysine mutation at residue 221 or by coexpression of the dominant-negative p300 HAT-deficient mutant (Chen et al., 2002). Inspection of the crystal structure of the RelA/p50-DNA complex reveals that Lys221 participates in a direct contact with the DNA backbone (Chen et al., 1998). Therefore, acetylation of Lys221 may result in a conformational change in the protein that promotes enhanced binding to the κB enhancer (Chen et al., 2002). A similar mechanism has been previously postulated for GATA-1, where acetylation in the DNA-binding domain enhances the DNA-binding activity of this transcription factor (Boyes et al., 1998). Acetylation of p50 also leads to increased DNA binding as demonstrated in streptavidin/biotin pull-down assays employing radiolabeled κB enhancer oligonucleotides. This change in the DNA-binding affinity of p50 is correlated with increased transcriptional activation of NF-κB on the HIV long terminal repeat (Furia et al., 2002).

3.4. Acetylation and IκBα binding

Acetylation of RelA also plays a key role in regulating its assembly with the IκBα inhibitor. In GST-IκBα pull-down studies, p300-mediated acetylation of RelA markedly diminishes the binding of RelA to GST-IκBα. Conversely, coexpression of HDAC3, which promotes deacetylation of RelA, enhances RelA binding to GST-IκBα. Studies of the various arginine for lysine substitution mutants of RelA reveal that Lys221 plays a key role in IκBα assembly (Chen et al., 2002). In the crystal structure of RelA-p50-IκBα (Huxford et al., 1998; Jacobs & Harrison, 1998), Lys221 directly interacts with Met279 present in the sixth ankyrin repeat of IκBα. Acetylation of Lys221 may thus result in a conformational change in RelA that both improves its affinity for the κB enhancer and disables its interaction with IκBα (Fig. 2). Alternatively, the more rapid off-rate of κB enhancer DNA binding displayed by RelA K221R may derive in part from its greater affinity for IκBα.

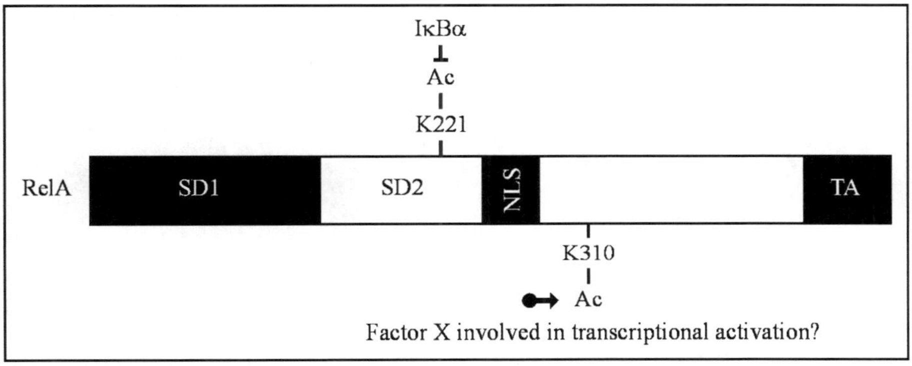

Fig. 2. Schematic representation of RelA indicating the major sites targeted by acetyltransferases. Acetylation of different lysine residues regulates distinct functions of RelA. Acetylation of Lys221 increases the DNA-binding affinity to the κB enhancer and prevents the association of RelA with IκBα. Acetylation of Lys310 enhances the full transactivation activity of RelA, although the mechanism is still unknown.

3.5. IκBα-dependent nuclear export of deacetylated RelA

When overexpressed in HeLa cells, GFP-RelA fusion proteins are mainly localized in the nuclear compartment. However, when coexpressed with HDAC3, the GFP-RelA fusion protein is almost exclusively found in the cytoplasm (Chen et al., 2001). Consistent with these results, a GFP-RelA fusion protein containing arginine for lysine substitutions at the three major acetylation sites principally localizes in the cytoplasm. HDAC3 expression induces cytoplasmic translocation of RelA. This effect is due to the deacetylation of RelA, which in turn promotes IκBα assembly and IκBα-dependent nuclear export of RelA. Consistent with this model, HDAC3 fails to promote RelA translocation into the cytoplasm. Moreover, when expressed in IκBα$^{-/-}$ MEFs, the GFP-RelA-KR mutant protein (all three lysines substituted with arginine) is predominantly located in the nucleus. Further, the cytoplasmic expression pattern of RelA-KR mutant appears to depend on the nuclear export signal (NES) present in IκBα. Expression of a NES-deficient mutant of IκBα produces a predominantly nuclear pattern of expression for the RelA-KR mutant in IκBα$^{-/-}$ MEFs. In contrast, wild-type IκBα promotes efficient translocation of GFP-RelA-KR to the cytoplasm in these cells. Acetylation may also regulate the subcellular localization of p50, as mutation of the acetylation sites at Lys440 and Lys441 apparently leads to a lack of nuclear entry (Chen et al., 2002; Furia et al., 2002). Taken together, these data show that reversible acetylation influences the subcellular localization of the NF-κB complex (Fig. 3).

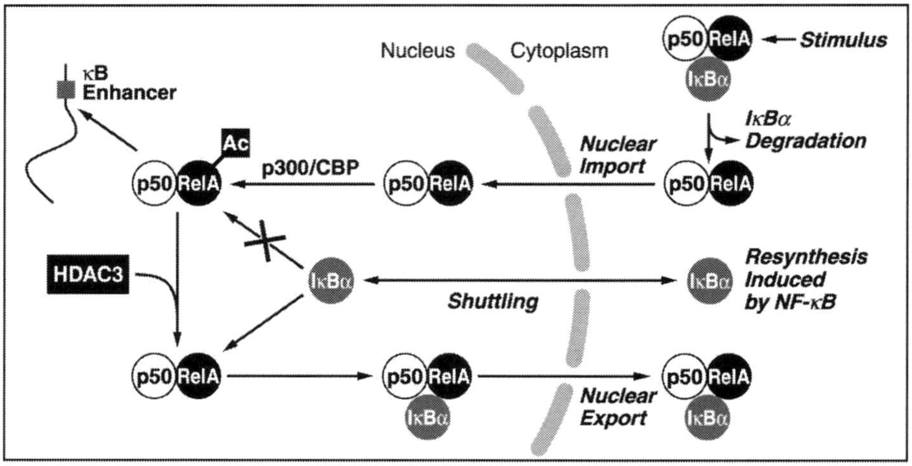

Fig. 3. Model indicating how the reversible acetylation of RelA forms an intranuclear molecular switch regulating the duration of the NF-κB response. HDAC3-mediated deacetylation of RelA promotes IκBα binding and IκBα-dependent export of the NF-κB complex to the cytoplasm. This deacetylation-controlled response leads to termination of the NF-κB transcriptional response and aids in re-establishing latent cytoplasmic forms of NF-κB bound to IκBα, thereby readying the cell to respond to the next NF-κB–inducing stimulus.

4. CONCLUSION AND FUTURE PERSPECTIVES

The above findings reveal key regulatory roles for both phosphorylation and acetylation in the biology of RelA. These posttranslational modifications affect many different biological properties of RelA, including DNA binding, transcriptional activity, assembly with the IκB inhibitors and subcellular localization of the RelA-containing complexes. However, many questions remain as to how these events are regulated. In this regard, it is intriguing to consider the possibility that acetylation of RelA may be contingent on prior phosphorylation of RelA. Precedence for such a scenario is found with p53. Phosphorylation of this tumor suppressor stimulates acetylation, likely by enhancing its association with CBP/p300 (Sakaguchi et al., 1998; Liu et al., 1999). A similar interdependent relationship between phosphorylation and acetylation has been described for the retinoblastoma tumor suppressor protein (pRb) (Chan et al., 2001). In the case of RelA, phosphorylation at Ser276 by PKA enhances the binding of p300/CBP (Zhong et al., 1998). RelA is also phosphorylated at multiple sites by different kinases within its C-terminal transactivation domain (Fig. 1), a region that participates in the interaction with p300/CBP (Zhong et al., 2002). Phosphorylation at one or more of these sites could thus trigger the recruitment of this coactivator, leading in turn to enhanced RelA acetylation. How the interplay between RelA and HDAC3 is regulated remains unknown. Future studies will help to further define the relationship between these intriguing reactions and how they contribute to the overall regulation of NF-κB action. It is interesting to consider the possibility that inhibition of one or both of these posttranslational modifications by small molecule antagonists could produce significant anti-inflammatory or immunosuppressive effects. The unknown in this scenario is whether the potential toxicity of such selective interference in the NF-κB pathway would be manageable.

5. REFERENCES

Ashburner BP, Westerheide SD & Baldwin AS (2001) The p65 (RelA) subunit of NF-κB interacts with the histone deacetylase (HDAC) corepressors HDAC1 and HDAC2 to negatively regulate gene expression. Mol. Cell. Biol. 21: 7065-7077.

Baeuerle PA, Lenardo M, Pierce JW & Baltimore D (1988) Phorbol-ester-induced activation of the NF-κB transcription factor involves dissociation of an apparently cytoplasmic NF-κB/inhibitor complex. Cold Spring Harb. Symp. Quant. Biol. 53: 789-789.

Baeuerle PA & Baltimore D (1996) NF-κB: ten years after. Cell 87: 13-20.

Bannister AJ & Miska EA (2000) Regulation of gene expression by transcription factor acetylation. Cell. Mol. Life Sci. 57: 1184-1192.

Barroga CF, Stevenson JK, Schwarz EM & Verma IM (1995) Constitutive phosphorylation of IκBα by casein kinase II. Proc. Natl. Acad. Sci. USA 92: 7637-7641.

Berger SL (1999) Gene activation by histone and factor acetyltransferases. Curr. Opin. Cell Biol. 11: 336-341.

Beyaert R, Cuenda A, Vanden Berghe W, Plaisance S, Lee JC, Haegeman G, Cohen P & Fiers W (1996) The p38/RK mitogen-activated protein kinase pathway regulates interleukin-6 synthesis response to tumor necrosis factor. EMBO J. 15: 1914-1923.

Birbach A, Gold P, Binder BR, Hofer E, de Martin R & Schmid JA (2002) Signaling molecules of the NF-κB pathway shuttle constitutively between cytoplasm and nucleus. J. Biol. Chem. 277: 10842-10851.

Bird TA, Schooley K, Dower SK, Hagen H & Virca GD (1997) Activation of nuclear transcription factor NF-κB by interleukin-1 is accompanied by casein kinase II-mediated phosphorylation of the p65 subunit. J. Biol. Chem. 272: 32606-32612.

Bonnard M, Mirtsos C, Suzuki S, Graham K, Huang J, Ng M, Itie A, Wakeham A, Shahinian A, Henzel WJ, Elia AJ, Shillinglaw W, Mak TW, Cao Z & Yeh WC (2000) Deficiency of T2K leads to apoptotic liver degeneration and impaired NF-κB-dependent gene transcription. EMBO J. 19: 4976-4985.

Boyes J, Byfield P, Nakatani Y & Ogryzko V (1998) Regulation of activity of the transcription factor GATA-1 by acetylation. Nature 396: 594-598.

Chakravarti D, Ogryzko V, Kao HY, Nash A, Chen H, Nakatani Y & Evans RM (1999) A viral mechanism for inhibition of p300 and PCAF acetyltransferase activity. Cell 96: 393-403.

Chan HM, Krstic-Demonacos M, Smith L, Demonacos C & La Thangue NB (2001) Acetylation control of the retinoblastoma tumour-suppressor protein. Nat. Cell Biol. 3: 667-674.

Chen FE, Huang DB, Chen YQ & Ghosh G (1998) Crystal structure of p50/p65 heterodimer of transcription factor NF-κB bound to DNA. Nature 391: 410-413.

Chen LF, Fischle W, Verdin E & Greene WC (2001) Duration of nuclear NF-κB action regulated by reversible acetylation. Science 293: 1653-1657.

Chen LF, Mu Y & Greene WC (2002) Acetylation of RelA at discrete sites regulates distinct nuclear functions of NF-κB. EMBO J. 21: 6539-6548.

Chrivia JC, Kwok RP, Lamb N, Hagiwara M, Montminy MR & Goodman RH (1993) Phosphorylated CREB binds specifically to the nuclear protein CBP. Nature 365: 855-859.

Diehl JA, Tong W, Sun G & Hannink M (1995) Tumor necrosis factor-α-dependent activation of a

RelA homodimer in astrocytes. Increased phosphorylation of RelA and MAD-3 precedes activation of RelA. J. Biol. Chem. 270: 2703-2707.

Eckner R, Arany Z, Ewen M, Sellers W & Livingston DM (1994) The adenovirus E1A-associated 300-kD protein exhibits properties of a transcriptional coactivator and belongs to an evolutionarily conserved family. Cold Spring Harb. Symp. Quant. Biol. 59: 85-95.

Fruman DA & Cantley LC (2002) Phosphoinositide 3-kinase in immunological systems. Semin. Immunol. 14: 7-18.

Furia B, Deng L, Wu K, Baylor S, Kehn K, Li H, Donnelly R, Coleman T & Kashanchi F (2002) Enhancement of nuclear factor-κB acetylation by coactivator p300 and HIV-1 Tat proteins. J. Biol. Chem. 277: 4973-4980.

Ganchi PA, Sun SC, Greene WC & Ballard DW (1993) A novel NF-κB complex containing p65 homodimers: implications for transcriptional control at the level of subunit dimerization. Mol. Cell. Biol. 13: 7826-7835.

Gerritsen ME, Williams AJ, Neish AS, Moore S, Shi Y & Collins T (1997) CREB-binding protein/p300 are transcriptional coactivators of p65. Proc. Natl. Acad. Sci. USA 94: 2927-2932.

Gillespie SK & Wasserman SA (1994) Dorsal, a Drosophila Rel-like protein, is phosphorylated upon activation of the transmembrane protein Toll. Mol. Cell. Biol. 14: 3559-3568.

Gu W & Roeder RG (1997) Activation of p53 sequence-specific DNA binding by acetylation of the p53 C-terminal domain. Cell 90: 595-606.

Gustin JA, Maehama T, Dixon JE & Donner DB (2001) The PTEN tumor suppressor protein inhibits tumor necrosis factor-induced nuclear factor κB activity. J. Biol. Chem. 276: 27740-27744.

Hamamori Y, Sartorelli V, Ogryzko V, Puri PL, Wu HY, Wang JY, Nakatani Y & Kedes L (1999) Regulation of histone acetyltransferases p300 and PCAF by the bHLH protein twist and adenoviral oncoprotein E1A. Cell 96: 405-413.

Hoeflich KP, Luo J, Rubie EA, Tsao MS, Jin O & Woodgett JR (2000) Requirement for glycogen synthase kinase-3β in cell survival and NF-κB activation. Nature 406: 86-90.

Huxford T, Huang DB, Malek S & Ghosh G (1998) The crystal structure of the IκBα/NF-κB complex reveals mechanisms of NF-κB inactivation. Cell 95: 759-770.

Imhof A, Yang XJ, Ogryzko VV, Nakatani Y, Wolffe AP & Ge H (1997) Acetylation of general transcription factors by histone acetyltransferases. Curr. Biol. 7: 689-692.

Jacobs MD & Harrison SC (1998) Structure of an IκBα/NF-κB complex. Cell 95: 749-758.

Johnson C, Van Antwerp D & Hope TJ (1999) An N-terminal nuclear export signal is required for the nucleocytoplasmic shuttling of IκBα. EMBO J. 18: 6682-6693.

Kane LP, Shapiro VS, Stokoe D & Weiss A (1999) Induction of NF-κB by the Akt/PKB kinase. Curr. Biol. 9: 601-604.

Koul D, Yao Y, Abbruzzese JL, Yung WK & Reddy SA (2001) Tumor suppressor MMAC/PTEN inhibits cytokine-induced NFκB activation without interfering with the IκB degradation pathway. J. Biol. Chem. 276: 11402-11408.

Krappmann D, Wulczyn FG & Scheidereit C (1996) Different mechanisms control signal-induced degradation and basal turnover of the NF-κB inhibitor IκBα in vivo. EMBO J. 15: 6716-6726.

Kuo MH, Zhou J, Jambeck P, Churchill ME & Allis CD (1998) Histone acetyltransferase activity of yeast Gcn5p is required for the activation of target genes in vivo. Genes Dev. 12: 627-639.

Leitges M, Sanz L, Martin P, Duran A, Braun U, Garcia JF, Camacho F, Diaz-Meco MT, Rennert PD & Moscat J (2001) Targeted disruption of the zetaPKC gene results in the impairment of the NF-κB pathway. Mol. Cell 8: 771-780.

Li CC, Dai RM, Chen E & Longo DL (1994a) Phosphorylation of NF-KB1-p50 is involved in NF-κB activation and stable DNA binding. J. Biol. Chem. 269: 30089-30092.

Li CC, Korner M, Ferris DK, Chen E, Dai RM & Longo DL (1994b) NF-κB/Rel family members are physically associated phosphoproteins. Biochem. J. 303: 499-506.

Li Q, Lu Q, Hwang JY, Buscher D, Lee K-F, Izpisua-Belmonte JC & Verma IM (1999) The IKK1-deficient mice exhibit abnormal development of skin and skeleton. Genes Dev. 13: 1322-1328.

Li Q, Estepa G, Memet S, Israel A & Verma IM (2000) Complete lack of NF-κB activity in IKK1 and IKK2 double-deficient mice: additional defect in neurulation. Genes Dev. 14: 1729-1733.

Lin R, Beauparlant P, Makris C, Meloche S & Hiscott J (1996) Phosphorylation of IκBα in the C-terminal PEST domain by casein kinase II affects intrinsic protein stability. Mol. Cell. Biol. 16: 1401-1409.

Liu L, Scolnick DM, Trievel RC, Zhang HB, Marmorstein R, Halazonetis TD & Berger SL (1999) p53 sites acetylated in vitro by PCAF and p300 are acetylated in vivo in response to DNA damage. Mol. Cell. Biol. 19: 1202-1209.

Marmiroli S, Bavelloni A, Faenza I, Sirri A, Ognibene A, Cenni V, Tsukada J, Koyama Y, Ruzzene M, Ferri A, Auron PE, Toker A & Maraldi NM (1998) Phosphatidylinositol 3-kinase is recruited to a specific site in the activated IL-1 receptor I. FEBS Lett. 438: 49-54.

Martinez-Balbas MA, Bauer UM, Nielsen SJ, Brehm A & Kouzarides T (2000) Regulation of E2F1 activity by acetylation. EMBO J. 19: 662-671.

Mayo MW, Madrid LV, Westerheide SD, Jones DR, Yuan XJ, Baldwin AS & Whang YE (2002) PTEN blocks tumor necrosis factor-induced NF-κB-dependent transcription by inhibiting the transactivation potential of the p65 subunit. J. Biol. Chem. 277: 11116-11125.

Mercurio F, Murray BW, Shevchenko A, Bennett BL, Young DB, Li JW, Pascual G, Motiwala A, Zhu H, Mann M & Manning AM (1999) IκB kinase (IKK)-associated protein 1, a common component of the heterogeneous IKK complex. Mol. Cell. Biol. 19: 1526-1538.

Mosialos G, Hamer P, Capobianco AJ, Laursen RA & Gilmore TD (1991) A protein kinase-A recognition sequence is structurally linked to transformation by p59v-rel and cytoplasmic retention of p68c-rel. Mol. Cell. Biol. 11: 5867-5877.

Mosialos G & Gilmore TD (1993) v-Rel and c-Rel are differentially affected by mutations at a consensus protein kinase recognition sequence. Oncogene 8: 721-730.

Na SY, Lee SK, Han SJ, Choi HS, Im SY & Lee JW (1998) Steroid receptor coactivator-1 interacts with the p50 subunit and coactivates nuclear factor κB-mediated transactivations. J. Biol. Chem. 273: 10831-10834.

Naumann M & Scheidereit C (1994) Activation of NF-κB in vivo is regulated by multiple phosphorylations. EMBO J. 13: 4597-4607.

Ng HH, Jeppesen P & Bird A (2000) Active repression of methylated genes by the chromosomal protein MBD1. Mol. Cell. Biol. 20: 1394-1406.

Perkins ND, Felzien LK, Betts JC, Leung K, Beach DH & Nabel GJ (1997) Regulation of NF-κB by cyclin-dependent kinases associated with the p300 coactivator. Science 275: 523-527.

Reddy SA, Huang JH & Liao WS (1997) Phosphatidylinositol 3-kinase in interleukin 1 signaling. Physical interaction with the interleukin 1 receptor and requirement in NFκB and AP-1 activation. J. Biol. Chem. 272: 29167-29173.

Sakaguchi K, Herrera JE, Saito S, Miki T, Bustin M, Vassilev A, Anderson CW & Appella E (1998) DNA damage activates p53 through a phosphorylation-acetylation cascade. Genes Dev. 12: 2831-2841.

Sakurai H, Chiba H, Miyoshi H, Sugita T & Toriumi W (1999) IκB kinases phosphorylate NF-κB p65 subunit on serine 536 in the transactivation domain. J. Biol. Chem. 274: 30353-30356.

Sartorelli V, Puri PL, Hamamori Y, Ogryzko V, Chung G, Nakatani Y, Wang JY & Kedes L (1999) Acetylation of MyoD directed by PCAF is necessary for the execution of the muscle program. Mol. Cell 4: 725-734.

Schmitz M, dos Santos Silva M & Baeuerle P (1995) Transactivation domain 2 (TA2) of p65 NF-κB. Similarity to TA1 and phorbol ester-stimulated activity and phosphorylation in intact cells. J. Biol. Chem. 270: 15576-15584.

Senftleben U & Karin M (2002) The IKK/NF-κB pathway. Crit. Care Med. 30: S18-26.

Sheppard KA, Rose DW, Haque ZK, Kurokawa R, McInerney E, Westin S, Thanos D, Rosenfeld MG, Glass CK & Collins T (1999) Transcriptional activation by NF-κB requires multiple coactivators. Mol. Cell. Biol. 19: 6367-6378.

Sizemore N, Leung S & Stark GR (1999) Activation of phosphatidylinositol 3-kinase in response to interleukin-1 leads to phosphorylation and activation of the NF-κB p65/RelA subunit. Mol. Cell. Biol. 19: 4798-4805.

Sizemore N, Lerner N, Dombrowski N, Sakurai H & Stark GR (2002) Distinct roles of the IκB

kinase α and β subunits in liberating nuclear factor κB (NF-κB) from IκB and in phosphorylating the p65 subunit of NF-κB. J. Biol. Chem. 277: 3863-3869.
Sterner DE & Berger SL (2000) Acetylation of histones and transcription-related factors. Microbiol. Mol. Biol. Rev. 64: 435-459.
Van Patten SM, Fletcher WH & Walsh DA (1986) The inhibitor protein of the cAMP-dependent protein kinase-catalytic subunit interaction. Parameters of complex formation. J. Biol. Chem. 261: 5514-5523.
Van Patten SM, Hotz A, Kinzel V & Walsh DA (1988) The inhibitor protein of the cyclic AMP-dependent protein kinase-catalytic subunit interaction. Composition of multiple complexes. Biochem. J. 256: 785-789.
Vanden Berghe W, Plaisance S, Boone E, De Bosscher K, Schmitz ML, Fiers W & Haegeman G (1998) p38 and extracellular signal-regulated kinase mitogen-activated protein kinase pathways are required for nuclear factor-κB p65 transactivation mediated by tumor necrosis factor. J. Biol. Chem. 273: 3285-3290.
Vanden Berghe W, De Bosscher K, Boone E, Plaisance S & Haegeman G (1999) The nuclear factor-κB engages CBP/p300 and histone acetyltransferase activity for transcriptional activation of the interleukin-6 gene promoter. J. Biol. Chem. 274: 32091-32098.
Vermeulen L, De Wilde G, Van Damme P, Vanden Berghe W & Haegeman G (2003) Transcriptional activation of the NF-κB p65 subunit by mitogen- and stress-activated protein kinase-1 (MSK1). EMBO J. 22: 1313-1324.
Wang D & Baldwin AS (1998) Activation of nuclear factor-κB-dependent transcription by tumor necrosis factor-α is mediated through phosphorylation of RelA/p65 on serine 529. J. Biol. Chem. 273: 29411-29416.
Wang D, Westerheide SD, Hanson JL & Baldwin AS (2000) Tumor necrosis factor α-induced phosphorylation of RelA/p65 on Ser529 is controlled by casein kinase II. J. Biol. Chem. 275: 32592-32597.
Werbajh S, Nojek I, Lanz R & Costas MA (2000) RAC-3 is a NF-κB coactivator. FEBS Lett. 485: 195-199.
Woodgett JR (2001) Judging a protein by more than its name: GSK-3. Sci. STKE 100: RE12.
Wu RC, Qin J, Hashimoto Y, Wong J, Xu J, Tsai SY, Tsai MJ & O'Malley BW (2002) Regulation of SRC-3 (pCIP/ACTR/AIB-1/RAC-3/TRAM-1) coactivator activity by IκB kinase. Mol. Cell. Biol. 22: 3549-3561.
Xiao G, Cvijic ME, Fong A, Harhaj EW, Uhlik MT, Waterfield M & Sun SC (2001) Retroviral oncoprotein Tax induces processing of NF-κB2/p100 in T cells: evidence for the involvement of IKKα. EMBO J. 20: 6805-6815.
Yin L, Wu L, Wesche H, Arthur CD, White JM, Goeddel DV & Schreiber RD (2001) Defective lymphotoxin-β receptor-induced NF-κB transcriptional activity in NIK-deficient mice. Science 291: 2162-2165.
Zhong H, SuYang H, Erdjument-Bromage H, Tempst P & Ghosh S (1997) The transcriptional activity of NF-κB is regulated by the IκB-associated PKAc subunit through a cyclic AMP-independent mechanism. Cell 89: 413-424.
Zhong H, Voll RE & Ghosh S (1998) Phosphorylation of NF-κB p65 by PKA stimulates transcriptional activity by promoting a novel bivalent interaction with the coactivator CBP/p300. Mol. Cell 1: 661-671.
Zhong H, May MS, Jimi E & Ghosh S (2002) The phosphorylation status of nuclear NF-κB determines its association with CBP/p300 or HDAC-1. Mol. Cell 9: 625-636.

6. AFFILIATION

Gladstone Institute of Virology and Immunology, POB 419100, Departments of Medicine, and Microbiology and Immunology, University of California, San Francisco, CA 94143, USA.

T. HUXFORD & G. GHOSH

THE STRUCTURAL BIOLOGY OF NF-κB

1. INTRODUCTION

The mechanisms underlying activation and DNA binding of the transcription factor NF-κB are central to our understanding of the cellular response to stress, development and apoptosis, among other key physiological processes (Baeuerle & Henkel, 1994; Siebenlist et al., 1994; Baeuerle & Baltimore, 1996; Baldwin, 1996; Ghosh et al., 1998; Gilmore, 1999; Silverman & Maniatis, 2001). Central to gene regulation by NF-κB is its release from inhibition by IκB and subsequent recognition of specific promoters of inducible genes. Three-dimensional structures of NF-κB proteins in complex with κB DNA or IκB inhibitors have provided new insights into the complexities of the NF-κB-signaling pathways and NF-κB-mediated gene regulation. In this review, we shall focus on the structural features of NF-κB/DNA and IκB/NF-κB complexes, as well as on the relation of these structures to biological function.

2. NF-κB PROTEINS

The mammalian Rel/NF-κB transcription factor family comprises five members, viz. p50, p52, p65 (RelA), c-Rel and RelB, which associate one with another to form active dimers. Each of these subunits shares an N-terminal segment of approximately 300 amino acids in length, known as the Rel homology region (RHR). Contained within the RHR are the amino acid sequences involved in subunit dimerization, DNA binding, nuclear translocation and binding to IκB.

2.1. Structural features of the NF-κB RHR

Three-dimensional x-ray crystal structures and limited proteolysis have revealed that the NF-κB RHR is organized into three relatively flexible segments (Ghosh et al., 1995; Müller et al., 1995; Cramer et al., 1997; Chen et al., 1998a,c; Cramer et al., 1999; Chen et al., 2000; Huang et al., 2001; Escalante et al., 2002). Two of the segments constitute separable domains that assume the Ig-like fold (Fig. 1). The N-terminal Ig-like domain conveys DNA-binding specificity to NF-κB. The C-terminal Ig-like domain is primarily responsible for subunit dimerization. Amino acids from the dimerization domain also participate in nonspecific contacts with DNA. The third independently flexible segment, which is composed

R. Beyaert (ed.), Nuclear Factor κB. Regulation and Role in Disease, 177-198.

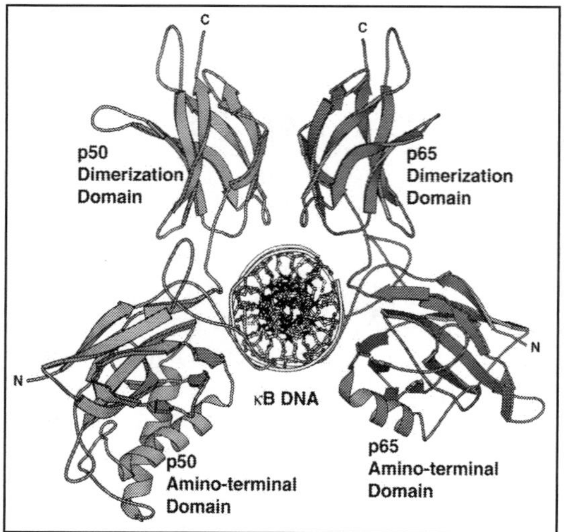

Fig. 1. Three-dimensional view of the NF-κB p50/p65 heterodimer bound to Ig-κB DNA, based on the x-ray crystal structure. The complex is viewed down the long axis of the DNA. In general, structures of all NF-κB/DNA complexes look very similar.

of approximately 30 C-terminal residues and includes the nuclear localization signal, does not assume an ordered structure in solution. Rel/NF-κB proteins are decorated with additional segments outside the RHR, displaying distinct physiological activities. For example, p65, c-Rel and RelB contain transcription activation domains at their C-termini. Although the C-termini of p50 and p52 extend some 80 residues beyond their RHR, these extensions are not known to take part in transcriptional activation functions. In addition to a C-terminal activation domain, RelB also contains 100 amino acids at its N-terminus (Ryseck et al., 1992; Dobrzanski et al., 1993). This region bears heptad leucine repeats characteristic of leucine zipper (LZ) proteins. The RelB LZ motif may play a role in activating transcription. The p50, p52, p65 and c-Rel subunits have smaller extensions at their N-termini.

2.2. *The NF-κB dimer interface*

Subunits of the Rel/NF-κB family form homo- and heterodimers in a combinatorial fashion. However, the stabilities of these dimers vary significantly to the degree that some of the possible subunit combinations have not been detected in cells. This suggests that the affinities of various dimers, such as RelB homodimer and RelB heterodimers with p65 and c-Rel, are too low to favor formation at cellular concentrations and in the presence of other NF-κB subunits with which they dimerize more readily (Ryseck et al., 1995). At higher protein concentrations these unfavorable dimers can be observed in vitro. The p50/p65 heterodimer is more stable than homodimers of p50 or p65 subunits. It is, therefore, not surprising that the p50/p65 heterodimer is also more abundant than the respective homodimers in cells.

The structural and biochemical bases behind selective NF-κB dimer formation for all five homodimers and the p50/p65 heterodimer have been determined. With the exception of the RelB homodimer, the interfaces of all other dimers are similar. Essentially, two symmetrical β-sheets stack next to one another forming a β-sandwich at the dimer interface (Fig. 2).

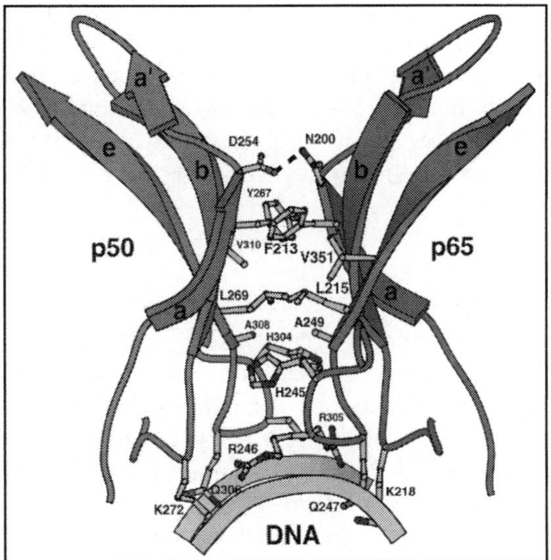

Fig. 2. Close-up view of the monomer/monomer interface of the p50 and p65 subunits in the NF-κB p50/p65/DNA complex (the orientation is the same as in Fig. 1). Shown are residues that are most important for stable dimer formation, as well as a few of the DNA-contacting residues emanating from the dimerization domains of p50 and p65.

Dimerization buries a relatively small solvent-accessible surface area. Thus even the most stable of the NF-κB dimers, the p50/p65 heterodimer and p50 homodimer, form with roughly high-nanomolar to low-micromolar binding affinities. X-ray crystallographic analyses reveal that roughly 14 conserved residues from each monomer contact the opposite subunit to form the dimer (Huang et al., 1997). Alanine scanning mutagenesis has revealed that only three of these 14 residues contribute significantly to p50 homodimer stability (Sengchanthalangsy et al., 1999), namely Tyr267, Leu269 and Val310. Together with another residue (Ala308), which tightly fits against the methyl groups of Leu269, these three residues form the hot spot of the interface (Hart et al., 2001). Leu269, Ala308 and Val310 are invariant throughout the Rel/NF-κB family. In p50, Tyr267 is the most critical residue in conferring the p50 homodimer stability. Interestingly, little change in p50 homodimer stability results when Tyr267 is replaced by hydrophobic amino acids longer than alanine. Besides these core residues, Asp254 also regulates p50 dimerization. Mutation of Asp254 to alanine enhances p50 dimer stability, suggesting that the close proximity of deprotonated aspartate-oxygen atoms from each of two subunits may exert electrostatic repulsion on the interacting subunits. The NF-κB subunits p65 and c-Rel both contain phenylalanine and asparagine at positions homologous to Tyr267 and Asp254 in p50. The hydroxyl group of p50 subunit Tyr267 forms several hydrogen bonds with the opposing subunit, which may explain why phenylalanine substitutions in p65 and c-Rel homodimers make these dimers less stable than the p50 homodimer. In the p50/p65 heterodimer, a hydrogen bond formed between Asp254 of p50 and the corresponding Asn200 of p65 appears to be at least partly responsible for conveying the highest overall stability to the heterodimer.

The RelB homodimer interface is unique among NF-κB family members. Superposition of RelB and p50 demonstrates that the two RelB subunits are separated by 8 Å and rotated by 30° relative to the p50 subunits. Remarkably, the amino acids occupying the hydrophobic core positions of the p50 homodimer are invariant in RelB. Therefore, it remains unclear as to why RelB homodimer formation is so unfavorable. The orientation of Tyr300 (equivalent to Tyr267 in p50) is significantly altered in RelB, which apparently prevents close association of the two RelB subunits. The structure does not clearly indicate why this is so. Although no structure of p50/RelB heterodimer is available, one can build a model based on monomers from p50 and RelB. Such a model shows that RelB can form heterodimers with p50 and p52.

2.3. Other structural differences among NF-κB subunits

The N-terminal domain of all NF-κB proteins contains an insertion that deviates from the standard Ig-fold. The role of this insert is yet to be identified. The NF-κB p50 subunit RHR contains a 58-residue long nonhomologous insertion within its NTD (Ghosh et al., 1995; Müller et al., 1995). Roughly half of these amino acids form two long helices which interact with one another, while the rest of the insert is unstructured. In p52, the insertion length is 42 residues (Cramer et al., 1997). As in p50, a part of these residues forms a helical bundle. In other family members the insert is much smaller, comprising only 25 residues. In p65 and c-Rel smaller helical motifs are formed by these segments (Chen et al., 1998c; Chen et al., 2000; Huang et al., 2001).

3. DNA BINDING BY NF-κB DIMERS

3.1. Basic features of NF-κB/DNA complexes

NF-κB dimers regulate expression of several hundreds of genes (Pahl, 1999). Each of these genes contains binding sites for NF-κB that are collective referred to as κB DNA. The DNA-binding properties of some of the most abundant and functionally most important NF-κB dimers, such as p50/p65 heterodimer and p50 homodimer, have been characterized biochemically. The DNA-bound x-ray crystal structures of several NF-κB dimers are known. Together these studies begin to define the rules that govern target DNA-site selection by NF-κB dimers (Fig. 3). The p50/p65 heterodimer binds to 10-bp DNA targets with the consensus sequence of 5'-GGGRN W TTCC-3' (where N is any base, R is any purine and W represents either A or T). The p50 subunit of the heterodimer binds to the 5' subsite consisting of 5 bp (5'-GGGRN-3'), whereas the p65 subunit binds to the 3' subsite comprising 4 bp (5'-TTCC-3'). Following this pattern of binding, the p50 homodimer prefers an 11-bp target DNA with two 5-bp half-sites separated by a central A/T base pair. Similarly, the p65 homodimer binds a 9-bp κB DNA with two 4-bp half-sites separated by the central A/T.

In NF-κB, DNA-contacting residues are contributed by protein loops connecting secondary structures. The most important of these loops, referred to as loop L1, contributes five base-contacting amino acids in p50 and four base-contacting residues in p65 (Fig. 4). His64 of p50 directly contacts the first guanine, which

		-5-4-3-2-1 0 +1+2+3+4
p50/p65	Consensus	GGGRN W TTCC
	IgκB (and HIVκB)	GGGAC T TTCC
	IFN-β	GGGAA A TTCC
	VCAM	GGGGT T TCCC
	TNFα	GGGAA A GCCC
	IL-2	GGGAT T TCAC
	M-CSF	GGGAA C TACC
	Urokinase	GGGAA A GTAC
		-4-3-2-1 0 +1+2+3+4
p65 and c-Rel	Consensus	GGAN W TTCC
	IL-8	GGAA T TTCC
	TF	GGAG T TTCC
	IL-2	AGAA A TTCC
	IFN-γ	GAAT T TTCC
	ISRE	TCAT T TTCC
		-5-4-3-2-1 0 +1+2+3+4+5
p50	Consensus	GGGRN W NYCCC
	MHC I	GGGGA A TCCCC
NF-AT	Consensus	GGAAAnnnn
	IL2	GGAAAaact
	IFN-γ	GGAAAtttt
	TNF-α	GGAGAaacc
	IL-8	GGAATttcc
STAT	Consensus	TTCNNNNGAA
	Ig Heavy chain	TTCCCAAGAA
	IL1-Ra	TTCCCAGGAA
	CD23	TACCTGAGAA

Fig. 3. Some physiological DNA targets and consensus sequences for the NF-κB, NF-AT and STAT families of transcription factors. For the NF-κB family, sequences of three different dimer classes are shown (nonconsensus bases are boxed).

explains the requirement of an extra G^{-5}:C^{-5} bp in the p50 half-site. In p65, as in c-Rel and RelB, this histidine is substituted with alanine. The $G^{-4}G^{-3}/C^{-4}C^{-3}$ bp at both half-sites forms the core of the DNA-binding element for both p50 and p65. Two conserved arginine side chains and a glutamic acid contact this core element. These side chains also interact one with another to position themselves in proper orientation for DNA recognition. This type of buttressing interaction between DNA-contacting protein side chains is common in many protein/DNA complexes and favors the recognition process by reducing the entropic penalty due to the side chain flexibility of the arginine amino acids. An invariant tyrosine amino acid is the last residue from the L1 loop that directly contacts κB DNA bases, doing so by forming van der Waals interactions with the edges of the internal pyrimidines. Mutational studies have suggested that the interactions between the tyrosine and DNA might be the most important for the binding affinity of NF-κB/DNA complexes (Liu et al., 1994). In addition, the hydroxyl group of tyrosine makes DNA backbone contact. All DNA-bound NF-κB complexes reveal that the position of this tyrosine side chain is fixed by the stacking of a series of interior residues, the majority of which are histidines.

In addition to loop L1, the loop L2 also contributes a lysine in the case of p50 or an arginine in the case of p65. In most of the DNA-bound NF-κB crystal structures, this basic side chain contacts the +1 bp. Some of the other residues that clearly play a role in DNA recognition and binding, are a pair of lysines corresponding to Lys144 and Lys145 in p50. The lysines project from the loop immediately N-terminal to the helical insertion. Modification of these lysines and Cys59 is known to significantly reduce DNA-binding affinity. In addition to DNA base-specific interactions, NF-κB dimers also employ amino acids projecting from loops in the N-terminal and C-terminal dimerization domains to contact the DNA nonspecifically through the sugar-phosphate backbone. Though nonspecific in nature, these interactions have a huge impact on overall DNA-binding affinity and they allow NF-κB to bind nonconsensus κB DNA-binding sites.

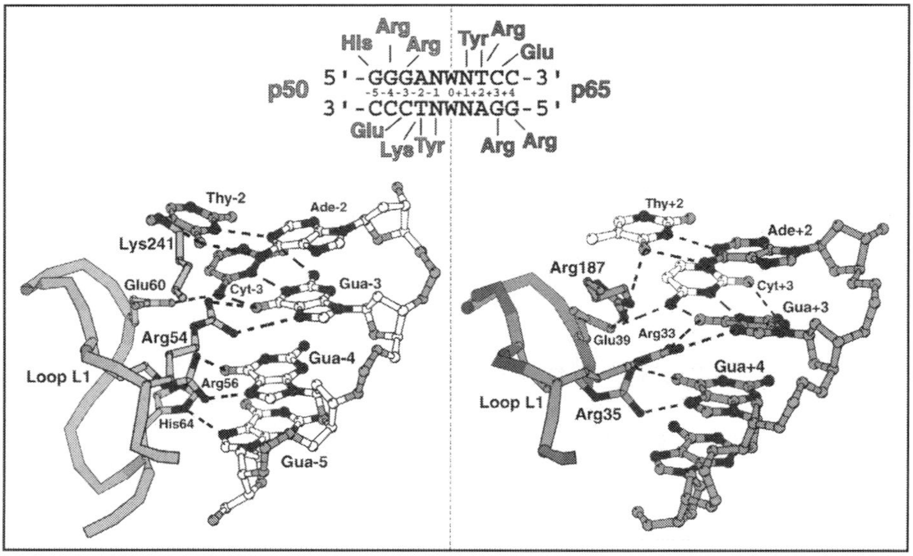

Fig. 4. Consensus κB-DNA target for the NF-κB p50/p65 heterodimer and specific base-contacting residues (top), as well as base-specific interactions between p50/p65 and DNA (bottom).

3.2. DNA-binding affinity and specificity

NF-κB p50/p65 heterodimer binds to physiological 10-bp κB DNA targets with affinities in the range of 0.5 to 5 nM (affinity defined as the equilibrium dissociation constant K_D) (Phelps et al., 2000). The p50 homodimer binds both its preferred 11-bp consensus and 10-bp sites with affinities ranging from 5 to 50 nM. NF-κB p50 homodimer is known to bind these sites in quiescent cells and repress transcription (Franzoso et al., 1992; Plaksin et al., 1993). Homodimers of p65 and c-Rel subunits bind their 9-bp cognate DNA sites with affinities between 10 and 100 nM. Although the affinities of the cognate complexes differ by over 100-fold, these dimers bind to noncognate sites with significantly lower affinities. For example, p50 homodimer and the p50/p65 heterodimer display low affinity toward 9-bp κB DNA sites, which are preferred by the p65 and c-Rel homodimer and the p65/c-Rel heterodimer. Conversely, p65 and c-Rel homodimers bind to 10-bp and 11-bp κB DNA with much lower affinities as compared to their 9-bp cognate sites. That the p50 homodimer and p50/p65 heterodimer would discriminate against 9-bp DNA targets is understandable due to the missing His64-G^{-5} contact. It is surprising that NF-κB p65 and c-Rel homodimers fail to stably bind 11-bp or 10-bp sequences, as the 9-bp consensus sequence is embedded within these longer DNA targets. It appears that the presence of a G:C bp at the -5 position is detrimental to p65 and c-Rel binding. It is also possible that higher binding affinities of p50 homodimer and p50/p65 heterodimer for the 10-bp κB DNA elements are not simply due to the contact between His64 and G^{-5}, but rather to the conformation of the DNA, which varies with the presence or absence of guanine in this position. This, in turn, could affect the stabilities of protein-DNA complexes. Intriguingly, a distinct DNA conformational variation has been observed between an HIV κB site (5'-GGGACTTTCC-3') and a mutant site, where the GGG sequence was replaced with CTC.

The binding affinity between NF-κB p50/p65 heterodimer and κB DNA from the immunoglobulin-κ light-chain gene (Ig-κB) is salt- and pH-dependent, but is not dependent on temperature (Phelps et al., 2000). The dissociation values referred to previously were all measured in 50 mM salt at pH 7.5. Upon studying the effects of salt on complex formation, it was determined that NF-κB binds to DNA almost nonspecifically at physiological salt concentrations. How then does NF-κB recognize specific DNA sequences in cells? One possibility is an increased DNA-binding affinity due to cooperative interactions between NF-κB and other neighboring transcriptional activators, coactivators or basal factors within the context of the enhanceosome. The lack of dependence on temperature suggests that the intrinsic enthalpy change upon NF-κB complex formation is negligible. Therefore, it appears that the binding process is entropically driven through the release of counterions and water molecules. We have further observed that many DNA-contacting residues are preorganized in the free state of the protein domains. Thus ordering of the amino acid side chains for DNA binding is not required, minimizing the loss in entropy due to local DNA-binding surface organization.

3.3. DNA-binding conformation and biological regulation

As a result of the modular architecture of the NF-κB RHR, alternative overall NF-κB conformations are observed upon binding to highly conserved κB DNA target sequences. It has been shown that even seemingly small changes in sequence can alter DNA dynamics and conformation. By virtue of the relative mobility of its ordered domains, NF-κB proteins respond to DNA structural variations with altered binding conformations. This principle is illustrated in the conformational differences between p50/p65 heterodimer bound to Ig-κB and IFN-κB DNA. Both target DNAs contain the same bases at critical positions for optimal NF-κB p50/p65 heterodimer recognition and binding. However, the overall conformations of these complexes differ significantly. In complex with NF-κB, the IFN-κB target DNA is straight, whereas the Ig-κB DNA is bent by 15° (Chen et al., 1998a). The N-terminal domains of both p50 and p65 subunits are required to rotate and to translate in order to make optimum contacts with the DNA. In other cases where target κB DNA sequences vary more significantly from the consensus, the degree of conformational variation becomes more pronounced. One such case is the p65 homodimer in complex with a consensus κB DNA target, in which one of the core G:C bp is replaced with an A:T (Chen et al., 1998c; Chen et al., 2000). As a result of this single bp change, the N-terminal domain of the p65 subunit bound to the nonconsensus half-site must rotate to avoid negative interactions between Glu39 and the O4 oxygen of the thymine base. The domain rotates by 20° and translates 7 Å in order to find an alternate binding surface upon the target DNA. As no change is observed within the C-terminal dimerization domains, such movement arises entirely as a result of the flexibly linked N-terminal domains of NF-κB.

Are the conformational differences exhibited by NF-κB dimers on different κB DNA physiologically meaningful? Whereas one can clearly envision functional differences due to the affinities and specificities of various κB DNA and NF-κB dimers, functional variation due to a conformational change is more difficult to prove. For example, the binding to 9-bp κB DNA sequences by the p65 and c-Rel homodimers as well as p65/c-Rel heterodimer, but not by p50 subunit-containing NF-κB dimers, can be explained by the respective binding affinities. It remains unclear, however, which NF-κB dimers regulate genes that contain the 9-bp κB DNA within their enhancers. Most current data are derived from transfection experiments. Such studies often fail to truly recreate in vivo the role of different NF-κB dimers in regulating specific κB DNA targets. It has been shown that expression of IL-2 and IL-12 is regulated by c-Rel, but not by p65 (Sanjabi et al., 2000). c-Rel-dependent expression of these interleukins is regulated at different levels: IL-2 expression by c-Rel is apparently regulated at the level of nuclear translocation of c-Rel, IL-12 expression is regulated by productive transcription complex formation by c-Rel. In the latter case, it is possible that only c-Rel can specifically recruit the other required transcriptional activators or coactivators, and form a productive enhanceosome complex upon the IL-12 promoter. In the case of the p50/p65 heterodimer, it is known that the IL-2Rα-κB and Ig-κB DNA sequences may not be swapped between their respective promoters without a loss of in vivo biological activity (Cross et al., 1989).

There are other well-characterized examples of DNA sequence-dependent conformational changes as a means of regulating gene expression. One such system are the POU domain proteins, which can bind to two different DNA targets and elicit different biological responses (Jacobson et al., 1997). As is the case with NF-κB, POU domain proteins are modular in design and can bind different DNA sequences with high affinity in different conformations.

3.4. NF-κB modification and biological regulation

Regulation of transcriptional activity by NF-κB dimers has been observed to involve phosphorylation in vivo (Kushner & Ricciardi, 1999). Hyperphosphorylation of the p50 subunit in response to phorbol ester and hemagglutinin has been shown to be important for enhanced DNA binding. In contrast, hypophosphorylation of p50 decreases DNA-binding affinity in adenovirus E1-transformed hooded Lister rat cells. Adenovirus exploits hyperphosphorylation of p50 to down-regulate the expression of the major histocompatibility complex I in an effort to evade cytotoxic T-cell-mediated lysis. The exact p50 subunit phosphorylation sites are yet to be identified.

Phosphorylation of the NF-κB p65 subunit has been studied in more detail (Zhong et al., 1998; Zhong et al., 2002). Both the p65 RHR and the transactivation domain contain sites of phosphorylation. Ser276, located within the dimerization domain, is reported to be phosphorylated by two different kinases: cAMP-dependent protein kinase (PKA) as well as mitogen- and stress-activated kinase 1 (MSK1). This phosphorylation has been shown to be essential for conferring full DNA binding and transactivation potential to p65. In its unphosphorylated state, the p65 subunit transactivation domain associates with its own RHR in an interaction that lowers DNA-binding affinity. Phosphorylation at Ser276 relieves this autoinhibition and enables p65 to bind κB DNA target sites. Phosphorylation at this site is also critical for the recruitment of coactivator protein CBP/p300. Substitution of p65 subunit Ser276 with alanine leads to lethality in mice due to massive liver apoptosis, a phenotype similar to the p65 subunit knockout animals.

How does phosphorylation of NF-κB subunits enhance DNA binding? Phosphorylation cannot increase DNA-binding affinity by a direct mechanism, since the addition of negative charge from the addition of a phosphate group can only provide repulsion to the highly negatively charged DNA backbone. Therefore, phosphorylation must affect DNA binding indirectly, perhaps by stabilizing positively charged residues that directly contact DNA. The Ser276 site of phosphorylation in p65 is located on a loop within the dimerization domain and is located near the DNA backbone. Phosphorylation of this side chain could reorganize some of the nearby positively charged residues, both within the same loop and on loop L1, in such a way that DNA binding by p65 is enhanced. Such a hypothesis awaits the structural characterization of NF-κB/DNA complexes with differing states of phosphorylation.

Acetylation of p65 has been reported to modulate its transcriptional activity. The exact sites of acetylation within the DNA-binding domain of p65 are not yet known. If the acetylation sites prove to be lysine amino acids that contact DNA, then acetylation in principle should repress transcription. However, it is known that acetyltransferases enhance transcription by modifying histones. Determination of the acetylation sites within p65 and the in vivo effects of p65 acetylation will clarify whether histone acetylation and p65 acetylation affect transcription in opposite ways.

3.5. Enhanceosome formation and NF-κB

The complex enhancer of a gene includes multiple binding sites for various transcription activators and is unique for specific genes. Some enhancers, such as the

HIV proximal promoter, T-cell receptor and IFNβ, are well characterized. The IFNβ enhancer is an approximately 55-bp DNA sequence located between -110 and -55 upstream of the transcriptional start site. It is both necessary and sufficient for virus-induced expression of the cytokine IFNβ. Located within this DNA element are binding sites for the NF-κB p50/p65 heterodimer, interferon regulatory factors 1 and 3 (IRF1/3) and c-Jun/ATF-2 heterodimer. The κB DNA is proximal and the AP1 site is distal to the transcription start site. Following viral infection, the inducible transcription factors bind their respective sites in a cooperative manner. The molecular nature of the cooperativity, the manner in which these factors interact with one another upon the DNA, is not known. However, this cooperative binding is further enhanced by coactivator CBP/p300 binding to the activator/DNA complex. Within the IFNβ enhancer, two binding sites for HMGI(Y) overlap the κB and AP1 sites. It is known that HMGI(Y) binds to these sites with high affinity in vitro. However, whether or not HMGI(Y) co-occupies these sites with NF-κB and c-Jun/ATF 2 is not clear. It is likely that HMGI(Y) plays a role in the initial binding steps of the activators. The remarkable feature of highly stable activator/enhancer DNA complexes is that the stability is afforded by multiple weak interactions among the different proteins and between proteins and DNA. These weak interactions allow for rapid dissociation of the complex in restoring the repressed transcriptional state.

3.6. NF-AT and STAT

Members of two other transcription factor families, namely 'nuclear factor of activated T-cells' (NF-AT) as well as 'signal transducer and activator of transcription' (STAT), also use Ig-like domains to bind their specific DNA targets (Darnell, 1997; Rao et al., 1997; Leonard & O'Shea, 1998). Like NF-κB, the functions of NF-AT and STAT are involved in lymphocyte activation and the immune response. Although the level of sequence identity between NF-AT and NF-κB is low, the two proteins share high structural homology. Most NF-AT (NFAT1-4) proteins are monomers in solution and bind DNA relatively weakly, often requiring other transcription factors for stable DNA binding. One of the family members, NF-AT5, is dimeric both in solution and bound to DNA. DNA target sites for NF-AT proteins bear partial similarity to κB DNA. The consensus NF-AT DNA-binding site is 5'-GGAAA-3', which is identical to the DNA half-site sequence recognized by NF-κB p65 and c-Rel subunits. Most DNA-contacting residues from NF-κB are preserved in NF-AT proteins. The protein-DNA interface of the monomer is remarkably similar to NF-κB half-complexes. The inability of NF-AT proteins to form dimers simply prevents them from binding DNA strongly. Indeed, NF-AT5, which does form homodimers, can bind asymmetric NF-AT-specific DNA sequences as well as the more symmetric p65/c-Rel-specific κB DNA sequences with high affinity.

Although STAT and NF-κB show only 10 to 15% sequence homology in their DNA-binding domains, their three-dimensional structures are strikingly similar

(Becker et al., 1998; Chen et al., 1998b). The consensus STAT DNA sequence is 5'-TTCC(C/G)GGAA-3', where the central C:G bp lies on a pseudo-twofold axis. This STAT consensus sequence resembles a κB DNA with a different half-site arrangement. NF-κB dimers cannot bind to this sequence in their normal DNA-binding mode. However, depending on the flanking DNA sequences, STAT DNA-binding sites can overlap with those of NF-κB and NF-AT proteins. Some STAT DNAs look like 5'-nngaattcc(C/G)ggaattcnn-3', and two NF-κB dimers may bind cooperatively to these sequences through the embedded lower-case κB DNA. The close resemblance in the DNA target sequences and overall architecture of the complexes suggests the intriguing possibility that at least some genes could be targeted by members of more than one of these three transcription factor families. It is possible that NF-κB, NF-AT and STAT proteins compete with one another for these common DNA targets or, more likely, that they each bind to these DNAs at different times in response to different signals.

4. IκB PROTEINS

4.1. Scope

IκB was identified as an activity that both inhibits NF-κB DNA binding in vitro and sequesters NF-κB to the cytoplasm of resting cells in vivo (Baeuerle & Baltimore, 1988; Zabel & Baeuerle, 1990). IκB proteins belong to the large and biologically diverse superfamily of proteins containing an ankyrin repeat domain (ARD). IκB proteins constitute a subfamily within the ARD superfamily by virtue of their higher sequence homology both within and outside the ARD. To date, eight different IκB proteins are known. These are IκBα (Zabel et al., 1993), IκBβ (Thompson et al., 1995), IκBε (Simeonidis et al., 1997; Whiteside et al., 1997), IκBγ (Inoue et al., 1992), p105 (Hatada et al., 1992; Liou et al., 1992), p100, IκBζ (Kitamura et al., 2000; Shiina et al., 2001; Yamazaki et al., 2001) and Bcl-3 (Wulczyn et al., 1992). The number of ankyrin repeats within these proteins varies between 6 and 7.

4.2. Structure

Ankyrin repeats are a 33-amino acid tandem-repeat motif (Michaely & Vennett, 1992; Bork, 1993). The repeating structural unit is composed of a helix-turn-helix connected to a β-hairpin turn by an extended segment (Fig. 5). The ARD is created as individual repeats stack in layers with a slight left-handed superhelical twist. Many ARD-containing proteins also exhibit a slight smooth bend that arises as a result of conserved small amino acid positions within the ankyrin consensus helix-turn-helix. The β-hairpin turns from each repeat project towards the concave surface in an orientation roughly perpendicular to the plane created by stacked helices and protrude like fingers (Groves & Barford, 1999; Sedgwick & Smerdon, 1999).

The IκB family proteins carry out overlapping, but distinct functions (Verma et al., 1995; Whiteside & Isräel, 1997). Some of their functional differences can be explained by the sequence differences within the ARD. IκBβ, for example, contains a nonhomologous insertion of 47 amino acids within its third ankyrin repeat. Other smaller insertions are also present within the ARD of other IκB family proteins. The ankyrin repeat insertions are accommodated within the extended segment which connects the second outer helix and the β-hairpin turn, and are therefore exposed at the convex outer surface of the proteins. The structural integrity of the stacked repeats is not disrupted by the presence of the insertions.

The x-ray crystal structures of three IκB family members, namely IκBα (Huxford et al., 1998; Jacobs & Harrison, 1998), IκBβ and Bcl-3 (Michel et al., 2001), have been determined. Whereas the IκBα and IκBβ structures were solved as complexes with NF-κB, the structure of Bcl-3 has been determined as a free protein. In each of these cases, the ARD structures closely resemble one another.

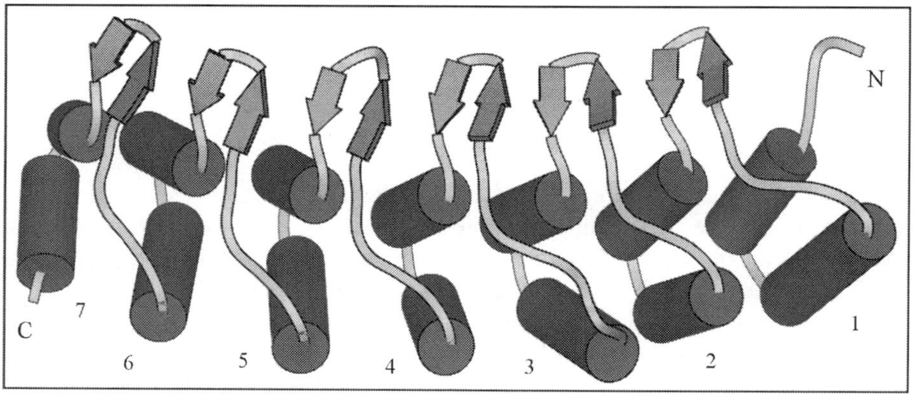

Fig. 5. Three-dimensional model of the ankyrin domain of Bcl-3. Helices in the convex face and β-hairpin turns on the concave face are shown as cylinders and arrows, respectively. The seven individual ankyrin repeats are numbered.

4.3. Classification

IκB represents a diverse protein family both with regard to structure and function. Indeed, all of these factors may not even function as NF-κB inhibitors. IκB proteins can be divided into three subgroups based on differences in their structures and functions.

IκBα, IκBβ and IκBε comprise the first subgroup. These three IκB proteins are the primary inhibitors of activating NF-κB dimers, such as the p50/p65 heterodimer and p65 homodimer. The ARD of these proteins comprises six ankyrin repeats flanked by an N-terminal signal-response region (SRR) and a C-terminal domain rich in proline, glutamic acid, serine and threonine (PEST). The SRR contains conserved serine sites of phosphorylation by the IκB kinase complex (IKK) and lysine residues that are polyubiquitinated prior to proteolytic degradation (Karin & Ben-Neriah, 2000). The acidic PEST sequence is common to many proteins that display rapid turnover in the cell (Rogers et al., 1986). It remains unclear whether the C-terminus of IκBε contains a functional PEST. Serine and threonine side chains within the PEST sequence of IκBα and IκBβ undergo constitutive phosphorylation by casein kinase II (Barroga et al., 1995; McElhinny et al., 1996; Schwarz et al., 1996). In contrast with the N-terminal SRR, the C-terminal PEST sequence of IκBα and IκBβ is also important for stable NF-κB binding (Ernst et al., 1995).

The second IκB family subgroup comprises p105 and p100. These two complicated proteins play dual cellular roles. First, p105 and p100 serve as precursors of the mature NF-κB subunits p50 and p52, respectively. The precursors are not processed entirely into NF-κB p50 and p52 subunit products, and the C-terminal half of the unprocessed molecules serves as inhibitor of other monomeric NF-κB subunits, including the p50 and p52 processed products. As IκB family members, p105 and p100 serve as nonspecific inhibitors of NF-κB. Furthermore, a separate gene identical to the C-terminus of p105 gives rise to the rare IκBγ molecule that shows affinity for NF-κB p50 homodimers.

Bcl-3 and IκBζ comprise the third subgroup. Contrary to the other IκB proteins, which are primarily cytoplasmic, Bcl-3 and IκBζ contain recognizable nuclear localization sequences and are observed in the nucleus. Their exact function within the nucleus, however, remains unclear. Bcl-3 interacts primarily with NF-κB p50 and p52 homodimers, and is known to exist in multiple phosphorylated forms. Bcl-3 may act both as a repressor and an activator of transcription, depending on its phosphorylation state.

5. IκB/NF-κB COMPLEX STRUCTURES

5.1. Structure of the IκBα/NF-κB p50/p65 heterodimer complex

The x-ray crystal structure of IκBα bound to the NF-κB p50/p65 heterodimer has been determined independently by two different laboratories (Huxford et al., 1998; Jacobs & Harrison, 1998). These two structures reveal the complex protein interface formed between IκBα and NF-κB, and provide a structural basis for the two primary inhibitory functions of IκBα, which are the regulation of subcellular localization and DNA binding of the NF-κB p50/p65 heterodimer (Fig. 6).

The structures of several NF-κB dimers bound to DNA showed that in its free form the dimeric NF-κB RHR has five segments flexibly linked to one another and dynamic in solution. These segments are the dimer platform composed of two closely packed C-terminal dimerization domains, two N-terminal domains and two NLS polypeptides. IκBα also has two binding elements, viz. the structured ARD and the flexible C-terminal PEST sequence. Thus it was clear from the beginning that formation of stable IκBα/NF-κB complexes requires the complicated ordering of multiple independent and flexible structural elements. Before elucidation of the IκBα/NF-κB structure, biochemical assays had shown that the N-terminal domain of NF-κB p50 was not involved in IκBα binding. Hence both complex structures were determined with a heterodimer from which the p50 N-terminal domain had been removed. The structures revealed a discontinuous protein-protein interface that can be divided into three parts. First, ankyrin repeats 1-2 of IκBα contact the NF-κB p65 subunit NLS polypeptide. Second, IκBα ankyrin repeats 3-4-5-6 and the C-terminal PEST contact the dimerization domains of both p50 and p65 subunits. Finally, the IκBα C-terminal PEST contacts the p65 subunit N-terminal domain.

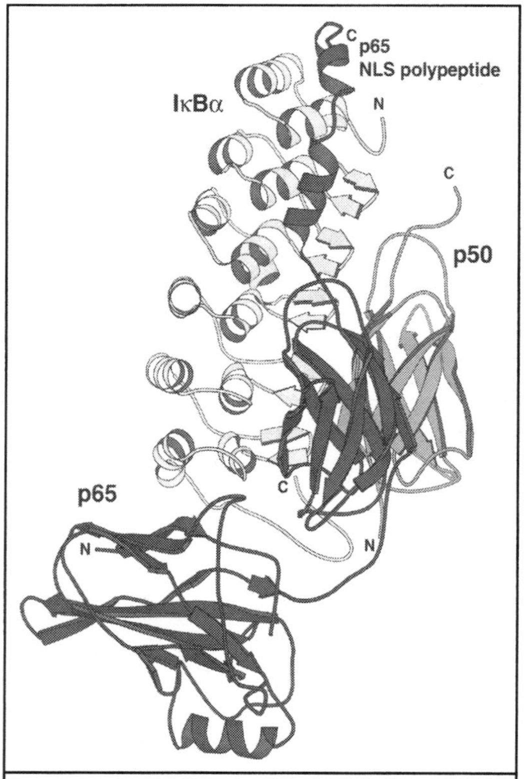

Fig. 6. Ribbon representation of IκBα in complex with the NF-κB p50/p65 heterodimer based on two published IκBα/NF-κB complex crystal structures. Note that the p50 NLS does not contact IκBα.

When bound to IκBα, the NF-κB p65 subunit NLS polypeptide forms two short helices oriented orthogonally with respect to each other. The second helix rests on top of the first IκBα ankyrin repeat and mediates multiple interactions. The first helix of the p65 NLS polypeptide ends with Lys301, Arg302, Lys303 and Arg304. These amino acids, three of which take part in ion-pairing interactions with acidic side chains of IκBα, constitute the basic NLS. The contacts in this region of the IκBα/NF-κB complex explain a mechanism for how the p65 subunit NLS is "masked" upon binding to IκBα. The finger-like protrusions from ankyrin repeats 3, 4 and 5 of IκBα make several interactions with the p50 subunit dimerization domain. This observation is surprising, as IκBα derives its high binding affinity by association with the p65 subunit in NF-κB dimers and fails to bind appreciably to the p50 homodimer. None of these interactions appears to be specific to p50, which implies that similar contacts could occur between IκBα and the NF-κB p65 homodimer.

The N-terminal domain of the NF-κB p65 subunit undergoes a dramatic conformational change upon binding to IκBα. A rotation of 171° and a translation of 38 Å are required to return this domain to its DNA-bound "open" conformation. This conformational change is stabilized by long-range electrostatic interactions between the positively charged DNA-binding surfaces of the p65 N-terminal domain and the acidic IκBα PEST, together with the uniquely acidic bottom portion of the p65 dimerization domain. Adoption of this "closed" IκB-bound conformation is likely to be further aided by constitutive IκBα PEST phosphorylation and renders NF-κB incapable of binding DNA.

5.2. Specificity of IκBα to NF-κB p65 subunits

Although NF-κB p50 and p65 subunits share a significant primary sequence homology, there are great sequence differences C-terminal to the basic NLS. The p65 subunit contains several hydrophobic amino acids within the second helix of the p65 NLS polypeptide that mediate hydrophobic contacts with the first ankyrin repeat of IκBα. These residues are absent in p50 and may explain the significantly lower binding affinity of IκBα to the NF-κB p50 homodimer. To test the hypothesis that the NF-κB p65 NLS polypeptide is in fact the primary element required for high affinity binding to NF-κB, the NLS polypeptides of p50 and p65 were swapped to create hybrid homodimers. Surprisingly, neither hybrid protein binds IκBα with near-native affinity. However, when two additional amino acids within the dimerization domain of p50 (Thr256 and Ala257 in mouse p50) are converted to corresponding residues of p65 (Asn202 and Ser203), the hybrid p50 homodimer with p65 NLS polypeptide displayed a significant recovery of IκBα-binding affinity. The IκBα/NF-κB complex crystal structures reveal that the p65 Asn202 and Ser203 are present on the face of the NF-κB dimerization platform opposite the complex protein-protein interface and that the nearest IκBα atom is located at a distance of nearly 20 Å. These results are supported by a prior yeast genetic screen for interactions of a system homologous to NF-κB in Drosophila (Lehming et al., 1995). Screening for mutants of the Drosophila NF-κB homologue Dorsal that escape regulation by the IκB homologue Cactus repeatedly identified Cys233 to Arg or Ser234 and Pro Dorsal mutations. Residues 233 and 234 in Dorsal are in a homologous position to Asn202 and Ser203 of p65. Taken together, these data point to a mechanism where the p65 subunit NLS polypeptide is a required element of high affinity interactions with IκBα. However, additional amino acid positions, some well outside of the complex interface, are also required in order to provide the proper environment and promote stable association of IκBα and NF-κB. Whether the effect of these distant amino acid positions concerns protein flexibility, protein folding or kinetics of complex formation remains to be determined.

5.3. IκBα/NF-κB complexes shuttle between the cytoplasm and nucleus

The most surprising result revealed by the two complementary IκBα/NF-κB p50/p65 complex structures was that the p50 NLS is exposed to solvent and fails to contact IκBα. This suggested that in quiescent cells the complex must be able to enter the nucleus by virtue of the free p50 NLS. By implementing a selective inhibitor of nuclear export, it was proven that IκBα/NF-κB complexes in fact shuttle between the nucleus and the cytoplasm (Johnson et al., 1999; Huang et al., 2000; Tam et al., 2000). The complex relies upon the free p50 NLS for nuclear import; the nuclear export sequences present in both IκBα and the activation domain of p65 are responsible for nuclear export (Malek et al., 2001). At equilibrium in resting cells, nuclear export is faster than import, which explains why under normal circumstances the IκBα/NF-κB complexes are seen predominantly within the cell cytoplasm.

It has been shown that the NF-κB p65 homodimer contains one masked and one free NLS in complex with IκBα. This suggests that this represents the general mode of binding in all physiological IκBα/NF-κB complexes. However, the question remains why IκBα/NF-κB complexes shuttle between the nucleus and the cytoplasm. At present the answer is not clear. However, it is known that expression of the IκBα gene falls under transcriptional control of NF-κB. NF-κB induction therefore results in newly synthesized IκBα, entering the nucleus and terminating NF-κB transcriptional activity by association with NF-κB. Thus it could simply be that the transcriptional repression function of IκBα requires it to shuttle between these two compartments. A second possibility is that the shuttling ability of the complex allows it to be ubiquitinated for degradation in the nucleus. Ubiquitination of phospho-IκBα could occur in the nucleus, as the IκBα-specific E3 ubiquitin-protein ligase receptor subunit is known to localize within the nucleus (Davis et al., 2002). The ability of IκBα/NF-κB complexes to transiently localize to the nucleus does not endow them with DNA-binding ability. Therefore it matters little whether these complexes reside in the cytoplasm or in the nucleus. The presence of IκBα alone assures transcriptional repression.

5.4. Regulation of NF-κB by IκBβ

IκBβ is another major inhibitor of NF-κB. Like IκBα, IκBβ is present in most cell types tested; nearly equivalent amounts of the activating NF-κB dimers are bound to IκBα and IκBβ. Both IκBβ and IκBα share similar specificity to p65 subunit-containing NF-κB dimers. However, IκBβ displays distinct functional differences from IκBα. First, IκBβ degradation is much slower than IκBα in response to most inducers (Thompson et al., 1995; Tran et al., 1997). Second, rather than shuttle between the nucleus and cytoplasm, IκBβ/NF-κB complexes are strictly cytoplasmic in quiescent cells (Malek et al., 2001; Tam & Sen, 2001). Third, in response to some inducers, such as IL-1 and viruses, IκBβ can enter the nucleus as complexes with NF-κB. Fourth, IκBβ is responsible for persistent NF-κB activation. Finally, IκBβ/NF-κB complexes are capable of binding κB DNA with low affinity. The mechanisms behind these functional differences are unclear.

We have recently determined the x-ray crystal structure of IκBβ in complex with the NF-κB p65 homodimer. We observe that in many ways this structure is similar to that of the IκBα/NF-κB heterodimer complex with two possible exceptions. First, the binding mode of the acidic C-terminal PEST is clearly different in these two complexes. The PEST of IκBβ does not contact any DNA-contacting residues of p65, which allows one to speculate on ternary complex formation between IκBβ/p65/DNA. Second, though one p65 NLS is bound in a manner analogous to that observed in the IκBα/NF-κB heterodimer complex, the second p65 NLS also participates in weak interactions with IκBβ. This weak binding is not likely to sufficiently block nuclear import of the complex. It is therefore possible that this complex represents the nuclear IκBβ/NF-κB complex, which was predicted by early biological studies.

5.5. Bcl-3 structure and function

The x-ray crystal structure of the ARD of Bcl-3 has recently been determined (Michel et al., 2001). This protein contains seven ankyrin repeats, each of which assumes the canonical ankyrin fold. Over half of the IκBα amino acids involved in specific contacts with NF-κB is conserved in Bcl-3. This suggests that Bcl-3 contacts NF-κB dimers with an orientation similar to that observed in the IκBα/NF-κB and IκBβ/NF-κB complexes. The structure of the Bcl-3 ARD differs from IκBα and IκBβ at the N- and C-terminus. At the N-terminus, the conserved acidic amino-acid side-chain positions, responsible for contacting the NF-κB p65 NLS in both the IκBα/NF-κB and IκBβ/NF-κB complex structures, assume an alternative conformation in Bcl-3. Although influenced by crystal packing forces, this altered conformation likely represents less stable binding between Bcl-3 ankyrin repeat 1 and the NLS of NF-κB dimers. At the C-terminus, Bcl-3 contains a seventh ankyrin repeat. In contrast to the IκBα C-terminal PEST, the seventh ankyrin repeat of Bcl-3 turns away from the NF-κB-binding interface. Additional basic and hydrophobic amino acids on the bottom surface of Bcl-3 could potentially mediate interactions with DNA. These structural findings are supporting two previous observations: first, two Bcl-3 monomers can bind to one NF-κB dimer (Wulczyn et al., 1992; Bundy & McKeithan, 1997); second, Bcl-3 can recognize and bind to NF-κB/DNA complexes (Bours et al., 1993; Fujita et al., 1993). Tertiary complex formation alters the NF-κB transcriptional signal.

5.6. Regulation of NF-κB by other IκB proteins

Little is known about how other IκB proteins regulate NF-κB function. It appears that these inhibitors play more restricted roles in a tissue- and cell type-specific manner. IκBε has been shown to be more specific for p65 and c-Rel homodimers than for the p65/c-Rel heterodimer (Simeonidis et al., 1997; Whiteside et al., 1997). The structural and biochemical basis for this specificity is unclear. The overall structure of NF-κB complexes with IκBε is expected to be similar to those observed with IκBα and IκBβ. However, some details of the complexes are clearly different. For example, the SRR of IκBε is much longer than either IκBα or IκBβ proteins.

The p105 protein is known to inhibit all NF-κB proteins, including p50 non-specifically (Rice et al., 1992; Mercurio et al., 1993). The mode of p105 binding to NF-κB can be understood from the binding of IκBγ and NF-κB p50 homodimers. IκBγ is the protein product of a separate gene which is identical to the C-terminal 606 residues of p105 (Inoue et al., 1992). IκBγ exists rarely in cells and therefore its physiological role is in doubt. In contrast, p105 is a clear functional inhibitor of NF-κB.

IκBγ binds strongly to the NF-κB p50 homodimer and the other p50 subunit-containing dimers, such as the p50/p65 heterodimer. A complex between the IκBγ and p50 homodimer is essentially equivalent to the complex between p50 and p105. Similarly, the IκBγ/p50/p65 complex is similar to the p65/p105 complex. One can envision that newly synthesized p105 forms homodimers of an unusual asymmetric complex, in which only one C-terminal ARD associates intramolecularly with the dimer formed by the N-terminal RHR dimer equivalent to the p50 homodimer. This leaves the second C-terminal ARD exposed to solvent, allowing it to be processed. The result is the formation of a stable p50/p105 complex. The mature processed p50 subunit within the p50/p105 complex can be exchanged with other NF-κB subunits such as p65 and c-Rel, leading to formation of the potent transactivating NF-κB heterodimers. Immediately C-terminal to the seventh ankyrin repeat in p105 is a death domain and a consensus IKK phosphorylation sequence (Heissmeyer et al., 2001). The precise structural and functional roles of these p105 C-terminal elements remain to be elucidated.

6. CONCLUSIONS

Since the determination of the first two NF-κB/DNA complex crystal structures in 1995, structural analysis of relevant NF-κB proteins and their complexes has been a valuable component in understanding NF-κB molecular biology. This will continue as structural biologists forge ahead with projects aimed at unraveling the secrets of multiple transcription factor association and enhanceosome assembly, processing of the NF-κB precursor proteins and activation of NF-κB by phosphorylation, ubiquitination and proteolysis of IκB.

7. ACKNOWLEDGMENTS

The authors wish to acknowledge C. Phelps for help with preparing figures and proofreading the text. Work in the authors' laboratory is supported by NIH/NCI.

8. REFERENCES

Baeuerle PA & Baltimore D (1988) IκB: a specific inhibitor of the NF-κB transcription factor. Science 242: 540-546.

Baeuerle PA & Baltimore D (1996) NF-κB: ten years after. Cell 87: 13-20.

Baeuerle PA & Henkel T (1994) Function and activation of NF-κB in the immune system. Annu.

Rev. Immunol. 12: 141-179.

Baldwin AS (1996) The NF-κB and IκB proteins: new discoveries and insights. Annu. Rev. Immunol. 14: 649-683.

Barroga CF, Stevenson JK, Schwarz EM & Verma IM (1995) Constitutive phosphorylation of IκBα by casein kinase II. Proc. Natl. Acad. Sci. USA 92: 7637-7641.

Becker S, Groner B & Müller CW (1998) Three-dimensional structure of the Stat3β homodimer bound to DNA. Nature 394: 145-151.

Bork P (1993) Hundreds of ankyrin-like repeats in functionally diverse proteins: mobile modules that cross phyla horizontally? Proteins 17: 363-374.

Bours V, Franzoso G, Azarenko V, Park S, Kanno T, Brown K & Siebenlist U (1993) The oncoprotein Bcl-3 directly transactivates through κB motifs via association with DNA-binding p50B homodimers. Cell 72: 729-739.

Bundy DL & McKeithan TW (1997) Diverse effects of Bcl-3 phosphorylation on its modulation of NF-κB p52 homodimer binding to DNA. J. Biol. Chem. 272: 33132-33139.

Chen FE, Huang D-B, Chen Y-Q & Ghosh G (1998a) Crystal structure of p50/p65 heterodimer of transcription factor NF-κB bound to DNA. Nature 391: 410-413.

Chen X, Vinkemeier U, Hao Y, Jeruzalmi D, Darnell JE & Kuriyan J (1998b) Crystal structure of a tyrosine phosphorylated Stat-1 dimer bound to DNA. Cell 93: 827-839.

Chen Y-Q, Ghosh S & Ghosh G (1998c) A novel DNA recognition mode by the NF-κB p65 homodimer. Nature Struct. Biol. 5: 67-73.

Chen Y-Q, Sengchanthalangsy LL, Hackett A & Ghosh G (2000) NF-κB p65 (RelA) homodimer uses distinct mechanisms to recognize DNA targets. Structure 8: 419-428.

Cramer P, Larson CJ, Verdine GL & Müller CW (1997) Structure of the human NF-κB p52 homodimer-DNA complex at 2.1 Å resolution. EMBO J. 16: 7078-7090.

Cramer P, Varrot A, Barillas-Mury C, Kafatos FC & Müller CW (1999) Structure of the specificity domain of the Dorsal homologue Gambif1 bound to DNA. Structure 7: 841-852.

Cross SL, Halden NF, Lenardo MJ & Leonard WJ (1989) Functionally distinct NF-κB binding sites in the immunoglobulin κ and IL-2 receptor α chain genes. Science 244: 466-469.

Darnell JE (1997) Stats and gene regulation. Science 277: 1630-1635.

Davis M, Hatzubai A, Andersen JS, Ben-Shushan E, Fisher GZ, Yaron A, Bauskin A, Mercurio F, Mann M & Ben-Neriah Y (2002) Pseudosubstrate regulation of the SCF/β-TrCP ubiquitin ligase by hnRNP-u. Genes Dev. 16: 439-451.

Dobrzanski P, Ryseck R-P & Bravo R (1993) Both N- and C-terminal domains of RelB are required for full transactivation: role of the N-terminal leucine zipper-like motif. Mol. Cell. Biol. 13: 1572-1582.

Ernst MK, Dunn LL & Rice NR (1995) The PEST-like sequence of IκBα is responsible for inhibition of DNA binding but not for cytoplasmic retention of c-Rel or RelA homodimers. Mol. Cell. Biol. 15: 872-882.

Escalante CR, Shen LY, Thanos D & Aggarwal AK (2002) Structure of NF-κB p65 (RelA) p50/p65 heterodimer bound to the PRDII DNA element from the interferon-β promoter. Structure 10: 383-391.

Franzoso G, Bours V, Park S, Tomita-Yamaguchi M, Kelly K & Siebenlist U (1992) The candidate oncoprotein Bcl-3 is an antagonist of p50/NF-κB-mediated inhibition. Nature 359: 339-342.

Fujita T, Nolan GP, Liou H-C, Scott ML & Baltimore D (1993) The candidate proto-oncogene Bcl-3 encodes a transcriptional coactivator that activates through NF-κB p50 homodimers. Genes Dev. 7: 1354-1363.

Ghosh G, Van Duyne G, Ghosh S & Sigler PB (1995) Structure of NF-κB p50 homodimer bound to a κB site. Nature 373: 303-310.

Ghosh S, May MJ & Kopp EB (1998) NF-κB and Rel proteins: evolutionarily conserved mediators of immune responses. Annu. Rev. Immunol. 16: 225-260.

Gilmore TD (1999) The Rel/NF-κB signal transduction pathway. Introduction. Oncogene 18: 6842-6844.

Groves MR & Barford D (1999) Topological characteristics of helical repeat proteins. Curr. Opin.

Struct. Biol. 9: 383-389.

Hart DJ, Speight RE, Sutherland JD & Blackburn JM (2001) Analysis of the NF-κB p50 dimer interface by diversity screening. J. Mol. Biol. 310: 563-575.

Hatada EN, Nieters A, Wulczyn FG, Naumann M, Meyer R, Nucifora G, McKeithan TW & Scheidereit C (1992) The ankyrin repeat domains of the NF-κB precursor p105 and the proto-oncogene Bcl-3 act as specific inhibitors of NF-κB DNA binding. Proc. Natl. Acad. Sci. 89: 2489-2493.

Heissmeyer V, Krappmann D, Hatada EN & Scheidereit C (2001) Shared pathways of IκB kinase-induced SCF/βTrCP-mediated ubiquitination and degradation for the NF-κB precursor p105 and IκBα. Mol. Cell. Biol. 21: 1024-1035.

Huang D-B, Chen Y-Q, Ruetsche M, Phelps CB & Ghosh G (2001) X-ray crystal structure of proto-oncogene product c-Rel bound to the CD28 response element of IL-2. Structure 9: 669-678.

Huang D-B, Huxford T, Chen Y-Q & Ghosh G (1997) The role of DNA in the mechanism of NF-κB dimer formation: crystal structures of the dimerization domains of the p50 and p65 subunits. Structure 5: 1427-1436.

Huang TT, Kudo N, Yoshida M & Miyamoto S (2000) A nuclear export signal in the N-terminal regulatory domain of IκBα controls cytoplasmic localization of inactive NF-κB/IκBα complexes. Proc. Natl. Acad. Sci. USA 97: 1014-1019.

Huxford T, Huang D-B, Malek S & Ghosh G (1998) The crystal structure of the IκBα/NF-κB complex reveals mechanisms of NF-κB inactivation. Cell 95: 759-770.

Inoue JI, Kerr LD, Kakizuka A & Verma IM (1992) IκBγ, a 70 kD protein identical to the C-terminal half of p110 NF-κB: a new member of the IκB family. Cell 68: 1109-1120.

Jacobs MD & Harrison SC (1998) Structure of an IκBα/NF-κB complex. Cell 95: 749-758.

Jacobson EM, Li P, Leon-Del-Rio A, Rosenfeld MG & Aggarwal AK (1997) Structure of Pit-1 POU domain bound to DNA as a dimer: unexpected arrangement and flexibility. Genes Dev. 11: 198-212.

Johnson C, Van Antwerp D & Hope TJ (1999) An N-terminal nuclear export signal is required for the nucleocytoplasmic shuttling of IκBα. EMBO J. 18: 6682-6693.

Karin M & Ben-Neriah Y (2000) Phosphorylation meets ubiquitination: the control of NF-κB activity. Annu. Rev. Immunol. 18: 621-663.

Kitamura H, Kanehira K, Okita K, Morimatsu M & Saito M (2000) MAIL, a novel nuclear IκB protein that potentiates LPS-induced IL-6 production. FEBS Lett. 485: 53-56.

Kushner DB & Ricciardi RP (1999) Reduced phosphorylation of p50 is responsible for diminished NF-κB binding to the major histocompatibility complex class I enhancer in adenovirus type 12-transformed cells. Mol. Cell. Biol. 19: 2169-2179.

Lehming N, McGuire S, Brickman JM & Ptashne M (1995) Interactions of a Rel protein with its inhibitor. Proc. Natl. Acad. Sci. USA 92: 10242-10246.

Leonard WJ & O'Shea JJ (1998). Jaks and Stats: biological implications. Annu Rev. Immunol. 16: 293-322.

Liou HC, Nolan GP, Ghosh S, Fujita T & Baltimore D (1992) The NF-κB p50 precursor, p105, contains an internal IκB-like inhibitor that preferentially inhibits p50. EMBO J. 11: 3003-3009.

Liu J, Sodeoka M, Lane WS & Verdine GL (1994) Evidence for a non-α-helical DNA-binding motif in the Rel homology region. Proc. Natl. Acad. Sci. USA 91: 908-912.

Malek S, Chen Y, Huxford T & Ghosh G (2001) IκBβ, but not IκBα, functions as a classical cyto-plasmic inhibitor of NF-κB dimers by masking both NF-κB nuclear localization sequences in resting cells. J. Biol. Chem. 276: 45225-45235.

McElhinny JA, Trushin SA, Bren GD, Chester N & Paya CV (1996) Casein kinase II phosphorylates IκBα at S283, S289, S293, and T291 and is required for its degradation in vitro. Mol. Cell. Biol. 16: 899-906.

Mercurio F, Didonato JA, Rosette C & Karin M (1993) P105 and p98 precursor proteins play an active role in NF-κB-mediated signal transduction. Genes Dev. 7: 705-718.

Michaely P & Vennett V (1992) The ANK repeat: a ubiquitous motif involved in macromolecular recognition. Trends Cell. Biol. 2: 127-130.

Michel F, Soler-Lopez M, Petosa C, Cramer P, Siebenlist U & Müller CW (2001) Crystal structure of the ankyrin repeat domain of Bcl-3: a unique member of the IκB protein family. EMBO J. 20: 6180-6190.

Müller CW, Rey FA, Sodeoka M, Verdine GL & Harrison SC (1995) Structure of the NF-κB p50 homodimer bound to DNA. Nature 373: 311-317.

Pahl HL (1999) Activators and target genes of Rel/NF-κB transcription factors. Oncogene 18: 6853-6866.

Phelps CB, Sengchanthalangsy LL, Malek S & Ghosh G (2000) Mechanism of κB DNA binding by Rel/NF-κB dimers. J. Biol. Chem. 275: 24392-24399.

Plaksin D, Baeuerle PA & Eisenbach L (1993) KBF1 (p50 NF-κB homodimer) acts as a repressor of H-2K-b gene expression in metastatic tumor cells. J. Exp. Med. 177: 1651-1662.

Rao A, Luo C & Hogan PG (1997) Transcription factors of the NF-AT family: regulation and function. Annu. Rev. Immunol. 15: 707-747.

Rice NR, Mackichan ML & Isräel A (1992) The precursor of NF-κB p50 has IκB-like functions. Cell 71: 243-253.

Rogers S, Wells R & Rechsteiner M (1986) Amino acid sequences common to rapidly degraded proteins: the PEST hypothesis. Science 234: 364-368.

Ryseck R-P, Novotny J & Bravo R (1995) Characterization of elements determining the dimerization properties of RelB and p50. Mol. Cell. Biol. 15: 3100-3109.

Ryseck RP, Bull P, Takamiya M, Bours V, Siebenlist U, Dobrzanski P & Bravo R (1992) RelB, a new Rel family transcription activator that can interact with p50-NF-κB. Mol. Cell. Biol. 12: 674-684.

Sanjabi S, Hoffmann A, Liou H-C, Baltimore D & Smale ST (2000) Selective requirement for c-Rel during IL-12 p40 gene induction in macrophages. Proc. Natl. Acad. Sci. USA 97: 12705-12710.

Schwarz EM, Van Antwerp D & Verma IM (1996) Constitutive phosphorylation of IκBα by casein kinase II occurs preferentially at serine 293: requirement for degradation of free IκBα. Mol. Cell. Biol. 16: 3554-3559.

Sedgwick SG & Smerdon SJ (1999) The ankyrin repeat: a diversity of interactions on a common structural framework. Trends Biochem. Sci. 24: 311-316.

Sengchanthalangsy LL, Datta S, Huang D-B, Anderson E, Braswell EH & Ghosh G (1999) Characterization of the dimer interface of transcription factor NF-κB p50 homodimer. J. Mol. Biol. 289: 1029-1040.

Shiina T, Morimatsu M, Kitamura H, Ito T, Kidou S, Matsubara K, Matsuda Y, Saito M & Syuto B (2001) Genomic organization, chromosomal localization, and promoter analysis of the mouse MAIL gene. Immunogenetics 53: 649-655.

Siebenlist U, Franzoso G & Brown K (1994). Structure, regulation and function of NF-κB. Annu. Rev. Immunol. 12: 405-455.

Silverman N & Maniatis T (2001) NF-κB signaling pathways in mammalian and insect innate immunity. Genes Dev. 15: 2321-2342.

Simeonidis S, Liang S, Chen G & Thanos D (1997) Cloning and functional characterization of mouse IκBε. Proc. Natl. Acad. Sci. USA 94: 14372-14377.

Tam WF, Lee LH, Davis L & Sen R (2000) Cytoplasmic sequestration of Rel proteins by IκBα requires Crm1-dependent nuclear export. Mol. Cell. Biol. 20: 2269-2284.

Tam WF & Sen R (2001) IκB family members function by different mechanisms. J. Biol. Chem. 276: 7701-7704.

Thompson JE, Phillips RJ, Erdjument-Bromage H, Tempst P & Ghosh S (1995) IκBβ regulates the persistent response in a biphasic activation of NF-κB. Cell 80: 573-582.

Tran K, Merika M & Thanos D (1997) Distinct functional properties of IκBα and IκBβ. Mol. Cell. Biol. 17: 5386-5399.

Verma IM, Stevenson K, Schwarz EW, Van Antwerp D & Miyamoto S (1995) Rel/NF-κB/IκB family: intimate tales of association and dissociation. Genes Dev. 9: 2723-2735.

Whiteside ST, Epinat J-C, Rice NR & Isräel A (1997) IκBε, a novel member of the IκB family, controls RelA and c-Rel NF-κB activity. EMBO J. 16: 1413-1426.

Whiteside ST & Isräel A (1997) IκB proteins: structure, function and regulation. Semin. Canc. Biol. 8: 75-82.

Wulczyn FG, Naumann M & Scheidereit C (1992) Candidate proto-oncogene Bcl-3 encodes a subunit-specific inhibitor of transcription factor NF-κB. Nature 358: 597-599.

Yamazaki S, Muta T & Takeshige K (2001) A novel IκB protein, IκBζ, induced by proinflammatory stimuli, negatively regulates NF-κB in the nuclei. J. Biol. Chem. 276: 27657-27662.

Zabel U & Baeuerle PA (1990) Purified human IκB can rapidly dissociate the complex of the NF-κB transcription factor with its cognate DNA. Cell 61: 255-266.

Zabel U, Henkel T, Dos Santos Silva M & Baeuerle PA (1993) Nuclear uptake control of NF-κB by MAD-3, an IκB protein present in the nucleus. EMBO J. 12: 201-211.

Zhong H, May MJ, Jimi E & Ghosh S (2002) The phosphorylation status of nuclear NF-κB determines its association with CBP/p300 or HDAC-1. Mol. Cell 9: 625-636.

Zhong H, Voll RE & Ghosh S (1998) Phosphorylation of NF-κB p65 by PKA stimulates transcriptional activity by promoting a novel bivalent interaction with the coactivator CBP/p300. Mol. Cell 1: 661-671.

9. AFFILIATION

University of California, San Diego, Department of Chemistry and Biochemistry, 9500 Gilman Drive, MC 0359, La Jolla, CA 92093, USA.

Y. YAMAMOTO, U.N. VERMA & R.B. GAYNOR

TOOLS TO INTERFERE WITH NF-κB ACTIVATION

1. INTRODUCTION

The NF-κB-signaling pathway has attracted widespread interest due to its fundamental role in the regulation of immune and inflammatory responses and in the control of cellular proliferation. A number of steps in the NF-κB pathway are potential targets for drug therapy to inhibit the activity of the pathway. In fact, some of the agents that inhibit these steps in the NF-κB pathway are being tested in clinical trials for the treatment of inflammatory disorders and cancer. A better understanding of the regulation of the NF-κB-signaling pathway has also led to a re-evaluation of the function of several conventional anti-inflammatory and anti-cancer drugs and provided new insights into the regulation of NF-κB. In this chapter, we review recent progress on key steps involved in the control of the NF-κB pathway and agents which interfere with NF-κB activation. The therapeutic potential of these agents in inhibiting the inflammatory response and increasing the efficacy of cancer therapy are also addressed.

NF-κB comprises a family of inducible transcription factors that are important regulators of the host innate and adaptive immune and inflammatory responses (Baldwin, 1996; Ghosh et al., 1998; Silverman & Maniatis, 2001). The effects of NF-κB are mediated by its ability to increase the expression of specific cellular genes. These genes encode at least 27 different cytokines (e.g. TNFα, IL-1, IL-6 and GM-CSF), chemokines (e.g. IL-8, MIP-1α, MCP1, RANTES and eotaxin), adhesion molecules (e.g. E-selectin, VCAM-1 and ICAM-1) as well as enzymes such as cyclooxygenase (COX)-2 and inducible nitric oxide synthase (iNOS). Some of these effectors, such as TNFα and IL-1β, can also directly activate the NF-κB pathway to establish a positive autoregulatory loop that can amplify the inflammatory response and increase the duration of this response. NF-κB plays an important role in the adaptive immune response by regulating the production of both receptors involved in immune recognition (e.g. MHC) and cytokines (e.g. IL-2, IL-12 and IFNγ) increasing the proliferation and differentiation of lymphocytes. In addition to activating the expression of genes involved in the control of the immune and inflammatory response, the NF-κB pathway is also a key mediator of antiapoptotic genes, such as c-IAP1, c-IAP2, TRAF1, TRAF2, the Bcl-2 homologues (A1/BFl-1, IEX-IL), Fas ligand and c-myc, that are critical molecules involved in the regulation of cellular proliferation and apoptosis. Finally, NF-κB is also critical in the control of the cell cycle, which is an important

R. Beyaert (ed.), Nuclear Factor κB. Regulation and Role in Disease, 199-219.

element in determining the degree of cellular apoptosis and proliferation. For example, NF-κB is an important factor in modulating the expression of cyclin D1, a positive regulator of G1-to-S phase progression.

The wide variety of cellular genes regulated by NF-κB indicate that this family of transcription factors serves as an important regulator of the host immune and inflammatory responses. The increased expression of NF-κB-inducible genes has been implicated in the pathogenesis of chronic inflammatory diseases such as asthma, rheumatoid arthritis and inflammatory bowel disease (IBD). In addition, altered NF-κB regulation may be involved in other diseases such as atherosclerosis, Alzheimer's disease and diabetes, in which the inflammatory response is at least partially involved in the pathogenesis. Constitutive activation of the NF-κB pathway has also been implicated in the pathogenesis of some human cancers. As we better understand the regulation of this pathway, the therapeutic potential for inhibiting this pathway has received increasing attention. Since the activation of the NF-κB pathway requires a number of unique events, it is not surprising that there are a number of different targets to inhibit the activation of this pathway.

The first potential therapeutic target in the NF-κB pathway described in this chapter are the NF-κB proteins themselves. The NF-κB proteins comprise a family that shares a 300-amino acid domain (designated as the Rel homology domain), which mediates the DNA binding, dimerization and nuclear transport of the NF-κB proteins. The NF-κB family consists of five members in mammals: Rel (c-Rel), RelA (p65), RelB, NF-κB1 (p50 and its precursor p105) and NF-κB2 (p52 and its precursor p100). Classically, the NF-κB DNA-binding compartment was defined as a p65/p50 heterodimer. Although many functionally active dimeric forms of these factors have been detected and shown to bind to consensus sequences in DNA and activate gene expression, homodimers including p50/p50 and p52/p52 can repress the gene expression of their target genes. Following the completion of cytokine stimulation, the active p65/p50 heterodimer is replaced by binding of the p50 homodimer/histone deacetylase (HDAC) complex to the promoter elements of NF-κB regulatory genes, which is also seen under unstimulated conditions.

In unstimulated cells, the NF-κB proteins are predominantly localized in the cytoplasm associated with a family of inhibitor proteins known as IκB (comprising IκBα, IκBβ, IκBε and IκBΘ). The IκB proteins are characterized by multiple ankyrin repeats which are critical for their interaction with NF-κB. The IκB proteins mask the nuclear localization signal on the NF-κB proteins, which maintains their cytoplasmic localization. The phosphorylation of IκBα (and its subsequent degradation) is a key step involved in the regulation of NF-κB complexes. A variety of stimuli (including cytokines such as TNFα and IL-1, phorbol esters, LPS, viral infection, the human T-cell leukemia virus type 1-transforming protein Tax, ultraviolet radiation and free radicals) results in a regulated process that leads both to the degradation of IκBα and the nuclear translocation of NF-κB.

The phosphorylation of the IκBα proteins is mediated by IκB kinases (IKK) whose activity is strongly induced by activators of the NF-κB pathway (Zandi & Karin, 1999). IKK kinase activity is associated with a high molecular weight complex containing at least two kinase subunits (IKKα and IKKβ) and the associated modulatory protein (IKKγ or NEMO). These kinases have 52% amino acid homology and a similar structure including kinase, leucine zipper and helix-helix domains; these kinases are able to form both homo- and heterodimers. Biochemical analysis and gene disruption studies of the IKK genes in mice indicate that IKKβ is the critical kinase involved in activating the NF-κB pathway, while IKKα likely plays an accessory role (Zandi & Karin, 1999). After treatment of cells with agents that stimulate the NF-κB pathway, the IKK complex can be activated by upstream kinases, such as NIK and MEKK1, which phosphorylate specific serine residues in the activation loop of each IKK subunit. The activated IKK complex phosphorylates the IκBα proteins on two closely spaced serine residues in the amino terminus of these proteins. Phosphorylation of IκBα leads to its ubiquitination on two amino-terminal lysine residues to result in its degradation by the 26S proteasome. NF-κB then translocates to the nucleus, where it binds to specific elements in the promoter regions of target genes to activate gene expression.

We further discuss inhibitors that prevent NF-κB-regulated gene expression by interfering with this activation process, including an IκBα mutant which acts as a dominant-negative phenotype, a 26S proteasome inhibitor, a peptide with a nuclear localization sequence that can antagonize NF-κB nuclear localization, an IKKβ peptide containing the NEMO-binding domain (NBD) that can inhibit NEMO association with IKKβ, a peptide with the Toll/IL-1R domain-containing adapter domain (TIRAP) and a heat-shock protein (Hsp)-90 inhibitor that can downmodulate the TNFα receptor complex.

NF-κB transcriptional activation can also be regulated by both phosphorylation and acetylation of the Rel family member p65. NF-κB transcriptional activity can be induced by protein kinase A-induced phosphorylation of the p65 subunit on serine residue 276 (Zhong et al., 1998). Phosphorylation of this residue facilitates the interaction of the coactivator protein CBP with p65 to enhance its transcriptional activity. TNFα treatment of cells can also induce p65 phosphorylation, but this phosphorylation occurs on serine residue 529 (Wang & Baldwin, 1998). IKKα is likely required for the TNFα-induced p65 phosphorylation that is mediated by activation of the phophatidylinositol-3' kinase and the serine/threonine kinase AKT. A possible explanation for the importance of p65 phosphorylation is that it can lead to the recruitment of CBP, which can acetylate p65 and increase its nuclear localization and subsequent transcriptional activation (Zhong et al., 1998). Deacetylation of p65 by HDACs can also play a critical role in NF-κB transcriptional activation by reducing its nuclear levels and increasing its interaction with IκBα.

Given the diverse processes involved in activating the NF-κB pathway, a better understanding of this pathway has led to studies to address the function of conventional anti-inflammatory and anticancer drugs in regulating the NF-κB pathway (Fig. 1).

2. ANTISENSE AND INTERFERING RNAs TO INHIBIT THE NF-κB PATHWAY

2.1. Scope

Recent advances in our understanding of the pathways that control transcription and translation have provided new techniques to inhibit cellular gene expression. Emerging oligonucleotide approaches are not only powerful tools for the study of NF-κB-regulated gene expression, but also provide potential avenues for broad clinical applications.

2.2. RelA antisense oligodeoxynucleotides (ODNs)

The use of antisense ODNs as selective inhibitors of gene expression is potentially a powerful tool for the disruption of specific cellular genes (Lebedeva & Stein, 2001). Antisense ODNs are short stretches of DNA that are complementary to a target mRNA. The ODNs selectively hybridize to their complementary RNA and lead to duplex formation, which likely prevents the ribosomal complex from translating mRNA and thus prevents the synthesis of the protein of interest.

The therapeutic application of NF-κB antisense ODNs has been explored in the treatment of IBD (Neurath et al., 1996; Neurath et al., 1998; Murano et al., 2000). Increased cytokine production by lymphocytes and macrophages is implicated in the pathogenesis of IBD, including both Crohn's disease and ulcerative colitis. The local administration of RelA antisense ODNs prevents the clinical and histological signs of colitis in trinitrobenzene sulfonic acid colitis in mice, which is an animal model of Crohn's disease (Neurath et al., 1996). The oral administration of RelA antisense ODNs can also effectively downregulate the production of inflammatory cytokines in the colonic mucosa and result in the inhibition of both clinical and histological evidence of a mouse model of ulcerative colitis (Murano et al., 2000). The selective downregulation of p65 in macrophages in patients suffering from IBD after a treatment with RelA antisense ODNs considerably reduces the production of proinflammatory cytokines, which indicates the efficiency of RelA antisense ODNs to modify NF-κB-regulated genes (Neurath et al., 1998). These results suggest the potential therapeutic utility of RelA antisense ODNs for the treatment of patients with IBD.

The application of antisense ODNs has also been investigated in the treatment of a variety of other disorders. In some cases, the studies have reached the stage of clinical trials, such as the treatment of non-Hodgkin lymphoma patients with bcl-2 antisense ODNs (Lebedeva & Stein, 2001). However, the development of antisense ODNs has faced some challenges as well. One of the most difficult is the ability of the mRNA target sequences, which form three-dimensional structures, to hybridize with antisense ODNs in cells. Therefore, the successful application of antisense ODNs depends on the accessibility of mRNA target sequences to the antisense ODNs.

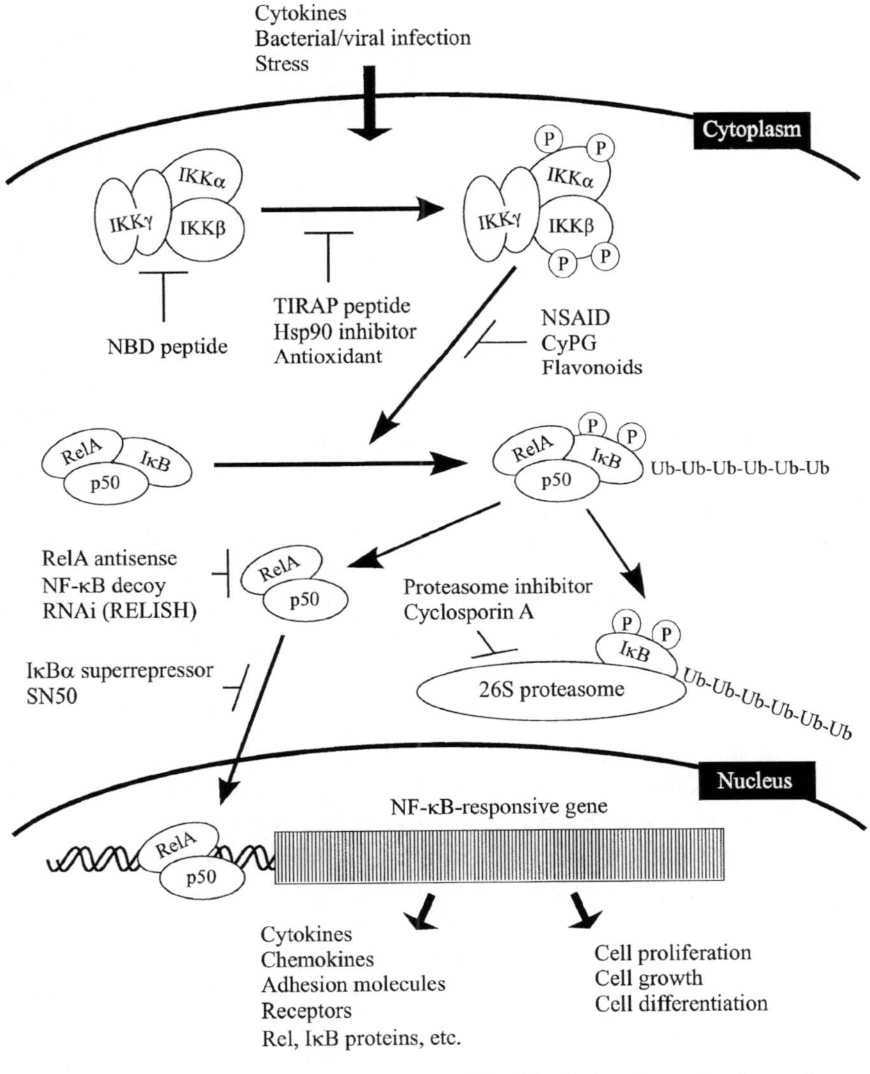

Fig. 1. Level of action of various inhibitors of the NF-κB signal transduction pathway.

2.3. *NF-κB decoy ODNs*

The transfection of synthetic double-stranded ODNs, referred to as decoy ODNs, has been reported to be a useful tool for silencing NF-κB transcription both under in vitro and in vivo conditions. This approach involves the introduction of decoy

ODNs containing consensus NF-κB DNA-binding sites, either by direct microinjection or by viral gene transfer. These ODNs render the endogenous NF-κB proteins incapable of subsequently binding to the promoter regions of NF-κB-regulated target genes. Thus the NF-κB decoy ODN approach potentially enables one to modulate NF-κB binding and its subsequent transcriptional regulation.

A first attempt to use NF-κB decoy ODNs for therapeutic purposes was in cardiovascular disorders, in which NF-κB activation during ischemia and reperfusion upregulates a variety of cytokines to result in myocardial dysfunction (Morishita et al., 1997). Introduction of NF-κB decoy ODNs into the coronary arteries reduces the myocardial infarct size in a rat model of acute coronary occlusion. This treatment also inhibits the expression of the NF-κB-regulated cytokine genes IL-6, IL-8, ICAM, VCAM and ELAM (Morishita et al., 1997). In an animal model of human giant cell myocarditis (Yokoseki et al., 2001), infusion of NF-κB decoy ODNs reduces the severity of experimental autoimmune myocarditis associated with a reduced expression of ICAM-1, iNOS, IL-2 and TNFα in the myocardium. In another study, transfection of NF-κB decoy ODNs in both the synovial cells derived from rheumatoid arthritis patients (Tomita et al., 2000) and within the joints of rats with a model of rheumatoid arthritis (Tomita et al., 1999) markedly suppresses IL-1β and TNFα expression. In the carrageenin-induced hind paw edema model of acute inflammation, NF-κB decoy ODNs dose-dependently inhibit edema formation in rat paw, which is associated with a reduced expression of COX-2 and iNOS (D'Acquisto et al., 2000). These studies suggest that NF-κB decoy ODNs may be useful to inhibit the expression of genes involved in the pathogenesis of immune and inflammatory diseases.

2.4. RNA interference

Recently, another promising oligonucleotide approach, termed RNA interference (RNAi), offers potential for specific gene silencing (Bass, 2000). In this technique, short double-stranded RNAs, produced by the annealing of sense and antisense RNAs homologous to the cellular mRNA of interest, are able to silence the expression of the corresponding cellular genes. In Drosophila, this approach inhibits the expression of the NF-κB-related protein Relish (Cornwell & Kirkpatrick, 2001). LPS-induced processing of the Relish protein, which results in release of the N-terminal Rel homology domain (RHD) from the C-terminal ankyrin repeats, leads to the subsequent nuclear translocation of the RHD. The decreased levels of Relish by the RNAi significantly reduce the production of antibacterial peptides induced by LPS. The efficacy of RNAi in cultured mammalian cells has been demonstrated using either synthesized 21-nucleotide short interfering RNAs (siRNAs) (Elbashir et al., 2001) or an expression vector that directs the synthesis of siRNA-like transcript (Brummelkamp et al., 2002). The combined depletion of IKKα and IKKβ proteins by treatment of cells with siRNA transfection has been shown to inhibit both TNFα and IL-1β-induced IκBα degradation and NF-κB DNA binding in gel-shift assays (Takaesu et al., 2003). These results indicate the future potential of RNAi as a therapy to decrease NF-κB-regulated transcription.

Although the above techniques are still early in their development as therapeutic tools, their clinical utility is quite interesting. However, a number of concerns remains. For example, systemic delivery of oligonucleotides is likely to result in their widespread uptake, causing potential nonspecific side effects. The development of efficient means to deliver oligonucleotides into specific targets may hold the key to the successful application of these technologies for future drug therapy.

3. INHIBITORS OF NF-κB NUCLEAR TRANSLOCATION

3.1. The IκBα superrepressor

The first evidence that the NF-κB pathway could be specifically inhibited, came from studies with a dominant-negative IκBα mutant (Heissmeyer et al., 2001; Horng et al., 2001). A variety of studies demonstrate that signal-induced phosphorylation and degradation of cytoplasmic IκBα is required for activation of the NF-κB pathway. Furthermore, IκBα proteins containing mutations in the IKK phosphorylation sites (serine residues 32 and 36) are not phosphorylated and, therefore, cannot be ubiquitinated and degraded by the proteasome. Mutations in these IκBα residues result in an IκBα mutant, known as a superrepressor, which has a dominant-negative phenotype and sequesters NF-κB in the cytoplasm to prevent activation of this pathway.

Not only does the IκBα superrepressor inhibit NF-κB-regulated gene expression, but this mutant also enhances the sensitivity of cells to apoptosis-inducing stimuli. For example, TNFα treatment of cells can activate the NF-κB pathway and also induce apoptosis (Van Antwerp et al., 1996; Wang et al., 1996). As prolonged NF-κB activation protects cells from undergoing apoptosis, blocking this pathway enhances cellular apoptosis. Expression of the IκBα superrepressor enhances the sensitivity to TNFα-induced apoptosis in Jurkat cells, that are relatively resistant to this process. Consistent with these findings, phosphatidylserine migration from the inner to the outer leaflet of plasma membrane, which is an early event in apoptosis, markedly increases in TNFα-treated Jurkat cells expressing the IκBα superrepressor (Van Antwerp et al., 1996). Expression of the IκBα superrepressor in human fibrosarcoma cells also enhances their sensitivity to apoptotic killing by ionizing radiation or the chemotherapeutic agent daunorubicin, each of which are able to activate the NF-κB pathway (Wang et al., 1996). In contrast, the sensitizing effects of the IκBα superrepressor are not seen in apoptosis induced by staurosporine, which induces cell death in an NF-κB-independent manner. These results indicate that the IκBα superrepressor can enhance cell killing by blocking activation of the NF-κB pathway.

The adenoviral delivery of the IκBα superrepressor to chemoresistant tumors in mice sensitizes these cells to undergo apoptosis in response to a treatment with either TNFα or the chemotherapeutic agent CPT-11, and can lead to tumor regression (Wang et al., 1999). The IκBα superrepressor also suppresses constitutive and TNFα-induced NF-κB activation in human head and neck carcinoma cells grown in tissue culture and reduces the growth of these tumors in SCID mice (Duffey et al., 1999). These results suggest that the IκBα superrepressor inhibits the expression of NF-κB-dependent genes, which leads to enhanced growth of squamous cell carcinomas. This evidence demonstrates an important role for NF-κB in the regulation of cellular proliferation and suggests that inhibition of the NF-κB pathway may be important in enhancing the efficacy of cancer chemotherapy.

3.2. 26S proteasome inhibitors prevent IκBα degradation and NF-κB activation

The activation of IKK and the subsequent phosphorylation and degradation of IκBα by the 26S proteasome is a key step in the nuclear translocation of NF-κB and subsequent NF-κB-regulated transcription. Given the fundamental role of the proteasome in the regulation of the NF-κB pathway, it provides a potential target for anticancer and anti-inflammatory therapy. A variety of natural and synthetic proteasome inhibitors has been studied, including PS-341 which is the first proteasome inhibitor to enter human trials for cancer therapy (Adams, 2002). Phase II clinical trials of PS-341 in patients with refractory multiple myeloma are in progress, but an ongoing study demonstrates that approximately 50% of these patients have either complete or partial decreases in the production of specific monoclonal proteins produced by the myeloma cells (Adams, 2002).

Several lines of evidence suggest that the anticancer properties of PS-341, at least partially result from its inhibitory effects on the NF-κB pathway (Berenson et al., 2001; Sunwoo et al., 2001; Tan & Waldmann, 2002). Pretreatment of multiple myeloma cells with PS-341 reduces the high constitutive levels of NF-κB found in these cells and enhances their killing by the chemotherapeutic agents (Berenson et al., 2001). In a mouse model of adult T-cell leukemia, PS-341 treatment in combination with anti-Tac, which is a humanized anti-IL-2α antibody, prolongs the survival of tumor-bearing mice (Tan & Waldmann, 2002). Similarly, the effects of PS-341 in mouse squamous cell carcinomas, exhibiting the highly constitutive NF-κB activity, suggest that PS-341 inhibition of NF-κB is responsible for the suppression of the growth and metastasis in squamous cell carcinoma-bearing mice (Sunwoo et al., 2001).

MG132 is another 26S proteasome inhibitor widely used to study NF-κB signaling. This agent enhances the sensitivity of cells to chemotherapeutic agents (Arlt et al., 2001). MG132 has been utilized to treat human carcinoma cell lines which have high levels of constitutive NF-κB and are resistant to the anticancer agent VP16, a topoisomerase-2 inhibitor. MG132 strongly enhances the apoptotic effects of VP16 on these chemotherapy-resistant human carcinoma cells.

Taken together, these studies support a potential role for inhibitors of the 26S proteasome as novel agents for the treatment of hematologic and solid tumor malignancies. Furthermore, combining proteasome inhibitors with chemotherapy may provide a means to overcome the high incidence of chemotherapy resistance. One concern in utilizing proteasome inhibitors to inhibit IκBα degradation is their specificity. For example, the phosphorylated serine residues in the C-terminus of the NF-κB precursor p105, which contain a conserved DSXXXDS destruction box motif, are recognized by the same SCF-βTrCP ubiquitin ligase complex mediating the ubiquitination of IκBα and its subsequent degradation by the 26S proteasome (Heissmeyer et al., 2001). The role of phosphorylation in this conserved motif of p105 has not yet been demonstrated. One model suggests that this phosphorylation is required for the generation of the p50 subunit of NFκB by proteolytic processing of p105. However, it has also been proposed that p50 is generated by a cotranslational processing mechanism. Although the mechanism of p50 production is controversial, it is possible that the inhibitory effects of proteasome inhibitors on the NF-κB activity partially result from the inhibition of p50 generation. Finally, proteasome function is also required for the degradation of a variety of other cellular proteins, including cyclins and β-catenin.

3.3. Peptides with nuclear localization sequences inhibit NF-κB activity

Translocation of the NF-κB heterodimer from the cytoplasm to the nucleus is a central paradigm in the regulation of the NF-κB pathway. Thus the development of inhibitors of NF-κB nuclear localization using recombinant peptides provides an approach that can mask the nuclear localization sequence (NLS) of NF-κB family members. This approach utilizes cell-penetrating peptides consisting of the NLS of the p50 NF-κB subunit, designated as SN50.

Introduction of SN50 into cells efficiently inhibits LPS- and TNFα-induced NF-κB nuclear translocation and reduces NF-κB DNA binding in cultured endothelial and monocytic cells (Lin et al., 1995). In human peripheral blood lymphocytes, SN50 treatment induces apoptosis without detectable activation of either caspase-1 or caspase-3 (Kolenko et al., 1999). SN50 selectively inhibits NF-κB DNA binding without affecting the binding of AP-1 and NF-AT, and its ability to induce apoptosis is inhibited by preactivation of T-cells to result in NF-κB translocation. SN50 also prevents GATA-3 transcription in primary mouse T-lymphocytes, which is critical for Th2 differentiation, cytokine production and allergic airway inflammation (Das et al., 2001). These results suggest that NF-κB is involved in Th2 differentiation by modulating the level of GATA-3 expression, which is supported by the observation of impaired GATA-3 expression in $p50^{-/-}$ cells. However, there is controversy regarding the specificity of SN50. For example, one study shows that SN50 inhibits the gene expression of IκBα and IL-2 in addition to the nuclear localization of NF-κB, AP-1, NFAT and STAT1 in Jurkat T-lymphocytes (Torgerson et al., 1998). These differences possibly represent a dose-dependent or cell type-dependent efficiency of SN50; however, it is also possible that SN50 competes with other cellular proteins generally involved in nuclear import.

3.4. Peptides with NMD and TIRAP inhibit NF-κB activity

Two other studies with cell-penetrating peptides were recently published. The first utilizes peptides corresponding to the region of IKKβ that binds to IKKγ/NMD (May et al., 2000) and the second uses TIRAP (Horng et al., 2001). NMD corresponds to the sequence of the C-terminus of IKKβ that facilitates its interaction with NEMO (May et al., 2000). NEMO association with the IKK complex is essential for NF-κB activation (in response to stimulation with TNFα and other cytokines) by facilitating the interaction of the IKK complex with the TNFα receptor complex, and the recruitment of IκBα into the IKK complex. NMD blocks the association of NEMO with IKKβ and inhibits TNFα-induced NF-κB activation and NF-κB-dependent gene expression. In a mouse model of acute inflammation, NMD ameliorates phorbol ester-induced ear edema and zymosan-induced peritonitis. These findings indicate the possibility of future use of NMD in the treatment of proinflammatory reactions.

The Toll-like receptors (TLRs) play a central role in the adaptive immune response by recognizing conserved products of microbial products (such as LPS) that activate NF-κB and other pathways (Horng et al., 2001). TLR and IL-1R share a homologous cytoplasmic domain named TIR, in addition to signaling components including the adapter protein MyD88, IL-1R-associated kinase (IRAK) and the TNF-R-associated factor 6 (TRAF6). TIRAP is a newly identified adapter protein associating with TLR4 only (Aderem & Ulevitch, 2000). Studies show that cell-penetrating peptides corresponding to TIRAP are capable of binding to TLR4 and inhibit LPS-induced NF-κB activation, but not IL-1β-induced responses. The TIRAP peptide also inhibits the production of IL-6 and IL-12 by dendritic cells in response to LPS. These data suggest that the TIRAP peptide is a specific inhibitor of LPS-induced activation of the NF-κB pathway.

3.5. The Hsp90 inhibitor

Hsp90 is one of the most abundant cellular proteins and is known to serve as a chaperone protein involved in maintaining protein stability and regulating cell signaling. Hsp90 has been identified as a specific target for the potential anticancer agent geldanamycin that binds to this chaperone protein and antagonizes its function (Neckers et al., 1999). The geldanamycin analogue 17-allylamino-17-desmethoxygeldanamycin (17-AAG) is currently in phase I clinical trials for cancer therapy. For example, one study demonstrates that 17-AAG inhibits the proliferation of breast cancer cell lines by arresting the cell cycle in the G1 phase (Munster et al., 2001). The mechanism, by which inhibition of Hsp90 causes apoptosis, remains unclear. However, a variety of studies (Devin et al., 2000; Lewis et al., 2000) may provide the possible answers to this question and the role of NF-κB in the process.

The initial step of TNFα-induced NF-κB activation is its binding to the TNF-R1 complex that consists of several components including the death domain kinase receptor-interacting protein (RIP) and the TNF-R-associated factor 2 (TRAF2). Geldanamycin treatment results in the degradation of RIP and subsequent inhibition of TNFα-induced IκBα degradation. These results suggest that Hsp90 facilitates TNFα-mediated NF-κB activation by modulating the stability of RIP (Lewis et al., 2000). This study also demonstrates that the degradation of RIP sensitizes cells to TNFα-induced apoptosis.

Knockout studies ($RIP^{-/-}$ and $TRAF2^{-/-}$ cells) show that both RIP and TRAF2 are essential for TNFα-induced IKK activation and that IKK requires the presence of RIP in the same complex for its kinase activity, while TRAF2 functions to recruit the IKK complex into the TNF-R1 complex (Devin et al., 2000). In another study, gel filtration analysis demonstrates that Hsp90 is an additional component of the IKK complex (Chen et al., 2002). Geldanamycin treatment decreases the mobility of the IKK complex and abolishes TNFα-induced IKK activation. These results indicate that the presence of Hsp90 in the IKK complex facilitates its kinase activity in response to TNFα stimulation. Geldanamycin treatment also inhibits TNFα-induced binding of IKK to TNF-R1. Thus Hsp90 plays an important role in TNFα-induced NF-κB activation, both by stabilizing the TNF-R1 complex and by enhancing IKK trafficking from the cytoplasm to the membrane. Currently, there is no direct evidence that connects the anticancer properties of the Hsp90 inhibitor with its ability to inhibit the NF-κB pathway. Further studies will be directed to the possibility that alteration in the TNF-R/IKK complex by treatment of cells with geldanamycin or 17-AAG is partly responsible for its therapeutic effects on cancer cells.

4. DIRECT CONTROL OF NF-κB DNA BINDING

4.1. HDAC inhibitors

Enzymes that have attracted considerable attention both as inhibitors of the NF-κB pathway and as therapeutic targets are HDACs. Their general role is to counteract the histone acetyltransferases (HATs), which acetylate the core histones coiled with DNA and thus generate a more open DNA conformation to enhance transcriptional activity (Khochbin et al., 2001). However, the anti-inflammatory properties (Leoni et al., 2002) and anticancer properties (Marks et al., 2001) of HDAC inhibitors may be independent of their ability to inhibit both the deacetylation of core histones and the NF-κB pathway (Ashburner et al., 2001; Chen et al., 2001; Zhong et al., 2002).

Administration of the HDAC inhibitor suberoylanilide hydroxamic acid (SAHA) reduces circulating levels of TNFα, IL-1β, IL-6 and IFNγ in mice stimulated with LPS. In human peripheral blood mononuclear cells, the mRNA levels of LPS-induced TNFα and IFNγ, but not of IL-1β or IL-8, are markedly decreased after a treatment with SAHA. Such data indicate that HDAC inhibitors can downregulate the gene expression of proinflammatory cytokines involved in the inflammatory response.

In contrast to these results, the HDAC inhibitor trichostatin A (TSA) increases TNFα-induced NF-κB transactivation and IL-8 production. These effects are associated with hyperacetylation of an NF-κB reporter gene and the IL-8 promoter (Ashburner et al., 2001). Two roles to explain the function of HDAC on NF-κB regulation have been proposed (Chen et al., 2001; Zhong et al., 2002). In one model, the acetylation of the p65 NF-κB subunit may be a key factor to prevent its binding to nuclear IκBα, which results in p65 nuclear export in an IκBα-dependent manner (Chen et al., 2001). The specific interaction between p65 and HDAC3 was demonstrated to enhance the binding of the p65/HDAC3 complex to IκBα. TSA inhibits the acetylation level of p65 and leads to the export of p65/HDAC3 complex. These results are consistent with a role for acetylation of p65 on NF-κB activation. The second model for HDAC regulation of the NF-κB pathway suggests that transcriptional repression of NF-κB by p50 homodimers is mediated by association of HDAC (Zhong et al., 2002). Chromatin immunoprecipitation assays demonstrated that p50 homodimers and HDAC1 form a complex on the promoter region of NF-κB-regulatory genes in unstimulated cells. TSA treatment inhibits the ability of the p50 homodimer/ HDAC1 complex to repress NF-κB transcription; this effect is not detected in $p50^{-/-}$ cells, which confirms a role for p50 and HDAC. Following cytokine stimulation, the p50 homodimer/HDAC1 complex is replaced with the p65/p50 heterodimer. Although questions, such as which isoform of HDAC binds to p65 or which of the two models is the most relevant for the control of the NF-κB transcriptional response, need to be investigated, these studies suggest an important role of HDACs in regulating the NF-κB pathway.

Several studies demonstrate that HDAC inhibitors lead to growth arrest, differentiation or apoptosis in a variety of transformed cells and tumor-bearing mice. Such inhibitors have entered clinical trials in patients with cancer (Leoni et al., 2002). In contrast, while NF-κB is known to regulate cellular growth properties as well as the host inflammatory and immune response, HDAC inhibitors have also been shown to upregulate NF-κB transcription. Given contradictory results, further experiments regarding the importance of acetylation on the regulation of the NF-κB pathway will be important in developing therapeutic applications of HDAC inhibitors.

5. CONVENTIONAL ANTI-INFLAMMATORY DRUGS ACT AS NF-κB INHIBITORS

5.1. Scope

Anti-inflammatory drugs, including both natural compounds and synthesized analogues, have been widely used in clinical medicine for years. However, the signaling pathways affected by agents have not been clearly evaluated. Recently a number of studies have demonstrated that anti-inflammatory agents can inhibit the NF-κB pathway.

5.2. Nonsteroidal anti-inflammatory drugs (NSAIDs)

NSAIDs include drugs of several structural classes sharing analgesic, antipyretic and anti-inflammatory activity (Vane, 1994). Besides their anti-inflammatory effects, other clinical effects seen with NSAIDs include a reduction in the risk of colorectal cancer and possibly other malignancies, a neuro-protective effect and reduction in the risk of Alzheimer's disease (Grilli et al., 1996). The most commonly accepted targets of NSAIDs mediating their anti-inflammatory properties are COXs, that are essential for the production of prostaglandins (PGs). However, the doses of NAIDs required for anti-inflammatory properties in patients with rheumatoid arthritis are much higher than those which inhibit the PG-mediated inflammatory response. Thus inhibition of the NF-κB pathway, which has been demonstrated in several studies, may be an attractive potential mechanism to explain many of these effects of NSAIDs.

Aspirin and salicylate are examples of NSAIDs whose effect are at least in part due to the inhibition of NF-κB (Kopp & Ghosh, 1994; Pierce et al., 1996; Yin et al., 1998; Yamamoto et al., 1999). Aspirin and sodium salicylate inhibit TNFα-induced mRNA expression of the adhesion molecules VCAM-1 and ICAM-1 in endothelial cells, as well as the surface expression of these molecules (Pierce et al., 1996). Such effects are independent of COX inhibition, as suggested by the fact that another NSAID, namely indomethacin (a potent inhibitor of PG synthesis), does not mediate these effects (Yin et al., 1998; Yamamoto et al., 1999). Treatment of endothelial monolayers with sodium salicylate also inhibits transendothelial migration of leukocytes (Pierce et al., 1996), suggesting that part of the anti-inflammatory properties of salicylates may be accounted for by inhibition of the NF-κB pathway. The mechanism of NF-κB inhibition was further elucidated in subsequent studies. Aspirin and sodium salicylate are able to inhibit ATP binding to IKKβ, thus impairing its kinase activity and decreasing IκBα phosphorylation, as well as the subsequent degradation and activation of the NF-κB pathway (Yin et al., 1998).

Sulindac, an NSAID structurally related to indomethacin, and its metabolites sulindac sulfide and sulindac sulfone are also potent inhibitors of the NF-κB pathway (Yamamoto et al., 1999). Inhibition of IKKβ activity by these agents is specific, as other evaluated kinases (including IKKα, p38, ERK2 and SAPK) are not inhibited by these agents. Sulindac, and to a lesser extent aspirin and sodium salicylate, can also induce apoptosis in HCT-15 colon cancer cells (Yamamoto et al., 1999). This particular cell line is defective in prostaglandin synthesis, providing further evidence that the proapoptotic response observed with these NSAIDs is independent of inhibition of prostaglandin synthesis (Yamamoto et al., 1999). In lung cancer cell lines, sulindac blocks TNFα-induced NF-κB nuclear translocation and DNA binding, and potentiates cell killing by TNFα. This suggests that NSAID-mediated blockage of NF-κB may have implications in cancer therapeutics by sensitizing tumor cells to chemotherapeutic agents (Berman et al., 2002).

Sulfasalazine, an anti-inflammatory agent widely used to treat inflammatory bowel disease, is cleaved into the NSAID moiety 5-amino-salicylic acid (5-ASA) and sulfapyridine after oral administration. Treatment of colonic epithelial cells with sulfasalazine prevents NF-κB activation induced by TNFα, LPS or phorbol esters (Wahl et al., 1997). Similar effects are not observed with 5-ASA or sulfapyridine. However, 5-ASA was demonstrated to inhibit both IKKα and IKKβ kinase activity and to prevent phosphorylation of IκBα in mouse colonic cells (Yan & Polk, 1999). Mesalamine, a related aminosalicylate compound, can block the phosphorylation of p65 without affecting IκB degradation, suggesting that NSAIDs might block the NF-κB pathway at multiple steps (Egan et al., 1999).

5.3. Immunosuppressive agents

Cyclosporin A (CsA) is an immunosuppressive agent used to prevent transplant rejection (Crabtree, 2001). The general mechanism of CsA action is to inhibit the activity of calcineurin, which is a calcium- and calmodulin-dependent serine-threonine phosphatase required for activation of the transcription factor NF-AT. The inhibition of calcineurin suppresses the NF-AT-regulated gene expression of IL-2, which is critical for the activation of T-lymphocytes. Unlike NF-AT, NF-κB activation does not require a calcium stimulus; however, overexpression of calcineurin can also activate the NF-κB pathway by enhancing the degradation of IκBα (Frantz et al., 1994). CsA acts as a noncompetitive inhibitor of the chymotrypsin-like activity of the 20S proteasome in vitro and suppresses LPS-induced IκB degradation in mouse macrophages by stabilizing the ubiquitinated forms of IκBα (Meyer et al., 1997). CsA also interferes with the degradation of IκBα following TPA/ionomycin stimulation in T-lymphocytes, without altering the IκBα phosphorylation (Marienfeld et al., 1997). In contrast, CsA does not influence the proteolytic processing of the NF-κB precursor p105, which also requires the 26S proteasome complex (Marienfeld et al., 1997). Overall, these results suggest that a target of CsA for inhibiting the NF-κB pathway may be the specific protease activity of the proteasome, rather than kinases or ubiquitin ligases which regulate the signal-induced phosphorylation and ubiquitination of IκB, respectively.

Thalidomide, an immunomodulator, has been demonstrated to exhibit anti-inflammatory and immunosuppressive properties due to modulation of various cytokine responses (including TNFα, IL-6 and IL-12) and chemokines. This agent is effective as an antineoplastic agent in certain malignancies, possibly by inhibition of angiogenesis. In endothelial cells, thalidomide prevents degradation of IκBα by inhibiting IKKβ kinase activity, which is consistent with its inhibitory effect on cytokine-induced NF-κB activation and on the expression of NF-κB-regulated genes (Keifer et al., 2001). Thalidomide also inhibits TNFα-, IL-1β- and H_2O_2-induced NF-κB activation in a wide variety of cells including T-lymphocytes, myeloid cells and epithelial cells, as assayed by a decreased p50/p65 or p50 homodimer DNA binding and NF-κB reporter gene activity

(Majumdar et al., 2002). Thus part of the immunosuppressive effect of thalidomide may be due to inhibition of NF-κB activation.

5.4. Cyclopentenone prostaglandins

Cyclopentenone prostaglandins (CyPGs) are naturally occurring anti-inflammatory prostaglandins which contain a cyclopentenone ring and which are derived from the metabolism of prostaglandin D_2 (Gilroy et al., 1999). These molecules are synthesized during the later phases of inflammation and are thought to play an important role in the resolution of inflammatory responses. Three distinct links of CyPG effects with the NF-κB pathway have been demonstrated. First, CyPGs may act as endogenous anti-inflammatory mediators by binding to peroxisome proliferation-activated receptor-γ (PPARγ) (Ricote et al., 1998). The treatment of peritoneal macrophages with the CyPG 15-deoxy-$\Delta^{12,14}$-prostaglandin J_2 (15d-PGJ_2) inhibits the expression of iNOS in a PPARγ-dependent manner. These results suggest that PPARγ interacts with CyPG to antagonize NF-κB-regulated gene expression (Ricote et al., 1998). Second, CyPGs can also directly inhibit the NF-κB pathway at multiple steps. 15d-PGJ_2 and PGA_1 inhibit IκBα degradation by inhibition of IKK activity due to direct modification of IKKβ at cysteine residue 179 within its activation loop (Rossi et al., 2000). The cysteine residues in the DNA-binding domain of p50 and p65 may also be targets of CyPG. Alkylation of these residues abolishes the inhibitory effects of 15d-PGJ_2 on NF-κB DNA binding, suggesting that direct modification of p50 and/or p65 by CyPG may be responsible for inhibition of NF-κB activation (Straus et al., 2000). Finally, it was proposed that NF-κB regulates both the onset and resolution of inflammation by modulating the gene expression of proinflammatory and anti-inflammatory mediators (Lawrence et al., 2001). For example, the DNA-binding activity of NF-κB in leukocytes from rat carrageenin pleurisy is associated with iNOS protein expression during the onset of inflammation, while the late phase of NF-κB activation is associated with the expression of COX2 directing the synthesis of anti-inflammatory CyPG (Lawrence et al., 2001). Taken together, the inhibition of NF-κB by CyPGs may be part of a negative feedback loop that contributes to resolution of inflammation.

5.5. Antioxidants and flavonoids

Antioxidants are implicated in preventing a variety of pathological states related to the generation of reactive oxygen species (ROS) by phagocytic leukocytes during chronic and acute inflammatory diseases or on environmental stresses (Berlett & Stadtman, 1997). The agents generating oxidative stress enhance the expression of proinflammatory genes regulated by NF-κB. In addition, several kinds of stimuli activating NF-κB can also increase the levels of intracellular ROS. However, both oxidative stress-induced activation of NF-κB and ROS generation mediated by NF-κB are cell type-specific. The question of how oxidative stresses may activate the NF-κB pathway is still unsolved. The potential of antioxidants to inhibit the NF-κB pathway has been demonstrated with several compounds. For example, treatment of rats with the antioxidant N-acetyl-L-cysteine (NAC) suppresses LPS-induced NF-κB activation and neutrophilic alveolitis in lung tissue (Blackwell et al., 1996). Vitamin C inhibits TNFα- and IL-1β-induced IKK phosphorylation of IκBα and subsequent NF-κB DNA binding in endothelial cell lines. This inhibitory effect of vitamin C is relieved by treatment with a p38 MAPK inhibitor, suggesting that vitamin C enhances the activity of p38 MAPK and apparently exerts a negative regulatory effect acting between the TNF receptor and IKK complex (Bowie & O'Neill, 2000).

Flavonoids are naturally occurring phenolic compounds found in plants that have been used as suppressors of inflammation, in addition to cancer chemoprevention and protection from vascular disease. Several studies demonstrate that flavonoids such as resveratrol may mediate their effects by inhibition of the NF-κB pathway (Tsai et al., 1999; Holmes-McNary & Baldwin, 2000). Resveratrol inhibits the expression of iNOS and NO generation in activated macrophages, which is consistent with its inhibitory effect on LPS-induced IκBα phosphorylation and subsequent IκBα degradation, as well as NF-κB DNA-binding activity (Tsai et al., 1999). Resveratrol also induces apoptosis in rat-1 cells by inhibition of ras-mediated activation of NF-κB (Holmes-McNary & Baldwin, 2000). These results suggest that the biological activities of flavonoids may be mediated by inhibition of the NF-κB pathway.

Aspirin and several antioxidants, including dietary supplements and herbs, are commonly used to reduce the risk of atherosclerosis, neurodegenerative disorders and cancer. Potential benefits of these agents may be derived by inhibition of the NF-κB pathway along one or multiple steps of the activation cascade. Other agents, such as immunosuppressors and immunomodulators, also target the NF-κB pathway; their inhibition may be responsible for both the therapeutic and side effects. A better understanding of the target specificity and serum levels of these agents required to inhibit the NF-κB pathway will provide more rational applications. The mechanism of their action and a structural insight may provide clues for designing new agents to inhibit the NF-κB pathway. The role of these agents, particularly of NSAIDs, in cancer therapeutics and chemoprevention needs to be further explored, as combinations of these agents with conventional chemotherapeutics appear to be promising.

6. CONCLUDING REMARKS AND PERSPECTIVES

With the growing understanding of the NF-κB pathway, NF-κB has become an increasingly attractive target for potential therapeutic interventions to treat inflammation and cancer. Indeed, the function of widely used anti-inflammatory agents has been shown to be due, at least in part, to inhibition of the NF-κB pathway. One concern about the inhibition of several of the components of the NF-κB pathway is the specificity of such drugs. For example, the proteasome has many other important functions besides IκB degradation, such as proteolytic processing of p100 and p105 precursor proteins and cyclins, degradation of β-catenin that transduces signals in the Wnt1/wingless pathway and degradation of the HIV-1 receptor CD4. Thus protease inhibition could potentially cause severe side effects. The discrimination of nuclear localization sequences (NLS) on NF-κB from other proteins with a similar consensus NLS may also be problematic. Specific inhibition of NF-κB with RelA antisense ODNs, NF-κB decoy ODNs or the superrepressor IκBα may lead to specific gene inhibition. However, the multiplicity of target genes regulated by NF-κB represents significant limitations in the use of these approaches. In addition to the problem of diverse functions of NF-κB, one unresolved challenge is the delivery of inhibitors to target genes in specific organs or tissues. The development of specific delivery systems to target these agents is a critical issue for their future application.

In this regard, the specific inhibition of upstream regulatory components of NF-κB, which function only under certain physiological conditions, might lead to the development of newer strategies against inflammation and cancer. Synthetic peptides such as TIRAP, which inhibits only LPS-induced NF-κB activation, may be one of these inhibitors. RNAi also has a great potential for gene therapy by inhibiting specific target proteins regulated by the NF-κB pathway. Another strategy to inhibit different aspects of the NF-κB pathway may be the identification of proapoptotic genes regulated by NF-κB. One of these examples is the recently identified gadd45β gene whose expression is induced in response to TNFα in an NF-κB-dependent manner. Expression of gadd45β leads to a reduced activation of the JNK pathway known to be a key regulator of cell death (De Smaele et al., 2001). Suppression of gadd45β by either synthetic peptides or RNAi may possibly enable cells to undergo apoptosis without causing any disruption of other genes regulated by the NF-κB pathway.

Besides the potential of new classes of anti-inflammatory and anticancer agents, inhibition of the NF-κB pathway may also provide a useful target to enhance the efficacy of cancer therapy. There is much evidence that NF-κB activity can protect normal cells from stimuli-induced apoptosis. Thus the application of NF-κB inhibitors may be a way to overcome the high incidence of resistance to chemotherapy and radiation therapy. Since the discovery of NF-κB more than a decade ago, NF-κB has attracted great interest because of its unique mechanism of regulation and the various functional consequences of its activity. The maintenance of appropriate levels of NF-κB activity in many tissues will likely be critical to prevent toxicity of NF-κB inhibitors. A better understanding of the mechanism that regulates the NF-κB pathway will provide opportunities for the development of novel treatments for inflammation and cancer.

7. REFERENCES

Adams J (2002) Proteasome inhibition: a novel approach to cancer therapy. Trends Mol. Med. 8: S49-54.

Aderem A & Ulevitch RJ (2000) Toll-like receptors in the induction of the innate immune response.

Nature 406: 782-787.

Arlt A, Vorndamm J, Breitenbroich M, Folsch UR, Kalthoff H, Schmidt WE & Schafer H (2001) Inhibition of NF-κB sensitizes human pancreatic carcinoma cells to apoptosis induced by etoposide (VP16) or doxorubicin. Oncogene 20: 859-868.

Ashburner BP, Westerheide SD & Baldwin AS (2001) The p65 (RelA) subunit of NF-κB interacts with the histone deacetylase (HDAC) corepressors HDAC1 and HDAC2 to negatively regulate gene expression. Mol. Cell. Biol. 21: 7065-7077.

Baldwin AS (1996) The NF-κB and IκB proteins: new discoveries and insights. Annu. Rev. Immunol. 14: 649-681.

Bass BL (2000) Double-stranded RNA as a template for gene silencing. Cell 101: 235-238.

Berenson JR, Ma HM & Vescio R (2001) The role of nuclear factor-κB in the biology and treatment of multiple myeloma. Semin. Oncol. 28: 626-633.

Berlett BS & Stadtman ER (1997) Protein oxidation in aging, disease, and oxidative stress. J. Biol. Chem. 272: 20313-20316.

Berman KS, Verma UN, Harburg G, Minna JD, Cobb MH & Gaynor RB (2002) Sulindac enhances tumor necrosis factor-α-mediated apoptosis of lung cancer cell lines by inhibition of nuclear factor-κB. Clin. Cancer Res. 8: 354-360.

Blackwell TS, Blackwell TR, Holden EP, Christman BW & Christman JW (1996) In vivo antioxidant treatment suppresses nuclear factor-κB activation and neutrophilic lung inflammation. J. Immunol. 157: 1630-1637.

Bowie AG & O'Neill LA (2000) Vitamin C inhibits NF-κB activation by TNF via the activation of p38 mitogen-activated protein kinase. J. Immunol. 165: 7180-7188.

Brummelkamp TR, Bernards R & Agami R (2002) A system for stable expression of short interfering RNAs in mammalian cells. Science 296: 550-553.

Chen G, Cao P & Goeddel DV (2002) TNF-induced recruitment and activation of the IKK complex require Cdc37 and Hsp90. Mol. Cell 9: 401-410.

Chen L, Fischle W, Verdin E & Greene WC (2001) Duration of nuclear NF-κB action regulated by reversible acetylation. Science 293: 1653-1657.

Cornwell WD & Kirkpatrick RB (2001) Cactus-independent nuclear translocation of Drosophila RELISH. J. Cell. Biochem. 82: 22-37.

Crabtree GR (2001) Calcium, calcineurin, and the control of transcription. J. Biol. Chem. 276: 2313-2316.

D'Acquisto F, Ialenti A, Ianaro A, Di Vaio R & Carnuccio R (2000) Local administration of transcription factor decoy oligonucleotides to nuclear factor-κB prevents carrageenin-induced inflammation in rat hind paw. Gene Ther. 7: 1731-1737.

Das J, Chen CH, Yang L, Cohn L, Ray P & Ray A (2001) A critical role for NF-κB in GATA3 expression and TH2 differentiation in allergic airway inflammation. Nat. Immunol. 2: 45-50.

De Smaele E, Zazzeroni F, Papa S, Nguyen DU, Jin R, Jones J, Cong R & Franzoso G (2001) Induction of gadd45β by NF-κB downregulates pro-apoptotic JNK signalling. Nature 414: 308-313.

Devin A, Cook A, Lin Y, Rodriguez Y, Kelliher M & Liu Z (2000) The distinct roles of TRAF2 and RIP in IKK activation by TNF-R1: TRAF2 recruits IKK to TNF-R1 while RIP mediates IKK activation. Immunity 12: 419-429.

Duffey DC, Chen Z, Dong G, Ondrey FG, Wolf JS, Brown K, Siebenlist U & Van Waes C (1999) Expression of a dominant-negative mutant inhibitor-κBα of nuclear factor-κB in human head and neck squamous cell carcinoma inhibits survival, proinflammatory cytokine expression, and tumor growth in vivo. Cancer Res. 59: 3468-3474.

Egan LJ, Mays DC, Huntoon CJ, Bell MP, Pike MG, Sandborn WJ, Lipsky JJ & McKean DJ (1999) Inhibition of interleukin-1-stimulated NF-κB RelA/p65 phosphorylation by mesalamine is accompanied by decreased transcriptional activity. J. Biol. Chem. 274: 26448-26453.

Elbashir SM, Harborth J, Lendeckel W, Yalcin A, Weber K & Tuschl T (2001) Duplexes of 21-nucleotide RNAs mediate RNA interference in cultured mammalian cells. Nature 411: 494-498.

Frantz B, Nordby EC, Bren G, Steffan N, Paya CV, Kincaid RL, Tocci MJ, O'Keefe SJ & O'Neill EA (1994) Calcineurin acts in synergy with PMA to inactivate IκB/MAD3, an inhibitor of NF-κB.

EMBO J. 13: 861-870.

Ghosh S, May MJ & Kopp EB (1998) NF-κB and Rel proteins: evolutionarily conserved mediators of immune responses. Annu. Rev. Immunol. 16: 225-260.

Gilroy DW, Colville-Nash PR, Willis D, Chivers J, Paul-Clark MJ & Willoughby DA (1999) Inducible cyclooxygenase may have anti-inflammatory properties. Nat. Med. 5: 698-701.

Grilli M, Pizzi M, Memo M & Spano P (1996) Neuroprotection by aspirin and sodium salicylate through blockade of NF-κB activation. Science 274: 1383-1385.

Heissmeyer V, Krappmann D, Hatada EN & Scheidereit C (2001) Shared pathways of IκB kinase-induced SCF(βTrCP)-mediated ubiquitination and degradation for the NF-κB precursor p105 and IκBα. Mol. Cell. Biol. 21: 1024-1035.

Holmes-McNary M & Baldwin AS (2000) Chemopreventive properties of trans-resveratrol are associated with inhibition of activation of the IκB kinase. Cancer Res. 60: 3477-3483.

Horng T, Barton GM & Medzhitov R (2001) TIRAP: an adapter molecule in the Toll signaling pathway. Nat. Immunol. 2: 835-841.

Keifer JA, Guttridge DC, Ashburner BP & Baldwin AS (2001) Inhibition of NF-κB activity by thalidomide through suppression of IκB kinase activity. J. Biol. Chem. 276: 22382-22387.

Khochbin S, Verdel A, Lemercier C & Seigneurin-Berny D (2001) Functional significance of histone deacetylase diversity. Curr. Opin. Genet. Dev. 11: 162-166.

Kolenko V, Bloom T, Rayman P, Bukowski R, Hsi E & Finke J (1999) Inhibition of NF-κB activity in human T lymphocytes induces caspase-dependent apoptosis without detectable activation of caspase-1 and -3. J. Immunol. 163: 590-598.

Kopp E & Ghosh S (1994) Inhibition of NF-κB by sodium salicylate and aspirin. Science 265: 956-959.

Lawrence T, Gilroy DW, Colville-Nash PR & Willoughby DA (2001) Possible new role for NF-κB in the resolution of inflammation. Nat. Med. 7: 1291-1297.

Lebedeva I & Stein CA (2001) Antisense oligonucleotides: promise and reality. Annu. Rev. Pharmacol. Toxicol. 41: 403-419.

Leoni F, Zaliani A, Bertolini G, Porro G, Pagani P, Pozzi P, Dona G, Fossati G, Sozzani S, Azam T, Bufler P, Fantuzzi G, Goncharov I, Kim SH, Pomerantz BJ, Reznikov LL, Siegmund B, Dinarello CA & Mascagni P (2002) The antitumor histone deacetylase inhibitor suberoylanilide hydroxamic acid exhibits antiinflammatory properties via suppression of cytokines. Proc. Natl. Acad. Sci. USA 99: 2995-3000.

Lewis J, Devin A, Miller A, Lin Y, Rodriguez Y, Neckers L & Liu ZG (2000) Disruption of hsp90 function results in degradation of the death domain kinase, receptor-interacting protein (RIP), and blockage of tumor necrosis factor-induced nuclear factor-κB activation. J. Biol. Chem. 275: 10519-10526.

Lin YZ, Yao SY, Veach RA, Torgerson TR & Hawiger J (1995) Inhibition of nuclear translocation of transcription factor NF-κB by a synthetic peptide containing a cell membrane-permeable motif and nuclear localization sequence. J. Biol. Chem. 270: 14255-14258.

Majumdar S, Lamothe B & Aggarwal BB (2002) Thalidomide suppresses NF-κB activation induced by TNF and H2O2, but not that activated by ceramide, lipopolysaccharides, or phorbol ester. J. Immunol. 168: 2644-2651.

Marienfeld R, Neumann M, Chuvpilo S, Escher C, Kneitz B, Avots A, Schimpl A & Serfling E (1997) Cyclosporin A interferes with the inducible degradation of NF-κB inhibitors, but not with the processing of p105/NF-κB1 in T cells. Eur. J. Immunol. 27: 1601-1609.

Marks PA, Richon VM, Breslow R & Rifkind RA (2001) Histone deacetylase inhibitors as new cancer drugs. Curr. Opin. Oncol. 13: 477-483.

May MJ, D'Acquisto F, Madge LA, Glockner J, Pober JS & Ghosh S (2000) Selective inhibition of NF-κB activation by a peptide that blocks the interaction of NEMO with the IκB kinase complex. Science 289: 1550-1554.

Meyer S, Kohler NG & Joly A (1997) Cyclosporine A is an noncompetitive inhibitor of proteasome activity and prevents NF-κB activation. FEBS Lett. 413: 354-358.

Morishita R, Sugimoto T, Aoki M, Kida I, Tomita N, Moriguchi A, Maeda K, Sawa Y, Kaneda Y,

Higaki J & Ogihara T (1997) In vivo transfection of cis element "decoy" against nuclear factor-κB binding site prevents myocardial infarction. Nat. Med. 3: 894-899.

Munster PN, Srethapakdi M, Moasser MM & Rosen N (2001) Inhibition of heat shock protein 90 function by ansamycins causes the morphological and functional differentiation of breast cancer cells. Cancer Res. 61: 2945-2952.

Murano M, Maemura K, Hirata I, Toshina K, Nishikawa T, Hamamoto N, Sasaki S, Saitoh O & Katsu K (2000) Therapeutic effect of intracolonically administered nuclear factor κB (p65) antisense oligonucleotide on mouse dextran sulphate sodium (DSS)-induced colitis. Clin. Exp. Immunol. 120: 51-58.

Neckers L, Schulte TW & Mimnaugh E (1999) Geldanamycin as a potential anti-cancer agent: its molecular target and biochemical activity. Invest. New Drugs 17: 361-373.

Neurath MF, Pettersson S, Meyer zum Buschenfelde KH & Strober W (1996) Local administration of antisense phosphorothioate oligonucleotides to the p65 subunit of NF-κB abrogates established experimental colitis in mice. Nat. Med. 2: 998-1004.

Neurath MF, Fuss I, Schurmann G, Pettersson S, Arnold K, Muller-Lobeck H, Strober W, Herfarth C & Buschenfelde KH (1998) Cytokine gene transcription by NF-κB family members in patients with inflammatory bowel disease. Ann. NY Acad. Sci. 859: 149-159.

Pierce JW, Read MA, Ding H, Luscinskas FW & Collins T (1996) Salicylates inhibit IκBα phosphorylation, endothelial-leukocyte adhesion molecule expression, and neutrophil transmigration. J. Immunol. 156: 3961-3969.

Ricote M, Li AC, Willson TM, Kelly CJ & Glass CK (1998) The peroxisome proliferator-activated receptor-γ is a negative regulator of macrophage activation. Nature 391: 79-82.

Rossi A, Kapahi P, Natoli G, Takahashi T, Chen Y, Karin M & Santoro MG (2000) Anti-inflammatory cyclopentenone prostaglandins are direct inhibitors of IκB kinase. Nature 403: 103-108.

Silverman N & Maniatis T (2001) NF-κB signaling pathways in mammalian and insect innate immunity. Genes Dev. 15: 2321-2342.

Straus DS, Pascual G, Li M, Welch JS, Ricote M, Hsiang CH, Sengchanthalangsy LL, Ghosh G & Glass CK (2000) 15-Deoxy-$\Delta^{12,14}$-prostaglandin J_2 inhibits multiple steps in the NF-κB signaling pathway. Proc. Natl. Acad. Sci. USA 97: 4844-4849.

Sunwoo JB, Chen Z, Dong G, Yeh N, Crowl Bancroft C, Sausville E, Adams J, Elliott P & Van Waes C (2001) Novel proteasome inhibitor PS-341 inhibits activation of nuclear factor-κB, cell survival, tumor growth, and angiogenesis in squamous cell carcinoma. Clin. Cancer Res. 7: 1419-1428.

Takaesu G, Surabhi RM, Park KJ, Ninomiya-Tsuji J, Matsumoto K & Gaynor RB (2003) TAK1 is critical for IκB kinase-mediated activation of the NF-κB pathway. J. Mol. Biol. 326: 105-115.

Tan C & Waldmann TA (2002) Proteasome inhibitor PS-341, a potential therapeutic agent for adult T-cell leukemia. Cancer Res. 62: 1083-1086.

Tomita T, Takeuchi E, Tomita N, Morishita R, Kaneko M, Yamamoto K, Nakase T, Seki H, Kato K, Kaneda Y & Ochi T (1999) Suppressed severity of collagen-induced arthritis by in vivo transfection of nuclear factor κB decoy oligodeoxynucleotides as a gene therapy. Arthritis Rheum. 42: 2532-2542.

Tomita T, Takano H, Tomita N, Morishita R, Kaneko M, Shi K, Takahi K, Nakase T, Kaneda Y, Yoshikawa H & Ochi T (2000) Transcription factor decoy for NFκB inhibits cytokine and adhesion molecule expressions in synovial cells derived from rheumatoid arthritis. Rheumatology 39: 749-757.

Torgerson TR, Colosia AD, Donahue JP, Lin YZ & Hawiger J (1998) Regulation of NF-κB, AP-1, NFAT, and STAT1 nuclear import in T lymphocytes by noninvasive delivery of peptide carrying the nuclear localization sequence of NF-κB p50. J. Immunol. 161: 6084-6092.

Tsai SH, Lin-Shiau SY & Lin JK (1999) Suppression of nitric oxide synthase and the down-regulation of the activation of NF-κB in macrophages by resveratrol. Br. J. Pharmacol. 126: 673-680.

Van Antwerp DJ, Martin SJ, Kafri T, Green DR & Verma IM (1996) Suppression of TNF-α-induced apoptosis by NF-κB. Science 274: 787-789.

Vane J (1994) Towards a better aspirin. Nature 367: 215-216.

Wahl C, Liptay S, Adler G & Schmid RM (1997) Sulfasalazine: a potent and specific inhibitor of NF-κB. J. Clin. Invest. 101: 1163-1174.

Wang CY, Mayo MW & Baldwin AS (1996) TNF- and cancer therapy-induced apoptosis: potentiation by inhibition of NF-κB. Science 274: 784-787.

Wang CY, Cusack JC, Liu R & Baldwin AS (1999) Control of inducible chemoresistance: enhanced anti-tumor therapy through increased apoptosis by inhibition of NF-κB. Nat. Med. 5: 412-417.

Wang D & Baldwin AS (1998) Activation of NF-κB-dependent transcription by TNFα is mediated through phosphorylation of RelA/p65 on serine 529. J. Biol. Chem. 273: 29411-29416.

Yamamoto Y, Yin M-J, Lin K-M & Gaynor RB (1999) Sulindac inhibits activation of the NF-κB pathway. J. Biol. Chem. 274: 27307-27314.

Yan F & Polk DB (1999) Aminosalicylic acid inhibits IκB kinase α phosphorylation of IκBα in mouse intestinal epithelial cells. J. Biol. Chem. 274: 36631-36636.

Yin M-J, Yamamoto Y & Gaynor RB (1998) The anti-inflammatory agents aspirin and salicylate inhibit the activity of IκB kinase-β. Nature 396: 77-80.

Yokoseki O, Suzuki J, Kitabayashi H, Watanabe N, Wada Y, Aoki M, Morishita R, Kaneda Y, Ogihara T, Futamatsu H, Kobayashi Y & Isobe M (2001) cis Element decoy against nuclear factor-κB attenuates development of experimental autoimmune myocarditis in rats. Circ. Res. 89: 899-906.

Zandi E & Karin M (1999) Bridging the gap: composition, regulation, and physiological function of the IκB kinase complex. Mol. Cell. Biol. 19: 4547-4551.

Zhong H, Voll RE & Ghosh S (1998) Phosphorylation of NF-κB p65 by PKA stimulates transcriptional activity by promoting a novel bivalent interaction with the coactivator CBP/p300. Mol. Cell 1: 661-671.

Zhong H, May MJ, Jimi E & Ghosh S (2002) The phosphorylation status of nuclear NF-κB determines its association with CBP/p300 or HDAC-1. Mol. Cell 9: 625-636.

8. AFFILIATION

University of Texas Southwestern Medical Center, Department of Medicine, Division of Hematology-Oncology, Harold Simmons Cancer Center, 5323 Harry Hines Boulevard, Dallas, TX 75390, USA.

K. DE BOSSCHER, W. VANDEN BERGHE
& G. HAEGEMAN

REGULATION OF NF-κB BY GLUCOCORTICOIDS

1. INTRODUCTION

Inflammation was first described in history by Cornelius Celsus (30 BC-38 AD) as a process that embraces "rubor et tumor cum calore et dolore", i.e. redness and swelling accompanied by heat and pain. Inflammation is the result of a series of enzymatic processes in the body. Cell membrane damage caused by a cut or scratch, for example, leads to the activation of phospholipases, which mediate the release of arachidonic acid. This metabolite is further processed by cyclooxygenases and lipoxygenases to produce the fever-causing prostaglandins, thromboxanes and leukotrienes. These fatty acid derivatives have a vasodilatory action, causing a higher blood flow to help attracting inflammatory cytokines and immune cells to the site of inflammation. This explains the accompanying symptoms of redness and swelling in inflamed tissue. The inflammation process is further mediated and controlled by the action of several messenger molecules called cytokines, chemokines and adhesion molecules, including TNF, IL-1, IL-2, IL-6, MCP-1, IL-8, GM-CSF, ICAM-1 and E-selectin (Cato & Wade, 1996; Barnes & Karin, 1997). These cytokines are produced by (and in turn activate) different surrounding cell types, such as fibroblasts, endothelial cells (lining blood vessels), macrophages and neutrophils, white blood cell components traveling through the bloodstream. Upon activation, the latter two cell types attach to the endothelial cell layer by a process called leukocyte rolling; they then migrate through the endothelial membrane (called diapedesis) to the place of inflammation. The infection is finally nipped in the bud by the local release of degrading enzymes, such as elastase and cathepsin, oxygen radicals and the action of phagocytosis ('eating' remaining debris by immune cells). Signaling by growth factors such as TGFβ promotes cell proliferation and contributes to the process of wound healing.

The dimeric transcription factor NF-κB (mainly p65 and p50) is now recognized as one of the most important regulators of proinflammatory gene expression. Cytokines, chemokines and adhesion molecules, which are upregulated during an inflammatory insult, contain responsive elements for NF-κB in their promoter region. The action of this transcription factor therefore represents a target for many new anti-inflammatory therapeutic strategies.

R. Beyaert (ed.), Nuclear Factor κB. Regulation and Role in Disease, 221-239.

2. GLUCOCORTICOID HORMONES

The hormone cortisone was isolated in the early 1930s from the adrenal glands. Its efficacy for the treatment of rheumatoid arthritis was empirically demonstrated in patients suffering from this debilitating disease. Due to their immunosuppressive effects, corticosteroids or glucocorticoids (GCs) are used to reduce organ rejection after transplantation and to treat autoimmune diseases, including multiple sclerosis, rheumatoid arthritis and inflammatory bowel disease. Other inflammatory diseases, for which GCs are part of the standard treatment, are systemic lupus erythematosus, sarcoidosis, asthma (Barnes & Adcock, 1997) and atopy (Beyaert, 1999). They are also used to treat brain edema, shock conditions and certain types of blood cancer (B- and T-cell lymphoma) (Jehn & Osborne, 1997), as well as conditions involving adrenal cortex insufficiency.

GCs inhibit leukocyte migration to sites of inflammation and thus reduce the general symptoms of inflammation (Cato & Wade, 1996). The synthesis and secretion of cortisol, the naturally occurring GC in humans, is subject to a negative feedback loop and under tight control by a careful balance between the adrenocorticotropin hormone, secreted by the pituitary gland in the brain, and the corticotropin hormone, secreted by the hypothalamus in a pulsatile and circadian way (Balsalobre et al., 2000).

GCs exert their functions by binding to the GC receptor (GR), a transcription factor that can regulate genes in a positive or negative way. Transcription factors are proteins that are able to bind the promoter regions of their target genes and that can modulate the rate of gene transcription. In a resting cell they can either be in the nucleus bound to DNA and waiting to be activated (e.g. by a phosphorylating signal), or they can reside in the cytoplasm kept in an inactive state before activation allows them to travel inside the nucleus. Just as NF-κB and its set of specific cytoplasmic inhibitor molecules, inactive GR is sequestered in the cytoplasm, although not via a specific inhibitor, but by its interaction with chaperoning molecules including Hsp90 and Hsp70.

Because of their lipophilic nature, ligands for GR, such as cortisol, corticosterone or the synthetic dexamethasone (DEX), can travel through the cell membrane, circumventing the need for membrane receptors to transmit their signal. Ligand binding induces a conformational change in GR, causing release of the interacting molecules and exposing its nuclear localization signal (a stretch of basic residues recognized by importin nuclear transport proteins). Once in the nucleus, GR can bind as a homodimer onto the GC response element (GRE), an imperfect palindromic recognition sequence GGTACAnnnTGTYCT (Y = T or C), and positively regulate steroid-responsive genes regulating metabolic homeostasis.

The GR protein consists of three main protein modules, an N-terminal heavily phosphorylated transactivating domain, a C-terminal dimerization and ligand-binding domain (LBD) and a central DNA-binding domain (DBD), of which eight out of nine cysteines form two tetrahedric structures in which a zinc ion is held.

This special structure mediates contact with the DNA; GR is therefore also called a zinc-finger protein. Although these three modules can work almost independently, there is a great overlap in certain functions. For example, transactivation functions are also found in the DBD and LBD, and nuclear translocation is not only found in the LBD, but also in the DBD.

GR can also influence gene expression indirectly. A recently recognized important function of activated GR is the inhibition of transcription of several cytokines and chemokines that are relevant in inflammatory diseases. In the last decade it has become evident that one of the main targets for GC-mediated cytokine gene suppression is NF-κB. Recent data, derived from elegant studies with dimerization-defective (and hence transactivation-defective, since only a dimeric GR recognizes the GRE consensus sequence) knocked-in GR mutant receptors in mice, demonstrated that the anti-inflammatory action of GCs solely arises from the 'negative' cross-talk of GR with NF-κB or AP-1 (Reichardt et al., 1998; Reichardt et al., 2001).

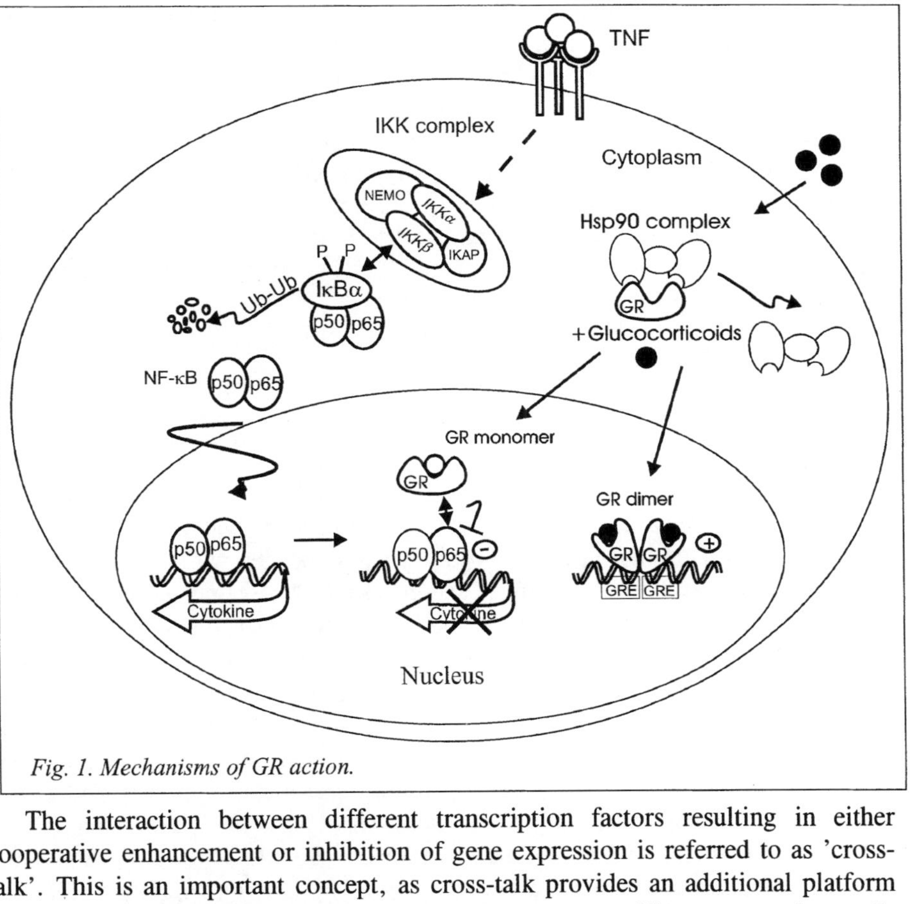

Fig. 1. Mechanisms of GR action.

The interaction between different transcription factors resulting in either cooperative enhancement or inhibition of gene expression is referred to as 'cross-talk'. This is an important concept, as cross-talk provides an additional platform of regulation to increase the number of possible gene-specific responses in a cell-specific manner. The long-term use of GCs in patients with chronic inflammatory disorders is, sadly enough, overshadowed by severe metabolic side effects, ascribed to the transactivating function of GR. Endogenous GCs protect the body from stress by regulating blood pressure levels, blood glucose levels, liver glycogen deposition and lipid metabolism. Consequently, a long-term treatment with GCs can result in diabetes, redistribution of fat and hypertension, but also HPA axis insufficiency, osteoporosis, skin and muscle atrophy, increased susceptibility to infections, cataract, peptic ulcers and a general retention of water due to a disturbed water household balance (because high levels of GCs can also activate the mineralocorticoid receptor) (Boumpas et al., 1993; Karin, 1998).

Besides metabolic actions, GCs also affect brain functions such as behavior and memory (De Kloet et al., 1998); long-term steroid use may therefore also lead to neuropsychiatric conditions. An additional problem is the fact that patients treated with GCs for long periods may develop resistance to a steroid-based therapy (Barnes et al., 1995). For all those reasons, the quest for 'better' anti-inflammatory drugs, separating the beneficial from the detrimental effects (so-called dissociated GCs) is a vigorous one. Progress has been made with the recent development and characterization of 'dissociating' GCs, which can separate to a certain extent transrepression from transactivation functions of GR (Resche-Rigon & Gronemeyer, 1998; Vanden Berghe et al., 1999b). In order to improve further therapies for inflammatory disorders, understanding of the molecular action mechanism of GCs is an absolute prerequisite. By no means a consensus has been reached with regard to the mechanism deployed by GCs in mediating proinflammatory gene repression. Different models aiming to explain this phenomenon have been advanced (Fig. 1).

3. CYTOPLASMIC MODELS

3.1. Upregulation of IκBα by GCs

One way GCs could repress NF-κB-driven gene expression is by sequestration of NF-κB in the cytoplasm. In the resting state, NF-κB activation is tightly controlled by its cytoplasmic inhibitor IκBα, which associates with NF-κB and prevents its migration to the nucleus. Two independent research groups proposed that GCs induced IκBα. This newly made IκBα would then travel to the nucleus to capture NF-κB from the DNA, ultimately resulting in retention of NF-κB in the cytoplasm and thus inhibition of cytokine gene expression (Auphan et al., 1995; Scheinman et al., 1995a). This mechanism was demonstrated for monocytes and T-cells, but other groups, including ours, did not find evidence for this mechanism in other cell lines comprising lung epithelial, fibroblasts and endothelial cells (Brostjan et al., 1996; De Bosscher et al., 1997; Hofmann et al., 1998; Adcock et al., 1999). Actually, in many cases GC-mediated repression occurred without a concomitant decrease in the DNA-binding capacity of NF-κB, as determined by gel retardation assays (Brostjan et al., 1996; De Bosscher et al.,

1997; Newton et al., 1998). On top of that, in some cases IκBα was found to be upregulated upon GC treatment, but quite unexpectedly also without apparent loss or decrease of NF-κB binding (Lezoualc'h et al., 1998; Wissink et al., 1998). It was recently shown for the ICAM-1 gene by in vivo footprinting analysis that GC repression happens without a change in conformation of the protein complex that binds to the NF-κB-binding site (Liden et al., 2000).

It seems that the upregulation of IκBα by GCs is a cell-specific phenomenon first observed for monocytes and T-lymphocytes (Auphan et al., 1995; Scheinman et al., 1995a). However, this observation is not unique or universal to immune cells. For instance, GCs do not upregulate IκBα in $CD4^+$ cells in vivo (Reichardt et al., 2001), whereas in cell types such as PC12, hepatocytes and breast carcinoma cells GCs do cause an increase in IκBα protein levels (De Vera et al., 1997; Ray & Searle, 1997; Lezoualc'h et al., 1998).

An interesting finding proving the point of cell specificity is that in the neuronal cortex of DEX-treated rats no IκBα was upregulated, whilst in the peripheral cells of the same animal IκBα levels were found to be enhanced (Unlap & Jope, 1997). Another example reflecting differences between different cell types are exemplified by the following observations. Elevated levels of IκBα were found in vascular endothelial tissue, but not in mononuclear cell infiltrates from GC-treated patients with Crohn's disease. Note that this result also seems to clash with the in vitro data, in which mainly T-cells seemed to respond to GCs with a higher IκBα protein level (Thiele et al., 1999). As an IκBα promoter construct was unresponsive to activation by DEX in L929sA cells, the actual mechanism by which GCs enhance IκBα levels in other cell types is even more puzzling (Vanden Berghe et al., 1999b).

Establishing the upregulation of IκBα by GCs is one thing, but the critical question is whether this effect can also be linked to the GC repression of NF-κB-driven genes. The answer is negative. The following observations have clearly pointed out that IκBα upregulation by GCs can be uncoupled from their anti-inflammatory cytokine gene-repressive effects. Mutational analysis of GR has shown that its DNA-binding capacity is dispensable for mediating transrepression on NF-κB-driven genes (Caldenhoven et al., 1995). This is confirmed in vivo by experiments using mice with a knocked-in dimerization-defective $GR^{dim/dim}$ mutant (A458T), which has lost DNA-binding and gene-activating properties (Reichardt et al., 1998). Vice versa, a GR mutant (S425G) defective in NF-κB-targeted gene repression was still shown to be capable of mediating enhanced IκBα synthesis (Tao et al., 2001). The repressive effects of GCs further remained apparent in the presence of the protein synthesis inhibitor cycloheximide, presenting evidence that novel protein synthesis is not required to inhibit NF-κB-driven gene expression (De Bosscher et al., 1997; Wissink et al., 1998). Finally, the use of 'dissociated' compounds lacking GR-transactivating capacities showed that GR-mediated transcription is not required for the inhibition of p65 transactivation and reciprocally GC analogues, which lack anti-inflammatory properties in vivo, could still upregulate IκBα (Heck et al., 1997; Vanden Berghe et al., 1999b).

The question remains why IκBα is upregulated by GCs in some and not in all cell types. One possible explanation could be a functional difference with regard to apoptosis. NF-κB has been reported to have an antiapoptotic function, as evidenced by the high level of liver apoptosis and subsequent death observed in p65-knockout embryos (Beg et al., 1995). Furthermore, NF-κB transcriptional activity has been implicated in cell cycle progression (Hinz et al., 1999; Kaltschmidt et al., 1999). In contrast, normal T-lymphoid and monocytic cells are sensitive to GC-induced apoptosis, a characteristic of GR used to its advantage in lymphoid cancers (Distelhorst, 2002). Now, T-cells of $GR^{dim/dim}$ mice were no longer subject to GC-mediated apoptosis, which argues for the need of gene-inductive effects by GR to mediate this event. IκBα induction was found in a GC-induced apoptosis-sensitive cell line, but not in resistant human leukemic T-cells (Ramdas & Harmon, 1998). Taken together, in lymphoid cells (which contain a high constitutive level of protective NF-κB) or other cells that have suffered too much damage, IκBα upregulation by GCs may ensure a functional apoptotic program to limit systemic immune responses that can otherwise lead to a lethal shock of the whole organism. In this respect, it is intriguing that the GR chaperone molecule Hsp90 has also been found as an additional component of death receptor-signaling complexes, such as IKK and RIP (Lewis et al., 2000; Chen et al., 2002).

3.2. Interference by other signal transduction pathways

A completely different mechanism by which GCs may exert part of their anti-inflammatory effects is the inhibition of signaling pathways that regulate inflammatory processes. One such example is the extracellular regulated kinase ERK-1,2, controlling the release of allergic mediators and induction of proinflammatory cytokine gene expression in mast cells. Recently the mechanism by which GCs inhibit ERK kinase activity was unraveled. This involves the increased expression and, crucially, at the same time a diminished proteosomal degradation of MAP kinase phosphatase-1 (Kassel et al., 2001). In other cell lines, however, such as L929sA mouse fibroblasts, GCs did not inhibit TNF-activated ERK activity (De Bosscher et al., 2001). The lack of blockage of MKP-1 degradation by GCs in these cells, as observed for NIH3T3 mouse fibroblasts (Kassel et al., 2001), is probably responsible for the differential outcome. The anti-inflammatory action of GCs may therefore not only be due to negative regulation by GR, but can also involve positive regulation by the receptor. Other examples of anti-inflammatory proteins upregulated by GCs include secretory leukocyte protease inhibitor, which protects healthy lung tissue from leukocytes activated during airway inflammation (Abbinante-Nissen et al., 1995), β-4-sulfoxide, which has a potent anti-inflammatory action on monocytes and macrophages (Young et al., 1999), as well as IL-1-receptor antagonist and lipocortin I and II, which are phospholipase inhibitor proteins (Barnes, 1998). Lipocortin upregulation by GCs may rather be a tissue- or cell-specific effect, since GCs did not affect its synthesis in L929sA fibroblasts and could therefore not be held responsible for the protective effects of DEX to TNF-mediated cytotoxicity (Beyaert et al., 1990). Along the same line, GC-mediated inhibition of c-Jun N-terminal kinase (JNK) activity leads to inhibition of c-Jun phosphorylation. This helps explaining the repressive action of GR towards the activity of AP-1, another transcription factor involved in proinflammatory gene expression (Caelles et al., 1997; Swantek et al., 1997; De Bosscher et al., 2001). Similar results have been found for the inhibition of p38 MAPK activity by GCs, although this seems again not a universal mechanism (De Bosscher et al., 2001; Lasa et al., 2001). The inhibition of p38 activation requires de novo protein synthesis (Lasa et al., 2001); in contrast, it was shown that GC-mediated repression of TNF-induced IL-6 mRNA occurred in the presence of cycloheximide, a protein synthesis inhibitor (De Bosscher et al., 1997). These differences may reflect subtle regulations by GR to enhance its repressive capacity over a longer period and again emphasize the diversity of regulatory possibilities GR has at hand to modulate cellular signaling events.

Another type of signaling cascade reported to have an effect on NF-κB and/or GR-driven transactivation is the protein kinase A (PKA) pathway. NF-κB-driven transcription is regulated by phosphorylation of RelA by PKA. Phosphorylation of Ser276 in p65 was shown to be essential for complex formation with the coactivator molecule CBP (cAMP response element binding (CREB)-binding protein) and subsequent stimulation of transactivation (Zhong et al., 1998). The catalytic subunit of PKA (PKAc) is also able to potentiate GR-dependent transcription.

Phosphorylation of a serine residue (Ser276) in the N-terminal Rel homology domain (RHD) by PKA was further demonstrated to be essential for p65-mediated repression of GR transactivation. Mutation of p65 at this conserved PKA phosphorylation site abolished the potential of p65 to repress GR (Doucas et al., 2000). Strikingly, an exclusively cytoplasmic variant of p65 (obtained by deleting the NLS of p65) was still capable to mediate transrepression of a GR-activated mouse mammary tumor virus promoter-driven reporter gene. From this result it was concluded that targeting GC-driven gene expression by p65 occurs in the cytoplasm, involving PKAc-dependent signaling as the molecular interface of this inhibition (Doucas et al., 2000). For the reciprocal mechanism controversy arises. A GR deletion variant (amino acids 589-697 deleted) with a predominant cytoplasmic localization would also still inhibit NF-κB-driven gene expression, arguing that a competition for PKAc would mediate mutual cross-coupling in the cytoplasm (Doucas et al., 2000). Mapping GR functions has, however, shown that there is more than one NLS in GR (Beato, 1989; McEwan et al., 1997). It can therefore not be ruled out that a minor proportion of this variant can still travel in and out of the nucleus, using the other NLS, and mediate gene repression in the nucleus.

Furthermore, Ser276 was found not to be a key player for repression of NF-κB activity by GR. In this case, mutation of Ser276 to a cysteine residue in a

Gal4-p65 fusion protein did not affect the ability of GR to block NF-κB-driven transcription (De Bosscher et al., 2000b). It becomes more and more clear that the reciprocal mechanisms of transrepression, i.e. the mutual antagonistic effects of NF-κB and GCs, do not necessarily use the same molecular mechanisms.

4. NUCLEAR MODELS

Direct binding of GR to DNA via a so-called 'nGRE', and as such negatively regulating gene expression, is quite a rare event. One example of this is the osteocalcin gene. The osteocalcin promoter region to which GR binds, is partially overlapping with the TATA box, occluding the build-up of a functional transcriptional complex (Meyer et al., 1997). Despite the ability of GCs to induce gene transcription, the major anti-inflammatory effects of GCs occur by repression of inflammatory and immune genes driven by NF-κB or AP-1, another mitogenic transcription factor (Adcock & Caramori, 2001). Intriguingly, the transrepressive relationship between GR and NF-κB appears to be mutual; so it seemed most logic at the time that both proteins would actively hinder each other's transactivating functions by no means other than a direct physical contact. The model of a direct interaction between GR and NF-κB was supported by several research groups for the last five years; however, only recently an actual physical interaction with endogenous proteins was demonstrated in A549 lung carcinoma cells (Adcock et al., 1997).

Some research groups have focused on mapping the involved functional domains in transrepression, albeit with sometimes conflicting results. For GR, domain swapping of its modular parts (N-terminal domain, DBD and LBD) with other members of the nuclear receptor superfamily, such as estrogen receptor (ER) and thyroid receptor (TR), demonstrated that the DBD was indispensable, not only for transactivation (as expected), but also for transrepression (Liden et al., 1997; Ray et al., 1997; Moras & Gronemeyer, 1998). To discern whether the DNA-binding function of GR per se was crucial for NF-κB transrepression point mutations in the P-box (= DNA-interacting N-terminal zinc finger) were performed. The outcome was that GR binding onto a classical GRE could be disrupted, but the transrepressive properties on NF-κB or AP-1 remained untouched (Heck et al., 1994; Caldenhoven et al., 1995; Helmberg et al., 1995). Another study, however, pointed to the other zinc finger, C-terminally placed, as a crucial determinant for NF-κB transrepression (Liden et al., 1997). The reason for this discrepancy remains undetermined, but may reside in the nature of the investigated NF-κB response elements (e.g. binding different NF-κB family members) in different cell lines. It is also possible that a different cofactor context or GR function-modulating chaperones in different cell lines contribute to these discrepancies. Recently it became apparent that the promoter context may codetermine whether or not a specific nuclear receptor can interfere with NF-κB activity (Amrani et al., 1999; De Bosscher et al., 2000a). Replacing the LBD with a nonrelated inert β-galactosidase moiety (to diminish unwanted effects of an incorrect folding of deletion variants) did not affect transrepression, arguing for a purely steric role of the LBD (Oro et al., 1988).

Conversely, also the domains of p65 involved in repression of GR activity have been mapped. Elaborate mutational analysis demonstrated that both the RHD, containing DNA-binding and dimerization functions, and the C-terminal domain of p65, harboring transactivation functions, are required for the repression of GR activity. However, in vitro a physical interaction was only observed between the RHD of p65 and GR (Scheinman et al., 1995b; Wissink et al., 1997).

A greater part of the gathered data points out that the protein-protein interaction between GR and NF-κB is most likely to occur in the nucleus. A first piece of evidence has been referred to above, namely the unchanged footprinting pattern of NF-κB bound to its response element in the GC-repressible ICAM promoter (Liden et al., 2000). Secondly, when p65 is fused to a DNA-binding yeast protein Gal4, this fusion protein is completely nuclear and now able to transactivate a GAL4-binding site containing a reporter gene. Importantly, GCs can inhibit the transactivation of Gal4-p65 to the same extent as wild-type p65, arguing that GC repression is a nuclear phenomenon (De Bosscher et al., 1997). The transactivating C-terminus of p65 has been shown to contact the general transcription factors TFIIB and TATA-binding protein in vitro (Schmitz et al., 1995); this binding could help stabilizing TFIID interactions to build up a functional promoter initiation complex (PIC) and to start gene transcription. The importance of the promoter context close to the TATA box sequences in mediating transrepression was exemplified by the finding of an NF-κB-driven gene that was no longer responsive to GC-mediated repression (De Bosscher et al., 2000b). Finding the sequences that determine this unexpected promoter specificity is an important issue. Another study exploring events around the start site of transcription came up with exciting new evidence that GR mediates repression by interfering with the phosphorylation of a serine residue in the C-terminal domain of RNA polymerase II, again without inhibiting the assembly of PIC (Nissen & Yamamoto, 2000).

Cofactors are nuclear proteins that are able to form a bridge between transcription factors and the basal transcription machinery, without contacting the DNA themselves (Horwitz et al., 1996). They are often associated with an enzymatic activity, either a histone acetylase (HAT) or a histone deacetylase (HDAC) activity, depending on their function as a coactivator or a corepressor, respectively. Histone acetylation is believed to be important for relaxing chromatin and favoring gene transcription, the opposite holding true for histone deacetylation (Wolffe & Pruss, 1996; Wolffe, 1997). This result postulates the existence of a novel type of corepressor associated with the LBD of GR, possibly a serine-2-phosphatase or a serine-2-kinase inhibitor (Nissen & Yamamoto, 2000).

A logic question now is whether a cytoplasmic model is in any way reconcilable with a nuclear model. Different findings, dependent on which cell lines are used, have sometimes led to seemingly conflicting results. Such observations should, however, not be contradictory, as nongenomic (cytoplasmic) and genomic (nuclear) mechanisms have recently been described to act independently on target gene expression (Falkenstein et al., 2000; Croxtall et al., 2002). Finally, it is quite acceptable that a different constitution in cofactor complexes or different allosteric conformations of target gene promoters may integrate all the signals coming from the cytoplasm with subtle differences, thus generating a slightly different mechanistic response by GCs (Hall et al., 2001; Hall et al., 2002; Katzenellenbogen & Katzenellenbogen, 2002; Shang & Brown, 2002). Also, a dosage effect or the duration of the proinflammatory insult to the cells may be of importance to the way GCs go about to inhibit the signaling mediated by NF-κB (Lawrence et al., 2001). Besides the transcriptional effects discussed here, important GC effects have also been detected at the posttranscriptional level, such as mRNA destabilization of proinflammatory genes (viz. iNOS, TNF, GM-CSF and COX-2) (Tobler et al., 1992; Delany et al., 1995; Chaudhary & Avioli, 1996; Lasa et al., 2001). Functional GRs must therefore be regarded as a context-dependent multi-targeting effector, rather than a mediator of repression via one exclusive pathway.

5. COFACTOR MODELS

As mentioned above, cofactor molecules provide an extra layer of transcriptional regulation in the nucleus (Katzenellenbogen & Katzenellenbogen, 2002; Shang & Brown, 2002). Coactivators are generally associated with a HAT activity, whereas corepressors are associated with a HDAC activity (Wolffe et al., 1997). The LBD of GR has been shown to interact, in a ligand-dependent way, with coactivators such as CBP/p300, GRIP1 and SRC-1 (Eggert et al., 1995; Onate et al., 1995; Chakravarti et al., 1996). The same coactivators have also been implicated in bridging other transcription factors, such as NF-κB and AP-1, to the basal transcription machinery (Kamei et al., 1996; Gerritsen et al., 1997; Perkins, 1997; Sheppard et al., 1998; Vanden Berghe et al., 1999a). Gene repression could therefore result from a competition between transcription factors for limiting amounts of coactivator molecules. A competition between p65 and GR for limiting amounts of CBP or SRC-1 was proposed to account for transrepression of NF-κB-dependent genes (McKay & Cidlowski, 1998; Sheppard et al., 1998). It is, however, difficult to understand how this mechanism would generate a specific transrepression of GC-repressible NF-κB- or AP-1-driven target genes only, since a great number of other transcription factors also utilize CBP/p300 or SRC-1 for enhancing their transactivation properties. These include CREB, ATF-2, MyoD, p53, Tax and STAT-2 (Horwitz et al., 1996; Xu et al., 1999), although preferences and differences have been noted. So it has been reported that NF-κB-mediated transactivation requires the presence of CBP, SRC-1 and p/CAF, using mainly the HAT activity of p/CAF, while CREB, for instance, uses CBP, p/CIP and p/CAF, but not SRC-1 (Korzus et al., 1998;

Sheppard et al., 1999). It may be that a constitution of different coactivator complexes provides a platform for the specific recognition and repression by GCs of certain transcription factor families only. Nevertheless, other experiments argue against a cofactor squelching model to explain mutual transrepression between NF-κB and GR. Increasing the coactivator levels in the cell by transient overexpression generates a general dose-responsive increase in gene expression levels of NF-κB-driven gene expression. However, upon activating GR the relative repression levels remain unaffected. Furthermore, in case of a competitive mechanism one would expect that the interaction between p65 and CBP would be diminished or completely lost upon interaction with activated GR. This has been proven not to be the case (De Bosscher et al., 2000a). Additional evidence diminishing the role of CBP in transrepression is offered by the fact that the GAL4-transactivating Gal4-p65 Ser276Cys mutant, leading to a defective CBP recruitment and subsequent loss of TNF inducibility, still demonstrated a functional repression by GCs (De Bosscher et al., 2000b). For the reciprocal mechanism, p65 mutation in the DNA-binding domain (but with the predicted coactivator recruitment domains intact) could no longer repress GC-mediated transactivation (Wissink et al., 1997). Finally, from the fact that dissociating ligands or GR point mutants can distinguish between transactivation and transrepression, it can be deduced that GR is not always associated with the same cofactor surrounding. In fact, a different steric conformation of GR could lie at the basis of this phenomenon, achieving transrepression when GR adopts a monomeric conformation as opposed to allowing transactivation of target genes, when GR is in a DNA-bound dimeric conformation (Lefstin & Yamamoto, 1998; Reichardt et al., 2001). These findings demonstrate an incompatibility with a general cofactor competition model.

It must be noted that all the experimental approaches, depending on overexpression and microinjection, overload the cell with transcriptional components, disregarding the influence of nuclear architecture and specific nuclear matrix targeting. The latter phenomenon is a concept that has recently gained importance. Territorial subdivision of transcription factor complexes in the nucleus (Doucas et al., 1999; Stenoien et al., 2000; Stewart & Crabtree, 2000) may explain how cofactors only target a certain gene in a designated compartment in the nucleus, whilst leaving the same factors associated with different genes in other compartments intact (Francastel et al., 2000; Hager et al., 2000; Lemon & Tjian, 2000). A specific nuclear matrix targeting signal has been described to include parts of the DBD and transactivation domains of GR (DeFranco & Guerrero, 2000).

Some members of the nuclear hormone receptor superfamily, such as retinoic/retinoid X acid receptors (RAR/RXR) and thyroid receptors (TR) are already bound to DNA in the absence of ligand. In this particular situation, corepressor complexes actively silence gene expression (Burke & Baniahmad, 2000). A HDAC-containing corepressor complex consisting of the components 'nuclear corepressor' (NcoR)/'silencing mediator of retinoid and thyroid receptors'

(SMRT), mSin3 and RDP3/HDAC1 is displaced by a HAT-containing coactivator complex comprising CBP, p/CAF and SRC-1 (Xu et al., 1999).

Co-crystal structures have revealed that helix 12 of the LBD of the related estrogen receptor (also harboring transrepressive properties) adopts a different conformation when bound to agonistic versus antagonistic estrogens (Brzozowski et al., 1998; Nichols et al., 1998). Antagonist-bound progesterone receptor and estrogen receptor were further found to interact in vitro with the corepressors NcoR and SMRT (Jackson et al., 1997; Lavinsky et al., 1998; Wagner et al., 1998; Zhang et al., 1998). Multiple ligands for nuclear receptors are thus capable of influencing the biological activity of the receptor by selectively affecting the recruitment of specific cofactor complexes. The role of each cofactor in vivo can be assessed by knockout models of the individual cofactors, such as for SRC-1 (Xu et al., 1998) or by cell reconstitution experiments (Lemon et al., 2001). Undoubtedly, GR will also recruit its own specific cofactor configuration to enable transactivation. The question is whether there is also a role for corepressor molecules in transrepression mechanisms between NF-κB and GR. GR recruitment of HDAC2 was shown to inhibit IL-1β-induced histone acetylation at specific lysine residues (Ito et al., 2000). Furthermore, this GR association with HDAC2 in vivo could be disrupted by the GR antagonist mifepristone (Ito et al., 2001). A deacetylase inhibitor trichostatin A (TSA) allegedly demonstrated involvement of HDAC activities in GR-transrepression mechanisms. However, the relative transrepression levels when comparing TNF + DEX + TSA with TNF + TSA were identical to the ones observed without TSA. De facto, these data alone do not allow to conclude that deacetylases are involved in the mechanism of transrepression between GR and NF-κB (Vanden Berghe et al., 2002). In addition, promoter responsivity to TSA does not necessarily reflect sensitivity to GCs, as both IL-8 and HIV promoter activities can be increased by TSA, whereas only IL-8 is responsive to GC repression (Vanden Berghe et al., 2002).

6. FUTURE PROSPECTS

It has already been mentioned that a GR-mediated serine phosphorylation switch of RNA polymerase II could form the basis of NF-κB and GR cross-talk mechanisms (Nissen & Yamamoto, 2000). It would be of utmost interest to investigate whether this event is promoter-specific or a more general characteristic of GC-mediated repression of NF-κB. It would furthermore be interesting to know whether this mechanism also accounts for the reciprocal mechanism. GR is able to block phosphorylation of CBP, which may be another or additional way by which this transcription factor modulates gene expression (Perissi et al., 1999). Involvement of this modification in cross-repression between NF-κB and GR remains to be explored.

As not only histone tails, but also nuclear receptors, NF-κB and cofactors can be acetylated or deacetylated, it will be interesting to learn how these various factor modifications can influence cross-talk mechanisms, in particular between GR and NF-κB. Similarly, other posttranslational modifications may exert their influence. Hormone-dependent histone 3- or histone 4-specific methyltransferases have now been characterized in transactivation (Ma et al., 2001; Wang et al., 2001). Again, not only histones, but also CBP/p300 are targeted by CARM1, causing a disturbance of interaction with the transcription factor CREB (Ma et al., 2001). A link between these enzymatic activities and GR/NF-κB interplay has not yet been established, but will probably be explored in the future.

As pointed out above, chromatin components can be modified by various enzymatic activities. Another quite novel concept is the so-called chromatin remodeling, referring to the alteration in the chromatin fiber structure of a particular nucleosome, or a series of adjacent nucleosomes. To this extent, liganded nuclear receptors may associate with remodeling components SWI/SNF and utilize large ATP-dependent complexes (Kniyamu et al., 2000; Urnov & Wolffe, 2001) to bring about these structural changes before any other modifications involved in transcription initiation may occur. Whether the transrepression of p65 by GR and/or the reciprocal transrepression of GR by p65 could work by influencing the chromatin-remodeling machinery needs further experimentation and confirmation in vivo.

Finally, we would like to stress that the models need not be exclusive. A direct protein-protein interaction would not rule out the involvement of cofactors or modulation by chromatin- or factor-modifying enzymatic activities, such as acetylation, methylation and ubiquitination. Understanding the subtle regulation and the precise contribution of each of these activities and parameters is of paramount importance to be able to design more specific anti-inflammatory strategies. The aim is to maintain the great effectiveness of GCs and to dispose of any detrimental side effects associated with their long-term usage.

7. REFERENCES

Abbinante-Nissen JM, Simpson LG & Leikauf GD (1995) Corticosteroids increase secretory leukocyte protease inhibitor transcript levels in airway epithelial cells. Am. J. Phys. 268: 601-606.

Adcock IM, Newton R & Barnes PJ (1997) NF-κB involvement in IL-1β-induction of GM-CSF and COX-2: inhibition by glucocorticoids does not require I-κB. Biochem. Soc. Trans. 25: 154s.

Adcock IM, Nasuhara Y, Stevens DA & Barnes PJ (1999) Ligand-induced differentiation of glucocorticoid receptor (GR) transrepression and transactivation: preferential targetting of NF-κB and lack of I-κB involvement. Br. J. Pharmacol. 127: 1003-1011.

Adcock IM & Caramori G (2001) Cross-talk between pro-inflammatory transcription factors and glucocorticoids. Immunol. Cell Biol. 79: 376-384.

Amrani Y, Lazaar AL & Panettieri RAJ (1999) Up-regulation of ICAM-1 by cytokines in human tracheal smooth muscle cells involves an NF-κB-dependent signaling pathway that is only partially sensitive to dexamethasone. J. Immunol. 163: 2128-2134.

Auphan N, DiDonato JA, Rosette C, Helmberg A & Karin M (1995) Immunosuppression by glucocorticoids: inhibition of NF-κB activity through induction of IκB synthesis. Science 270: 286-290.

Balsalobre A, Brown SA, Marcacci L, Tronche F, Kellendonk C, Reichardt HM, Schütz G & Schnibler U (2000) Resetting of circadian time in peripheral tissues by glucocorticoid signalling. Science 289: 2344-2347.

Barnes PJ, Greening AP & Crompton GK (1995) Glucocorticoid resistance in asthma. Am. J. Respir. Crit. Care Med. 152: S125-140.

Barnes PJ & Adcock IM (1997) NF-κB: a pivotal role in asthma and a new target for therapy. Trends Pharmacol. Sci. 18: 46-50.

Barnes PJ & Karin M (1997) Nuclear factor-κB: a pivotal transcription factor in chronic inflammatory diseases. N. Engl. J. Med 336: 1066-1071.

Barnes PJ (1998) Anti-inflammatory actions of glucocorticoids: molecular mechanisms. Clin. Sci. Colch. 94: 557-572.

Beato M (1989) Gene regulation by steroid hormones. Cell 56: 335-344.

Beg AA, Sha WC, Bronson RT, Ghosh S & Baltimore D (1995) Embryonic lethality and liver degeneration in mice lacking the RelA component of NF-κB. Nature 376: 167-170.

Beyaert R, Suffys P, Van Roy F & Fiers W (1990) Inhibition by glucocorticoids of tumor necrosis factor-mediated cytotoxicity. Evidence against lipocortin involvement. FEBS Lett. 262: 93-96.

Beyaert R (1999) NF-κB as an emerging target in atopy. Emerg. Ther. Targets 3: 213-228.

Boumpas DT, Chrousos GP, Wilder RL, Cupps TR & Balow JE (1993) Glucocorticoid therapy for immune-mediated diseases: basic and clinical correlates. Ann. Intern. Med. 119: 1198-1208.

Brostjan C, Anrather J, Csizmadia V, Stroka D, Soares M, Bach FH & Winkler H (1996) Glucocorticoid-mediated repression of NFκB activity in endothelial cells does not involve induction of IκBα synthesis. J. Biol. Chem. 271: 19612-19616.

Brzozowski AM, Pike ACW, Dauter Z, Hubbard RE, Bonn T, Engstrom O, Ohman L, Greene GL, Gustafsson JA & Carlquist M (1998) Molecular basis of agonism and antagonism in the oestrogen receptor. Nature 389: 753-758.

Burke LJ & Baniahmad A (2000) Corepressors. FASEB J. 14: 1876-1888.

Caelles C, Gonzalez Sancho JM & Muñoz A (1997) Nuclear hormone receptor antagonism with AP-1 by inhibition of the JNK pathway. Genes Dev. 11: 3351-3364.

Caldenhoven E, Liden J, Wissink S, van de Stolpe A, Raaijmakers J, Koenderman L, Okret S, Gustafsson JA & van der Saag PT (1995) Negative cross-talk between RelA and the glucocorticoid receptor: a possible mechanism for the antiinflammatory action of glucocorticoids. Mol. Endocrinol. 9: 401-412.

Cato AC & Wade E (1996) Molecular mechanisms of anti-inflammatory action of glucocorticoids. BioEssays 18: 371-378.

Chakravarti D, LaMorte VJ, Nelson MC, Nakajima T, Schulman IG, Juguilon H, Montminy M & Evans RM (1996) Role of CBP/p300 in nuclear receptor signalling. Nature 383: 99-103.

Chaudhary LR & Avioli LV (1996) Regulation of interleukin-8 gene expression by interleukin-1β, osteotropic hormones, and protein kinase inhibitors in normal human bone marrow stromal cells. J. Biol. Chem. 271: 16591-16596.

Chen G, Cao P & Goeddel DV (2002) TNF-induced recruitment and activation of the IKK complex require Cdc37 and Hsp90. Mol. Cell 9: 401-410.

Croxtall JD, van Hal PT, Choudhury Q, Gilroy DW & Flower RJ (2002) Different glucocorticoids vary in their genomic and non-genomic mechanism of action in A549 cells. Br. J. Pharmacol. 135: 511-519.

De Bosscher K, Schmitz ML, Vanden Berghe W, Plaisance S, Fiers W & Haegeman G (1997) Glucocorticoid-mediated repression of nuclear factor-κB-dependent transcription involves direct interference with transactivation. Proc. Natl. Acad. Sci. USA 94: 13504-13509.

De Bosscher K, Vanden Berghe W & Haegeman G (2000a) Mechanisms of anti-inflammatory action and of immunosuppression by glucocorticoids: negative interference of activated glucocorticoid receptor with transcription factors. J. Neuroimmunol. 109: 16-22.

De Bosscher K, Vanden Berghe W, Vermeulen L, Plaisance S, Boone E & Haegeman. G (2000b) Glucocorticoids repress NF-kB-driven genes by disturbing the interaction of p65 with the basal transcription machinery, irrespective of coactivator levels in the cell. Proc. Natl. Acad. Sci. USA 97: 3919-3924.

De Bosscher K, Vanden Berghe W & Haegeman G (2001) Glucocorticoid repression of AP-1 is not mediated by competition for nuclear coactivators. Mol. Endocrinol. 15: 219-227.

De Kloet ER, Vreugdenhil E, Oitzl MS & Joels M (1998) Brain corticosteroid receptor balance in health and disease. Endocr. Rev. 19: 269-301.
De Vera ME, Taylor BS, Wang Q, Shapiro RA, Billiar TR & Geller DA (1997) Dexamethasone suppresses iNOS gene expression by upregulating I-κBα and inhibiting NF-κB. Am. J. Physiol. 273: g1290-1296.
DeFranco DB & Guerrero J (2000) Nuclear matrix targeting of steroid receptors: specific signal sequences and adaptor proteins. Crit. Rev. Eukaryot. Gene Expr. 10: 39-44.
Delany AM, Gabbitas BY & Canalis E (1995) Cortisol downregulates osteoblast α1 (I) procollagen mRNA by transcriptional and posttranscriptional mechanisms. J. Cell. Biochem. 57: 488-494.
Distelhorst CW (2002) Recent insights into the mechanism of glucocorticosteroid-induced apoptosis. Cell Death Differ. 9: 6-19.
Doucas V, Tini M, Egan DA & Evans RM (1999) Modulation of CREB binding protein function by the promyelocytiv (PML) oncoprotein suggests a role for nuclear bodies in hormone signaling. Proc. Natl. Acad. Sci. USA 96: 2627-2632.
Doucas V, Shi Y, Miyamoto S, West A, Verma I & Evans RM (2000) Cytoplasmic catalytic subunit of protein kinase A mediates cross-repression by NF-κB and the glucocorticoid receptor. Proc. Natl. Acad. Sci. USA 97: 11893-11898.
Eggert M, Mows CC, Tripier D, Arnold R, Michel J, Nickel J, Schmidt S, Beato M & Renkawitz R (1995) A fraction enriched in a novel glucocorticoid receptor-interacting protein stimulates receptor-dependent transcription in vitro. J. Biol. Chem. 270: 30755-30759.
Falkenstein E, Tillmann HC, Christ M, Feuring M & Wehling M (2000) Multiple actions of steroid hormones. A focus on rapid, nongenomic effects. Pharmacol. Rev. 52: 513-556.
Francastel C, Schubeler D, Martin DI & Groudine M (2000) Nuclear compartmentalization and gene activity. Nat. Rev. Mol. Cell Biol. 1: 137-143.
Gerritsen ME, Williams AJ, Neish AS, Moore S, Shi Y & Collins T (1997) CREB-binding protein/p300 are transcriptional coactivators of p65. Proc. Natl. Acad. Sci. USA 94: 2927-2932.
Hager GL, Lim CS, Elbi C & Baumann CT (2000) Trafficking of nuclear receptors in living cells. J. Steroid Biochem. Mol. Biol. 74: 249-254.
Hall JM, Couse JF & Korach KS (2001) The multifaceted mechanisms of estradiol and estrogen receptor signaling. J. Biol. Chem. 276: 36869-36872.
Hall JM, McDonnell DP & Korach KS (2002) Allosteric regulation of estrogen receptor structure, function, and coactivator recruitment by different estrogen response elements. Mol. Endocrinol. 16: 469-486.
Heck S, Kullmann M, Gast A, Ponta H, Rahmsdorf HJ, Herrlich P & Cato AC (1994) A distinct modulating domain in glucocorticoid receptor monomers in the repression of activity of the transcription factor AP-1. EMBO J. 13: 4087-4095.
Heck S, Bender K, Kullmann M, Gottlicher M, Herrlich P & Cato AC (1997) IκBα downregulation of NF-κB activity by glucocorticoid receptor. EMBO J. 16: 4698-4707.
Helmberg A, Auphan N, Caelles C & Karin M (1995) Glucocorticoid-induced apoptosis of human leukemic cells is caused by the repressive function of the glucocorticoid receptor. EMBO J. 14: 452-460.
Hinz M, Krappmann D, Eichten A, Heder A, Scheidereit C & Strauss M (1999) NF-κB function in growth control: regulation of cyclin D expression and G0/G1-to-S-phase transition. Mol. Cell. Biol. 19: 2690-2698.
Hofmann TG, Hehner SP, Bacher S, Dröge W & Schmitz ML (1998) Various glucocorticoids differ in their ability to induce gene expression, apoptosis and to repress NF-κB-dependent transcription. FEBS Lett. 441: 441-446.
Horwitz KB, Jackson TA, Bain DI, Richer JK, Takimoto GK & Tung L (1996) Nuclear receptor coactivators and corepressors. Mol. Endocrinol. 10: 1167-1177.
Ito K, Barnes PJ & Adcock IM (2000) Glucocorticoid receptor recruitment of histone deacetylase 2 inhibits IL-1β-induced histone H4 acetylation on Lysines 8 and 12. Mol. Cell. Biol. 20: 6891-6903.
Ito K, Jazrawi E, Cosio B, Barnes PJ & Adcock IM (2001) p65-activated histone acetyltransferase activity is repressed by glucocorticoids: mifepristone fails to recruit HDAC2 to the p65-HAT

complex. J. Biol. Chem. 276: 30208-30215.

Jackson TA, Richer JK, Bain DL, Takimoto GS, Tung L & Horwitz KB (1997) The partial agonist activity of antagonist-occupied steroid receptors is controlled by a novel hinge domain-binding coactivator L7/SPA and the corepressors N-CoR or SMRT. Mol. Endocrinol. 11: 693-705.

Jehn BM & Osborne BA (1997) Gene regulation associated with apoptosis. Crit. Rev. Eukaryot. Gene Expr. 7: 179-193.

Kaltschmidt B, Kaltschmidt C, Hehner SP, Dröge W & Schmitz ML (1999) Repression of NF-κB impairs HeLa cell proliferation by functional interference with cell cycle checkpoint regulators. Oncogene 18: 3213-3225.

Kamei Y, Xu L, Heinzel T, Torchia J, Kurokawa R, Gloss B, Lin SC, Heyman RA, Rose DW, Glass CK & Rosenfeld MG (1996) A CBP integrator complex mediates transcriptional activation and AP-1 inhibition by nuclear receptors. Cell 85: 403-414.

Karin M (1998) New twists in gene regulation by glucocorticoid receptor: is DNA binding dispensable? Cell 93: 487-490.

Kassel O, Sancono A, Krätzschmar J, Kreft B, Stassen M & Cato CB (2001) Glucocorticoids inhibit MAP kinase via increased expression and decreased degradation of MKP-1. EMBO J. 20: 7108-7116.

Katzenellenbogen BS & Katzenellenbogen JA (2002) Biomedicine. Defining the "S" in SERMs. Science 295: 2380-2381.

Kniyamu HK, Fryer CJ, Horwitz KB & Archer TK (2000) The mouse mammary tumor virus promoter adopts distinct chromatin structures in hulan breast cancer cells with and without glucocorticoid receptor. J. Biol. Chem. 275: 20061-20068.

Korzus E, Torchia J, Rose DW, Xu L, Kurokawa R, McInerney EM, Mullen TM, Glass CK & Rosenfeld MG (1998) Transcription factor-specific requirements for coactivators and their acetyltransferase functions. Science 279: 703-707.

Lasa M, Brook M, Saklatvala J & Clark RA (2001) Dexamethasone destabilizes cyclooxygenase 2 mRNA by inhibiting mitogen-activated protein kinase p38. Mol. Cell. Biol. 21: 771-780.

Lavinsky RM, Jepsen K, Heinzel T, Torchia J, Mullen TM, Schiff R, Del Rio AL, Ricote M, Ngo S, Gemsch J, Hilsenbeck SG, Osborne CK, Glass CK, Rosenfeld MG & Rose DW (1998) Diverse signaling pathways modulate nuclear receptor recruitment of N-CoR and SMRT complexes. Proc. Natl. Acad. Sci. USA 95: 2920-2925.

Lawrence T, Gilroy DW, Colville-Nash PR & Willoughby DA (2001) Possible new role for NF-κB in the resolution of inflammation. Nat. Med. 7: 1291-1297.

Lefstin JA & Yamamoto KR (1998) Allosteric effects of DNA on transcriptional regulators. Nature 392: 885-888.

Lemon B & Tjian T (2000) Orchestrated response: a symphony of transcription factors for gene control. Genes Dev. 14: 2551-2569.

Lemon B, Inouye C, King DS & Tjian R (2001) Selectivity of chromatin-remodelling cofactors for ligand-activated transcription. Nature 414: 924-928.

Lewis J, Devin A, Miller A, Lin Y, Rodriguez Y, Neckers L & Liu ZG (2000) Disruption of hsp90 function results in degradation of the death domain kinase, receptor-interacting protein (RIP), and blockage of tumor necrosis factor-induced nuclear factor-κB activation. J. Biol. Chem. 275: 10519-10526.

Lezoualc'h F, Sagara Y, Holsboer F & Behl C (1998) High constitutive NF-κB activity mediates resistance to oxidative stress in neuronal cells. J. Neurosci. 18: 3224-3232.

Liden J, Delaunay F, Rafter I, Gustafsson J & Okret S (1997) A new function for the C-terminal zinc finger of the glucocorticoid receptor. Repression of RelA transactivation. J. Biol. Chem. 272: 21467-21472.

Liden J, Rafter I, Truss M, Gustafsson J & Okret S (2000) Glucocorticoid effects on NF-κB binding in the transcription of the ICAM-1 gene. Biochem. Biophys. Res. Commun. 273: 1008-1014.

Ma H, Baumann CT, Li H, Strahl BD, Rice R, Jelinek MA, Aswad DW, Allis CD & Stallcup MR (2001) hormone-dependent, CARM1-directed, arginine-specific methylation of histone H3 on a steroid-regulated promoter. Curr. Biol. 11: 1981-1985.

McEwan IJ, Wright AP & Gustafsson JA (1997) Mechanism of gene expression by the glucocorticoid receptor: role of protein-protein interactions. BioEssays 19: 153-160.

McKay LI & Cidlowski JA (1998) Cross-talk between nuclear factor-κB and the steroid hormone receptors: mechanisms of mutual antagonism. Mol. Endocrinol. 12: 45-56.

Meyer T, Carlstedt Duke J & Starr DB (1997) A weak TATA box is a prerequisite for glucocorticoid-dependent repression of the osteocalcin gene. J. Biol. Chem. 272: 30709-30714.

Moras D & Gronemeyer H (1998) The nuclear receptor ligand-binding domain: structure and function. Curr. Opin. Cell Biol. 10: 384-391.

Newton R, Hart LA, Stevens DA, Bergmann M, Donnelly LE, Adcock IM & Barnes PJ (1998) Effect of dexamethasone on interleukin-1β-(IL-1β)-induced nuclear factor-κB (NF-κB) and κB-dependent transcription in epithelial cells. Eur. J. Biochem. 254: 81-89.

Nichols M, Rientjes JM & Stewart AF (1998) Different positioning of the ligand-binding domain helix 12 and the F domain of the estrogen receptor accounts for functional differences between agonists and antagonists. EMBO J. 17: 765-773.

Nissen RM & Yamamoto KR (2000) The glucocorticoid receptor inhibits NFκB by interfering with serine-2 phosphorylation of the RNA polymerase II carboxy-terminal domain. Genes Dev. 14: 2314-2329.

Onate SA, Tsai SY, Tsai MJ & O'Malley BW (1995) Sequence and characterization of a coactivator for the steroid hormone receptor superfamily. Science 270: 1354-1357.

Oro AE, Hollenberg SM & Evans RM (1988) Transcriptional inhibition by a glucocorticoid receptor-β-galactosidase fusion protein. Cell 55: 1109-1114.

Perissi V, Dasen JS, Kurokawa R, Wang ZY, Korzus E, Rose DW, Glass CK & Rosenfeld MG (1999) Factor-specific modulation of CREB-binding protein acetyltransferase activity. Proc. Natl. Acad. Sci. USA 96: 3652-3657.

Perkins ND (1997) Achieving transcriptional specificity with NF-kB. Int. J. Biochem. Cell Biol. 29: 1433-1448.

Ramdas J & Harmon JM (1998) Glucocorticoid-induced apoptosis and regulation of NF-κB activity in human leukemic T cells. Endocrinology 139: 3813-3821.

Ray KP & Searle N (1997) Glucocorticoid inhibition of cytokine-induced E-selectin promoter activation. Biochem. Soc. Trans. 25: 189s.

Ray P, Ghosh SK, Zhang DH & Ray A (1997) Repression of interleukin-6 gene expression by 17 β-estradiol: inhibition of the DNA-binding activity of the transcription factors NF-IL6 and NF-κB by the estrogen receptor. FEBS Lett. 409: 79-85.

Reichardt HM, Kaestner KH, Tuckermann J, Kretz O, Wessely O, Bock R, Gass P, Schmid W, Herrlich P, Angel P & Schutz G (1998) DNA binding of the glucocorticoid receptor is not essential for survival. Cell 93: 531-541.

Reichardt HM, Tuckermann JP, Göttlicher M, Vujic M, Weih F, Angel P, Herrlich P & Schütz G (2001) Repression of inflammatory responses in the absence of DNA binding by the glucocorticoid receptor. EMBO J. 20: 7168-7173.

Resche-Rigon M & Gronemeyer H (1998) Therapeutic potential of selective modulators of nuclear receptor action. Curr. Opin. Chem. Biol. 2: 501-507.

Scheinman R, Cogswell PC, Lofquist A & Baldwin AS (1995a) Role of transcriptional activation of IκBα in mediation of immunosuppression by glucocorticoids. Science 270: 283-286.

Scheinman RI, Gualberto A, Jewell CM, Cidlowski JA & Baldwin AS (1995b) Characterization of mechanisms involved in transrepression of NF-κB by activated glucocorticoid receptors. Mol. Cell. Biol. 15: 943-953.

Schmitz ML, Stelzer G, Altmann H, Meisterernst M & Baeuerle PA (1995) Interaction of the COOH-terminal transactivation domain of p65 NF-κB with TATA-binding protein, transcription factor IIB, and coactivators. J. Biol. Chem. 270: 7219-7226.

Shang Y & Brown M (2002) Molecular determinants for the tissue specificity of SERMs. Science 295: 2465-2468.

Sheppard K-A, Phelps KM, Williams AJ, Thanos D, Glass CK, Rosenfeld MG, Gerritsen ME & Collins T (1998) Nuclear integration of glucocorticoid receptor and nuclear factor-κB signaling by

CREB-binding protein and steroid receptor coactivator-1. J. Biol. Chem. 273: 29291-29294.
Sheppard K-A, Rose DW, Haque ZK, Kurokawa R, McInerney E, Westin S, Thanos D, Rosenfeld MG, Glass CK & Collins T (1999) Transcriptional activation by NF-κB requires multiple coactivators. Mol. Cell. Biol. 19: 6367-6378.
Stenoien DL, Mancini MG, Patel K, Allegretto EA, Smith CL & Mancini MA (2000) Subnuclear trafficking of estrogen receptor α and steroid receptor coactivator-1. Mol. Endocrinol. 15: 518-534.
Stewart S & Crabtree GR (2000) Transcription. Regulation of the regulators. Nature 408: 46-47.
Swantek JL, Cobb MH & Geppert TD (1997) Jun N-terminal kinase/stress-activated protein kinase (JNK/SAPK) is required for lipopolysaccharide stimulation of tumor necrosis factor α (TNF-α) translation: glucocorticoids inhibit TNF-α translation by blocking JNK/SAPK. Mol. Cell. Biol. 17: 6274-6282.
Tao Y, Williams-Skipp C & Scheinman RI (2001) Mapping of glucocorticoid receptor DNA binding domain surfaces contributing to transrepression of NF-κB and induction of apoptosis. J. Biol. Chem. 276: 2329-2332.
Thiele K, Bierhaus A, Autschbach F, Hofmann M, Stremmel W, Thiele H, Ziegler R & Nawroth PP (1999) Cell specific effects of glucocorticoid treatment on the NF-κBp65/IκBα system in patients with Crohn's disease. Gut 45: 693-704.
Tobler A, Meier R, Seitz M, Dewald B, Baggiolini M & Fey MF (1992) Glucocorticoids downregulate gene expression of GM-CSF, NAP-1/IL-8, and IL-6, but not of M-CSF in human fibroblasts. Blood 79: 45-51.
Unlap MT & Jope RS (1997) Dexamethasone attenuates NF-κB DNA binding activity without inducing IκB levels in rat brain in vivo. Brain Res. Mol. Brain Res. 45: 83-89.
Urnov FD & Wolffe AP (2001) A necessary good: nuclear hormone receptors and their chromatin template. Mol. Endocrinol. 15: 1-16.
Vanden Berghe W, De Bosscher K, Boone E, Plaisance S & Haegeman G (1999a) The nuclear factor-κB engages CBP/p300 and histone acetyltransferase activity for transcriptional activation of the interleukin-6 gene promoter. J. Biol. Chem. 274: 32091-32098.
Vanden Berghe W, Francesconi E, De Bosscher K, Resche-Rigon M & Haegeman G (1999b) Dissociated glucocorticoids with anti-inflammatory potential repress interleukin-6 gene expression by a nuclear factor-κB-dependent mechanism. Mol. Pharmacol. 56: 797-806.
Vanden Berghe W, De Bosscher K, Vermeulen L, De Wilde G & Haegeman G (2002) Induction and repression of NF-κB-driven inflammatory genes, in: Recent Advances in Glucocorticoid Receptor Action (Cato A, Schaecke H & Asadullah K, eds), Ernst Schering Research Foundation Workshop, Vol. 40, Springer, Heidelberg, p. 233-278.
Wagner BL, Norris JD, Knotts TA, Weigel NL & McDonnell DP (1998) The nuclear corepressors NCoR and SMRT are key regulators of both ligand- and 8-bromo-cyclic AMP-dependent transcriptional activity of the human progesterone receptor. Mol. Cell. Biol. 18: 1369-1378.
Wang H, Huang ZQ, Xia L, Feng Q, Erdjument-Bromage H, Strahl BD, Briggs SD, Allis CD, Wong J, Tempst P & Zhang Y (2001) Methylation of histone H4 at arginine 3 facilitating transcriptional activation by nuclear hormone receptors. Science 293: 853-857.
Wissink S, van Heerde EC, Schmitz ML, Kalkhoven E, van der Burg B, Baeuerle PA & van der Saag PT (1997) Distinct domains of the RelA NF-κB subunit are required for negative cross-talk and direct interaction with the glucocorticoid receptor. J. Biol. Chem. 272: 22278-22284.
Wissink S, van Heerde EC, van der Burg B & van der Saag PT (1998) A dual mechanism mediates repression of NF-κB activity by glucocorticoids. Mol. Endocrinol. 12: 355-363.
Wolffe A & Pruss D (1996) Targeting chromatin disruption: transcription regulators that acetylate histones. Cell 84: 817-819.
Wolffe AP (1997) Transcriptional control. Sinful repression. Nature 387: 16-17.
Wolffe AP, Wong J & Pruss D (1997) Activators and repressors: making use of chromatin to regulate transcription. Genes Cells 2: 291-302.
Xu J, Qiu Y, DeMayo FJ, Tsai S, Tsai M & O'Malley BW (1998) Partial hormone resistance in mice with disruption of the steroid receptor coactivator-1 (SRC-1) gene. Science 279: 1922-1925.
Xu L, Glass CK & Rosenfeld MG (1999) Coactivator and corepressor complexes in nuclear receptor

function. Curr. Opin. Genet. Dev. 9: 140-147.

Young JD, Lawrence AJ, MacLean AG, Leung BP, McInnes IB, Canas B, Pappin DJC & Stevenson RD (1999) Thymosin β sulfoxide is an anti-inflammatory agent generated by monocytes in the presence of glucocorticoids. Nat. Med. 5: 1424-1427.

Zhang X, Jeyakumar M, Petukhov S & Bagchi MK (1998) A nuclear receptor corepressor modulates transcriptional activity of antagonist-occupied steroid hormone receptor. Mol. Endocrinol. 12: 513-524.

Zhong H, Voll RE & Ghosh S (1998) Phosphorylation of NF-κB p65 by PKA stimulates transcriptional activity by promoting a novel bivalent interaction with the coactivator CBP/p300. Mol. Cell 1: 661-671.

8. AFFILIATION

Ghent University and Flanders Interuniversity Institute for Biotechnology, Department for Molecular Biomedical Research, Unit of Eukaryotic Expression and Signal Transduction, K.L. Ledeganckstraat 35, 9000 Gent, Belgium.

X. LI & G.R. STARK

GENETIC ANALYSIS OF NF-κB-DEPENDENT SIGNALING PATHWAYS IN MAMMALIAN CELLS

1. GENERATION OF THE FORWARD GENETIC SYSTEM FOR ANALYZING NF-κB-DEPENDENT SIGNALING PATHWAYS

1.1. Forward genetics

Reverse genetics is used to study the function of a known gene product by knockout or replacement. In contrast, forward genetics is used to identify an unknown gene, based on its function. In using forward genetics, one can knockout or gain gene expression by random mutagenesis or by introducing expression libraries, followed by selection based on function (Stark & Gudkov, 1999). As a result, a collection of mutant cell lines and cDNA reagents will be generated, which can be used to define detailed mechanisms for specific components in a particular signaling pathway. The application of forward genetics to mammalian cells has led to significant advances in understanding signaling in response to interferons (Darnell et al., 1994) and has led to an appreciation of the role of Janus kinases (JAKs) in many signaling pathways (Velazquez et al., 1992; Stark et al., 1998). The forward genetic approach has also been successfully applied to study the TGFβ receptor (Hocevar & Howe, 1996) and T-cell receptor-mediated signaling (Goldsmith & Weiss, 1987; Serafini et al., 1995; Williams et al., 1998). While biochemical studies and reverse genetics have tremendously advanced our understanding of the NF-κB-dependent signaling pathways, the ways in which a forward genetic approach has also been used to study these pathways will be discussed in this chapter.

1.2. Setting up a cell line with NF-κB-dependent dual-drug selection

The first step in generating a forward genetic system to study signal transduction in a mammalian cell line is to set up a selection or screen both for and against the signal-induced expression of marker genes controlled by the upstream element of a gene responding to induction. The markers used originally to select, viz. guanine phosphoribosyltransferase (gpt), or to sort, viz. the cell-surface proteins cd2 or cd4, have been extremely useful, but have significant limitations. Lethal selection with 6-thioguanine in cells carrying a signal-regulated gpt gene is

R. Beyaert (ed.), Nuclear Factor κB. Regulation and Role in Disease, 241-264.

convenient, but works well only in a situation in which the promoter gives very low basal expression, for example the interferon-responsive 6-16 promoter (Pellegrini et al., 1989). Use of the fluorescence-activated cell sorter to separate cells physically on the basis of surface expression of cd2 or cd4 is less attractive than lethal selection because of the time and expense involved, and because each sort usually gives an enrichment of only ~100-fold. An even more serious concern is that mutants in IL-1 signaling that we isolated in this way were metastable, reverting to a wild-type phenotype at a low rate, making complementation very difficult. These problems have recently been improved by lethal selection based on the expression of herpesvirus thymidine kinase (TK) (Li et al., 1999c). With this enzyme, it is possible to use concentrations of gancyclovir (GCV) that allow survival of unresponsive mutant cells with a significant basal level of TK expression, but that still kill wild-type cells with a highly induced level. In other words, one can now work with most promoters that express low basal level of the genes that they control.

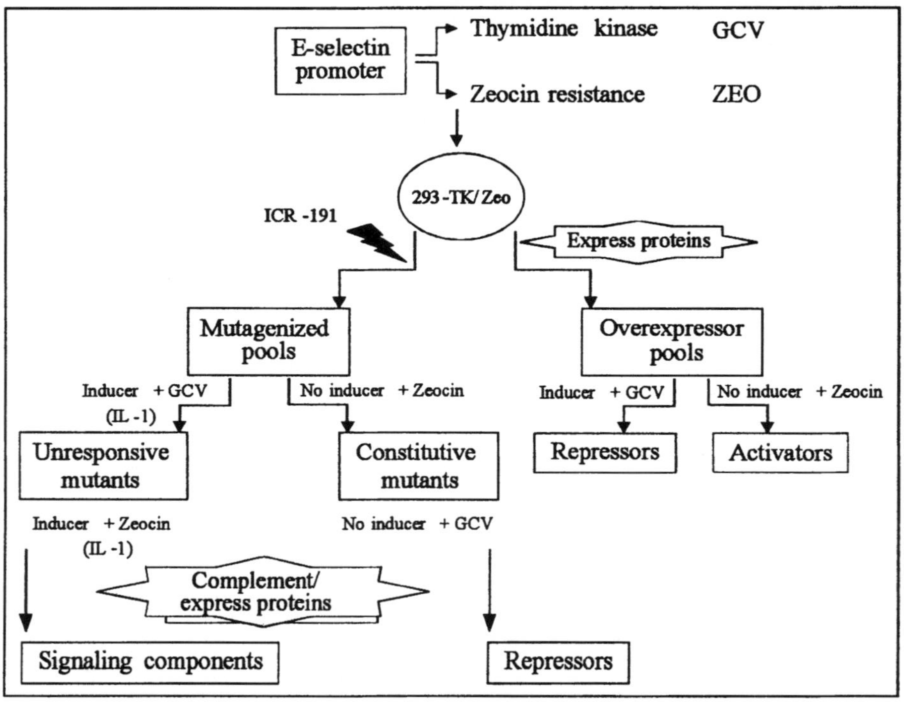

Fig. 1. Setting up a cell line with NF-κB-dependent dual-drug selection.

To establish a cell line with NF-κB-dependent drug selection, an NF-κB-containing upstream region (-730 +52) of the human E-selectin gene, also called ELAM-1 (Schindler & Baichwal, 1994; Whitley et al., 1994), was cloned upstream of TK and a zeocin resistance gene (Zeo). Both the E-selectin/TK and E-selectin/Zeo constructs were cotransfected into 293 cells, in which the E-selectin promoter has a low basal activity and is highly induced in response to signals that activate NF-κB (e.g. IL-1). The transfected cells were then selected for clones that have NF-κB-dependent TK and Zeo gene expression, using IL-1 to activate NF-κB. A clone was obtained that survives in GCV, dies completely in GCV plus IL-1, dies in Zeo and survives in Zeo plus IL-1. This parental clone (293-TK/Zeo) was used either to obtain mutant cell lines or to perform expression cloning of signaling components in NF-κB-dependent signaling pathways (Fig. 1). To obtain mutants, 293-TK/Zeo cells were mutagenized with the frameshift mutagen ICR-191 to generate mutagenized pools, which were screened for unresponsive mutants to a particular stimulus by selection in inducer (e.g. IL-1) plus GCV, and for constitutive mutants (in which NF-κB is constitutively activated) by selection in zeocin. The missing components in the isolated mutants were searched for by complementation with individual cDNAs of known components or by using retroviral cDNA libraries, followed by drug selections: inducer plus zeocin for complementing the unresponsive mutants and GCV for the constitutive mutants. The 293-TK/Zeo cells were also used to isolate cDNAs encoding proteins that constitutively activate or repress NF-κB activation. Retroviral cDNA libraries were directly introduced into 293-TK/Zeo cells to establish the expression pools, which were then selected in zeocin for cDNAs encoding NF-κB activators and in inducer plus GCV for NF-κB repressors.

2. IL-1-UNRESPONSIVE MUTANTS

2.1. The IL-1-mediated signaling pathway

IL-1 plays a central role in mediating a variety of inflammatory responses, exerting its biological activities mainly by activating the transcription of effector genes in target cells (Adachi et al., 1998). The transcription factors activated by IL-1 include NF-κB, AP-1 and ATF (O'Neill, 1995; Barnes & Karin, 1997; O'Neill, 1997) (Fig. 2). IL-1-mediated signaling begins when IL-1 binds to the receptor complex, composed of the type 1 receptor (IL-1R) and the receptor-accessory protein (IL-1RAcp) (Greenfeder et al., 1995; Huang et al., 1997; Korherr et al., 1997). The cytosolic proteins MyD88 (Lord et al., 1990; Wesche et al., 1997; Adachi et al., 1998) and Tollip (Burns et al., 2000) are then recruited to the receptor complex, where they function as adaptors, recruiting IL-1R-associated kinase (IRAK) in turn. IRAK, a serine-threonine kinase, is phosphorylated at the receptor complex and then presumably dissociates from the receptor complex to interact with TRAF6 (Cao et al., 1996a-b; Li et al., 1999c; Li et al., 2001). Phosphorylated IRAK is eventually ubiquitinated and degraded (Yamin & Miller, 1997). Recently, a member of the MAP kinase kinase kinase family (TAK1) and two proteins that bind to it (TAB1 and TAB2) have also been implicated in IL-1 signaling (Ninomiya-Tsuji et al., 1999; Takaesu et al., 2000). TAB2 is membrane-bound in untreated cells and translocates to the cytosol upon stimulation with IL-1, where it functions as an adapter, linking TRAF6 to TAK1 and TAB1, and thereby activating TAK1 (Takaesu et al., 2000). Activated TAK1 is suggested to activate the NF-κB-inducing kinase (NIK) and IκB kinase (IKK) cascade, leading to NF-κB activation (Malinin et al., 1997; Wang et al., 2001). However, the generation of NIK-deficient mice reveals that NIK is required for lymphotoxin-β signaling, but is dispensable for IL-1- and TNFα-dependent signaling (Matsushima et al., 2001; Yin et al., 2001). Thus a NIK-like molecule might be involved in the IL-1-signaling pathway.

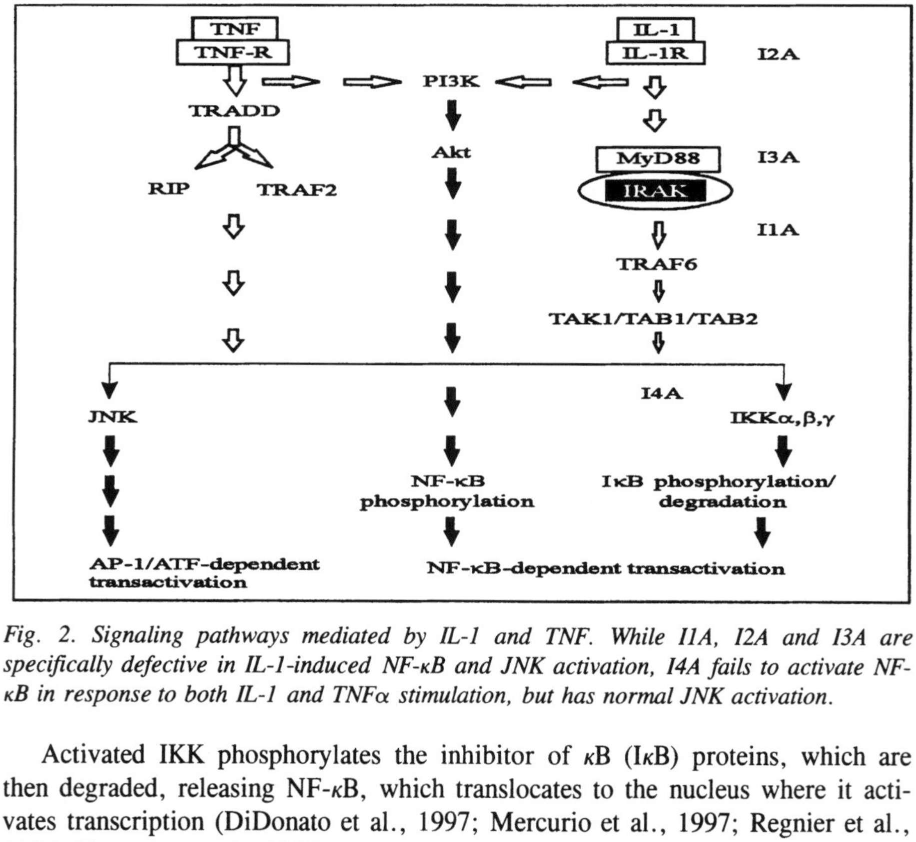

Fig. 2. Signaling pathways mediated by IL-1 and TNF. While I1A, I2A and I3A are specifically defective in IL-1-induced NF-κB and JNK activation, I4A fails to activate NF-κB in response to both IL-1 and TNFα stimulation, but has normal JNK activation.

Activated IKK phosphorylates the inhibitor of κB (IκB) proteins, which are then degraded, releasing NF-κB, which translocates to the nucleus where it activates transcription (DiDonato et al., 1997; Mercurio et al., 1997; Regnier et al., 1997; Woronicz et al., 1997; Zandi et al., 1997; Yamaoka et al., 1998; Li et al., 1999a-b; Li et al., 2000a; Rudolph et al., 2000; Hu et al., 2001). Through a parallel signaling pathway, activated IKK can also lead to phosphorylation of NF-κB, which is necessary for its transcriptional activity (Zhong et al., 1997; Wang & Baldwin, 1998; Zhong et al., 1998; Sizemore et al., 1999; Madrid et al., 2000; Wang et al., 2000; Madrid et al., 2001; Silverman & Maniatis, 2001). Activated TAK1 has also been implicated in the IL-1-induced activation of JNK and other MAP kinases (Ninomiya-Tsuji et al., 1999), leading to phosphorylation and activation of ATF and AP-1, thereby also activating transcription.

2.2. Mutagenesis and selection

The forward genetic approach was chosen to study IL-1-signaling pathways, namely by random mutagenesis to generate IL-1-unresponsive cell lines lacking specific components of the pathways. To generate recessive mutants lacking a single protein essential for a particular signaling pathway, one must usually inactivate both alleles of the relevant gene. The frequency of this highly improbable event can be confined to a reasonable range (10^{-6} to 10^{-7}) by repeated applications of a potent chemical mutagen such as ICR-191. Fortunately, the mutants so obtained do not often have defects in other genes in the same pathways, since they can almost always be complemented by a single cDNA (Stark et al., 1998).

To obtain IL-1-unresponsive mutants, the 293-TK/Zeo cells described above were expanded to pools of 10^7 cells each. The cells were then subjected to five rounds of mutagenesis (50-70% lethality each round) with ICR-191, an intercalating agent that induces frameshift mutations. The cells were then selected in GCV plus IL-1. Since clones from the same mutagenized pool may well be siblings, only one mutant clone from each pool was used for further study. Five mutant clones defective in IL-1-induced NF-κB activation were selected (Li et al., 1999c). Whereas IL-1 failed to activate NF-κB substantially in all five mutant clones (I1A, I2A, I2B, I3A and I4A), TNF activated NF-κB in all of these mutant clones except I4A (the mutant clones were named according to their complementation groups as described below). To confirm these results, an E-selectin promoter-driven luciferase plasmid was transfected into each mutant cell line. Both IL-1 and TNFα upregulated this promoter in wild-type cells (293-TK/Zeo). However, IL-1-induced promoter activity was absent in all five mutant clones, while their TNFα response was intact, except for I4A (Fig. 3A).

2.3. Dominance and complementation

Cells from each of the initial five mutant clones were made puromycin-resistant with pBabePURO and fused with hygromycin-resistant 293-TK/Zeo cells (transfected with pSV2hyg). After selection with both drugs, IL-1-induced activation of NF-κB was restored in all heterokaryons, indicating that the mutations are all recessive. To assign complementation groups, a puromycin-resistant population from each mutant clone was fused with a hygromycin-resistant population from each mutant, and IL-1-induced NF-κB activation was examined in the hybrid populations. Clones I2A and I2B were placed in the same complementation group, as IL-1-induced NF-κB activation was not restored in the heterokaryons (data not shown). I1A, I3A and I4A are in different complementation groups, since IL-1-induced NF-κB is restored in their respective hybrids (data not shown; the nomenclature for mutant clone I1A is I = IL-1-unresponsive, 1 = complementation group 1, A = independent isolate A). The isolation of four different complementation groups strongly suggests that we obtained mutant cell lines defective in different signaling components.

2.4. *Characterization of the mutants*

All four mutant clones (I1A, I2A, I3A and I4A) were assayed with antibodies against the known signaling components IL-1R1, IL-1RAcP, MyD88, Tollip, IRAK, IRAK2, IRAK4, TRAF6, NIK, TAK1, TAB1, TAB2, IKKα, IKKβ and IKKγ (Fig. 2). I1A cells lack IRAK (Li et al., 1999c; Fig. 3B), a serine-threonine kinase that is recruited to the IL-1R complex after IL-1 stimulation (Fig. 2). This result was confirmed by Northern analysis, showing that IRAK mRNA is also absent in I1A cells (Fig. 3C). IRAK-deficient I1A cells have been extremely useful in elucidating the structure/function of IRAK in IL-1 signaling, which will be discussed in detail below (Li et al., 2001; Qian et al., 2001).

No known component was missing in the other three mutant clones (I2A, I3A and I4A). IL-1-induced JNK activation was completely absent in I1A, I2A and I3A, whereas JNK activation was normal in I4A cells (Li et al., 1999c). Thus, while IRAK and the I2A- and I3A-defective components are required for activation of both NF-κB and JNK, I4A is deficient in a component only needed for NF-κB activation. In response to IL-1, IRAK, TAK1, TAB1 and TAB2 are phosphorylated in wild-type cells. Interestingly, none of these components were phosphorylated after IL-1 stimulation in I2A and I3A, whereas the signal-induced IRAK, TAK1, TAB1 and TAB2 modification was normal in I4A cells (unpublished data). These results indicate that the defects in the I2A and I3A mutant clones are upstream of IRAK, whereas I4A is defective in a component downstream of IRAK and TAK1. I2A and I3A displayed differential phenotypes when they were assayed for their IL-18 response compared to IL-1. While transfection of the IL-18R subunits rendered wild-type 293-TK/Zeo cells responsive to IL-18, both IL-1 and IL-18 failed to induce luciferase activity in the mutants I1A, I3A and I4A (Wald et al., 2001). In contrast, luciferase was greatly induced upon IL-18 treatment in I2A cells. These results show that IRAK as well as the I3A- and I4A-defective components are required for both IL-1 and IL-18 signaling, whereas the I2A-defective component is necessary for IL-1-dependent, but not for IL-18-dependent signaling. I2A cells may be defective in a component of the IL-1R that has not yet been identified or in a protein that interacts directly with the receptor. Western analysis pointed out that the two known IL-1R subunits (IL-1R1 and IL-1RAcp) and the adaptor protein MyD88 are present in I2A cells. As ICR-191, the mutagen used to generate the mutant cell lines (Li et al., 1999c), generates frameshifts instead of point mutations, it is unlikely that there is simply a point mutation in IL-1R1 or IL-1RAcp.

3. STUDY OF IRAK IN IRAK-DEFICIENT I1A CELLS

3.1. *Loss of IRAK accounts for the phenotype of I1A cells*

As presented above, E-selectin promoter activity induced by IL-1, as well as NF-κB and JNK activation were abolished in I1A cells deficient in IRAK (Fig. 3A).

To determine whether IRAK is able to complement the defect in I1A cells, TK-driven IRAK cDNA was cotransfected transiently with E-selectin/Luc into these cells. E-selectin-1 promoter activity induced by IL-1 was restored in I1A cells transfected with IRAK (Fig. 3D). NF-κB and JNK activation induced by IL-1 was also restored in I1A cells stably transfected with IRAK. These results indicate that IRAK is required for NF-κB and JNK activation induced by IL-1, and that the loss of IRAK protein accounts for the phenotype of I1A cells (Li et al., 1999c).

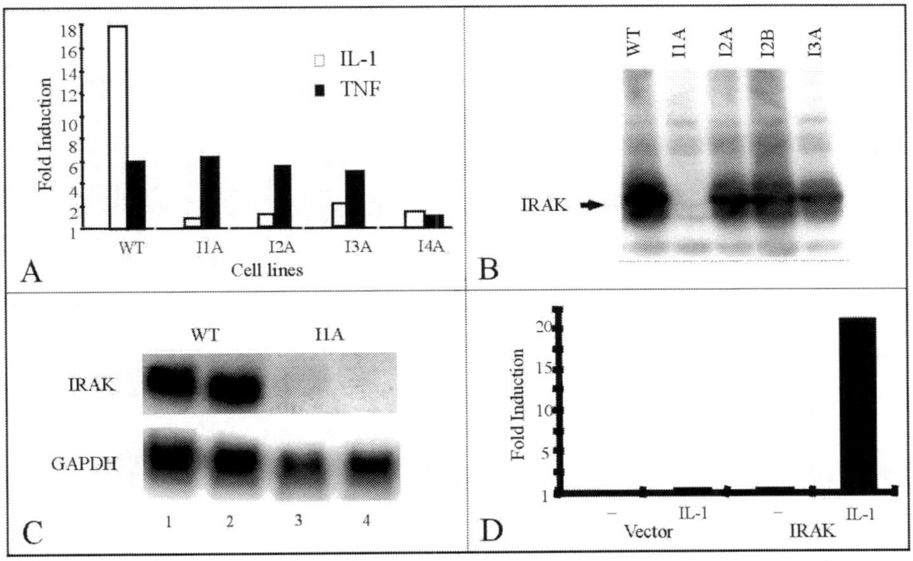

Fig. 3. IL-1-unresponsive mutants. (A) E-selectin luciferase reporter assay for IL-1-unresponsive mutants. E-selectin/Luc was transiently transfected into 293-TK/Zeo cells (WT) and IL-1-unresponsive mutants. After 36 h, the cells were either left untreated or stimulated for 4 h more with IL-1 or TNFα. Luciferase activities were normalized to β-galactosidase. (B) Clone I1A lacks IRAK protein. Extracts were made from 293-TK/Zeo (WT) and IL-1-unresponsive mutants, followed by Western analysis with anti-IRAK. (C) Clone I1A lacks IRAK mRNA. Total RNA made from 293-TK/Zeo cells (WT) and mutant I1A was analyzed by the Northern procedure with IRAK cDNA as probe. (D) IRAK complements I1A cells. I1A cells were cotransfected with E-selectin/Luc and TK promoter-driven IRAK expression vector. After 36 h, the cells were either left untreated or stimulated for 4 h more with IL-1. Luciferase activities were normalized to β-galactosidase.

3.2. IRAK functions between MyD88 and TRAF6

Ectopic expression of MyD88 strongly induces NF-κB activation, even in the absence of IL-1 (Muzio et al., 1997; Wesche et al., 1997). However, the constitutive activation of the E-selectin-1 promoter in response to overexpression of MyD88 was not observed in IRAK-null I1A cells. In I1A cells which had been stably transfected with IRAK, the effect of MyD88 was restored. Thus MyD88 cannot interact with downstream components in the pathway in the absence of IRAK. Overexpression of TRAF6 can also constitutively induce NF-κB activation (Cao et al., 1996b). However, when TRAF6 was cotransfected with E-selectin/Luc into either wild-type 293-TK/Zeo cells or I1A cells, the promoter activity in the absence of IL-1 was increased, indicating that TRAF6 interacts with components of the signaling pathway downstream of IRAK. These results reveal that IRAK functions between MyD88 and TRAF6 (Li et al., 1999c).

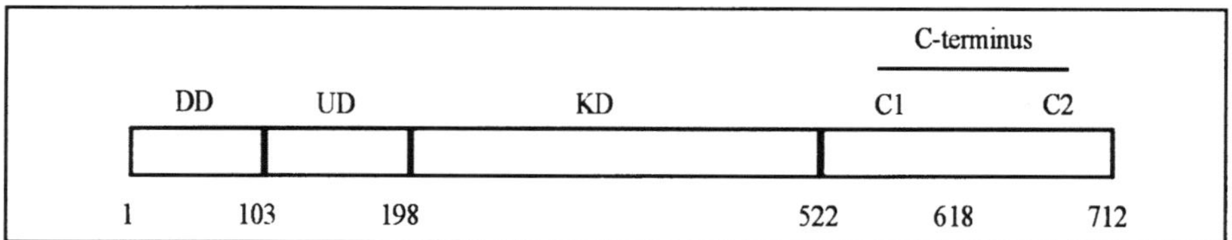

	ELAM-1/Luc	NF-κB	Phosphorylation Degradation	JNK	TAK1/TAB1-2 Phosphorylation
dDD	-	-	-	-	-
dUD	-	-	-	+	-
dKD	+	+	−	+	+
dC1	+	+	+	+	+
dC2	+	+	+	+	+
dC1+C2	-	-	-	-	-

Fig. 4. Structure/function analysis of IRAK in I1A cells transfected with IRAK constructs. IRAK deletion constructs were transfected into IRAK-null I1A cells to analyze their effect on various IL-1-induced events. +, wild-type effect; -, abolished effect; DD, death domain; UD, N-proximal undetermined region; KD, kinase domain; C1/C2, C-terminal region.

3.3. Structure/function relationship of IRAK in IL-1-induced NF-κB and JNK activation

IRAK-null I1A cells have been used to study the structure/function relationship of IRAK in IL-1-induced NF-κB and JNK activation (Li et al., 2001). IRAK is a multidomain protein (DiDonato et al., 1997; Hu et al., 2001) containing an N-terminal death domain (DD, residues 1-103), followed by a domain of unknown function (UD, residues 104-198), a kinase domain (KD, residues 199-522) and a two-part C-terminal domain, also of unknown function (residues 523-618 for C1 and residues 619-712 for C2) (Fig. 4). To address the role of the kinase activity of IRAK, its adenosine triphosphate-binding site was mutated (K239A) or the entire KD was deleted (dKD). Both IRAK-K239A and IRAK-dKD complemented all defects in IL-1-dependent signaling in mutant I1A cells, indicating that the kinase activity of IRAK is not necessary for these functions (Goldsmith & Weiss, 1987; Greenfeder et al., 1995) (Fig. 4). Analysis of other IRAK deletion mutants showed that both DD and C1/C2 of IRAK are required for NF-κB and JNK activation, but that UD was necessary only for NF-κB (not JNK) activation (Greenfeder et al., 1995). Furthermore, IL-1-induced phosphorylation of IRAK is required to activate NF-κB, but not JNK (Greenfeder et al., 1995). Taken together, these results suggest strongly that the IL-1-induced signaling pathways leading to NF-κB and JNK activation diverge at or upstream of IRAK.

3.4. IRAK functions as an organizer and transporter in IL-1-dependent signaling

Whereas IRAK is required for IL-1-induced activation of NF-κB and JNK, the kinase activity of IRAK is dispensable for its function, suggesting that IRAK mediates signaling by protein-protein interactions. IRAK-null I1A cells have also been utilized to investigate the function of IRAK in the sequential formation and activation of IL-1-induced signaling complexes (Qian et al., 2001; Y. Jiang & X. Li, unpublished data).

IRAK is recruited to the IL-1R and hyperphosphorylated after IL-1 stimulation (Fig. 5). While TRAF6 was also recruited to the receptor in wild-type cells, TRAF6 failed to interact with the receptor in IRAK-null I1A cells. Transfection of IRAK into I1A cells restored the recruitment of TRAF6 to IL-1R. These data indicate that IRAK is required for the recruitment of TRAF6 to the receptor. IL-1R/IRAK/TRAF6 is designated as complex I.

Assays in IRAK-deficient 1IA cells show that interaction of TRAF6 with TAK1 and TAB2 on the membrane is also IRAK-dependent. IRAK probably brings TRAF6 from the receptor complex I to form the membrane-associated complex II with TAK1, TAB1 and TAB2, which are preassociated on the membrane before stimulation. The formation of complex II leads to phosphorylation of TAK1 and TAB2 on the membrane. IRAK, but not its kinase activity, was shown to be required for IL-1-induced phosphorylation of TAK1, TAB1 and TAB2. Complex III, comprising TRAF6/TAK1/TAB1/TAB2, is then dissociated from IRAK and translocated to the cytosol. The formation of complex III and its interaction with additional cytosolic factors lead to activation of TAK1, which in turn activates IKK by an unknown mechanism, resulting in NF-κB activation. Using IRAK-deficient I1A cells, it has been shown that IRAK, but not its kinase activity, is required for translocation of complex III from the membrane to the cytosol and activation of TAK1 kinase activity in the cytosol. Phosphorylated IRAK remains on the membrane and eventually is ubiquitinated and degraded. The data show that IRAK plays a critical role in mediating the association and dissociation of IL-1-induced signaling complexes, functioning as an organizer and transporter protein in the IL-1-dependent pathway.

3.5. Activation of IRAK and TAK1 must involve other kinases

After IL-1 stimulation, the majority of IRAK translocates to the IL-1R complex, where it becomes multiphosphorylated, followed by ubiquitination and degradation by proteosomes (Yamin & Miller, 1997). K239A and dKD were still phosphorylated in I1A cells upon IL-1 stimulation, showing that IRAK must become phosphorylated by another kinase. As IRAK-K239A complements I1A cells just as well as wild-type IRAK, it is likely that the phosphorylation of IRAK by another kinase plays a more important role in signaling than autophosphorylation. While TAK1 and TAB2 are phosphorylated on the membrane in response to IL-1 stimulation, the kinase activity of TAK1 is only activated in the cytosol. These results indicate that phosphorylation of TAK1 on the membrane is probably not due to its autophosphorylation.

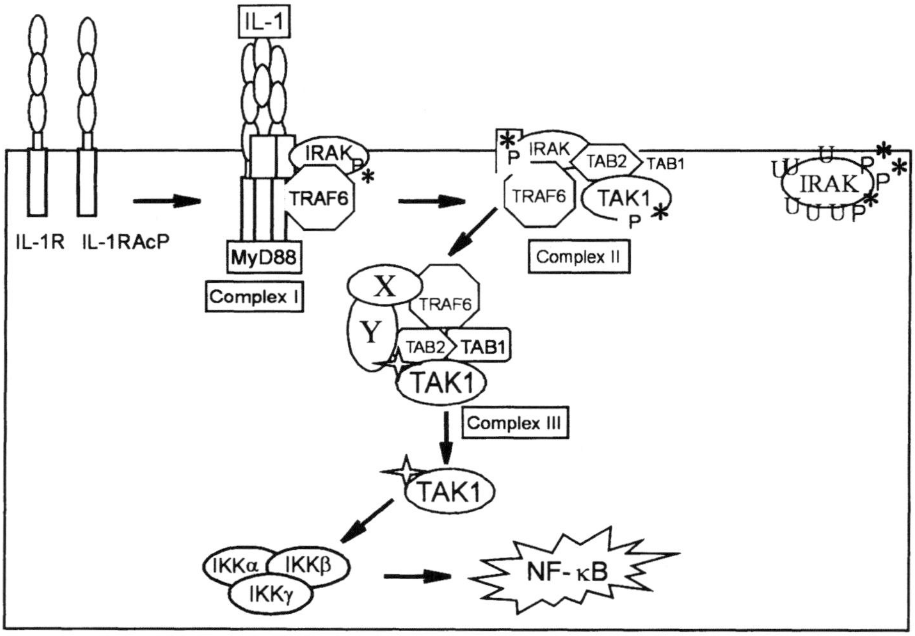

Fig. 5. Revised model for IL-1-dependent signaling (P, phosphorylation; U, ubiquitination; X and Y, unknown components).*

The fact that TAK1 and TAB2 are also phosphorylated in IRAK-deficient I1A cells restored with a kinase-dead IRAK mutant indicates that IRAK is not the kinase that phosphorylates TAK1 either. IRAK4, another member of the IRAK family, has recently been shown to be required for IL-1-mediated signaling (Li et al., 2002; Suzuki et al., 2002). However, IRAK4 interacts only with complex I, but not with complex II after IL-1 stimulation (Y. Jiang & X. Li, unpublished results); this indicates that, while IRAK4 might be an IRAK kinase, it cannot be the kinase that phosphorylates the components in complex II. Taken together, these results suggest that there must be another membrane-bound kinase that is responsible for IL-1-induced phosphorylation of TAK1 and TAB2 on the membrane.

3.6. *The function of IRAK in IL-1/Toll receptor-mediated signaling pathways*

The IL-1/Toll receptor superfamily includes the two IL-1R subunits, the two IL-18R subunits, the Toll-like receptors (TLRs), so far comprising ten members (Poltorak et al., 1998; Takeuchi et al., 1999; Underhill et al., 1999a-b; Sebastiani et al., 2000; Takeuchi et al., 2000; Alexopoulou et al., 2001), and the single Ig IL-1R-related molecule (SIGIRR) (Thomassen et al., 1999). The defining feature of a member of the superfamily is a Toll-like domain on the cytoplasmic side of the receptor. Because of the similarity in their cytoplasmic domains, some of the receptors in this superfamily have been shown to use signaling components that function with the IL-1R, including IRAK. Using IRAK-deficient I1A cells, it was clearly shown that IRAK is required for IL-18-mediated and TLR2-mediated NF-κB and JNK activation. In IRAK-deficient macrophages, LPS-stimulated TNFα production is impaired. IRAK has also been implicated in TLR9-mediated NF-κB activation in response to CpG DNA stimulation. However, recent studies with IRAK-deficient I1A cells show that IRAK is not involved in TLR3-mediated NF-κB and JNK activation in response to dsRNA stimulation (Y. Jiang & X. Li, unpublished data), indicating that the receptor-proximal signaling events are very different for TLR3 as compared to other members in this superfamily.

4. MUTANTS THAT DO NOT SIGNAL FROM ONCOGENIC Ki-Ras TO NF-κB

Several reports have indicated that NF-κB can be activated by oncoproteins, including oncogenic forms of Ras (Finco et al., 1997; Rayet & Gelinas, 1999; Jo et al., 2000). NF-κB activity is required for Ras to induce cellular transformation, and inhibition of NF-κB blocks oncogenic Ras-dependent cellular transformation (Finco et al., 1997; Jo et al., 2000). The exact mechanism by which oncogenic Ras mediates activation of NF-κB, still remains uncertain. Oncogenic Ras activates both IκB degradation and also a second independent pathway necessary for full activation of NF-κB, the activation of NF-κB itself by serine phosphorylation of the p65 NF-κB subunit (Finco et al., 1997). Induction of the transactivation potential of the p65 NF-κB subunit, by both oncogenic Ras and IL-1β, is mediated through a signal transduction pathway that involves phosphoinositol-3 kinase (PI3K), the serine/threonine kinase Akt/PKB and the IKKs, leading to phosphorylation and transactivation of p65 NF-κB (Sizemore et al., 1999; Madrid et al., 2000; Madrid et al., 2001; Sizemore et al., 2002).

A forward genetic approach was also used to identify currently unknown positive regulators that function to regulate NF-κB activity in response to oncogenic Ki-Ras (N. Sizemore & G.R. Stark, unpublished results). To isolate mutant cell lines defective in Ras-mediated NF-κB activation, mutant Ki-Ras was transfected into 293-TK/Zeo cells. The resulting cell line RasC6 has constitutively activated NF-κB, survives in Zeo and dies in GCV. After mutagenesis with ICR-

191, several pools of cells were selected in GCV to obtain mutants in which Ras-dependent activation of NF-κB is lost. The mutants have different phenotypes and belong to four different complementation groups. The phenotype is presumably due to loss of positive components necessary for Ras to activate NF-κB.

5. CONSTITUTIVE MUTANTS

5.1. Deregulation of NF-κB activation

Given that NF-κB regulates the expression of many growth factors, cytokines and other factors that mediate stress responses, cell proliferation, cell death and cell cycle progression, its activity and level must be delicately balanced in any given cell type to ensure normal cell proliferation. It is not surprising, therefore, that deregulation of the NF-κB pathway has been associated with various pathological conditions, including virus infection, radiation damage, septic shock, athero-sclerosis, acute inflammation and cancer (Siebenlist et al., 1994; Baeuerle & Baltimore, 1996). Loss of normal regulation of NF-κB is becoming more apparent as a major pathway that contributes to the deregulated growth, resistance to apoptosis and propensity to metastasize observed in many cancers (Rayet & Gelinas, 1999; Baldwin, 2001). Inappropriate constitutive activation of pathways that stimulate NF-κB are potential targets for cancer therapy (Biswas et al., 2001). Thus the discovery of novel negative regulators of NF-κB activation, as well as full understanding of how these negative regulators function, is important.

Table 1. Constitutive mutants.

Mutagenized pool	*Constitutive mutants (No.)*	*Mutants selected for further analysis*
C6P1	12	C6P1Z1 and Z12
C6P2	2	C6P2Z1 and Z2
C8P1	5	Z3 and Z23
C8P4	5	Z5
C18P3	1	Z13

C6, C8 and C18 are three different initial clones. P1-P4 are independently mutagenized pools. Z1–Z23 are zeocin-resistant clones.

5.2. Isolation of constitutive mutants

Using the selection strategy to isolate mutant cell lines that have constitutive NF-κB activity, it is most likely to isolate cells in which a negative regulator has been inactivated. To obtain such constitutive mutant cell lines, mutagenized pools of 293-TK/Zeo cells were subjected to selection in Zeo (50 μg/ml). Constitutive activation of NF-κB renders the cells resistant to Zeo and sensitive to GCV. Several Zeo-resistant clones were selected from each pool. A total of eight mutants from five pools was selected for further study (Table 1; S.S. Sathe &

G.R. Stark, unpublished data). Every mutant clone revealed constitutive DNA-binding activity of NF-κB (Fig. 6A). Constitutive activation of NF-κB results in constitutive activation of the luciferase reporter in mutants, but not in parental cells. The constitutive mutants were analyzed by Northern and Western procedures for expression of several known negative regulators, including PICOT (Witte et al., 2000), SODD (Jiang et al., 1999), PTEN (Li et al., 1997) as well as IκBα, IκBβ and IκBε (Brockman et al., 1995). All of these proteins were present, indicating that the constitutive activation of NF-κB in the mutants is due to mutations in other genes.

5.3. Dominance and complementation

Whereas mutants Z12 and Z23 show partial dominance, the rest of the mutants is recessive. The constitutive mutants fall into at least three complementation groups (S.S. Sathe & G.R. Stark, unpublished data). Pairwise complementation analysis between mutants P2Z1, Z2 and Z5 was done by fusing a puromycin-resistant population of one mutant with a hygromycin-resistant population of the other. The heterokaryons were tested for constitutive NF-κB activation. Constitutive activation of NF-κB is greatly reduced or abolished in the heterokaryons, indicating that mutants C6P2Z1, Z2 and Z5 are in different complementation groups. The rest of the mutants (P1Z1, Z3 and Z13) has not yet been assigned to a complementation group.

5.4. Biochemical analysis of constitutive mutants

Members of the IκB family of proteins are the best studied negative regulators of NF-κB. To check their status in the constitutive mutants, extracts of parental and mutant clones were immunoblotted and probed with IκBα,β antibodies. As NF-κB upregulates IκBα levels, the cells were treated with cycloheximide for 3 h before lysing them. As shown in Fig. 6B, there is constitutive degradation of IκBα in all the mutants, and IκBβ is constitutively degraded in all, except Z12 (S.S. Sathe & G.R. Stark, unpublished data). In order to test whether IκB kinase was activated in the constitutive mutants, it was immunoprecipitated; an in vitro kinase assay was performed using IκBα (1-54) as a substrate (S.S. Sathe & G.R. Stark, unpublished data). Among the seven constitutive mutants tested, IKK is activated only in Z12 cells (Fig. 6C). In the rest of the mutants, degradation of IκB is probably not due to phosphorylation of Ser32 and Ser36, suggesting that alternative mechanisms exist that lead to IκB degradation and subsequent NF-κB activation. In mutant Z12, NF-κB is constitutively active as a result of constitutive IKK activity. As many pathways leading to NF-κB activation also lead to JNK activation, the constitutive mutants were tested for JNK activity by Western analysis using anti-pJNK. JNK was activated in Z12 cells, but not in the other mutants tested. Mutant Z12 may either lack a negative regulator upstream of IKK and JNK or may constitutively secrete a ligand such as IL-1 (Arlt et al., 2002).

To test for the constitutive activation of other kinases that have been implicated in NF-κB-dependent signaling, Western transfers of the constitutive mutants were probed with several phospho-specific antibodies (S.S. Sathe & G.R. Stark, unpublished data). Akt is a downstream target for PI3K and activates NF-κB in response to stimuli such as TNF and IL-1. Immunoblots of the constitutive mutants were probed with anti-pAkt. Parental cells contain high levels of endogenous Akt. However, an additional, more slowly migrating band, presumably a superphosphorylated or acetylated form of Akt appearing in IL-1-treated cells, was seen in mutants Z2, Z12 and Z3, implying that the PI3K/Akt pathway is constitutively activated in these three mutants, placing the missing negative regulator upstream of Akt.

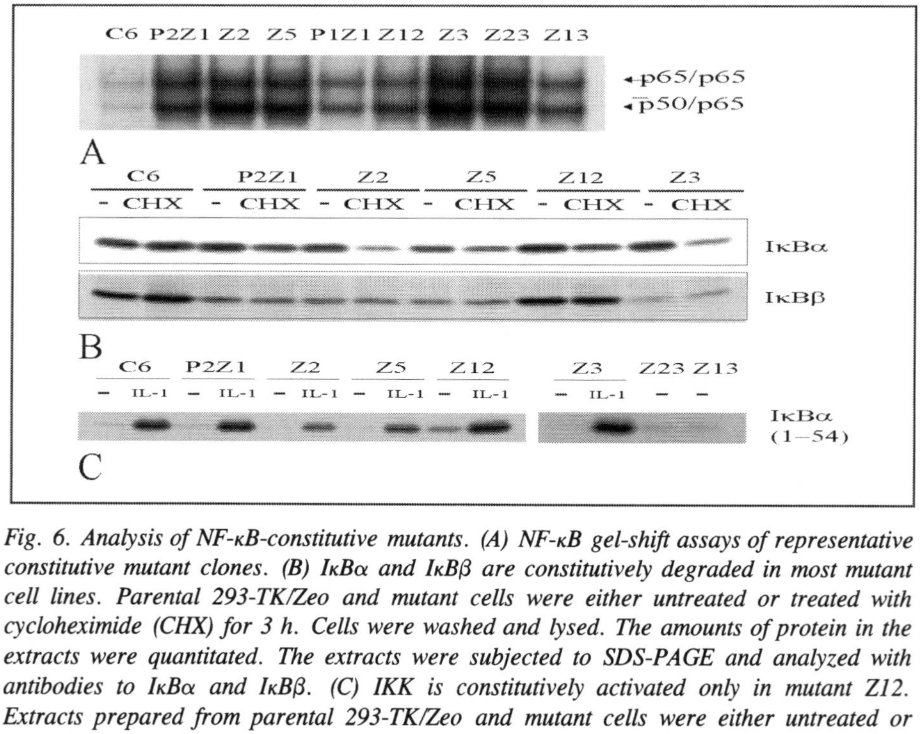

Fig. 6. Analysis of NF-κB-constitutive mutants. (A) NF-κB gel-shift assays of representative constitutive mutant clones. (B) IκBα and IκBβ are constitutively degraded in most mutant cell lines. Parental 293-TK/Zeo and mutant cells were either untreated or treated with cycloheximide (CHX) for 3 h. Cells were washed and lysed. The amounts of protein in the extracts were quantitated. The extracts were subjected to SDS-PAGE and analyzed with antibodies to IκBα and IκBβ. (C) IKK is constitutively activated only in mutant Z12. Extracts prepared from parental 293-TK/Zeo and mutant cells were either untreated or treated with IL-1 for 15 min, followed by immunoprecipitation with anti-IKKα overnight at 4°C and an in vitro kinase assay using GST-tagged IκBα (residues 1-54) as a substrate.

Mitogen-activated 90-kDa ribosomal S6 kinase ($p90^{rsk1}$), a serine-threonine kinase, is involved in transducing signals induced by stimuli that activate Ras-MAP kinase cascades. It is phosphorylated and subsequently activated by the extracellularly regulated kinases $p44^{erk1}$ and $p42^{erk2}$. $p90^{rsk1}$ has been reported to phosphorylate IκB on serine 32, both in vitro and in vivo; moreover, $p90^{rsk1}$ is required for TPA-induced phosphorylation and subsequent degradation of IκBα (Schouten et al., 1997). In order to determine whether these two kinases are activated in the constitutive mutants, cell extracts were separated by SDS-PAGE, after which the Western transfers were probed with anti-p90^{rsk1} and anti-pErk (p42/44). Constitutive activation of these two kinases was observed in mutants Z12, Z3 and Z13, but not in parental controls.

While complementation analysis has placed the mutants into at least three groups, biochemical analysis has assigned them to at least five different categories, depending on the kinases activated (Table 2). These observations indicate that multiple pathways and mechanisms lead to IκB degradation and NF-κB activation. The activation of different pathways in different constitutive mutants implies that different negative regulators are likely to be mutated in each of them.

Table 2. Constitutive activation of kinases in different constitutive mutant clones.

Cell type	*IKK*	*JNK*	*Akt*	*p90^{rskI}*	*ERK*
Z12	↑	↑	↑	↑	↑
Z2	—	—	↑	—	—
Z3	—	—	↑	↑	↑
Z13	—	—	—	↑	↑
P2Z1, P1Z1, Z5, Z23	—	—	—	—	—

Dashes mean that the kinase was not constitutively active.

6. COMPLEMENTATION

6.1. Complementation of mutants with retroviral expression libraries

Although the above biochemical analyses help to characterize the mutants, the ultimate goal is to complement them. If the mutants are not complemented with a known cDNA, expression cDNA libraries are not needed. The success of complementation with a library demands a high efficiency of introducing genes into the mutant cells, appropriate expression of the transfected gene and tight selection of the truly complemented cells against the background. Retroviral cDNA expression libraries have previously been successfully used to complement mutant cell lines (Yamaoka et al., 1998). To complement IL-1-unresponsive mutants and NF-κB constitutive mutants, a retroviral cDNA library from parental 293 cells was generated in the retroviral vector pBabeX, a derivative of the Moloney murine leukemia virus-based vector pBabe-puro (Morgenstern & Land, 1990). IL-1-unresponsive cells infected with retroviruses were selected in Zeo plus IL-1 for restoration of the responsiveness to IL-1, whereas infected constitutive mutants were selected in GCV. The cDNA inserts were amplified by PCR from the selected clones, using viral vector primers. The PCR products were cloned into an expression vector and reintroduced into the mutant cell line to confirm complementation. So far, this approach has had very limited success for complementing IL-1-unresponsive mutants. Several cDNA clones were obtained by complementation of the constitutive mutants, and their characterization is ongoing.

6.2. *New approaches*

6.2.1. *Regulation of expression levels in cDNA libraries*

Complemented clones are often not obtained because the level of expression of the complementing protein does not match the normal level and is deleterious. cDNA libraries in a retroviral vector with a drug-regulatable promoter are now being constructed so that one can at least attempt complementation at different expression levels.

6.2.2. *Insertional mutagenesis*

This strategy would avoid functional complementation entirely, marking the mutational event and directly identifying the affected gene. Briefly, cells will be heavily mutagenized with ICR-191 to obtain a pseudo-haploid population. Mutants due solely to chemical mutagenesis will be removed by selection. Insertional mutations will be introduced into these cells, for example by using a retroviral vector. The inserted sequences will be flanked by lox P-sites so that, when a putative mutant is obtained, the insert can be excised with Cre recombinase, which should revert the mutant phenotype. Genomic sequences flanking the insertion site will be cloned and the human genome database will be used to identify the affected gene.

7. EXPRESSION CLONING

7.1. *Identification of NF-κB activators*

While 293-TK/Zeo cells were used successfully to isolate IL-1-unresponsive mutant cell lines and constitutive mutants (Li et al., 1999c), the same cells can also be utilized to clone cDNAs whose expression leads to constitutive activation of NF-κB, applying zeocin for selection (Fig. 1). A retroviral cDNA library made from human keratinocyte HaCat cells was introduced into 293-TK/Zeo cells and then selected infected cells in zeocin. cDNA inserts were recovered by RT-PCR from the obtained clones (Li et al., 2000b). One of the cDNAs encodes a 60-kDa polypeptide (574 amino acids) named NF-κB activator-1 (Act1). Simultaneously, Act1 was also cloned (and named CIKS) by yeast two-hybrid screening, based on its interaction with IKKγ (Leonardi et al., 2000).

Overexpression of Act1 in 293 cells leads to constitutive activation of both NF-κB and JNK. Act1 activates NF-κB through the TAK1- and IKK-signaling cascade (Y. Qian & X. Li, unpublished data). The major sites of Act1 expression are epithelial cells of the respiratory and gastrointestinal tract, and its expression is augmented by proinflammatory mediators (X. Li, unpublished data). Act1 does not have any enzymatic domain, but does contain a helix-loop-helix at the N-terminus and a coiled-coil at the C-terminus. In addition, Act1 contains two TRAF-binding sites, namely EEESE (residues 38-42) and EERPA (residues 333-337).

7.2. *Act1 functions in CD40-mediated signaling*

7.2.1. *Endogenous Act1 is recruited to CD40 upon stimulation with CD40L*

Since Act1 contains several putative TRAF-binding sites, it was examined for its interaction with TRAFs. Flag-tagged TRAFs 1 to 6 were transfected into 293 cells already expressing HA-tagged Act1. While HA-tagged Act1 was immunoprecipitated with TRAF3, weak interactions of TRAF1 and TRAF5 with Act1 were also observed (Y. Qian & X. Li, unpublished data). Since TRAF3 and TRAF5 participate in signaling pathways mediated by CD40, LTβR and several other members of the TNF-R superfamily, the above result suggests that Act1 may play a role in one or more pathways mediated by the same receptors. To test the role of Act1 in the CD40-mediated pathway, Flag-tagged CD40 was transfected into 293 cells containing HA-tagged Act1. Interestingly, while Act1 interacted weakly with CD40 in untreated cells, stimulation with CD40L greatly induced the interaction with CD40 (Y. Qian & X. Li, unpublished data). Since the above experiment was done with overexpressed Act1 and CD40, it was critical to examine whether endogenous Act1 can also interact with the stimulated CD40 receptor. Since Act1 is well expressed in epithelial cells of the respiratory and gastrointestinal tracts, and since Act1 expression is augmented by proinflammatory mediators (X. Li, unpublished data), epithelial cell lines were used as models to study the functions of Act1. Western analysis show that both CD40 and Act1 are expressed at high levels in human intestinal (HT29) and cervical (HeLa) cells (Fig. 7A). Importantly, endogenous Act1 does indeed interact with CD40, both in HeLa and HT29 cells stimulated with CD40L (Fig. 7B). The interaction is transient, peaking at 5 min and disappearing after 30 min. Taken together, the above data clearly show that Act1 is recruited to CD40 upon stimulation with CD40L, strongly suggesting that Act1 plays a critical role in CD40-mediated signaling (Y. Qian & X. Li, unpublished data).

7.2.2. *Expression of Act1 confers CD40L-induced NF-κB activation in Act1-null C33A cells and protects them from CD40-mediated apoptosis*

While NF-κB is strongly activated by CD40L in HT29 and HeLa cells where Act1 is highly expressed, and is slightly activated in A549 cells, which express low level of Act1, CD40L failed to activate NF-κB in C33A cells that do not express Act1 at all (Fig. 7A-C). It is clear that the level of CD40L-induced NF-κB activation correlates well with the levels of Act1 in these cells, suggesting that Act1 might be required for CD40-mediated NF-κB activation. To test this hypothesis, Act1 was expressed in natural Act1-null C33A cells and tested for its ability to make these cells sensitive to CD40L-induced NF-κB activation. Stable transfection of pCMV-Act1 into C33A cells did render them susceptible to CD40L-induced NF-κB activation (Fig. 7D; Y. Qian & X. Li, unpublished data), showing that Act1 can mediate this process.

Previous studies showed that activation of CD40 can lead to apoptosis in some epithelial cells (Young et al., 1998). CD40L activates apoptosis in C33A or

A549, but not in HT29 or HeLa cells (Fig. 7E). These results show a correlation between expression of Act1 and protection against apoptosis, which is likely to have resulted from Act1-mediated NF-κB activation upon stimulation with CD40L. Importantly, expression of Act1 in Act1-null C33A cells protects them from CD40L-induced apoptosis (Fig. 7E; Y. Qian & X. Li, unpublished data).

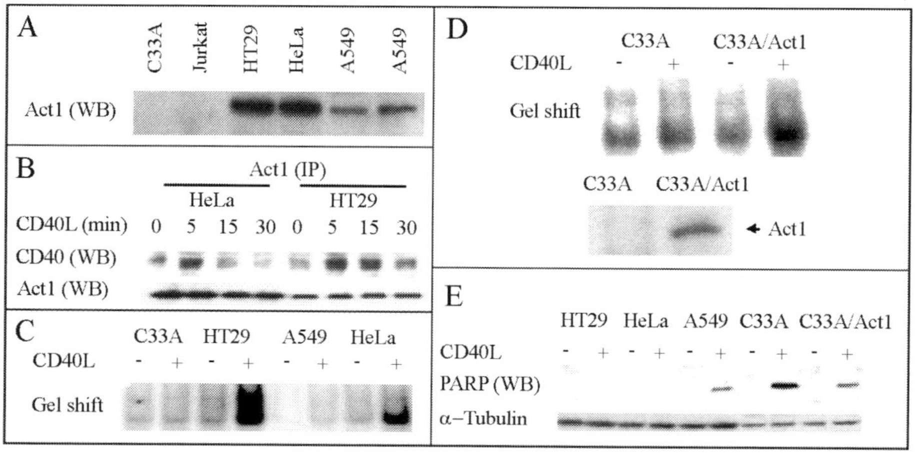

Fig. 7. Role of Act1 in CD40-mediated signaling. (A) Expression of Act1. Extracts of different cell lines were used for Western analysis with anti-Act1. (B) Recruitment of Act1 to CD40 after CD40L stimulation. Extracts from HeLa and HT29 cells, untreated (0) or treated with CD40L for different times, were immunoprecipitated with anti-Act1 and probed with anti-CD40 and anti-Act1. (C) NF-κB gel-shift assays. Extracts of C33A, HT29, A549 and HeLa cells, untreated (-) or treated with CD40L (+), were used for gel-shift assays. (D) Act1 confers CD40-mediated NF-κB activation. Extracts of C33A cells and C33A cells stably transfected with pCMV-Act1 (C33/Act1), untreated (-) or treated with CD40L (+), were used for gel-shift assays. Extracts of C33A and C33A/Act1 cells were used for Western analysis with anti-Act1. (E) Act1 protects cells from CD40L-induced apoptosis. Extracts from different cell lines, either untreated (-) or treated with CD40L (+) for 24 h, were used for Western analysis with an antibody recognizing cleaved poly(ADP-ribose) polymerase (PARP).

The work presented here leads to a working model for the role of Act1 in the CD40 pathway. Act1 functions as a receptor-proximal adapter involved in CD40-mediated NF-κB activation and inhibits CD40L-induced apoptosis in epithelial cells. Act1 is activated by its recruitment to the CD40 receptor through its interaction with TRAF proteins. Activated Act1 then links TRAFs to the TAK1/IKK signalsome, mediating the activation of NF-κB, and to an unknown MAP kinase complex to activate JNK. The Act1-mediated activation of NF-κB protects cells from apoptosis. These Act1-mediated signaling events in the CD40 pathway are regulated by the various interactions of Act1 with different TRAFs. The relative levels of Act1 and TRAFs may play critical roles in the life-or-death decisions of epithelial cells when stimulated with CD40L.

7.3. Identification of NF-κB repressors

While 293-TK/Zeo cells have been used successfully to isolate cDNAs encoding proteins that constitutively activate NF-κB, these cells should also be very useful in identifying NF-κB repressors. Retroviral cDNA libraries can be directly introduced into 293-TK/Zeo cells to establish expression pools, which are then selected in inducer (e.g. IL-1) plus GCV for NF-κB repressors.

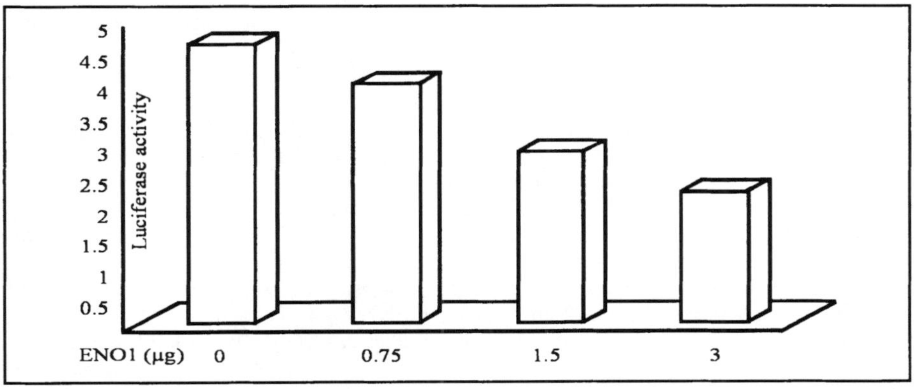

Fig. 8. ENO1 inhibits IL-1-stimulated NF-κB promoter activity in the RasC6 cell line. Cells were transiently cotransfected with increasing amounts of pCR 3.1 ENO1 and 1 μg of the NF-κB-dependent reporter E-selectin/Luc, and analyzed after 48 h. Cells were incubated with IL-1 for 4 h. Luciferase activity was normalized for efficiency of transfection.

This approach has been made to identify currently unknown regulators that function to inhibit NF-κB activity in response to oncogenic Ki-Ras (N. Sizemore & G.R. Stark, unpublished data). To isolate negative regulators of Ras signaling to NF-κB, the Ras C6 parental was infected with a retroviral library to overexpress proteins and selected in GCV to clone Ras/NF-κB repressors. Three independently obtained clones were complemented with the same cDNA. These clones were GCV-resistant and constitutive NF-κB activation was repressed. The cDNA encodes a variant form of enolase-1 (ENO1), a known transcriptional repressor that maps to a chromosomal breakpoint frequently found in human cancers. ENO1 shares 97% similarity with Myc promoter-binding protein-1 (MBP-1), which binds to and negatively regulates the c-myc P2 promoter (Feo et al., 2000). The complementing ENO1 cDNA codes for a truncated protein identical in sequence and length to the reported alternatively translated form of ENO1 (Feo et al., 2000). Therefore, the truncated ENO1 will function and be localized identically with MBP-1. Structural analysis of α-enolase, the protein product of the ENO1 gene, has revealed that the N-terminal part is necessary for its function as a glycolytic enzyme, but not for its ability to repress the c-myc P2 promoter (Subramanian & Miller, 2000). The isolated complementing ENO1 cDNA codes only for the portion of α-enolase that is responsible for its ability to repress the c-myc P2 promoter. The complementing ENO1 cDNA was subcloned into the vector pCR 3.1 under control of the CMV promoter and increasing amounts were transiently cotransfected with an NF-κB-dependent luciferase reporter into the RasC6 cell line. IL-1-induced activation was inhibited with increasing amounts of ENO1 (Fig. 8).

8. REFERENCES

Adachi O, Kawai T, Takeda K, Matsumoto M, Tsutsui H, Sakagami M, Nakanishi K & Akira S (1998) Targeted disruption of the MyD88 gene results in loss of IL-1- and IL-18-mediated function. Immunity 9: 143-150.

Alexopoulou L, Holt AC, Medzhitov R & Flavell RA (2001) Recognition of double-stranded RNA and activation of NF-κB by Toll-like receptor 3. Nature 413: 732-738.

Arlt A, Vorndamm J, Muerkoster S, Yu H, Schmidt WE, Folsch UR & Schafer H (2002) Autocrine production of interleukin 1β confers constitutive nuclear factor κB activity and chemoresistance in pancreatic carcinoma cell lines. Cancer Res. 62: 910-916.

Baeuerle PA & Baltimore D (1996) NF-κB: ten years after. Cell 87: 13-20.

Baldwin AS (2001) Control of oncogenesis and cancer therapy resistance by the transcription factor NF-κB. J. Clin. Invest. 107: 241-246.

Barnes PJ & Karin M (1997) Nuclear factor-κB: a pivotal transcription factor in chronic inflammatory diseases. N. Engl. J. Med. 336: 1066-1071.

Biswas DK, Dai SC, Cruz A, Weiser B, Graner E & Pardee AB (2001) The nuclear factor κB (NF-κB): a potential therapeutic target for estrogen receptor negative breast cancers. Proc. Natl. Acad. Sci. USA 98: 10386-10391.

Brockman JA, Scherer DC, McKinsey TA, Hall SM, Qi X, Lee WY & Ballard DW (1995) Coupling of a signal response domain in IκBα to multiple pathways for NF-κB activation. Mol. Cell. Biol. 15: 2809-2818.

Burns K, Clatworthy J, Martin L, Martinon F, Plumpton C, Maschera B, Lewis A, Ray K, Tschopp J & Volpe F (2000) Tollip, a new component of the IL-1RI pathway, links IRAK to the IL-1 receptor. Nat. Cell Biol. 2: 346-351.

Cao Z, Henzel WJ & Gao X (1996a) IRAK: a kinase associated with the interleukin-1 receptor. Science 271: 1128-1131.

Cao Z, Xiong J, Takeuchi M, Kurama T & Goeddel DV (1996b) TRAF6 is a signal transducer for interleukin-1. Nature 383: 443-446.

Darnell JE, Kerr IM & Stark GR (1994) Jak-STAT pathways and transcriptional activation in response to IFNs and other extracellular signaling proteins. Science 264: 1415-1421.

DiDonato JA, Hayakawa M, Rothwarf DM, Zandi E & Karin M (1997) A cytokine-responsive IκB kinase that activates the transcription factor NF-κB. Nature 388: 548-554.

Feo S, Arcuri D, Piddini E, Passantino R, Giallongo A (2000) ENO1 gene product binds to the c-myc promoter and acts as a transcriptional repressor: relationship with Myc promoter-binding protein 1 (MBP-1). FEBS Lett. 473: 47-52.

Finco TS, Westwick JK, Norris JL, Beg AA, Der CJ & Baldwin AS (1997) Oncogenic Ha-Ras-induced signaling activates NF-κB transcriptional activity, which is required for cellular transformation. J. Biol. Chem. 272: 24113-24116.

Goldsmith MA & Weiss A (1987) Isolation and characterization of a T-lymphocyte somatic mutant with altered signal transduction by the antigen receptor. Proc. Natl. Acad. Sci. USA 84: 6879-6883.

Greenfeder SA, Nunes P, Kwee L, Labow M, Chizzonite RA & Ju G (1995) Molecular cloning and characterization of a second subunit of the interleukin 1 receptor complex. J. Biol. Chem. 270: 13757-13765.

Hocevar BA & Howe PH (1996) Isolation and characterization of mutant cell lines defective in transforming growth factor β signaling. Proc. Natl. Acad. Sci. USA 93: 7655-7660.

Hu Y, Baud V, Oga T, Kim KI, Yoshida K & Karin M (2001) IKKα controls formation of the epidermis independently of NF-κB. Nature 410: 710-714.

Huang J, Gao X, Li S & Cao Z (1997) Recruitment of IRAK to the interleukin 1 receptor complex requires interleukin 1 receptor accessory protein. Proc. Natl. Acad. Sci. USA 94: 12829-12832.

Jiang Y, Woronicz JD, Liu W & Goeddel DV (1999) Prevention of constitutive TNF receptor 1 signaling by silencer of death domains. Science 283: 543-546.

Jo H, Zhang R, Zhang H, McKinsey TA, Shao J, Beauchamp RD, Ballard DW & Liang P (2000) NF-κB is required for H-ras oncogene induced abnormal cell proliferation and tumorigenesis. Oncogene 19: 841-849.

Korherr C, Hofmeister R, Wesche H & Falk W (1997) A critical role for interleukin-1 receptor accessory protein in interleukin-1 signaling. Eur. J. Immunol. 27: 262-267.

Leonardi A, Chariot A, Claudio E, Cunningham K & Siebenlist U (2000) CIKS, a connection to IκB kinase and stress-activated protein kinase. Proc. Natl. Acad. Sci. USA 97: 10494-10499.

Li J, Yen C, Liaw D, Podsypanina K, Bose S, Wang SI, Puc J, Miliaresis C, Rodgers L, McCombie R, Bigner SH, Giovanella BC, Ittmann M, Tycko B, Hibshoosh H, Wigler MH & Parsons R (1997) PTEN, a putative protein tyrosine phosphatase gene mutated in human brain, breast, and prostate cancer. Science 275: 1943-1947.

Li Q, Estepa G, Memet S, Israel A & Verma IM (2000a) Complete lack of NF-κB activity in IKK1 and IKK2 double-deficient mice: additional defect in neurulation. Genes Dev. 14: 1729-1733.

Li Q, Lu Q, Hwang JY, Buscher D, Lee KF, Izpisua-Belmonte JC & Verma IM (1999a) IKK1-deficient mice exhibit abnormal development of skin and skeleton. Genes Dev. 13: 1322-1328.

Li Q, Van Antwerp D, Mercurio F, Lee KF & Verma IM (1999b) Severe liver degeneration in mice lacking the IκB kinase 2 gene. Science 284: 321-325.

Li S, Strelow A, Fontana EJ & Wesche H (2002) IRAK-4: A novel member of the IRAK family with the properties of an IRAK-kinase. Proc. Natl. Acad. Sci. USA 99: 5567-5572.

Li X, Commane M, Burns C, Vithalani K, Cao Z & Stark GR (1999c) Mutant cells that do not respond to interleukin-1 (IL-1) reveal a novel role for IL-1 receptor-associated kinase. Mol. Cell. Biol. 19: 4643-4652.

Li X, Commane M, Jiang Z & Stark GR (2001) IL-1-induced NFκB and c-Jun N-terminal kinase (JNK) activation diverge at IL-1 receptor-associated kinase (IRAK). Proc. Natl. Acad. Sci. USA 98: 4461-4465.

Li X, Commane M, Nie H, Hua X, Chatterjee-Kishore M, Wald D, Haag M & Stark GR (2000b) Act1, an NF-κB-activating protein. Proc. Natl. Acad. Sci. USA 97: 10489-10493.

Lord KA, Hoffman-Liebermann B & Liebermann DA (1990) Nucleotide sequence and expression of a cDNA encoding MyD88, a novel myeloid differentiation primary response gene induced by IL6. Oncogene 5: 1095-1097.

Madrid LV, Mayo MW, Reuther JY & Baldwin AS (2001) Akt stimulates the transactivation potential of the RelA/p65 Subunit of NF-κB through utilization of the IκB kinase and activation of the mitogen-activated protein kinase p38. J. Biol. Chem. 276: 18934-18940.

Madrid LV, Wang CY, Guttridge DC, Schottelius AJ, Baldwin AS & Mayo MW (2000) Akt suppresses apoptosis by stimulating the transactivation potential of the RelA/p65 subunit of NF-κB. Mol. Cell. Biol. 20: 1626-1638.

Malinin NL, Boldin MP, Kovalenko AV & Wallach D (1997) MAP3K-related kinase involved in NF-κB induction by TNF, CD95 and IL-1. Nature 385: 540-544.

Matsushima A, Kaisho T, Rennert PD, Nakano H, Kurosawa K, Uchida D, Takeda K, Akira S & Matsumoto M (2001) Essential role of nuclear factor (NF)-κB-inducing kinase and inhibitor of κB (IκB) kinase α in NF-κB activation through lymphotoxin β receptor, but not through tumor necrosis factor receptor I. J. Exp. Med. 193: 631-636.

Mercurio F, Zhu H, Murray BW, Shevchenko A, Bennett BL, Li J, Young DB, Barbosa M, Mann M, Manning A & Rao A (1997) IKK-1 and IKK-2: cytokine-activated IκB kinases essential for NF-κB activation. Science 278: 860-866.

Morgenstern JP & Land H (1990) Advanced mammalian gene transfer: high titre retroviral vectors with multiple drug selection markers and a complementary helper-free packaging cell line. Nucleic Acids Res. 18: 3587-3596.

Muzio M, Ni J, Feng P & Dixit VM (1997) IRAK (Pelle) family member IRAK-2 and MyD88 as proximal mediators of IL-1 signaling. Science 278: 1612-1615.

Ninomiya-Tsuji J, Kishimoto K, Hiyama A, Inoue J, Cao Z & Matsumoto K (1999) The kinase TAK1 can activate the NIK-IκB as well as the MAP kinase cascade in the IL-1 signalling pathway. Nature 398: 252-256.

O'Neill LA (1995) Towards an understanding of the signal transduction pathways for interleukin 1. Biochim. Biophys. Acta 1266: 31-44.

O'Neill LA (1997) Molecular mechanisms underlying the actions of the pro-inflammatory cytokine interleukin 1. Royal Irish Academy Medal Lecture. Biochem. Soc. Trans. 25: 295-302.

Pellegrini S, John J, Shearer M, Kerr IM & Stark GR (1989) Use of a selectable marker regulated by α interferon to obtain mutations in the signaling pathway. Mol. Cell. Biol. 9: 4605-4612.

Poltorak A, He X, Smirnova I, Liu MY, Huffel CV, Du X, Birdwell D, Alejos E, Silva M, Galanos C, Freudenberg M, Ricciardi-Castagnoli P, Layton B & Beutler B (1998). Defective LPS signaling in C3H/HeJ and C57BL/10ScCr mice: mutations in Tlr4 gene. Science 282: 2085-2088.

Qian Y, Commane M, Ninomiya-Tsuji J, Matsumoto K & Li X (2001) IRAK-mediated translocation of TRAF6 and TAB2 in the interleukin-1-induced activation of NFκB. J. Biol. Chem. 276: 41661-41667.

Rayet B & Gelinas C (1999) Aberrant rel/nfkb genes and activity in human cancer. Oncogene 18: 6938-6947.

Regnier CH, Song HY, Gao X, Goeddel DV, Cao Z & Rothe M (1997) Identification and characterization of an IκB kinase. Cell 90: 373-383.

Rudolph D, Yeh WC, Wakeham A, Rudolph B, Nallainathan D, Potter J, Elia AJ & Mak TW (2000) Severe liver degeneration and lack of NF-κB activation in NEMO/IKKγ-deficient mice. Genes Dev. 14: 854-862.

Schindler U & Baichwal VR (1994) Three NF-κB binding sites in the human E-selectin gene required for maximal tumor necrosis factor α-induced expression. Mol. Cell. Biol. 14: 5820-5831.

Schouten GJ, Vertegaal AC, Whiteside ST, Israel A, Toebes M, Dorsman JC, van der Eb AJ & Zantema A (1997) IκBα is a target for the mitogen-activated 90 kDa ribosomal S6 kinase. EMBO J. 16: 3133-3144.

Sebastiani G, Leveque G, Lariviere L, Laroche L, Skamene E, Gros P & Malo D (2000) Cloning and characterization of the murine toll-like receptor 5 (Tlr5) gene: sequence and mRNA expression studies in Salmonella-susceptible MOLF/Ei mice. Genomics 64: 230-240.

Serafini AT, Lewis RS, Clipstone NA, Bram RJ, Fanger C, Fiering S, Herzenberg LA & Crabtree GR (1995) Isolation of mutant T lymphocytes with defects in capacitative calcium entry. Immunity 3: 239-250.

Siebenlist U, Franzoso G & Brown K (1994) Structure, regulation and function of NF-κB. Annu. Rev. Cell Biol. 10: 405-455.

Silverman N & Maniatis T (2001) NF-κB signaling pathways in mammalian and insect innate immunity. Genes Dev. 15: 2321-2342.

Sizemore N, Lerner N, Dombrowski N, Sakurai H & Stark GR (2002) Distinct roles of the IκB kinase α and β subunits in liberating nuclear factor κB from IκB and in phosphorylating the p65 subunit of NF-κB. J. Biol. Chem. 277: 3863-3869.

Sizemore N, Leung S & Stark GR (1999) Activation of phosphatidylinositol 3-kinase in response to interleukin-1 leads to phosphorylation and activation of the NF-κB p65/RelA subunit. Mol. Cell. Biol. 19: 4798-4805.

Stark GR & Gudkov AV (1999) Forward genetics in mammalian cells: functional approaches to gene discovery. Hum. Mol. Genet. 8: 1925-1938.

Stark GR, Kerr IM, Williams BR, Silverman RH & Schreiber RD (1998) How cells respond to interferons. Annu. Rev. Biochem. 67: 227-264.

Subramanian A & Miller DM (2000) Structural analysis of α-enolase. Mapping the functional domains involved in down-regulation of the c-myc protooncogene. J. Biol. Chem. 275: 5958-5965.

Suzuki N, Suzuki S, Duncan GS, Millar DG, Wada T, Mirtsos C, Takada H, Wakeham A, Itie A, Li S, Penninger JM, Wesche H, Ohashi PS, Mak TW & Yeh WC (2002) Severe impairment of interleukin-1 and Toll-like receptor signalling in mice lacking IRAK-4. Nature 416: 750-756.

Takaesu G, Kishida S, Hiyama A, Yamaguchi K, Shibuya H, Irie K, Ninomiya-Tsuji J & Matsumoto K (2000) TAB2, a novel adaptor protein, mediates activation of TAK1 MAPKKK by linking TAK1 to TRAF6 in the IL-1 signal transduction pathway. Mol. Cell 5: 649-658.

Takeuchi O, Hoshino K, Kawai T, Sanjo H, Takada H, Ogawa T, Takeda K & Akira S (1999) Differential roles of TLR2 and TLR4 in recognition of gram-negative and gram-positive bacterial cell wall components. Immunity 11: 443-451.

Takeuchi O, Kaufmann A, Grote K, Kawai T, Hoshino K, Morr M, Muhlradt PF & Akira S (2000) Cutting edge: preferentially the R-stereoisomer of the mycoplasmal lipopeptide macrophage-activating lipopeptide-2 activates immune cells through a toll-like receptor 2- and MyD88-dependent signaling pathway. J. Immunol. 164: 554-557.

Thomassen et al, 1999 Thomassen E, Renshaw BR & Sims JE (1999) Identification and characterization of SIGIRR, a molecule representing a novel subtype of the IL-1R superfamily. Cytokine 11: 389-399.

Underhill DM, Ozinsky A, Hajjar AM, Stevens A, Wilson CB, Bassetti M & Aderem A (1999a) The Toll-like receptor 2 is recruited to macrophage phagosomes and discriminates between pathogens. Nature 401: 811-815.

Underhill DM, Ozinsky A, Smith KD & Aderem A (1999b) Toll-like receptor-2 mediates mycobacteria-induced proinflammatory signaling in macrophages. Proc. Natl. Acad. Sci. USA 96: 14459-14463.

Velazquez L, Fellous M, Stark GR & Pellegrini S (1992) A protein tyrosine kinase in the interferon α/β signaling pathway. Cell 70: 313-322.

Wald D, Commane M, Stark GR & Li X (2001) IRAK and TAK1 are required for IL-18-mediated signaling. Eur. J. Immunol. 31: 3747-3754.

Wang C, Deng L, Hong M, Akkaraju GR, Inoue J & Chen ZJ (2001) TAK1 is a ubiquitin-dependent kinase of MKK and IKK. Nature 412: 346-351.

Wang D & Baldwin AS (1998) Activation of nuclear factor-κB-dependent transcription by tumor necrosis factor-α is mediated through phosphorylation of RelA/p65 on serine 529. J. Biol. Chem. 273: 29411-29416.

Wang D, Westerheide SD, Hanson JL & Baldwin AS (2000) Tumor necrosis factor α-induced phosphorylation of RelA/p65 on Ser529 is controlled by casein kinase II. J. Biol. Chem. 275: 32592-32597.

Wesche H, Henzel WJ, Shillinglaw W, Li S & Cao Z (1997) MyD88: an adapter that recruits IRAK to the IL-1 receptor complex. Immunity 7: 837-847.

Whitley MZ, Thanos D, Read MA, Maniatis T & Collins T (1994) A striking similarity in the organization of the E-selectin and β interferon gene promoters. Mol. Cell. Biol. 14: 6464-6475.

Williams BL, Schreiber KL, Zhang W, Wange RL, Samelson LE, Leibson PJ & Abraham RT (1998) Genetic evidence for differential coupling of Syk family kinases to the T-cell receptor: reconstitution studies in a ZAP-70-deficient Jurkat T-cell line. Mol. Cell. Biol. 18: 1388-1399.

Witte S, Villalba M, Bi K, Liu Y, Isakov N & Altman A (2000) Inhibition of the c-Jun N-terminal kinase/AP-1 and NF-κB pathways by PICOT, a novel protein kinase C-interacting protein with a thioredoxin homology domain. J. Biol. Chem. 275: 1902-1909.

Woronicz JD, Gao X, Cao Z, Rothe M & Goeddel DV (1997) IκB kinase-β: NF-κB activation and complex formation with IκB kinase-α and NIK. Science 278: 866-869.

Yamaoka S, Courtois G, Bessia C, Whiteside ST, Weil R, Agou F, Kirk HE, Kay RJ & Israel A (1998) Complementation cloning of NEMO, a component of the IκB kinase complex essential for NF-κB activation. Cell 93: 1231-1240.

Yamin TT & Miller DK (1997) The interleukin-1 receptor-associated kinase is degraded by proteasomes following its phosphorylation. J. Biol. Chem. 272: 21540-21547.

Yin L, Wu L, Wesche H, Arthur CD, White JM, Goeddel DV & Schreiber RD (2001) Defective lymphotoxin-β receptor-induced NF-κB transcriptional activity in NIK-deficient mice. Science 291: 2162-2165.

Young LS, Eliopoulos AG, Gallagher NJ & Dawson CW (1998) CD40 and epithelial cells: across the great divide. Immunol. Today 19: 502-506.

Zandi E, Rothwarf DM, Delhase M, Hayakawa M & Karin M (1997) The IκB kinase complex (IKK) contains two kinase subunits, IKKα and IKKβ, necessary for IκB phosphorylation and NF-κB activation. Cell 91: 243-252.

Zhong H, SuYang H, Erdjument-Bromage H, Tempst P & Ghosh S (1997) The transcriptional activity of NF-κB is regulated by the IκB-associated PKAc subunit through a cyclic AMP-independent mechanism. Cell 89: 413-424.

Zhong H, Voll RE & Ghosh S (1998) Phosphorylation of NF-κB p65 by PKA stimulates transcriptional activity by promoting a novel bivalent interaction with the coactivator CBP/p300. Mol. Cell 1: 661-671.

9. AFFILIATIONS

The Cleveland Clinic Foundation, Lerner Research Institute, Department of Immunology (X. Li) and Department of Molecular Biology (G.R. Stark), 9500 Euclid Avenue, Cleveland, OH 44195, USA.

N. SUZUKI, M. BONNARD, S. SUZUKI & W.-C. YEH

DELINEATION OF NF-κB SIGNALING BY GENE TARGETING

1. INTRODUCTION

Active NF-κB transcription factors are composed of dimeric combinations of members of the Rel family proteins. The Rel family proteins include RelA (p65), RelB, c-Rel, NF-κB1 (p105, the precursor for p50) and NF-κB2 (p100, the precursor for p52) (Karin & Ben-Neriah, 2000). Among these, RelA and c-Rel contain potent transcriptional activation domains. Thus the transcriptionally active forms of NF-κB are dimers of c-Rel/p50, c-Rel/RelA, RelA/RelA and p50/RelA (the most abundant form of NF-κB) (Ghosh et al., 1998). In contrast, homodimers of p50 and p52 are thought to be transcriptionally repressive. In resting cells, NF-κB dimers are normally sequestered in the cytoplasm via binding to IκB proteins. The IκB family includes IκBα, IκBβ, IκBγ, IκBε and Bcl-3. The IκB proteins mask the nuclear localization signal of NF-κB, preventing its translocation to the nucleus. A variety of stimuli, including proinflammatory cytokines and products of viral and bacterial infections, induce the serine phosphorylation of IκB proteins. Phosphorylated IκB proteins become targets for ubiquitination and proteosome-mediated degradation, a process that releases NF-κB and promotes its translocation to the nucleus. In the nucleus, NF-κB binds to its specific κB DNA motif in the promoters of target genes and activates their transcription (Ghosh et al., 1998; Karin & Ben-Neriah, 2000) (Fig. 1).

Inducible phosphorylation of IκB is carried out primarily by the IκB kinase (IKK) complex (Delhase et al., 1999). The IKK complex comprises two catalytic subunits, viz. IKKα (IKK1) and IKKβ (IKK2), as well as a modulator protein called IKKγ or NEMO (NF-κB essential modulator) (Zandi et al., 1997; Rothwarf et al., 1998). Other kinases, namely IKKε (also named IKK-i) (Shimada et al., 1999) and T2K (Bonnard et al., 2000) [also called TBK1 (Pomerantz & Baltimore, 1999) or NAK (Tojima et al., 2000)], can phosphorylate IκB and have been implicated in the activation of NF-κB-directed gene transcription. The precise roles of these enzymes and how they might coordinate different signaling pathways are critical to our understanding of the activation of NF-κB.

Many signaling molecules upstream of the IKK complex have been implicated in regulating NF-κB activation based on the results of overexpression and biochemical studies. Interestingly, most of these signaling molecules can be mapped

R. Beyaert (ed.), Nuclear Factor κB. Regulation and Role in Disease, 265-294.

to three key signaling cascades that transduce signals downstream of three major cell surface receptor families associated with NF-κB activation: (1) the T-cell receptor (TCR)-signaling pathway; (2) the interleukin-1 receptor (IL-1R)/Toll-like receptor (TLR) pathway; (3) the tumor necrosis factor receptor (TNF-R) pathway. This observation suggests that certain mediators may be pathway-specific, and that the cell requires flexible molecular mechanisms to trigger the activation of the IKK complex and NF-κB. On the other hand, molecules such as TRAF6 (Cao et al., 1996b) and receptor-interacting protein 2 (RIP2) (McCarthy et al., 1998), mediate NF-κB signaling downstream of two or more receptor systems. Studies of these proteins are particularly helpful in elucidating the unique and common characteristics of different receptor-signaling complexes.

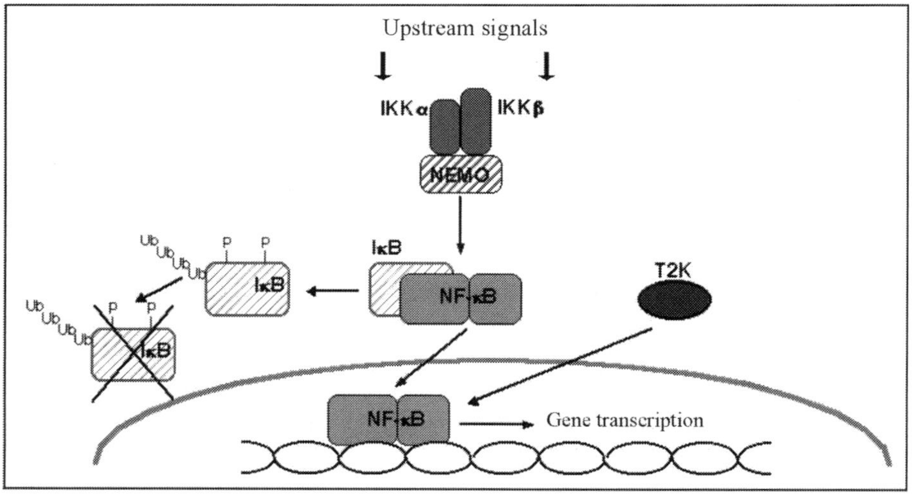

Fig. 1. In the nucleus, NF-κB binds to its specific κB DNA motif in the promoters of target genes and activates their transcription.

In assigning physiological functions to signaling proteins associated with NF-κB activation, two additional factors should be noted. First, some proteins, depending on their positions in the signaling cascade, are also mediators in other downstream signaling events that are independent of NF-κB, e.g. stress kinase activation. These molecules may serve to promote cross-talk between different signaling pathways. Secondly, whether or not NF-κB activation occurs is governed by a balance of signaling molecules that have positive or negative effects. This feature is particularly evident in the complex pathways of TNF-R signaling. These interesting aspects of NF-κB signaling, if properly explored, may assist us in devising strategies to modulate the impact of NF-κB activation as well as biological effects mediated by other pathways.

In this chapter we focus our discussion on the functions of a wide range of molecules associated with NF-κB signaling as revealed by gene-targeting studies.

We review the relevant gene knockouts starting with animals deficient in various NF-κB subunits or IκB proteins, progressing through mice lacking components of the IKK complex or IKK-related proteins, and concluding with mutants deleted for genes specifying individual cytoplasmic signaling proteins required in the TCR-, IL-1R/TLR- or TNF-R-signaling pathways. While knockouts of the cell surface receptors themselves can sometimes be informative, we did not include studies of those mutants in this review because the phenotypes observed likely represent a combination of effects on signaling cascades in addition to the NF-κB pathway.

2. NF-κB SUBUNITS

2.1. RelA

Mice deficient in RelA exhibit a surprisingly severe phenotype in that they die between 14.5 and 16 days of embryogenesis (E14.5-16) as a result of massive hepatocyte apoptosis (Beg et al., 1995b). $RelA^{-/-}$ fibroblasts and macrophages are hypersensitive to TNFα-induced apoptosis in vitro (Beg & Baltimore, 1996). These results provided the first hint that NF-κB might oppose apoptosis induced by TNF stimulation. Interestingly, elimination of TNF-R1 or TNF rescues the embryonic lethality of $RelA^{-/-}$ mice (Doi et al., 1999; Alcamo et al., 2001), suggesting that the pathology of apoptotic liver degeneration can be attributed to TNF signaling.

Mice deficient in both RelA and TNF-R1 ($RelA^{-/-}TNF\text{-}R1^{-/-}$ mice) are born healthy. Such animals are, however, highly susceptible to bacterial infections and they die within a few weeks after birth (Alcamo et al., 2001). In these animals, the development of T- and B-lymphocytes appears to be normal (Doi et al., 1997; Alcamo et al., 2002). However, $RelA^{-/-}$ T-cells exhibit an impaired proliferation in response to either concanavalin A (ConA) or an anti-CD3 antibody with or without anti-CD28. Likewise, the proliferative responses of $RelA^{-/-}$ B-cells to lipopolysaccharide (LPS) or anti-IgM are reduced compared to wild-type controls. Furthermore, the T-dependent antibody responses of B-cells are defective in $RelA^{-/-}TNF\text{-}R1^{-/-}$ animals (in comparison with $RelA^{+/+}TNF\text{-}R1^{-/-}$ controls), such that basal IgA and IgG1 levels are reduced in lethally irradiated mice reconstituted with RelA-deficient fetal liver (Doi et al., 1997; Alcamo et al., 2002).

At the gene level, RelA was reported to be essential in murine embryonic fibroblasts (MEFs) for TNF-mediated induction of the known NF-κB target genes IκBα and GM-CSF (Beg et al., 1995a). This result was particularly striking in that it showed that NF-κB dimers containing other Rel family members cannot compensate for a lack of RelA in the activation of these genes. Studies of the specificity of various Rel family proteins in the regulation of target genes involved in inflammation, immune responses and protection against apoptosis will be of great interest.

2.2. *c-Rel*

Mice lacking c-Rel are viable and grossly normal, indicating that c-Rel, unlike RelA, is not essential for embryonic survival (Kontgen et al., 1995). Lymphocyte development and hematopoiesis are also normal in c-Rel$^{-/-}$ mice. However, c-Rel-deficient animals exhibit defects in T- and B-lymphocyte proliferation and in the synthesis of IgG1 and IgG2a. c-Rel$^{-/-}$ T-cells stimulated with anti-CD3 with or without anti-CD28 produce significantly lower levels of IL-2 than did wild-type controls. While addition of exogenous IL-2 can rescue the proliferation of c-Rel$^{-/-}$ T-cells, neither IL-2, IL-4, IL-5 nor IL-6 can restore B-cell proliferation, indicating that c-Rel-mediated regulation of T- and B-cell activation affects different genes. Interestingly, κB DNA-binding activity can still be induced in c-Rel$^{-/-}$ B-cells. The relevant complexes were composed predominantly of RelB/p50 dimers and, to a lesser extent, RelA/p50 dimers (Kontgen et al., 1995).

2.3. *RelB*

Like c-Rel$^{-/-}$ mice, RelB$^{-/-}$ mice are viable and appear normal at birth. However, 2-6 weeks after birth, the mutant mice fall ill and some die prematurely (Weih et al., 1995). Although the development of mature T- and B-lymphocytes appears normal, RelB-deficient mice develop a severe and complex phenotype that includes thymic atrophy, splenomegaly and myeloid hyperplasia of the bone marrow (Weih et al., 1995). The mutant animals also display an absence of secondary lymphoid tissues and an apparent loss of dendritic cells (DC), correlating with a dysfunctional immune response (Burkly et al., 1995; Weih et al., 1995). Thymic medullary epithelial cells are lacking in RelB-deficient mice and negative selection of T-cells is defective (Burkly et al., 1995; DeKoning et al., 1997). RelB-deficient mice also produce excess neutrophils and macrophages (Burkly et al., 1995), and exhibit multi-organ inflammation that has been attributed to overproduction of cytokines in non-lymphoid tissues (Weih et al., 1995; DeKoning et al., 1997). The RelB target genes responsible for this phenotype remain unknown. At the gene level, expression of p50, p52, RelA, c-Rel and IκBα is unaltered in RelB$^{-/-}$ thymocytes, and κB binding of p50/RelA dimers is normal (Weih et al., 1995). These observations indicate that the role of RelB in thymocytes cannot be compensated for by other members of the Rel family (Weih et al., 1995).

2.4. *NF-κB1 (p50)*

NF-κB1-deficient mice appear healthy and fertile, but are more prone to infection than wild-type mice (Sha et al., 1995). Although lymphocyte populations appear normal, basal serum immunoglobulin levels are reduced in NF-κB1$^{-/-}$ mice due to a defect in isotype switching. Proliferation of NF-κB1$^{-/-}$ B-cells in response to LPS or anti-IgM stimulation is also impaired. Immune responses to certain bacterial pathogens are defective, while responses to murine encephalomyocarditis virus are enhanced. This result reflects a dependence on different cell populations and cytokines for the clearance of bacteria as opposed to viruses. In this context, NF-κB1$^{-/-}$ macrophages show impaired responses to TNF or IL-1 stimulation, but levels of IFNβ are upregulated in similarly treated NF-κB1$^{-/-}$ fibroblasts. NF-κB1 may therefore have positive regulatory effects on some genes, but negative regulatory effects on others (Sha et al., 1995).

RelA$^{-/-}$NF-κB1$^{-/-}$ animals die with liver apoptosis at E12 (Horwitz et al., 1997), considerably earlier than RelA$^{-/-}$ animals. Lethally irradiated hosts reconstituted with RelA$^{-/-}$NF-κB1$^{-/-}$ fetal liver precursors show evidence of severe defects in lymphocyte development, including decreased populations of B-cell precursors (pro-B-cells and pre-B-cells), reduced numbers of immature and mature B-cells, and impaired T-cell development (Horwitz et al., 1997). The phenotype of RelA$^{-/-}$NF-κB1$^{-/-}$ mice demonstrates that these Rel family members have critical, but redundant roles in lymphocyte development.

2.5. NF-κB2 (p52)

Mice lacking NF-κB2 are viable and show no gross developmental abnormalities. However, they do exhibit a complex defect in the microarchitecture of the spleen and lymph nodes (Caamano et al., 1998; Franzoso et al., 1998; Poljak et al., 1999). Follicular dendritic cells are lacking, numbers of marginal metallophilic macrophages are decreased, and the B-cell compartment is markedly reduced (Poljak et al., 1999). NF-κB2$^{-/-}$ B-cells have mildly compromised proliferative responses, and undergo normal maturation and immunoglobulin class switching (Caamano et al., 1998). However, antibody responses to both T-dependent and T-independent antigens in vivo are impaired (Franzoso et al., 1998). NF-κB2$^{-/-}$ T-cells, which have normal proliferative responses, show enhanced production of IL-2 and GM-CSF, suggesting that NF-κB2 negatively regulates the expression of these cytokine genes (Franzoso et al., 1998). Further studies into the roles of NF-κB1 and NF-κB2 in the positive and negative regulation of genes involved in hematopoiesis and lineage commitment will be critical to understand their contributions to NF-κB-dependent transcription.

3. IκB FAMILY PROTEINS

3.1. IκBα

Mice deficient in IκBα are viable and appear normal at birth. However, these mutants cease to grow after three days and are severely runty. They also suffer from widespread dermatitis and extensive granulopoiesis, and typically die by 7-10 days of age (Beg et al., 1995b; Klement et al., 1996). Although IκBα$^{-/-}$ MEFs exhibit a minimal constitutive NF-κB activation, the hematopoietic tissues of IκBα$^{-/-}$ mice display enhanced levels of nuclear NF-κB (as assayed by EMSA).

The phenotypic features of IκB$\alpha^{-/-}$ mice are at least partly accounted for by the increased level of NF-κB activation in hematopoietic tissues, as mice lacking both IκBα and NF-κB1 show a significant delay in the onset of abnormalities (Beg et al., 1995b).

IκB$\alpha^{-/-}$ thymocytes show elevated levels of mRNA for NF-κB target genes such as G-CSF, MIP-2 and VCAM-1 (Beg et al., 1995b). Similarly, enhanced levels of TNF have been observed in the skin of IκB$\alpha^{-/-}$ animals. TNF-mediated induction of κB DNA binding is comparable in wild-type and IκB$\alpha^{-/-}$ fibroblasts, but the nuclear localization of NF-κB is prolonged in IκB$\alpha^{-/-}$ cells. This observation suggests that IκBα may play a role in repressing NF-κB activation (Beg et al., 1995b; Klement et al., 1996). Further analysis of IκB$\alpha^{-/-}$ mice will be important for understanding the regulation of NF-κB activation in response to different stimuli in different cell types. In this context, it will be also of interest to assess the physiological effects of mutations in other IκB isoforms.

3.2. Bcl-3

In contrast to other IκB family members, Bcl-3 appears to act as a positive regulator of NF-κB activity. Bcl-3 preferentially interacts with NF-κB1 and NF-κB2 homodimers. Consistent with the synergistic effect of these proteins on transcription, Bcl-3-deficient mice share phenotypic defects with NF-κB1$^{-/-}$ and NF-κB2$^{-/-}$ animals. Like NF-κB1-deficient mice, Bcl-3$^{-/-}$ mice are viable, but are abnormally susceptible to infections and have difficulty in clearing pathogens such as Listeria monocytogenes (Franzoso et al., 1997; Schwarz et al., 1997). Like NF-κB2-deficient mice, Bcl-3$^{-/-}$ mice exhibit defects in the microarchitecture of secondary lymphoid organs, and spleens of Bcl-3$^{-/-}$ animals are devoid of germinal centers (Franzoso et al., 1997; Schwarz et al., 1997; Poljak et al., 1999). Bcl-3$^{-/-}$ mice also have a defect in the formation of B-cell follicles and are unable to form follicular dendritic cell networks upon antigenic challenge (Poljak et al., 1999). Furthermore, T-dependent antibody responses and T_H1 responses are impaired in Bcl-3-deficient animals (Franzoso et al., 1997; Schwarz et al., 1997). These phenotypes are enhanced in Bcl-3$^{-/-}$NF-κB2$^{-/-}$ mice (Schwarz et al., 1997).

4. THE IKK COMPLEX

4.1. IKKα

IKKα-deficient mice display a phenotype strikingly different from those of IKK$\beta^{-/-}$ and NEMO$^{-/-}$ mice (see below). IKK$\alpha^{-/-}$ animals die within a few hours of birth showing abnormal limb outgrowth and craniofacial deformities. These defects are most likely due to excessive proliferation of the basal layer of the epidermis (Hu et al., 1999; Takeda et al., 1999). κB DNA binding in IKK$\alpha^{-/-}$ keratinocytes in response to TNF or IL-1 stimulation is comparable to that observed in wild-type keratinocytes (Takeda et al., 1999; Hu et al., 2001). These results revealed the unexpected finding that IKKα plays an important role in epidermal differentiation, possibly through an NF-κB-independent pathway (Hu et al., 2001).

IKKα may also play an important and unique role in B-cell maturation and responses. Lethally irradiated hosts reconstituted with IKK$\alpha^{-/-}$ fetal liver show reduced numbers of mature B-cells, and IKK$\alpha^{-/-}$ B-cells exhibit impaired responses to mitogens as well as increased sensitivity to apoptosis (Kaisho et al., 2001a; Senftleben et al., 2001a). In addition, the formation of germinal centers in response to immunization with DNP-KLH is impaired in IKK$\alpha^{-/-}$ chimeras (Senftleben et al., 2001a). Importantly, processing of the NF-κB2 from p100 to p52 is defective in IKK$\alpha^{-/-}$ B-cells and reduced levels of NF-κB2 are observed in these cells (Senftleben et al., 2001a). While LPS-induced κB DNA binding is detectable in IKK$\alpha^{-/-}$ B-cells, it is variably reduced according to the circumstance (Kaisho et al., 2001a; Senftleben et al., 2001a). NF-κB1- and NF-κB2-containing complexes are also reduced in IKK$\alpha^{-/-}$ B-cells (Kaisho et al., 2001a; Senftleben et al., 2001a). Further studies of IKKα-deficient animals and cells will be needed to determine if IKKα plays different roles in the regulation of NF-κB in different cell types.

4.2. IKKβ

IKK$\beta^{-/-}$ mice die at E14.5 with massive apoptotic degeneration of the liver, a phenotype that is reminiscent of RelA$^{-/-}$ mice. In addition, like RelA$^{-/-}$ MEFs, IKK$\beta^{-/-}$ MEFs are hypersensitive to TNFα-induced apoptosis (Li et al., 1999; Tanaka et al., 1999). Degradation of IκBα and IκBβ (Li et al., 1999; Tanaka et al., 1999), as well as κB DNA binding in response to stimulation with proinflammatory cytokines are reduced in IKK$\beta^{-/-}$ cells. As was true for RelA$^{-/-}$ mice, deletion of TNF-R1 rescues the embryonic lethality and liver apoptosis associated with IKKβ deficiency (Li et al., 1999). IKKβ also plays an important role in lymphocyte development and activation. Lethally irradiated Rag$^{-/-}$ mice reconstituted with IKK$\beta^{-/-}$ fetal liver cells show an absence of peripheral blood lymphocytes, dramatically reduced numbers of thymocytes and increased thymic apoptosis. These defects are significantly attenuated in IKK$\beta^{-/-}$TNF-R1$^{-/-}$ animals, indicating that IKKβ is required to protect developing lymphocytes from TNFα-induced apoptosis (Senftleben et al., 2001b). Taken together, these results emphasize that the transcription of genes required to protect against TNF-induced apoptosis depends on NF-κB activation, and NF-κB activation in response to proinflammatory cytokines depends on IKKβ.

IKK$\beta^{-/-}$TNF-RI$^{-/-}$ thymocytes stimulated with anti-CD3, ConA or PMA plus ionomycin display impaired proliferation and defective κB DNA-binding activity. It is therefore possible that, in addition to its role in facilitating antiapoptotic NF-κB activation, IKKβ influences the regulation of NF-κB target genes involved in thymocyte proliferation (Senftleben et al., 2001b). It will be of great interest to identify which IKKβ target genes protect against apoptosis and which promote lymphocyte proliferation.

4.3. NEMO/IKKγ

NEMO/IKKγ-deficient mice closely resemble $RelA^{-/-}$ and $IKK\beta^{-/-}$ mice, and thus greatly differ from $IKK\alpha^{-/-}$ mice. $NEMO^{-/-}$ mice die at E12.5-13 due to liver apoptosis; the limb defects seen in $IKK\alpha^{-/-}$ mice are not present. $NEMO^{-/-}$ MEFs are hypersensitive to TNF-induced apoptosis (Makris et al., 2000; Rudolph et al., 2000; Schmidt-Supprian et al., 2000) and NF-κB activation in response to TNF, IL-1 or LPS is completely blocked. This inhibition is associated with impaired phosphorylation and degradation of IκBα and reduced κB DNA binding. Thus, at least in response to proinflammatory cytokines or LPS, NEMO is essential for NF-κB activation. Consistent with the immunological defects observed in other mice lacking Rel family members, lymphocytes cannot be reconstituted in vivo from NEMO-deficient ES cells (Schmidt-Supprian et al., 2000).

NEMO-deficient mice may have a human counterpart. NEMO is an X-linked gene. In humans, deletion of NEMO/IKKγ has been linked to the genetic disorder named incontinentia pigmenti, which is lethal in males and causes variable abnormalities of the skin, hair, teeth, nails and nervous system in affected females (Smahi et al., 2000). Interestingly, female $NEMO^{+/-}$ mice develop a condition analogous to the human disease with a similar pattern of inheritance (Makris et al., 2000; Schmidt-Supprian et al., 2000). Apoptosis of keratinocytes in these animals is increased, further highlighting the role of NF-κB target genes in protecting against apoptosis (Makris et al., 2000; Schmidt-Supprian et al., 2000).

5. OTHER IKK-RELATED KINASES

T2K (also called TBK1 or NAK) was originally isolated as a kinase associated with TRAF2 in a protein complex that also includes TANK (Cheng & Baltimore, 1996). T2K shares some sequence homology with IKKα and IKKβ, particularly in the kinase domains. T2K can phosphorylate IκBα in vitro, but at only one of two critical serine residues (Ser36). T2K may also activate IKKβ directly, and thus influence the IKK complex. T2K-deficient mice die at E14.5 due to liver apoptosis. Similar to mice lacking RelA or IKKβ, the embryonic lethality and liver apoptosis of $T2K^{-/-}$ mice can be rescued by the elimination of TNF-R1 (Bonnard et al., 2000). However, unlike $RelA^{-/-}$ and $IKK\beta^{-/-}$ cells, $T2K^{-/-}$ thymocytes and MEFs are not sensitive to TNF-induced apoptosis, suggesting that T2K may be critical for protection against apoptosis only in certain cell types.

Interestingly, phosphorylation of IκBα, degradation of IκBα, IκBβ and IκBε, and κB binding in response to either TNF or IL-1 stimulation are comparable in wild-type and $T2K^{-/-}$ MEFs. However, cytokine-induced NF-κB-dependent gene expression is significantly impaired in fibroblasts lacking T2K. TNF-induced endogenous expression of TLR-2 and c-IAP1 are reduced in T2K$^{-/-}$ fibroblasts compared to wild-type controls. These results suggest that the role of T2K in NF-κB-dependent transcription may lie outside events leading to IκB degradation (Bonnard et al., 2000). Further studies of T2K-deficient animals will determine if T2K plays different roles in the regulation of NF-κB in different cell types, or in response to stimuli other than proinflammatory cytokines. It will also be interesting to characterize the molecular mechanism by which T2K modulates the activation of NF-κB.

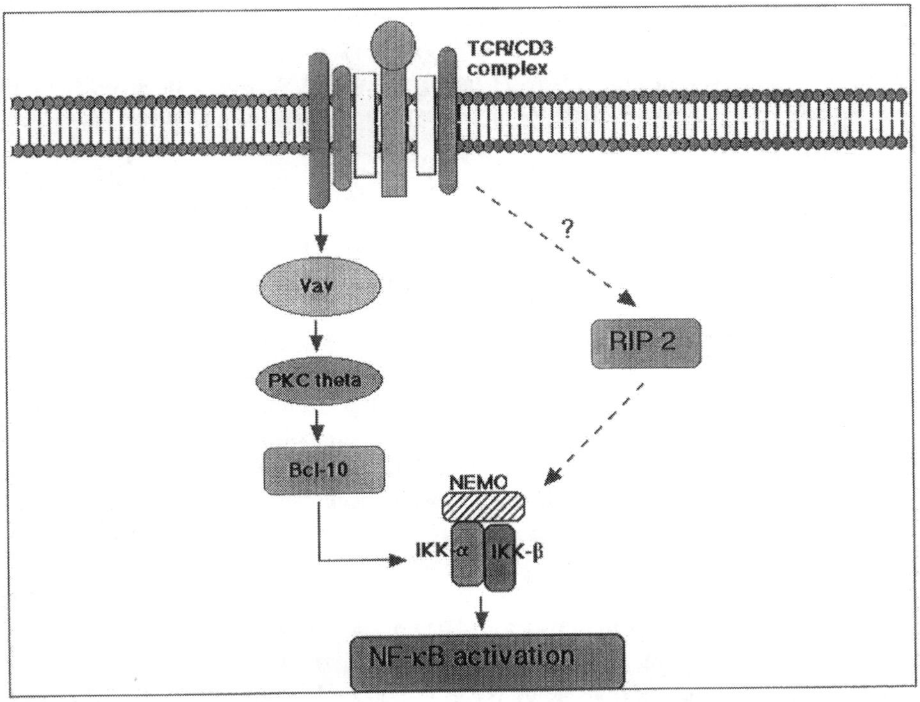

Fig. 2. Molecules whose functions have been linked to TCR-induced NF-κB activation following gene-targeting studies.

6. THE TCR-SIGNALING PATHWAY

6.1. Scope

Upon recognition of foreign antigens, TCRs transduce signals leading to clonal expansion and differentiation of T-effector cells (Voll & Ghosh, 1999). Activation of NF-κB is crucial for these events. Though a plethora of signaling proteins associated with the TCR complex has been characterized, the precise molecular pathway leading to IKK complex activation and NF-κB-dependent gene transcription remains unclear. We focus on four molecules whose functions were linked to TCR-induced NF-κB activation by the results of gene-targeting studies (Fig. 2).

6.2. Vav-1

Vav proteins are GTP/GDP exchange factors required for the activation of Rho family GTPases (Costello et al., 1999). Vav-1 is predominantly expressed in tissues and cells of hematopoietic origin. The related protein Vav-2 is structurally related to Vav-1 (Schuebel et al., 1996), but is more widely expressed. Upon TCR engagement, Vav-1 is rapidly tyrosine-phosphorylated and carries out functions that cannot be replaced by Vav-2 activity (Doody et al., 2000). Vav-1$^{-/-}$ T-cells cannot proliferate in response to TCR stimulation because they fail to initiate expression of the IL-2 gene and secretion of the IL-2 protein. Importantly, NF-κB activation induced by TCR engagement is defective in Vav-1$^{-/-}$ lymphocytes. Vav-1 is also required for thymocyte development, activation-dependent actin polymerization, intracellular calcium flux and activation of mitogen-activated protein kinases (MAPKs) induced by TCR stimulation (Costello et al., 1999). In contrast, activation of the tyrosine kinases Lck and ZAP-70, two key mediators of TCR signaling, appears to be normal in Vav-1$^{-/-}$ lymphocytes (Costello et al., 1999).

6.3. Protein kinase Cθ (PKCθ)

Following TCR activation, PKCθ (Baier et al., 1993) interacts with Vav-1 and is rapidly recruited to the site of TCR clustering at the cell membrane. This observation suggests a potential role for PKCθ in TCR signaling (Arendt et al., 2002). Indeed, the highest levels of PKCθ expression occur in mature T-lymphocytes, and PKCθ$^{-/-}$ mature T-cells have severely impaired proliferative responses to stimulation with anti-CD3 with or without anti-CD28, and to PMA plus ionomycin (Sun et al., 2000). Mature PKCθ$^{-/-}$ T-lymphocytes show a striking defect in anti-CD3-mediated NF-κB activation, but normal signaling through the MAPK and stress-induced c-Jun N-terminal kinase (JNK) pathways (Sun et al., 2000). Activation of the transcription factor NF-AT induced by anti-CD3 cross-linking is also normal in mature PKCθ$^{-/-}$ T-lymphocytes (Sun et al., 2000). However, the involvement of PKCθ in NF-κB activation is remarkably specific, depending on both the cell type and the stimulus. For example, unlike mature PKCθ$^{-/-}$ T-cells, PKCθ$^{-/-}$ thymocytes show an intact activation of NF-κB in response to anti-CD3 cross-linking. Thymocyte development is also normal in PKCθ-deficient mice. With respect to stimuli, PKCθ is required for NF-κB activation induced by anti-CD3 or PMA plus ionomycin, but is dispensable for NF-κB activation induced by proinflammatory cytokines. IκB degradation is defective in PKCθ$^{-/-}$ T-lymphocytes, indicating that PKCθ likely acts upstream of the IKK complex in the NF-κB-signaling cascade (Arendt et al., 2002). PKCθ is also required for activation of the transcription factor AP-1 in mature T-cells stimulated with PMA or PMA plus ionomycin, which is consistent with the role of this kinase in transducing signals received from the intracellular messenger diacylglycerol (Sun et al., 2000).

6.4. *Bcl-10*

Bcl-10 is a CARD domain-containing protein identified from the t(1;14)(p22;q32) breakpoint in MALT lymphomas (Zhang et al., 1999). Overexpression studies have suggested that Bcl-10 mediates signals leading both to apoptosis and NF-κB activation (Zhang et al., 1999). However, studies of Bcl-$10^{-/-}$ mice do not support a physiological role for Bcl-10 in apoptosis. Rather, Bcl-10 has a surprisingly specific function in lymphocyte activation. Bcl-$10^{-/-}$ mice are severely immunodeficient, and have reduced basal concentrations of all immunoglobulins. Upon viral challenge, Bcl-$10^{-/-}$ mice show impaired primary and memory immune responses, and a failure in immunoglobulin class switching (Ruland et al., 2001).

Bcl-10-deficient lymphocytes have a defect in NF-κB activation that is specific for TCR and B-cell receptor signaling. Both Bcl-$10^{-/-}$ T- and B-lymphocytes fail to proliferate in response to antigen receptor stimulation, but calcium flux and MAPK activation pathways are normal. Moreover, there are no defects in TNF- or IL-1-induced NF-κB activation in Bcl-$10^{-/-}$ cells. Bcl-10 is required for IκBα phosphorylation and degradation induced by PMA plus ionomycin, suggesting that Bcl-10 acts upstream of IKK complex activation (Ruland et al., 2001). It is possible that in T-cells, PKCθ and Bcl-10 collaborate in a linear signaling pathway that regulates NF-κB activation leading to lymphocyte proliferation (McAllister-Lucas et al., 2001). Interestingly, while PKCθ also controls AP-1 activation in response to PMA (Baier-Bitterlich et al., 1996), Bcl-10 is dispensable for this process. It remains to be determined whether the same or a different isoform of PKC and Bcl-10 participate in a similar cascade in B-cells.

About one-third of Bcl-$10^{-/-}$ embryos develop exencephaly associated with a defect in neural tube closure (Ruland et al., 2001). Neural tube closure involves the orchestration of multiple events, and the role of Bcl-10 in this process remains unclear. However, it is intriguing that IKK$\alpha^{-/-}$IKK$\beta^{-/-}$ mice demonstrate a neural phenotype similar to that of exencephalic Bcl-$10^{-/-}$ embryos. It is possible that Bcl-10 and the IKK complex function in an NF-κB-mediated pathway required for nervous system development, perhaps by regulating developmental apoptosis. Insights into the multiple functions of Bcl-10 and its potential for oncogenesis (Shih et al., 2001) may be gained from studies of several proteins recently identified as interacting with Bcl-10, including MALT1 (Akagi et al., 1999), CARD10 (Bimp1) (Wang et al., 2001), CARD11 and CARD14 (Bertin et al., 2001).

6.5. *RIP2*

RIP2 is a serine/threonine kinase similar to RIP, but RIP2 has a CARD domain at its C-terminus (McCarthy et al., 1998), whereas RIP has a death domain (Kelliheretal.,1998). RIP2 has been shown to associate with a variety of other CARD domain-containing molecules, suggesting that this kinase may function in a variety of signaling cascades (McCarthy et al., 1998). Besides, overexpression of RIP2 induces activation of both NF-κB and JNK (McCarthy et al., 1998). The complex nature of the phenotype of RIP2-deficient mice also suggests that RIP2 may play a role in multiple signaling pathways (Chin et al., 2002; Kobayashi et al., 2002). With respect to TCR signaling, RIP2$^{-/-}$ T-cells show severe reductions in proliferation, IL-2 production, and NF-κB activation in response to anti-CD3 cross-linking, but normal responses to stimulation with ConA or PMA plus ionomycin (Kobayashi et al., 2002). Study of the defect in NF-κB signaling shows that RIP2 acts upstream of the phosphorylation and degradation of IκB (Kobayashi et al., 2002). Interestingly, RIP2-deficient mice are highly susceptible to infection by Listeria monocytogenes, an intracellular pathogen that requires a T-cell-mediated response for clearance (Chin et al., 2002). However, RIP2 is also required for TLR-mediated signaling and for IL-12 signaling that transduces Stat4 activation and T_H1 differentiation (Chin et al., 2002). RIP2 is therefore a key signal mediator linking innate and adaptive immunity.

7. THE IL-1R/TLR-SIGNALING PATHWAY

7.1. Scope

The innate immune response is initiated by the recognition of pathogen-associated molecular patterns (PAMPs) by a set of germline-encoded receptors known as TLRs (O'Neill, 2000; Akira et al., 2001; Kimbrell & Beutler, 2001). To date, five mammalian TLRs have been associated with specific pathogen-associated ligands: TLR2, TLR3, TLR4, TLR5 and TLR9 receive and transmit signals in response to lipoproteins/peptidoglycans (PGN), viral double-stranded RNA (dsRNA), LPS, bacterial flagellin and bacterial (unmethylated CpG) DNA, respectively (Akira et al., 2001). Based on the homology of their cytoplasmic domains, TLRs are related to members of the IL-1 receptor family (IL-1R and IL-18R) (O'Neill, 2000). The cytoplasmic domains of TLRs and IL-1R are also homologous to domains in the plant R gene product, giving these regions their name of "TIR" domains (O'Neill, 2000). TIR domain-containing receptors play crucial roles in both innate and adaptive immunity in mammals, and the TIR-signaling cascades are implicated in a number of human diseases including sepsis. Remarkably, the signal transduction mechanisms employed by TIRs (Fig. 3) show a significant conservation from invertebrates to vertebrates (O'Neill, 2000; Akira et al., 2001; Kimbrell & Beutler, 2001).

7.2. MyD88

Upon stimulation of IL-1R or at least some TLRs, the TIR-containing adapter protein MyD88 is recruited to and associates directly with the receptor complex via interaction of the TIR domains (Medzhitov et al., 1998). MyD88 also contains a death domain that can associate with the death domains of IL-1R-associated kinase (IRAK) family members (Wesche et al., 1997). Downstream signal transduction inducing NF-κB activation and proinflammatory cytokine production then ensues. MyD88-deficient mice have defects in cytokine production and NF-κB activation induced by IL-1 and the majority of known TLR ligands. MyD88$^{-/-}$ mice are resistant to LPS-induced septic shock, but succumb to bacterial infection (e.g. Staphylococcus aureus), consistent with a critical role of MyD88 in mounting inflammatory responses (Kawai et al., 1999). In vivo studies have confirmed that MyD88 is required for TLR2- and TLR9-mediated NF-κB activation (Hacker et al., 2000).

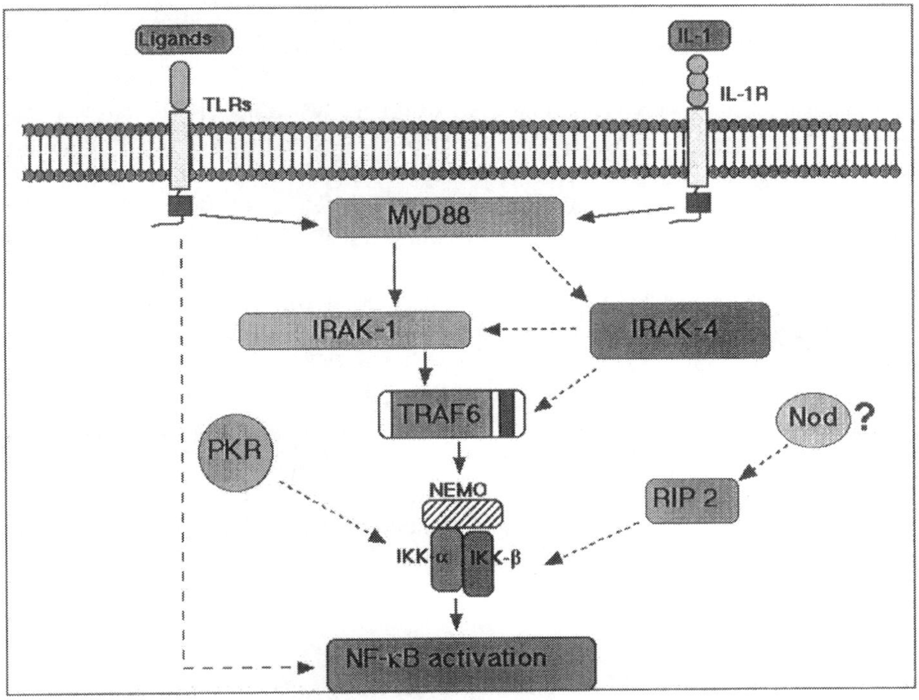

Fig. 3. The signal transduction mechanisms employed by TIRs.

However, MyD88 does not control all TLR-signaling pathways. For example, LPS stimulation of TLR4 still activates NF-κB signaling in MyD88$^{-/-}$ cells (Alexopoulou et al., 2001). LPS-induced DC maturation is also normal in the absence of MyD88 (Kaisho et al., 2001b). MyD88-deficient macrophages fail to produce proinflammatory cytokines upon stimulation with LPS, but LPS-induced NF-κB, JNK and p38 MAPK activation are only slightly delayed in these cells (Alexopoulou et al., 2001). Furthermore, IRF3 activation and induction of IP-10 and GARG16 are all normal in MyD88-deficient cells (Kawai et al., 2001). These results suggest that different TLRs may transduce signals via distinct pathways. TLR signaling that is MyD88-independent may be mediated at least in part by Mal, a MyD88-like adapter protein (Fitzgerald et al., 2001).

7.3. *IRAK-1*

IRAK-1 is the prototypical member of the IRAK family of serine/threonine kinases (Cao et al., 1996a). Upon receptor stimulation, IRAK-1 is recruited to MyD88, possibly through the adapter Tollip (Burns et al., 2000), and is rapidly phosphorylated. However, IRAK-1 is not the sole molecule transducing MyD88 signals, since IRAK-1-deficient mice show impaired, but not abolished responses to IL-1 or LPS stimulation (Kanakaraj et al., 1998; Thomas et al., 1999). Activation of NF-κB and JNK induced by IL-1 or IL-18 is also only mildly inhibited in IRAK-$1^{-/-}$ cells (Adachi et al., 1998). Thus IRAK-1 is important for optimal signal transduction mediated by the IL-1R family and perhaps by TLRs, but other cellular proteins are able to compensate for an absence of IRAK-1 and take over the majority of its in vivo functions. For example, two other IRAK family members, IRAK-2 and IRAK-M, display similar functions to IRAK-1 in tissue culture models (Muzio et al., 1997; Wesche et al., 1999).

Genetic analyses have shown that activated IRAKs transduce downstream signals via the adapter protein TRAF6 (Burns et al., 1998). The latter in turn activates TAK1 through a protein complex that also includes TAB1 and TAB2 (Qian et al., 2001; Takaesu et al., 2001). Studies of gene-targeted mice and cells have revealed that IRAK-1 is required for IL-1-induced activation of TAK1, which subsequently activates NF-κB through the IKK complex (Takaesu et al., 2001). In IRAK-1-deficient cells, both the translocation of TAB2 from the cell membrane to the cytoplasm and the association of TAB2 with TRAF6 are abrogated. IL-1-mediated stimulation of a mutant 293 cell line lacking IRAK-1 also demonstrated a defect in the translocation of TAB2 and TRAF6 to the cytoplasm, as well as a failure in the formation of the TAK1-activating complex (Takaesu et al., 2001).

7.4. *IRAK-4*

IRAK-4 shares the domain structure of other IRAKs and interacts with IRAK-1 and TRAF6 upon IL-1R/TLR stimulation (Li et al., 2002). Expression of kinase-inactive mutants of IRAK-4 in cells inhibits IL-1-induced NF-κB activation even in the presence of IRAK-1, suggesting an important role for IRAK-4 in IL-1 signaling. Gene-targeting studies have confirmed that IRAK-4 is indispensable for the responses of animals and cells to IL-1 and various TLR ligands. IRAK-4-deficient animals are completely resistant to a lethal dose of LPS (Suzuki et al., 2002). In addition, animals lacking IRAK-4 are severely impaired in their defense against viral and bacterial infections (Suzuki et al., 2002). These results indicate that the in vivo function of IRAK-4 cannot be replaced by other IRAK family members.

Activation of NF-κB is dramatically reduced in IRAK-$4^{-/-}$ cells stimulated with IL-1, PGN, dsRNA or bacterial DNA (Suzuki et al., 2002). However, as was true in MyD$88^{-/-}$ cells, LPS-induced NF-κB signaling is not completely abolished in IRAK-4$^{-/-}$ cells. The kinetics of NF-κB activation are similarly delayed in both knockouts, but whereas the peak DNA-binding activity by NF-κB is reduced in IRAK-4$^{-/-}$ cells, it is normal in MyD88$^{-/-}$ cells (Schmidt-Supprian et al., 2000; Suzuki et al., 2002). Much will be learned from a comparative study of IRAK-4$^{-/-}$ and MyD88$^{-/-}$ cells. In particular, the role of IRAK-4 in MyD88-independent DC maturation and signaling pathways merits investigation. Additional comparisons among IRAK family members may elucidate important functional interrelationships that are required for proximal molecular signaling downstream of IL-1R/TLR.

7.5. TRAF6

TRAF6 is the only TRAF family member involved in IL-1R/TLR signaling (Cao et al., 1996b). TRAF6-deficient cells completely lose their ability to respond to IL-1 and LPS and show defects in NF-κB activation and nitric oxide production (Lomaga et al., 1999). However, TRAF6 is also involved in other pathways, such as signals initiated by engagement of CD40, receptor activator of NF-κB (RANK), and other receptors that govern animal growth and survival (Lomaga et al., 1999; Lomaga et al., 2000).

7.6. PKR

The dsRNA-dependent protein kinase PKR is a serine/threonine kinase whose expression is induced by interferon (IFN) (Yang et al., 1995). In the presence of dsRNA, PKR is activated by autophosphorylation and mediates downstream signals leading to NF-κB activation and induction of IFN-regulatory factor 1 (IRF-1) (Kumar et al., 1997). In response to dsRNA, PKR$^{-/-}$ MEFs fail to activate NF-κB and cannot produce IFNα. However, although PKR$^{-/-}$ mice are more susceptible than the wild-type to virus infections, IFNα and IFNβ production appears normal in spleens of PKR$^{-/-}$ mice injected with dsRNA (Yang et al., 1995; Abraham et al., 1999). These results suggest that PKR may play a role in recognizing and transmitting signals induced by a virus or its dsRNA, but the functional relationships between PKR and the TLR-mediated signaling cascade remain to be determined.

7.7. RIP2

RIP2 was introduced above as a CARD domain-containing serine/threonine kinase that plays a role in TCR signaling (Chin et al., 2002). However, expression of RIP2 is inducible by LPS, and RIP2 is capable of activating NF-κB and JNK (McCarthy et al., 1998). These observations raise the possibility that RIP2 may be involved in IL-1R/TLR signaling. Indeed, cytokine production by RIP2-deficient cells is reduced in response to stimulation with LPS, PGN or dsRNA, but not to stimulation with bacterial DNA. RIP2 therefore likely acts downstream of certain TLRs, but not others. RIP2-deficient mice are mildly resistant to LPS-induced endotoxic shock (Chin et al., 2002; Kobayashi et al., 2002). NF-κB activation induced by LPS is somewhat reduced in RIP2-deficient macrophages, but is not completely abolished (Kobayashi et al., 2002).

Interestingly, RIP2 is required for activation of NF-κB by overexpression of NOD1 or NOD2 (Chin et al., 2002). NOD1 and NOD2 are intracellular CARD domain-containing proteins derived from genes with significant homology to disease resistance genes in plants (Inohara et al., 2000; Ogura et al., 2001). The corresponding plant proteins are important mediators of innate immunity in these organisms. In addition, the leucine-rich repeat regions of NOD proteins are similar to the extracellular domains of the TLRs, suggesting that these regions may determine the specificity of pathogen detection. Taken together, these observations imply that the NOD proteins may act as intracellular TLRs that receive and transmit signals induced by microbial pathogens, and that RIP2 may be an important effector downstream of various NOD proteins (Inohara et al., 2000; Ogura et al., 2001). Interestingly, cells lacking RIP2 are hyporesponsive to IL-18 stimulation, but only slightly impaired in IL-1-induced signaling (Chin et al., 2002; Kobayashi et al., 2002). These results suggest that the NOD-RIP2-signaling pathway of NF-κB activation may be distinct from the MyD88-IRAK-TRAF6 cascade. Further studies of cells from various knockout animals will help clarify this issue.

Fig. 4. Signaling pathways of the TNF-R superfamily members.

8. THE TNF-R-SIGNALING PATHWAY

8.1. Scope

TNF-R superfamily members can be categorized into two classes based on their cytoplasmic domains and associated signaling complexes (Baker & Reddy, 1998). The first class consists of the 'death receptors' Fas, TNF-R1, DR3, DR4, DR5 and DR6 (Strater & Moller, 2000). These molecules contain death domains in their intracellular portions that are essential for the induction of apoptosis. Fas, DR4 and DR5 associate directly with the adapter protein FADD, which in turn recruits and activates caspase-8 (Strater & Moller, 2000). Caspase-8 is an upstream caspase that triggers the apoptosis execution cascade (Varfolomeev et al., 1998). Signaling initiated by TNF-R1, DR3 or DR6 engagement involves an additional adapter protein called TNF-R-associated death domain (TRADD) that directly binds to these receptors (Hsu et al., 1995). Recruitment of FADD and caspase-8 by TRADD also results in the execution of the cell death program (Hsu et al., 1996). However, if TRADD recruits TRAF2 and RIP to TNF-R1, signaling leading to NF-κB activation is initiated (Fig. 4) (Hsu et al., 1996). NF-κB activation is also the end result of stimulation of members of the second class of TNF-R family members (Baker & Reddy, 1998). TNF-R2, CD30, CD40 and RANK do not contain death domains and instead associate with various TRAF proteins to signal for NF-κB activation (Bradley & Pober, 2001b). Studies of gene-targeted mice deficient in one or more molecules in TNF-R-related signaling pathways have revealed much about the induction of immune and inflammatory responses, and about pathways leading to apoptosis and cell survival. These results are considered in the following sections, along with the negative feedback regulation of TRAF1, TRAF2 and A20 expression operating in the TNF-R-signaling cascade (Song et al., 1996) (Fig. 5).

8.2. TRAF1

TRAF1 is unique among TRAF family members in that it lacks a RING finger domain and is highly expressed in activated lymphocytes (Rothe et al., 1994). TRAF1 has been implicated as a transducer of signals triggered by the engagement of many TNF-R superfamily members, including TNF-R1 and TNF-R2 (Rothe et al., 1995). Surprisingly, TRAF1-deficient T-cells exhibit increased proliferation when stimulated by anti-CD3 cross-linking. Moreover, activated T-cells from TRAF1$^{-/-}$ mice show a heightened response to TNF stimulation that results in further proliferation and hyperactivation of both the NF-κB and JNK pathways (Tsitsikov et al., 2001). This hypersensitivity to TNF is mediated only by TNF-R2 and not by TNF-R1, despite the fact that TRAF1 associates with both TNF receptors. In vivo, skin tissues of TRAF1-deficient mice are hypersensitive to TNF-induced necrosis (Tsitsikov et al., 2001). Expression of TRAF1 is inducible upon TNF stimulation at least in certain cell types, and the recruitment of TRAF1 to the signaling complex may downregulate signals triggered by the original TNF stimulation.

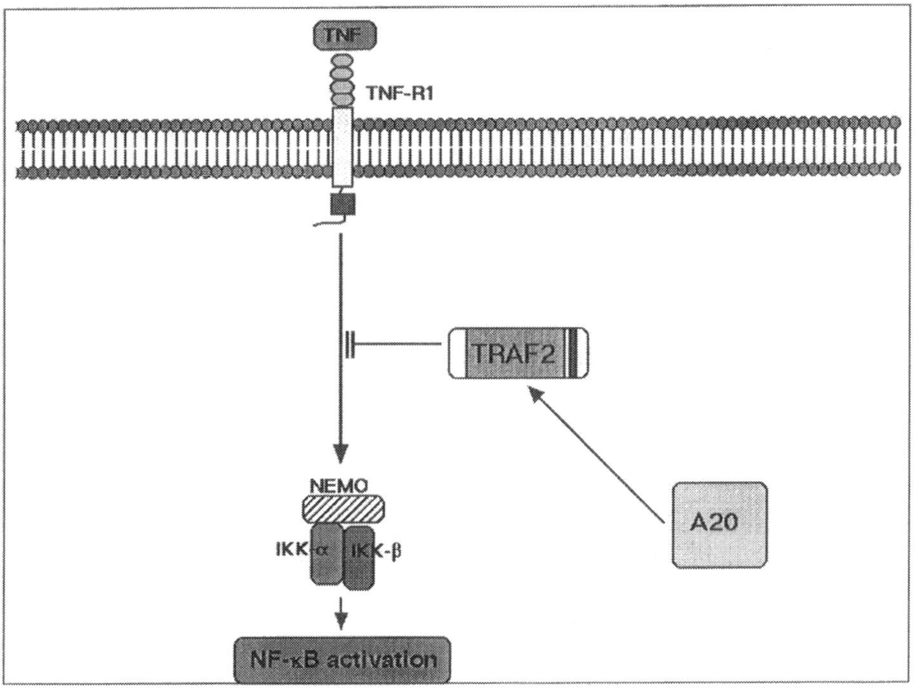

Fig. 5. Negative feedback regulation of TRAF2 and A20 expression operating in the TNF receptor-signaling cascade.

8.3. TRAF2

Overexpression studies have indicated that TRAF2 plays a critical role in transducing signals initiated by TNF-R1, TNF-R2 and CD40 engagement (Tewari & Dixit, 1996). Indeed, biochemical examination of TRAF2$^{-/-}$ cells shows a severe reduction in TNF-mediated JNK activation (Yeh et al., 1997). However, gene-targeting studies have revealed a very complex picture of TRAF2 involvement in TNF-R signaling. Roughly half of TRAF2-deficient animals die at E14.5 with massive liver apoptosis (Yeh et al., 1997), a phenotype strikingly similar to those of RelA$^{-/-}$, IKK$\beta^{-/-}$ and NEMO$^{-/-}$ mice in which NF-κB activation is abrogated. Furthermore, several types of TRAF2-deficient cells, including thymocytes, hematopoietic progenitors and MEFs, are highly sensitive to TNF-induced cell death.

However, biochemical analyses have demonstrated that TRAF2 deficiency has only a mild effect on TNF-induced NF-κB activation, in that the kinetics of NF-κB activation are delayed and the intensity of NF-κB DNA-binding activity is slightly reduced in mutant MEFs (Yeh et al., 1997). These results suggest, first that TRAF2-independent pathways of NF-κB activation exist, and second that TRAF2 mediates an NF-κB-independent signal induced by TNF that protects cells from apoptosis.

TRAF2-deficient mice that survive birth are runty, devoid of fat deposits and muscle mass, and gradually waste away until death. Thymocytes, B-cell precursors and peripheral lymphocytes are all depleted in moribund mutant mice and elevated basal levels of serum TNF are present. Significantly, TRAF2$^{-/-}$TNF-R1$^{-/-}$ and TRAF2$^{-/-}$TNF$^{-/-}$ mutants are viable and generally healthy, indicating that much of the pathology in TRAF2$^{-/-}$ mice is due to unregulated effects of TNF (Nguyen et al., 1999). TRAF2-deficient macrophages are hypersensitive to TNF stimulation and produce copious quantities of inflammatory cytokines and nitric oxide, molecules considered to be targets of NF-κB signaling.

In contrast, studies of TRAF2$^{-/-}$TNF-R1$^{-/-}$ B-cells stimulated by anti-CD40 cross-linking showed that TRAF2 is required for CD40-mediated NF-κB activation and B-cell proliferation (Nguyen et al., 1999). Together, these results imply distinct functions for TRAF2 in mediating signals from different members of the TNF-R superfamily. TRAF2 may also play a role in anchoring the negative regulation of TNF signal transduction that appears to be exerted at later time points during TNF stimulation.

8.4. TRAF3

TRAF3 was originally identified as a signal transducer that interacts with CD40 and mediates NF-κB activation (Cheng et al., 1995). TRAF3-deficient mice exhibit a surprisingly severe phenotype in that these mice become increasingly runty, display progressive hypoglycemia and leukocyte depletion, and die by 10 days of age (Xu et al., 1996). However, TRAF3$^{-/-}$ fetal liver cells (unlike TRAF2$^{-/-}$ cells) can reconstitute all hematopoietic lineages in lethally irradiated mice, arguing against an intrinsic defect of survival in TRAF3-deficient cells. TRAF3$^{-/-}$ reconstituted mice are able to respond normally to a T-independent antigen, but are impaired in their immune responses to a T-dependent antigen (Xu et al., 1996). Despite the circumstances of its original isolation, TRAF3 is likely dispensable for CD40-mediated signaling. TRAF3$^{-/-}$ B-cells stimulated with CD40 ligand show normal upregulation of CD23 and a wild-type proliferative response.

8.5. TRAF5

In vitro studies have implicated TRAF5 in NF-κB activation induced by engagement of the TNF-R superfamily members CD27, CD30, CD40 and lymphotoxin-β receptor (LTβR) (Ishida et al., 1996; Nakano et al., 1996). However, activation of NF-κB induced by CD27 or CD40 is normal in TRAF5-deficient mice (Nakano et al., 1999). A mild defect in the CD40 response is observed in TRAF5$^{-/-}$ B-cells, in that proliferation and upregulation of the surface markers CD23, CD54, CD80, CD86 and Fas is slightly reduced. CD27-mediated costimulatory signaling is impaired in $TRAF5^{-/-}$ T-cells (Nakano et al., 1999). NF-κB activation is normal in TRAF5-deficient cells stimulated with TNF, but cells lacking both TRAF2 and TRAF5 show a more severe impairment of NF-κB activation than TNF-stimulated $TRAF2^{-/-}$ cells (Tada et al., 2001). This result suggests that TRAF2 and TRAF5 have redundant roles in TNF-stimulated NF-κB activation.

8.6. *TRAF6*

In addition to playing a role in IL-1R/TLR signaling, TRAF6 is implicated in various TNF-R-signaling pathways (Cao et al., 1996b). Among the TRAFs, TRAF6 shows the least homology to the prototypical TRAF domain sequence. Since the TRAF domain dictates receptor interaction, this divergence allows TRAF6 to bind to a site on the TNF receptors that is distinct from that bound by other TRAFs (Bradley & Pober, 2001a). For example, TRAF6 binds to a proximal site in the cytoplasmic region of CD40, whereas TRAF2 binds to a distal region. Indeed, $TRAF6^{-/-}$ cells, like $TRAF2^{-/-}$ cells, are impaired in CD40-induced NF-κB activation and cell proliferation, suggesting that the cooperation of TRAF2 and TRAF6 is essential for the normal function of this receptor system (Lomaga et al., 1999).

One striking phenotype displayed by $TRAF6^{-/-}$ mice is osteopetrosis (Lomaga et al., 1999). A severe impairment of osteoclast function results in defects in bone resorption, remodeling and tooth eruption. Both the differentiation and activation of osteoclasts require signals initiated by the binding of osteoprotegerin ligand (OPGL) to its cognate receptor RANK (Kong & Penninger, 2000). Indeed, TRAF6 is an essential mediator for RANK-induced signaling (Lomaga et al., 1999). Another interesting phenotype displayed by about half of $TRAF6^{-/-}$ embryos is exencephaly due to a failure in neural tube closure (Lomaga et al., 2000), a phenotype reminiscent of Bcl-$10^{-/-}$ embryos. Although there is no apparent overlap between TRAF6 and Bcl-10 in the signaling systems we have discussed, studies of these two proteins in the neural tissue will help determine their potential functional relationship.

8.7. *RIP*

RIP is a death domain-containing adapter protein originally identified as interacting with Fas (Stanger et al., 1995). However, RIP is also recruited by TRADD to TNF-R1 upon ligand stimulation, and analysis of RIP-deficient mice has shown that RIP primarily plays a role in TNF-induced NF-κB activation (Kelliher et al., 1998). $RIP^{-/-}$ mice appear normal at birth, but fail to thrive, dying at 1-3 days of age with extensive apoptosis in both lymphoid and adipose tissues. RIP-deficient cells are highly sensitive to TNF-induced apoptosis. TNF stimulation fails to activate NF-κB in RIP-deficient cells, but, unlike $TRAF2^{-/-}$

cells, $RIP^{-/-}$ cells show normal activation of JNK (Kelliher et al., 1998). These results demonstrate that both TRAF2 and RIP play key roles in mediating the survival effects of TNF, but likely do so through distinct signaling pathways initiated from the receptor-proximal protein complex.

8.8. A20

A20 was first identified as a gene whose expression is rapidly induced after TNF stimulation (Opipari et al., 1990). A20 is a cytoplasmic zinc finger-containing protein that interacts with TRAF1 and TRAF2 (Song et al., 1996). Overexpression studies have shown that A20 can inhibit NF-κB activation and TNF-mediated cell death (Song et al., 1996). Studies of A20-deficient mice have further demonstrated that a critical physiological function of A20 is to negatively regulate TNF signaling (Lee et al., 2000). While A20 is not required for embryonic survival, $A20^{-/-}$ mice are runty, develop severe multi-organ inflammation and die prematurely. These mutant animals are abnormally susceptible to sublethal doses of LPS or TNF, and die within 2 hours of injection. A20-deficient cells exhibit prolonged NF-κB activation in response to TNF stimulation and are more sensitive to TNF-induced apoptosis (Lee et al., 2000).

These results suggest that A20 is required to limit signaling induced by TNF, particularly NF-κB activation. After its induction, A20 may be recruited to the TNF-R-signaling complex through interaction with TRAF proteins or NEMO/IKKγ. Binding of A20 to the complex may inhibit further transduction of positive signals. A20 interacts with two other proteins, ABIN-1 and ABIN-2, which also inhibit NF-κB activation (Klinkenberg et al., 2001). Further studies of the physiological functions of A20, ABIN-1 and ABIN-2 will expand our understanding of mechanisms underlying the negative feedback regulation of TNF-induced NF-κB activation.

8.9. NF-κB-inducing kinase (NIK)

NIK was originally implicated as an important molecule transducing signals from proinflammatory cytokines to NF-κB activation (Malinin et al., 1997). Overexpression and biochemical studies indicated that NIK is a key link between the TRAF proteins and the IKK complex (Woronicz et al., 1997). However, $NIK^{-/-}$ MEFs display wild-type biological responses, and NF-κB activation induced by TNF or IL-1 is normal (Yin et al., 2001). Interestingly, $NIK^{-/-}$ mice lack all lymph nodes, a defect in lymph organogenesis very similar to that in mice lacking LTβR (Benedict et al., 2001). Indeed, $NIK^{-/-}$ cells stimulated by LTβR engagement fail to transcribe κB-dependent genes such as IκBα and MCP-1 (Yin et al., 2001). Intriguingly, LTβR-induced NF-κB translocation into nuclei and κB DNA binding are intact in $NIK^{-/-}$ cells (Yin et al., 2001). Like T2K, NIK appears to be required not for the degradation of IκB or translocation of NF-κB, but for the activity of NF-κB in transactivating target genes.

8.10. MEKK3

The MEK protein kinases, like T2K and NIK, are members of the mitogen-activated protein kinase kinase kinase (MAP3K) family (Hagemann & Blank, 2001). Overexpression of MEKK proteins results in the activation of the IKK complex (Madrid et al., 2000). While the roles of MEKK1 and MEKK2 in NF-κB signaling remain controversial (Garrington et al., 2000; Xia et al., 2000), disruption of MEKK3 severely impairs TNF-induced activation of NF-κB (Yang et al., 2001). MEKK3-deficient embryos die at E10.5-11 just as the fetal liver starts to develop. MEKK3$^{-/-}$ cells are highly sensitive to TNF-induced apoptosis, a phenotype consistent with impaired NF-κB signaling. Furthermore, MEKK3 has been shown to interact with RIP and can directly phosphorylate IKK (Yang et al., 2001). MEKK3 may play a role in proinflammatory cytokine-induced NF-κB activation by providing a link between RIP and the IKK complex.

8.11. GSK3β

GSK3β is an inhibitory component of the Wnt-signaling pathway important for embryonic development (Rubinfeld et al., 1996). In frogs and zebrafish, GSK3 kinase homologues are required for the definition of the embryonic axes (Dierick & Bejsovec, 1999). Surprisingly, GSK3β-deficient mice show a specific and limited phenotype of fetal liver apoptosis similar to that observed in mice with NF-κB activation defects (Hoeflich et al., 2000). GSK3β$^{-/-}$ MEFs are highly susceptible to TNF-induced apoptosis, consistent with the observation that wild-type fibroblasts treated with the GSK3 inhibitor lithium are abnormally sensitive to TNF killing (Hoeflich et al., 2000). Interestingly, TNF-induced IκB phosphorylation and degradation are normal in GSK3β$^{-/-}$ cells (Hoeflich et al., 2000). However, DNA-binding activity was somewhat reduced and NF-κB activation was diminished in GSK3β$^{-/-}$ cells, leading to a severe defect in transcriptional activation of a target reporter gene (Hoeflich et al., 2000; Pomerantz & Baltimore, 2000).

8.12. Atypical protein kinase C (PKCζ)

PKCζ has also been implicated in TNF-induced NF-κB activation (Sanz et al., 2000). PKCζ can associate with RIP through the adapter protein p62 (Sanz et al., 2000). Deficiency of PKCζ severely impairs cellular responses to TNF that depend on the transcriptional activity of NF-κB (Leitges et al., 2001). However, IKK activation is normal in PKCζ-deficient MEFs as for cells lacking T2K or GSK3β, and NF-κB DNA binding in response to TNF is only mildly reduced in PKCζ$^{-/-}$ cells (Leitges et al., 2001). RelA phosphorylation is impaired in PKCζ$^{-/-}$ cells, suggesting that PKCζ is involved in an IKK-independent pathway of NF-κB activation. PKCζ$^{-/-}$ mice do not exhibit fetal liver apoptosis or embryonic lethality, but instead display a reduced number of Peyer's patches. Consistent with an NF-κB-signaling defect, PKCζ $^{-/-}$ MEFs are hypersensitive to TNF-induced apoptosis (Leitges et al., 2001).

9. CONCLUSION

In this chapter we have described the contributions of knockout mice to the elucidation of NF-κB signaling. As more genes are studied using this technique, we can expect to 'complete' the signaling pathways established by other methodologies. In many cases, overexpression and biochemical approaches can accurately predict the physiological functions of proteins that are later confirmed by knockout phenotypes. Where a conflict exists, as arises when a phenotype predicted by overexpression studies is absent, the physiological function of the protein of interest can still be reconciled by concluding redundancy. While the gene-targeting approach is limited in that it shows only the nonredundant functions of a given molecule, it has the added value of revealing unexpected phenotypes derived from a previously unsuspected feature or function of the protein of interest. Studies of animals mutated for genes encoding the NF-κB-signaling molecules involved in three major signaling systems and those lacking proteins involved in the NF-κB execution machinery itself have confirmed or discovered the existence of major players that regulate inflammatory responses, apoptosis and cell survival. Our next mission in this field of study is to utilize these cornerstones to identify more important signaling proteins in each signaling cascade and to construct a blueprint of the regulatory network that controls NF-κB. Determining the specificity and commonality of different signaling pathways will enable us to devise therapeutic strategies in the near future to combat infection, autoimmunity and cancer, diseases which are all linked to deregulation of NF-κB signaling.

10. REFERENCES

Abraham N, Stojdl DF, Duncan PI, Methot N, Ishii T, Dube M, Vanderhyden BC, Atkins HL, Gray DA, McBurney MW, Koromilas AE, Brown EG, Sonenberg N & Bell JC (1999) Characterization of transgenic mice with targeted disruption of the catalytic domain of the double-stranded RNA-dependent protein kinase, PKR. J. Biol. Chem. 274: 5953-5962.

Adachi O, Kawai T, Takeda K, Matsumoto M, Tsutsui H, Sakagami M, Nakanishi K & Akira S (1998) Targeted disruption of the MyD88 gene results in loss of IL-1- and IL-18-mediated function. Immunity 9: 143-150.

Akagi T, Motegi M, Tamura A, Suzuki R, Hosokawa Y, Suzuki H, Ota H, Nakamura S, Morishima Y, Taniwaki M & Seto M (1999) A novel gene, MALT1 at 18q21, is involved in t(11;18)(q21;q21) found in low-grade B-cell lymphoma of mucosa-associated lymphoid tissue. Oncogene 18: 5785-5794.

Akira S, Takeda K & Kaisho T (2001) Toll-like receptors: critical proteins linking innate and acquired immunity. Nat. Immunol. 2: 675-680.

Alcamo E, Mizgerd JP, Horwitz BH, Bronson R, Beg AA, Scott M, Doerschuk CM, Hynes RO & Baltimore D (2001) Targeted mutation of TNF receptor I rescues the RelA-deficient mouse and reveals a critical role for NF-κB in leukocyte recruitment. J. Immunol. 167: 1592-1600.

Alcamo E, Hacohen N, Schulte LC, Rennert PD, Hynes RO & Baltimore D (2002) Requirement for the NF-κB family member RelA in the development of secondary lymphoid organs. J. Exp. Med. 195: 233-244.

Alexopoulou L, Holt AC, Medzhitov R & Flavell RA (2001) Recognition of double-stranded RNA and activation of NF-κB by Toll-like receptor 3. Nature 413: 732-738.

Arendt CW, Albrecht B, Soos TJ & Littman DR (2002) Protein kinase C-θ: signaling from the center of the T-cell synapse. Curr. Opin. Immunol. 14: 323-330.

Baier G, Telford D, Giampa L, Coggeshall KM, Baier-Bitterlich G, Isakov N & Altman A (1993) Molecular cloning and characterization of PKCθ, a novel member of the protein kinase C (PKC) gene family expressed predominantly in hematopoietic cells. J. Biol. Chem. 268: 4997-5004.

Baier-Bitterlich G, Uberall F, Bauer B, Fresser F, Wachter H, Grunicke H, Utermann G, Altman A & Baier G (1996) Protein kinase C-θ isoenzyme selective stimulation of the transcription factor complex AP-1 in T lymphocytes. Mol. Cell. Biol. 16: 1842-1850.

Baker SJ & Reddy EP (1998) Modulation of life and death by the TNF receptor superfamily. Oncogene 17: 3261-3270.

Beg AA, Sha WC, Bronson RT & Baltimore D (1995a) Constitutive NF-κB activation, enhanced granulopoiesis, and neonatal lethality in IκBα-deficient mice. Genes Dev. 9: 2736-2746.

Beg AA, Sha WC, Bronson RT, Ghosh S & Baltimore D (1995b) Embryonic lethality and liver degeneration in mice lacking the RelA component of NF-κB. Nature 376: 167-170.

Beg AA & Baltimore D (1996) An essential role for NF-κB in preventing TNF-α-induced cell death. Science 274: 782-784.

Benedict CA, Banks TA, Senderowicz L, Ko M, Britt WJ, Angulo A, Ghazal P & Ware CF (2001) Lymphotoxins and cytomegalovirus cooperatively induce interferon-β, establishing host-virus detente. Immunity 15: 617-626.

Bertin J, Wang L, Guo Y, Jacobson MD, Poyet JL, Srinivasula SM, Merriam S, DiStefano PS & Alnemri ES (2001) CARD11 and CARD14 are novel caspase recruitment domain (CARD)/membrane-associated guanylate kinase (MAGUK) family members that interact with BCL10 and activate NF-κB. J. Biol. Chem. 276: 11877-11882.

Bonnard M, Mirtsos C, Suzuki S, Graham K, Huang J, Ng M, Itie A, Wakeham A, Shahinian A, Henzel WJ, Elia AJ, Shillinglaw W, Mak TW, Cao Z & Yeh WC (2000) Deficiency of T2K leads to apoptotic liver degeneration and impaired NF-κB-dependent gene transcription. EMBO J. 19: 4976-4985.

Bradley JR & Pober JS (2001a) Tumor necrosis factor receptor-associated factors (TRAFs). Oncogene 20: 6482-6491.

Bradley JR & Pober JS (2001b) Tumor necrosis factor receptor-associated factors (TRAFs). Oncogene 20: 6482-6491.

Burkly L, Hession C, Ogata L, Reilly C, Marconi LA, Olson D, Tizard R, Cate R & Lo D (1995) Expression of relB is required for the development of thymic medulla and dendritic cells. Nature 373: 531-536.

Burns K, Martinon F, Esslinger C, Pahl H, Schneider P, Bodmer JL, Di Marco F, French L & Tschopp J (1998) MyD88, an adapter protein involved in interleukin-1 signaling. J. Biol. Chem. 273: 12203-12209.

Burns K, Clatworthy J, Martin L, Martinon F, Plumpton C, Maschera B, Lewis A, Ray K, Tschopp J & Volpe F (2000) Tollip, a new component of the IL-1RI pathway, links IRAK to the IL-1 receptor. Nat. Cell Biol. 2: 346-351.

Caamano JH, Rizzo CA, Durham SK, Barton DS, Raventos-Suarez C, Snapper CM & Bravo R (1998) Nuclear factor (NF)-κB2 (p100/p52) is required for normal splenic microarchitecture and B cell-mediated immune responses. J. Exp. Med. 187: 185-196.

Cao Z, Henzel WJ & Gao X (1996a) IRAK: a kinase associated with the interleukin-1 receptor. Science 271: 1128-1131.

Cao Z, Xiong J, Takeuchi M, Kurama T & Goeddel DV (1996b) TRAF6 is a signal transducer for interleukin-1. Nature 383: 443-446.

Cheng G, Cleary AM, Ye ZS, Hong DI, Lederman S & Baltimore D (1995) Involvement of CRAF1, a relative of TRAF, in CD40 signaling. Science 267: 1494-1498.

Cheng G & Baltimore D (1996) TANK, a co-inducer with TRAF2 of TNF- and CD 40L-mediated NF-κB activation. Genes Dev. 10: 963-973.

Chin AI, Dempsey PW, Bruhn K, Miller JF, Xu Y & Cheng G (2002) Involvement of receptor-interacting protein 2 in innate and adaptive immune responses. Nature 416: 190-194.

Costello PS, Walters AE, Mee PJ, Turner M, Reynolds LF, Prisco A, Sarner N, Zamoyska R & Tybulewicz VL (1999) The Rho-family GTP exchange factor Vav is a critical transducer of T cell receptor signals to the calcium, ERK, and NF-κB pathways. Proc. Natl. Acad. Sci. USA 96: 3035-3040.

DeKoning J, DiMolfetto L, Reilly C, Wei Q, Havran WL & Lo D (1997) Thymic cortical epithelium is sufficient for the development of mature T cells in relB-deficient mice. J. Immunol. 158: 2558-2566.

Delhase M, Hayakawa M, Chen Y & Karin M (1999) Positive and negative regulation of IκB kinase activity through IKKβ subunit phosphorylation. Science 284: 309-313.

Dierick H & Bejsovec A (1999) Cellular mechanisms of wingless/Wnt signal transduction. Curr. Top. Dev. Biol. 43: 153-190.

Doi TS, Takahashi T, Taguchi O, Azuma T & Obata Y (1997) NF-κB RelA-deficient lymphocytes: normal development of T cells and B cells, impaired production of IgA and IgG1 and reduced proliferative responses. J. Exp. Med. 185: 953-961.

Doi TS, Marino MW, Takahashi T, Yoshida T, Sakakura T, Old LJ & Obata Y (1999) Absence of tumor necrosis factor rescues RelA-deficient mice from embryonic lethality. Proc. Natl. Acad. Sci. USA 96: 2994-2999.

Doody GM, Billadeau DD, Clayton E, Hutchings A, Berland R, McAdam S, Leibson PJ & Turner M (2000) Vav-2 controls NFAT-dependent transcription in B- but not T-lymphocytes. EMBO J. 19: 6173-6184.

Fitzgerald KA, Palsson-McDermott EM, Bowie AG, Jefferies CA, Mansell AS, Brady G, Brint E, Dunne A, Gray P, Harte MT, McMurray D, Smith DE, Sims JE, Bird TA & O'Neill LA (2001) Mal (MyD88-adapter-like) is required for Toll-like receptor-4 signal transduction. Nature 413: 78-83.

Franzoso G, Carlson L, Scharton-Kersten T, Shores EW, Epstein S, Grinberg A, Tran T, Shacter E, Leonardi A, Anver M, Love P, Sher A & Siebenlist U (1997) Critical roles for the Bcl-3 oncoprotein in T cell-mediated immunity, splenic microarchitecture, and germinal center reactions. Immunity 6: 479-490.

Franzoso G, Carlson L, Poljak L, Shores EW, Epstein S, Leonardi A, Grinberg A, Tran T, Scharton-Kersten T, Anver M, Love P, Brown K & Siebenlist U (1998) Mice deficient in nuclear factor (NF)-κB/p52 present with defects in humoral responses, germinal center reactions, and splenic microarchitecture. J. Exp. Med. 187: 147-159.

Garrington TP, Ishizuka T, Papst PJ, Chayama K, Webb S, Yujiri T, Sun W, Sather S, Russell DM, Gibson SB, Keller G, Gelfand EW & Johnson GL (2000) MEKK2 gene disruption causes loss of cytokine production in response to IgE and c-Kit ligand stimulation of ES cell-derived mast cells. EMBO J. 19: 5387-5395.

Ghosh S, May MJ & Kopp EB (1998) NF-κB and Rel proteins: evolutionarily conserved mediators of immune responses. Annu. Rev. Immunol. 16: 225-260.

Hacker H, Vabulas RM, Takeuchi O, Hoshino K, Akira S & Wagner H (2000) Immune cell activation by bacterial CpG-DNA through myeloid differentiation marker 88 and tumor necrosis factor receptor-associated factor (TRAF)6. J. Exp. Med. 192: 595-600.

Hagemann C & Blank JL (2001) The ups and downs of MEK kinase interactions. Cell. Signal. 13: 863-875.

Hoeflich KP, Luo J, Rubie EA, Tsao MS, Jin O & Woodgett JR (2000) Requirement for glycogen synthase kinase-3β in cell survival and NF-κB activation. Nature 406: 86-90.

Horwitz BH, Scott ML, Cherry SR, Bronson RT & Baltimore D (1997) Failure of lymphopoiesis after adoptive transfer of NF-κB-deficient fetal liver cells. Immunity 6: 765-772.

Hsu H, Xiong J & Goeddel DV (1995) The TNF receptor 1-associated protein TRADD signals cell death and NF-κB activation. Cell 81: 495-504.

Hsu H, Shu HB, Pan MG & Goeddel DV (1996) TRADD-TRAF2 and TRADD-FADD interactions define two distinct TNF receptor 1 signal transduction pathways. Cell 84: 299-308.

Hu Y, Baud V, Delhase M, Zhang P, Deerinck T, Ellisman M, Johnson R & Karin M (1999) Abnormal morphogenesis but intact IKK activation in mice lacking the IKKα subunit of IκB kinase. Science 284: 316-320.

Hu Y, Baud V, Oga T, Kim KI, Yoshida K & Karin M (2001) IKKα controls formation of the epidermis independently of NF-κB. Nature 410: 710-714.

Inohara N, Koseki T, Lin J, del Peso L, Lucas PC, Chen FF, Ogura Y & Nunez G (2000) An induced proximity model for NF-κB activation in the Nod1/RICK and RIP signaling pathways. J. Biol. Chem. 275: 27823-27831.

Ishida TK, Tojo T, Aoki T, Kobayashi N, Ohishi T, Watanabe T, Yamamoto T & Inoue J (1996) TRAF5, a novel tumor necrosis factor receptor-associated factor family protein, mediates CD40 signaling. Proc. Natl. Acad. Sci. USA 93: 9437-9442.

Kaisho T, Takeda K, Tsujimura T, Kawai T, Nomura F, Terada N & Akira S (2001a) IκB kinase α is essential for mature B cell development and function. J. Exp. Med. 193: 417-426.

Kaisho T, Takeuchi O, Kawai T, Hoshino K & Akira S (2001b) Endotoxin-induced maturation of MyD88-deficient dendritic cells. J. Immunol. 166: 5688-5694.

Kanakaraj P, Schafer PH, Cavender DE, Wu Y, Ngo K, Grealish PF, Wadsworth SA, Peterson PA, Siekierka JJ, Harris CA & Fung-Leung WP (1998) Interleukin (IL)-1 receptor-associated kinase (IRAK) requirement for optimal induction of multiple IL-1 signaling pathways and IL-6 production. J. Exp. Med. 187: 2073-2079.

Karin M & Ben-Neriah Y (2000) Phosphorylation meets ubiquitination: the control of NF-κB activity. Annu. Rev. Immunol. 18: 621-663.

Kawai T, Adachi O, Ogawa T, Takeda K & Akira S (1999) Unresponsiveness of MyD88-deficient mice to endotoxin. Immunity 11: 115-122.

Kawai T, Takeuchi O, Fujita T, Inoue J, Muhlradt PF, Sato S, Hoshino K & Akira S (2001) Lipopolysaccharide stimulates the MyD88-independent pathway and results in activation of IFN-regulatory factor 3 and the expression of a subset of lipopolysaccharide-inducible genes. J. Immunol. 167: 5887-5894.

Kelliher MA, Grimm S, Ishida Y, Kuo F, Stanger BZ & Leder P (1998) The death domain kinase RIP mediates the TNF-induced NF-κB signal. Immunity 8: 297-303.

Kimbrell DA & Beutler B (2001) The evolution and genetics of innate immunity. Nat Rev Genet 2: 256-267.

Klement JF, Rice NR, Car BD, Abbondanzo SJ, Powers GD, Bhatt PH, Chen CH, Rosen CA & Stewart CL (1996) IκBα deficiency results in a sustained NF-κB response and severe widespread dermatitis in mice. Mol. Cell. Biol. 16: 2341-2349.

Klinkenberg M, Van Huffel S, Heyninck K & Beyaert R (2001) Functional redundancy of the zinc fingers of A20 for inhibition of NF-κB activation and protein-protein interactions. FEBS Lett. 498: 93-97.

Kobayashi K, Inohara N, Hernandez LD, Galan JE, Nunez G, Janeway CA, Medzhitov R & Flavell RA (2002) RICK/Rip2/CARDIAK mediates signalling for receptors of the innate and adaptive immune systems. Nature 416: 194-199.

Kong YY & Penninger JM (2000) Molecular control of bone remodeling and osteoporosis. Exp. Gerontol. 35: 947-956.

Kontgen F, Grumont RJ, Strasser A, Metcalf D, Li R, Tarlinton D & Gerondakis S (1995) Mice lacking the c-rel proto-oncogene exhibit defects in lymphocyte proliferation, humoral immunity, and interleukin-2 expression. Genes Dev. 9: 1965-1977.

Kumar A, Yang YL, Flati V, Der S, Kadereit S, Deb A, Haque J, Reis L, Weissmann C & Williams BR (1997) Deficient cytokine signaling in mouse embryo fibroblasts with a targeted deletion in the PKR gene: role of IRF-1 and NF-κB. EMBO J. 16: 406-416.

Lee EG, Boone DL, Chai S, Libby SL, Chien M, Lodolce JP & Ma A (2000) Failure to regulate TNF-induced NF-κB and cell death responses in A20-deficient mice. Science 289: 2350-2354.

Leitges M, Sanz L, Martin P, Duran A, Braun U, Garcia JF, Camacho F, Diaz-Meco MT, Rennert

PD & Moscat J (2001) Targeted disruption of the ζPKC gene results in the impairment of the NF-κB pathway. Mol. Cell 8: 771-780.

Li Q, Van Antwerp D, Mercurio F, Lee KF & Verma IM (1999) Severe liver degeneration in mice lacking the IκB kinase 2 gene. Science 284: 321-325.

Li S, Strelow A, Fontana EJ & Wesche H (2002) IRAK-4: A novel member of the IRAK family with the properties of an IRAK-kinase. Proc. Natl. Acad. Sci. USA 99: 5567-5572.

Lomaga MA, Yeh WC, Sarosi I, Duncan GS, Furlonger C, Ho A, Morony S, Capparelli C, Van G, Kaufman S, van der Heiden A, Itie A, Wakeham A, Khoo W, Sasaki T, Cao Z, Penninger JM, Paige CJ, Lacey DL, Dunstan CR, Boyle WJ, Goeddel DV & Mak TW (1999) TRAF6 deficiency results in osteopetrosis and defective interleukin-1, CD40, and LPS signaling. Genes Dev. 13: 1015-1024.

Lomaga MA, Henderson JT, Elia AJ, Robertson J, Noyce RS, Yeh WC & Mak TW (2000) Tumor necrosis factor receptor-associated factor 6 (TRAF6) deficiency results in exencephaly and is required for apoptosis within the developing CNS. J. Neurosci. 20: 7384-7393.

Madrid LV, Wang CY, Guttridge DC, Schottelius AJ, Baldwin AS & Mayo MW (2000) Akt suppresses apoptosis by stimulating the transactivation potential of the RelA/p65 subunit of NF-κB. Mol. Cell. Biol. 20: 1626-1638.

Makris C, Godfrey VL, Krahn-Senftleben G, Takahashi T, Roberts JL, Schwarz T, Feng L, Johnson RS & Karin M (2000) Female mice heterozygous for IKKγ/NEMO deficiencies develop a dermatopathy similar to the human X-linked disorder incontinentia pigmenti. Mol. Cell 5: 969-979.

Malinin NL, Boldin MP, Kovalenko AV & Wallach D (1997) MAP3K-related kinase involved in NF-κB induction by TNF, CD95 and IL-1. Nature 385: 540-544.

McAllister-Lucas LM, Inohara N, Lucas PC, Ruland J, Benito A, Li Q, Chen S, Chen FF, Yamaoka S, Verma IM, Mak TW & Nunez G (2001) Bimp1, a MAGUK family member linking protein kinase C activation to Bcl10-mediated NF-κB induction. J. Biol. Chem. 276: 30589-30597.

McCarthy JV, Ni J & Dixit VM (1998) RIP2 is a novel NF-κB-activating and cell death-inducing kinase. J. Biol. Chem. 273: 16968-16975.

Medzhitov R, Preston-Hurlburt P, Kopp E, Stadlen A, Chen C, Ghosh S & Janeway CA (1998) MyD88 is an adaptor protein in the hToll/IL-1 receptor family signaling pathways. Mol. Cell 2: 253-258.

Muzio M, Ni J, Feng P & Dixit VM (1997) IRAK (Pelle) family member IRAK-2 and MyD88 as proximal mediators of IL-1 signaling. Science 278: 1612-1615.

Nakano H, Oshima H, Chung W, Williams-Abbott L, Ware CF, Yagita H & Okumura K (1996) TRAF5, an activator of NF-κB and putative signal transducer for the lymphotoxin-β receptor. J. Biol. Chem. 271: 14661-14664.

Nakano H, Sakon S, Koseki H, Takemori T, Tada K, Matsumoto M, Munechika E, Sakai T, Shirasawa T, Akiba H, Kobata T, Santee SM, Ware CF, Rennert PD, Taniguchi M, Yagita H & Okumura K (1999) Targeted disruption of Traf5 gene causes defects in CD40- and CD27-mediated lymphocyte activation. Proc. Natl. Acad. Sci. USA 96: 9803-9808.

Nguyen LT, Duncan GS, Mirtsos C, Ng M, Speiser DE, Shahinian A, Marino MW, Mak TW, Ohashi PS & Yeh WC (1999) TRAF2 deficiency results in hyperactivity of certain TNFR1 signals and impairment of CD40-mediated responses. Immunity 11: 379-389.

Ogura Y, Inohara N, Benito A, Chen FF, Yamaoka S & Nunez G (2001) Nod2, a Nod1/Apaf-1 family member that is restricted to monocytes and activates NF-κB. J. Biol. Chem. 276: 4812-4818.

O'Neill L (2000) The Toll/interleukin-1 receptor domain: a molecular switch for inflammation and host defence. Biochem. Soc. Trans. 28: 557-563.

Opipari AW, Boguski MS & Dixit VM (1990) The A20 cDNA induced by tumor necrosis factor α encodes a novel type of zinc finger protein. J. Biol. Chem. 265: 14705-14708.

Poljak L, Carlson L, Cunningham K, Kosco-Vilbois MH & Siebenlist U (1999) Distinct activities of p52/NF-κB required for proper secondary lymphoid organ microarchitecture: functions enhanced by Bcl-3. J. Immunol. 163: 6581-6588.

Pomerantz JL & Baltimore D (1999) NF-κB activation by a signaling complex containing TRAF2, TANK and TBK1, a novel IKK-related kinase. EMBO J. 18: 6694-6704.

Pomerantz JL & Baltimore D (2000) Signal transduction. A cellular rescue team. Nature 406: 26-27, 29.

Qian Y, Commane M, Ninomiya-Tsuji J, Matsumoto K & Li X (2001) IRAK-mediated translocation of TRAF6 and TAB2 in the interleukin-1-induced activation of NFκB. J. Biol. Chem. 276: 41661-41667.

Rothe M, Wong SC, Henzel WJ & Goeddel DV (1994) A novel family of putative signal transducers associated with the cytoplasmic domain of the 75 kDa tumor necrosis factor receptor. Cell 78: 681-692.

Rothe M, Pan MG, Henzel WJ, Ayres TM & Goeddel DV (1995) The TNFR2-TRAF signaling complex contains two novel proteins related to baculoviral inhibitor of apoptosis proteins. Cell 83: 1243-1252.

Rothwarf DM, Zandi E, Natoli G & Karin M (1998) IKK-γ is an essential regulatory subunit of the IκB kinase complex. Nature 395: 297-300.

Rubinfeld B, Albert I, Porfiri E, Fiol C, Munemitsu S & Polakis P (1996) Binding of GSK3β to the APC-β-catenin complex and regulation of complex assembly. Science 272: 1023-1026.

Rudolph D, Yeh WC, Wakeham A, Rudolph B, Nallainathan D, Potter J, Elia AJ & Mak TW (2000) Severe liver degeneration and lack of NF-κB activation in NEMO/IKKγ-deficient mice. Genes Dev. 14: 854-862.

Ruland J, Duncan GS, Elia A, del Barco Barrantes I, Nguyen L, Plyte S, Millar DG, Bouchard D, Wakeham A, Ohashi PS & Mak TW (2001) Bcl10 is a positive regulator of antigen receptor-induced activation of NF-κB and neural tube closure. Cell 104: 33-42.

Sanz L, Diaz-Meco MT, Nakano H & Moscat J (2000) The atypical PKC-interacting protein p62 channels NF-κB activation by the IL-1-TRAF6 pathway. EMBO J. 19: 1576-1586.

Schmidt-Supprian M, Bloch W, Courtois G, Addicks K, Israel A, Rajewsky K & Pasparakis M (2000) NEMO/IKKγ-deficient mice model incontinentia pigmenti. Mol. Cell 5: 981-992.

Schuebel KE, Bustelo XR, Nielsen DA, Song BJ, Barbacid M, Goldman D & Lee IJ (1996) Isolation and characterization of murine vav2, a member of the vav family of proto-oncogenes. Oncogene 13: 363-371.

Schwarz EM, Krimpenfort P, Berns A & Verma IM (1997) Immunological defects in mice with a targeted disruption in Bcl-3. Genes Dev. 11: 187-197.

Senftleben U, Cao Y, Xiao G, Greten FR, Krahn G, Bonizzi G, Chen Y, Hu Y, Fong A, Sun SC & Karin M (2001a) Activation by IKKα of a second, evolutionary conserved, NF-κB signaling pathway. Science 293: 1495-1499.

Senftleben U, Li ZW, Baud V & Karin M (2001b) IKKβ is essential for protecting T cells from TNFα-induced apoptosis. Immunity 14: 217-230.

Sha WC, Liou HC, Tuomanen EI & Baltimore D (1995) Targeted disruption of the p50 subunit of NF-κB leads to multifocal defects in immune responses. Cell 80: 321-330.

Shih LY, Fu JF, Shurtleff SA, Morris SW & Downing JR (2001) Lack of BCL10 mutations in multiple myeloma and plasma cell leukemia. Genes Chromosomes Cancer 30: 402-406.

Shimada T, Kawai T, Takeda K, Matsumoto M, Inoue J, Tatsumi Y, Kanamaru A & Akira S (1999) IKK-i, a novel lipopolysaccharide-inducible kinase that is related to IκB kinases. Int. Immunol. 11: 1357-1362.

Smahi A, Courtois G, Vabres P, Yamaoka S, Heuertz S, Munnich A, Israel A, Heiss NS, Klauck SM, Kioschis P, Wiemann S, Poustka A, Esposito T, Bardaro T, Gianfrancesco F, Ciccodicola A, D'Urso M, Woffendin H, Jakins T, Donnai D, Stewart H, Kenwrick SJ, Aradhya S, Yamagata T, Levy M, Lewis RA & Nelson DL (2000) Genomic rearrangement in NEMO impairs NF-κB activation and is a cause of incontinentia pigmenti. Nature 405: 466-472.

Song HY, Rothe M & Goeddel DV (1996) The tumor necrosis factor-inducible zinc finger protein A20 interacts with TRAF1/TRAF2 and inhibits NF-κB activation. Proc. Natl. Acad. Sci. USA 93: 6721-6725.

Stanger BZ, Leder P, Lee TH, Kim E & Seed B (1995) RIP: a novel protein containing a death domain that interacts with Fas/APO-1 (CD95) in yeast and causes cell death. Cell 81: 513-523.

Strater J & Moller P (2000) Expression and function of death receptors and their natural ligands in

the intestine. Ann. NY Acad. Sci. 915: 162-170.
Sun Z, Arendt CW, Ellmeier W, Schaeffer EM, Sunshine MJ, Gandhi L, Annes J, Petrzilka D, Kupfer A, Schwartzberg PL & Littman DR (2000) PKC-θ is required for TCR-induced NF-κB activation in mature but not immature T lymphocytes. Nature 404: 402-407.
Suzuki N, Suzuki S, Duncan GS, Millar DG, Wada T, Mirtsos C, Takada H, Wakeham A, Itie A, Li S, Penninger JM, Wesche H, Ohashi PS, Mak TW & Yeh WC (2002) Severe impairment of interleukin-1 and Toll-like receptor signalling in mice lacking IRAK-4. Nature 416: 750-756.
Tada K, Okazaki T, Sakon S, Kobarai T, Kurosawa K, Yamaoka S, Hashimoto H, Mak TW, Yagita H, Okumura K, Yeh WC & Nakano H (2001) Critical roles of TRAF2 and TRAF5 in tumor necrosis factor-induced NF-κB activation and protection from cell death. J. Biol. Chem. 276: 36530-36534.
Takaesu G, Ninomiya-Tsuji J, Kishida S, Li X, Stark GR & Matsumoto K (2001) Interleukin-1 (IL-1) receptor-associated kinase leads to activation of TAK1 by inducing TAB2 translocation in the IL-1 signaling pathway. Mol. Cell. Biol. 21: 2475-2484.
Takeda K, Takeuchi O, Tsujimura T, Itami S, Adachi O, Kawai T, Sanjo H, Yoshikawa K, Terada N & Akira S (1999) Limb and skin abnormalities in mice lacking IKKα. Science 284: 313-316.
Tanaka M, Fuentes ME, Yamaguchi K, Durnin MH, Dalrymple SA, Hardy KL & Goeddel DV (1999) Embryonic lethality, liver degeneration, and impaired NF-κB activation in IKK-β-deficient mice. Immunity 10: 421-429.
Tewari M & Dixit VM (1996) Recent advances in tumor necrosis factor and CD40 signaling. Curr. Opin. Genet. Dev. 6: 39-44.
Thomas JA, Allen JL, Tsen M, Dubnicoff T, Danao J, Liao XC, Cao Z & Wasserman SA (1999) Impaired cytokine signaling in mice lacking the IL-1 receptor-associated kinase. J. Immunol. 163: 978-984.
Tojima Y, Fujimoto A, Delhase M, Chen Y, Hatakeyama S, Nakayama K, Kaneko Y, Nimura Y, Motoyama N, Ikeda K, Karin M & Nakanishi M (2000) NAK is an IκB kinase-activating kinase. Nature 404: 778-782.
Tsitsikov EN, Laouini D, Dunn IF, Sannikova TY, Davidson L, Alt FW & Geha RS (2001) TRAF1 is a negative regulator of TNF signaling. enhanced TNF signaling in TRAF1-deficient mice. Immunity 15: 647-657.
Varfolomeev EE, Schuchmann M, Luria V, Chiannilkulchai N, Beckmann JS, Mett IL, Rebrikov D, Brodianski VM, Kemper OC, Kollet O, Lapidot T, Soffer D, Sobe T, Avraham KB, Goncharov T, Holtmann H, Lonai P & Wallach D (1998) Targeted disruption of the mouse caspase 8 gene ablates cell death induction by the TNF receptors, Fas/Apo1, and DR3 and is lethal prenatally. Immunity 9: 267-276.
Voll RE & Ghosh S (1999) Role of NF-κB in T-lymphocyte development. Cold Spring Harb. Symp. Quant. Biol. 64: 485-490.
Wang L, Guo Y, Huang WJ, Ke X, Poyet JL, Manji GA, Merriam S, Glucksmann MA, DiStefano PS, Alnemri ES & Bertin J (2001) Card10 is a novel caspase recruitment domain/membrane-associated guanylate kinase family member that interacts with BCL10 and activates NF-κB. J. Biol. Chem. 276: 21405-21409.
Weih F, Carrasco D, Durham SK, Barton DS, Rizzo CA, Ryseck RP, Lira SA & Bravo R (1995) Multiorgan inflammation and hematopoietic abnormalities in mice with a targeted disruption of RelB, a member of the NF-κB/Rel family. Cell 80: 331-340.
Wesche H, Henzel WJ, Shillinglaw W, Li S & Cao Z (1997) MyD88: an adapter that recruits IRAK to the IL-1 receptor complex. Immunity 7: 837-847.
Wesche H, Gao X, Li X, Kirschning CJ, Stark GR & Cao Z (1999) IRAK-M is a novel member of the Pelle/interleukin-1 receptor-associated kinase (IRAK) family. J. Biol. Chem. 274: 19403-19410.
Woronicz JD, Gao X, Cao Z, Rothe M & Goeddel DV (1997) IκB kinase-β: NF-κB activation and complex formation with IκB kinase-α and NIK. Science 278: 866-869.
Xia Y, Makris C, Su B, Li E, Yang J, Nemerow GR & Karin M (2000) MEK kinase 1 is critically required for c-Jun N-terminal kinase activation by proinflammatory stimuli and growth factor-induced cell migration. Proc. Natl. Acad. Sci. USA 97: 5243-5248.

Xu Y, Cheng G & Baltimore D (1996) Targeted disruption of TRAF3 leads to postnatal lethality and defective T-dependent immune responses. Immunity 5: 407-415.

Yang J, Lin Y, Guo Z, Cheng J, Huang J, Deng L, Liao W, Chen Z, Liu Z & Su B (2001) The essential role of MEKK3 in TNF-induced NF-κB activation. Nat. Immunol. 2: 620-624.

Yang YL, Reis LF, Pavlovic J, Aguzzi A, Schafer R, Kumar A, Williams BR, Aguet M & Weissmann C (1995) Deficient signaling in mice devoid of double-stranded RNA-dependent protein kinase. EMBO J. 14: 6095-6106.

Yeh WC, Shahinian A, Speiser D, Kraunus J, Billia F, Wakeham A, de la Pompa JL, Ferrick D, Hum B, Iscove N, Ohashi P, Rothe M, Goeddel DV & Mak TW (1997) Early lethality, functional NF-κB activation, and increased sensitivity to TNF-induced cell death in TRAF2-deficient mice. Immunity 7: 715-725.

Yin L, Wu L, Wesche H, Arthur CD, White JM, Goeddel DV & Schreiber RD (2001) Defective lymphotoxin-β receptor-induced NF-κB transcriptional activity in NIK-deficient mice. Science 291: 2162-2165.

Zandi E, Rothwarf DM, Delhase M, Hayakawa M & Karin M (1997) The IκB kinase complex (IKK) contains two kinase subunits, IKKα and IKKβ, necessary for IκB phosphorylation and NF-κB activation. Cell 91: 243-252.

Zhang Q, Siebert R, Yan M, Hinzmann B, Cui X, Xue L, Rakestraw KM, Naeve CW, Beckmann G, Weisenburger DD, Sanger WG, Nowotny H, Vesely M, Callet-Bauchu E, Salles G, Dixit VM, Rosenthal A, Schlegelberger B & Morris SW (1999) Inactivating mutations and overexpression of BCL10, a caspase recruitment domain-containing gene, in MALT lymphoma with t(1;14)(p22;q32). Nat. Genet. 22: 63-68.

11. AFFILIATIONS

University of Toronto, Department of Medical Biophysics, and Advanced Medical Discovery Institute, University Health Network, 620 University Avenue, Suite 706, Toronto M5G 2C1, Canada.

E. ANDREAKOS, I. UDALOVA, S. SACRE
& B.M. FOXWELL

THE ROLE OF NF-κB IN INFLAMMATORY DISEASES

1. MOLECULAR AND CELLULAR BASIS OF INFLAMMATION AND ITS ROLE IN PATHOLOGY

The term inflammation was used as early as 1600 BC in Egyptian papyrus writings to describe a complex series of events induced by a wound or an infection. These events can be recognized by the redness, swelling, fever and pain observed, and are now known to reflect the vasodilation, increased capillary permeability and influx of phagocytic cells. Thus, under most circumstances, inflammation is a normal physiological and very beneficial response, employed by our organism to fight infectious agents and protect us from disease. Inflammation involves the activation of both the innate and adaptive branches of the immune system and is under the control of a large number of cellular and molecular components. Major cellular components include B-cells, T-cells, dendritic cells and macrophages (Table 1).

Major molecular components involved in inflammation comprise cytokines, adhesion molecules, acute phase proteins, kinins, prostaglandins/leukotrienes, nitric oxide (NO) and inducible NO synthase (iNOS), and histamine. Cytokines are small proteins or glycoproteins that act as local messengers in inflammation and immunity, but also in many other important biological processes such as cell growth, development, repair and fibrosis. Cytokines are considered to play a central role in the development of inflammatory responses, as they regulate both the recruitment and activation of the cellular components, as well as the expression of the molecular components of inflammation. Their own expression, in turn, is subject to control by the other molecular components, thus setting up a regulatory loop that can either reduce or extend the persistence of the inflammatory response. Adhesion molecules are essential for the adhesion, activation and transmigration of leukocytes to inflammatory sites as well as the costimulation and activation process of T- and B-cells upon recognition of antigen (Ley, 2001). Acute phase proteins are serum proteins released in response to tissue damage. C-reactive protein, in particular, is produced by the liver and binds to the C-polysaccharide cell wall component of bacteria and fungi, activating complement, mediating phagocytosis and increasing pathogen clearance (Mortensen, 2001).

R. Beyaert (ed.), Nuclear Factor κB. Regulation and Role in Disease, 295-323.

Kinins are small peptides, found in blood plasma in an inactive form, that become activated in response to tissue injury to cause vasodilation and increased capillary permeability (Couture et al., 2001). Some kinins such as bradykinin, also stimulate pain receptors in the skin. Finally, prostaglandins and leukotrienes are eicosanoid lipid mediators derived from phospholipase-released arachidonic acid that are involved in numerous homeostatic and inflammatory functions, including vasodilation, chemotaxis, fever and pain (Funk, 2001). NO is also a mediator of inflammation that is important in bronchodilation, leukocyte adhesion and extravasation to tissues and antimicrobial activity through its oxidant derivatives (Bogdan, 2001). Production of NO is under the control of two constitutive NO synthase enzymes and one inducible NO synthase enzyme, which renders its study very difficult.

Table 1. Major functions of immune system cells.

Cell type		*Function*
Lymphoid	Helper T-lymphocytes	Stimulate B-lymphocyte growth and differentiation, and activate macrophages
	Cytotoxic T-lymphocytes	Kill virally infected or tumor cells and activate macrophages
	B-lymphocytes	Present antigen to T-lymphocytes and produce antibodies
	Large granular lymphocytes or natural killer (NK) cells	Involved in the early responses to virally infected or tumor cells and mediate antibody-dependent cellular cytotoxicity
Myeloid	Mononuclear phagocytes (monocytes/macrophages)	Phagocytose foreign particles and dead cells, present antigen to T-lymphocytes, induce fibroblast and vascular endothelial cell proliferation important in tissue repair, and recruit other inflammatory cells
	Neutrophils	Phagocytose foreign particles and dead cells
	Eosinophils	Phagocytose IgE-coated particles and mediate immediate-type hypersensitivity reactions
	Basophils/mast cells	Involved in IgE-mediated hypersensitivity
	Platelets	Mediate blood clotting
	Dendritic cells	Present antigen to T-lymphocytes
Endothelial		Regulate the recruitment of effector cells and the blood-clotting process

Although inflammation is a very beneficial response, prolonged or deregulated inflammation can lead to extensive tissue damage, disability or even death. A number of human diseases have been attributed to sustained inflammatory responses and include rheumatoid arthritis, inflammatory bowel disease, inflammatory lung diseases (asthma and fibrosing alveolitis), inflammatory skin diseases (psoriasis and incontinentia pigmenti), atherosclerosis, diabetes and neurodegenerative diseases such as Alzheimer's disease and multiple sclerosis. Understanding what regulates the inflammatory response in these conditions may lead to novel therapeutics of increased efficacy and reduced toxicity.

2. NF-κB IN THE CONTROL OF MOLECULAR FACTORS OF INFLAMMATION

In the past, most studies of inflammation were focusing on the role of mediators, such as cytokines, adhesion molecules, kinins and prostaglandins, in the development of the inflammatory response. However, in recent years, the development of powerful molecular biology tools has made it possible to study not only how distinct mediators regulate each other, but also which signaling pathways and transcription factors are involved and how they interact with each other. Central to the development of the inflammatory response was found to be the transcription factor NF-κB.

Table 2. Genes involved in inflammation and regulated by NF-κB.

Function	*Gene*
Activation of immune cells and inflammation	TNFα, IL-1, IL-2, IL-6, IL-12, IL-17, IL-18, IFNγ, GM-CSF
Chemotaxis/recruitment of immune cells	IL-8, MIP-1α/β, MCP-1, Gro-α/β/γ, eotaxin, ICAM-1, VCAM-1, E-selectin
Angiogenesis/neovascularization	VEGF
Vasodilation/increased capillary permeability	B1 kinin receptor, iNOS, COX-2, PGE2, PLA_2
Tissue damage	C-reactive protein, hsp32, hsp70
Tissue remodeling	MMP-1, MMP-3, MMP-9, MMP-13
Survival and proliferation	TRAF1/2, c-IAP1/2, XIAP, A1/Bfl-1, bcl-x_L, c-Myc, cyclin D

NF-κB is essential for host defense and inflammatory responses to microbial and viral infections (Barnes & Karin, 1997; Ghosh et al., 1998). Mouse models deficient in different NF-κB subunits have demonstrated the pivotal role that NF-κB plays in lymphoid organ development, immunity and inflammation (Beg et al., 1995; Sha et al., 1995; Weih et al., 1997; Franzoso et al., 1998). In response to pathogens, coordinated activation of NF-κB occurs in almost every cell type known and includes cells involved in immunity such as neutrophils, macrophages, lymphocytes, endothelial cells and epithelial cells. NF-κB activation, in turn, induces the expression of numerous inflammatory mediators such as cytokines, adhesion molecules, chemokines, cyclooxygenase 2 (COX-2), nitric oxide and others that contain NF-κB-binding sites in their promoters (Table 2) (Pahl, 1999). As many of these inflammatory mediators such as TNFα and IL-1 can also activate NF-κB, a positive regulatory loop is created that maintains the activation of NF-κB and amplifies the inflammatory response (Karin, 1999). NF-κB is tightly regulated and provides an important mechanism by which the fine balance between beneficial immune responses to fight infections and pathological immune responses that lead to tissue damage are achieved. Here possible mechanisms of highly specific NF-κB-dependent regulation of gene expression during inflammation are discussed.

Homodimers and heterodimers of five distinct Rel proteins form the family of transcription factors collectively referred to as NF-κB. All five proteins are characterized by the presence of a common Rel homology domain responsible for DNA binding and dimerization. Nevertheless there are two distinct classes within the family. The first consists of p65, RelB and c-Rel proteins that are synthesized as mature proteins and encompass the 'transactivating domain'. The second group includes p50 and p52 proteins that first are synthesized as large precursors p105 and p100, respectively, and then undergo proteolytic processing. Neither p50 nor p52 protein contain the transactivating domain. This diversity of NF-κB complexes lays the first basis for specificity of gene regulation. Indeed, there is growing body of evidence that the p50/p50 homodimer (lacking transactivating potential) exerts an inhibitory effect on gene transcription (Schmitz & Baeuerle, 1991; Franzoso et al., 1992; Franzoso et al., 1993; Udalova et al., 2000), possibly due to a competition with p50/p65 for binding to DNA. Overexpression of p50 blocked lipopolysaccharide (LPS)-induced transcription from TNFα promoter reporter constructs, suggesting that this factor is an inhibitor of the TNFα gene (Baer et al., 1998).

It has been also observed that expression of IFNβ, which is transcriptionally regulated by NF-κB in response to a viral infection, is increased in p50 knockout mice (Sha et al., 1995). Which NF-κB dimer is more likely to interact with a given promoter depends both on cell-specific protein distribution and DNA-binding specificity of different NF-κB sites, which significantly differ between the dimers (Kunsch et al., 1992; Udalova et al., 2002a).

p65 and c-Rel-containing NF-κB dimers are held in the cytoplasm in an inactive form by specific inhibitors, termed as inhibitors of NF-κB (IκB) proteins such as IκBα, IκBβ and IκBε, which undergo rapid ubiquitin-dependent degradation after the cell becomes exposed to a stimulus. Tissue-specific distribution of IκB molecules together with their different preference for NF-κB dimers may constitute another level of fine gene regulation by virtue of inhibiting specific NF-κB subsets in different cells. Various IκB proteins target different NF-κB species: IκBα and IκBβ bind preferentially to p50/p65 and p50/c-Rel heterodimers (Thompson et al., 1995b), IκBε associates only with homo- or heterodimeric complexes containing p65 and c-Rel (Simeonidis et al., 1997; Whiteside et al., 1997), and Bcl-3, another member of the IκB family, interacts specifically with p50 and p52 homodimers (Franzoso et al., 1992; Wulczyn et al., 1992). Bcl-3 can form ternary complexes with DNA-bound p52/p52, and enables it to activate transcription directly (Bours et al., 1993; Fujita et al., 1993), while interaction of this unusual member of the IκB family with DNA-bound p50/p50 is believed to induce its dissociation from DNA, allowing other NF-κB species to bind instead (Franzoso et al., 1992; Franzoso et al., 1993; Pan & McEver,

1995). Direct phosphorylation of Rel proteins constitutes another mode of NF-κB regulation (Schmitz et al., 2001). It has been suggested that Rel phosphorylation is critical for full NF-κB activation and that this can occur independently of IκB degradation. Both protein kinase A (PKA)-mediated and casein kinase II (CKII)-mediated phosphorylation of p65 have been demonstrated (Zhong et al., 1997; Wang et al., 2000). Such integration of different signaling pathways allows for activation of distinct Rel proteins to take place in response to different proinflammatory signals.

Upstream kinase cascades that result in IκB phosphorylation and degradation constitute another level of control of NF-κB activation as their activity is tightly regulated. Central to this cascade is the IκB kinase complex (IKK), a large multi-subunit kinase complex that comprises IKKα, IKKβ and IKKγ, and that is the point of convergence of multiple signals (e.g. microbial and viral products, cytokines and T-cell receptor engagement) leading to IκB phosphorylation and NF-κB activation (Karin, 1999). The spectrum of adaptor proteins and kinases that signal downstream from the receptors to the IKK complex differ between microbial (Toll-like receptor), cytokine (e.g. TNF receptors) and T-cell receptor-signaling pathways. Many of the signaling molecules in these pathways are yet to be identified. However, recent studies identified a common molecule, namely receptor-interacting protein kinase (RIP1/RIP2), which is involved in signaling from Toll-like, T-cell and TNF receptors, thus integrating signals for both innate and adaptive immune responses during the inflammation (Fig. 1) (Kelliher et al., 1998; Chin et al., 2002; Kobayashi et al., 2002).

A different mechanism of NF-κB RelB/p52 translocation into the nucleus has recently emerged. RelB is associated with p100 precursor in nonstimulated cells. p105 and p100-containing complexes are held in the cytoplasm by the C-terminal IκB-like domain of the proteins, which is proteolytically processed by ubiquitin-dependent partial degradation (Karin & Ben-Neriah, 2000). Whether the processing of p105 to p50 is inducible remains in question (Belich et al., 1999; Ciechanover et al., 2001; Dumitru et al., 2000). However, there is clear evidence for regulated processing of p100 (Senftleben et al., 2001; Xiao et al., 2001). Lymphotoxins α and β (LTα, LTβ) are essential in both lymphoid organ development and inflammation (Ruddle, 1999). Signaling from the LTβ receptor (LTβR) activates the MAP3K protein NIK, which in turn activates IKKα, leading to phosphorylation, ubiquitination and degradation of the C-terminal domain of p100. While the canonical NF-κB pathway, which is essential for innate immunity (Senftleben et al., 2001), is mostly dependent on IKKβ (Li et al., 1999a-b), this alternative pathway depends on IKKα and is mostly involved in lymphoid organ development and adaptive immunity (Senftleben et al., 2001). At the same time, signaling through the LTβR has also been shown to potentially use IKKβ to induce IκBα phosphorylation and NF-κB activation (Smith et al., 2001). It remains to be determined whether two different NF-κB pathways activate distinct subsets of genes and whether the proinflammatory genes are likely to be regulated by one selected pathway.

A different mechanism of NF-κB RelB/p52 translocation into the nucleus has recently emerged. RelB is associated with the p100 precursor in nonstimulated cells. p105- and p100-containing complexes are held in the cytoplasm by the C-terminal IκB-like domain of the proteins, which is proteolytically processed by ubiquitin-dependent partial degradation (Karin & Ben-Neriah, 2000).

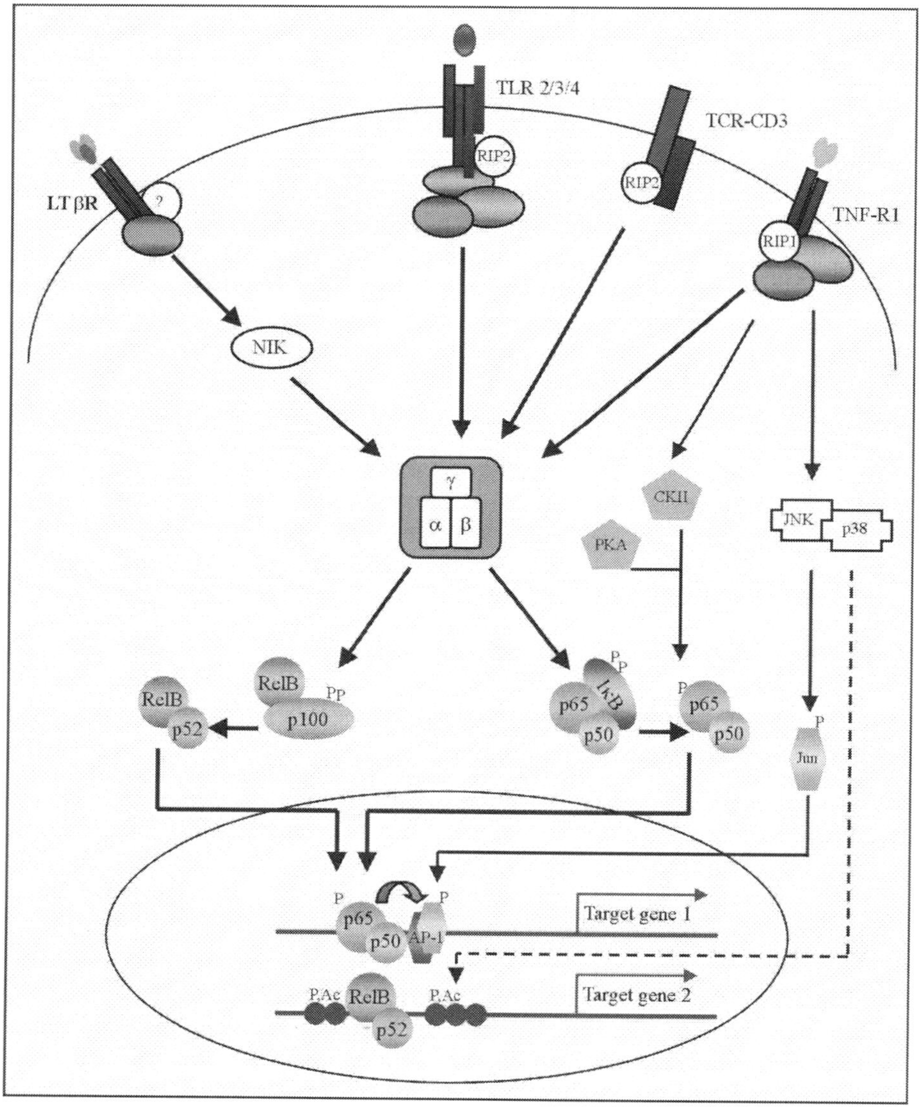

Fig. 1. Integration and specificity in NF-κB-dependent activation of the proinflammatory genes. In this model different proinflammatory stimuli are recognized by different receptors at the cell surface: TLR (for microbial lipoproteins, LPS, dsRNA), TNF-R1 and LTβR (for cytokines TNFα and LTαLTβ$_2$ trimer) and TCR (for T-cell-specific signaling). They

activate intracellular signaling cascades through distinct sets of adaptor proteins and kinases, with RIP1/2 integrating signals between the cascades. The IKK complex congregates the pathways. The activated IKK complex phosphorylates IκB, mostly by IKKβ, leading to IκB ubiquitination and degradation and the release of free NF-κB p50/p65. Its translocation into the nucleus is accompanied by PKA or CKII-dependent phosphorylation of p65. LTβR-NIK signaling is the only example to date of acting through IKKα subunit of the IKK, resulting in phosphorylation and degradation of p100 and nuclear translocation of NF-κB RelB/p52. In the nucleus p50/p65 and RelB/p52 complexes (and other NF-κB/Rel dimers not described in this simplified model) find specific DNA-binding sites in the promoters of inflammatory mediators (cytokines, adhesion molecules, nitric oxide, macrophage proteins, etc.) based on DNA sequence preferences, association with other transcription factors (e.g. AP-1) and high-order chromatin structure. The chromatin structure of the promoters can either be immediately accessible to NF-κB binding (e.g. the TNFα promoter) or require stimulus-dependent modification of histones (e.g. p38-dependent phosphorylation and acetylation of H3 histone in the IL-6 promoter). Binding of a specific NF-κB complex and its interaction with promoter-specific transcription factors leads to recruitment of coactivators and chromatin remodeling factors and results in highly specific signal-dependent activation of inflammatory gene transcription.

Whether the processing of p105 to p50 is inducible, remains in question (Belich et al., 1999; Dumitru et al., 2000; Ciechanover et al., 2001). However, there is clear evidence for regulated processing of p100 (Senftleben et al., 2001; Xiao et al., 2001). Lymphotoxin α and β (LTα/LTβ) are essential in both lymphoid organ development and inflammation (Ruddle, 1999). Signaling from the LTβR activates the MAP3K protein NIK, which in turn activates IKKα, leading to phosphorylation, ubiquitination and degradation of the C-terminal domain of p100.

While the canonical NF-κB pathway, which is essential for innate immunity (Senftleben et al., 2001), is mostly dependent on IKKβ (Li et al., 1999a-b), this alternative pathway depends on IKKα and is mostly involved in lymphoid organ development and adaptive immunity (Senftleben et al., 2001). At the same time, signaling through the LTβR has also been shown to potentially use IKKβ to induce IκBα phosphorylation and NF-κB activation (Smith et al., 2001). It remains to be determined whether two different NF-κB pathways activate distinct subsets of genes and whether the proinflammatory genes are likely to be regulated by one selected pathway.

Another mode of specificity in NF-κB-dependent gene activation lies in its ability to orchestrate gene expression in concert with other transcription factors. Eukaryotic genes are regulated largely through the cooperative interactions of distinct transcription factors and their assembly into multicomponent enhancer complexes (Carey, 1998). The repertoire of proteins NF-κB interacts with determines the timing, level and duration of gene expression. For instance, the production of IFNβ in response to virus induction requires the correct assembly of NF-κB p50/p65, ATF-2, c-Jun, IRF-3, IRF-7 and HMG-I(Y) into enhanceosome (Thanos & Maniatis, 1995), in which both the composition and the orientation of transcription complexes are essential (Thanos & Maniatis, 1995; Falvo et al.,

2000). Moreover, such assembly leads to ordered recruitment of coactivator molecules, histone acetylases and chromatin remodeling factors and general transcription machinery, that determines activation time and duration of the gene transcription (Agalioti et al., 2000). DNA-binding studies reveal a requirement for NF-κB and a small group of other transcriptional activators for maximal levels of adhesion molecules during inflammatory responses. The organization of the cytokine-inducible element in the E-selectin promoter is remarkably similar to the IFNβ gene in that both promoters require NF-κB, ATF-2 and HMG-I(Y) (Whitley et al., 1994), whereas vascular cell adhesion molecule-1 (VCAM-1) expression is induced through interactions of NF-κB with IRF-1 and HMG-I(Y), which also depends on constitutively present SP-1 (Neish et al., 1995a-b). NF-κB physically associates with the transcription factor NF-IL-6 in synergistic activation of transcription of the inflammatory cytokine IL-6 (Matsusaka et al., 1993), and together with AP-1 of IL-8 (Roebuck, 1999).

The ability of NF-κB to physically interact with another ubiquitous transcription factor AP-1, whose transcriptional induction follows a distinct signal transduction pathway involving phosphorylation of c-Jun kinase (JNK) (Chang & Karin, 2001), is of particular importance, as it may play a general role in regulation of the inflammatory response by simultaneous activation of the majority of inflammatory mediators (Stein et al., 1993). Indeed, many of the inflammatory genes require these two transcription factors working cooperatively, including VCAM-1 (Ahmad et al., 1998), IL-8 (Roebuck, 1999), COX-2 (Allport et al., 2000), macrophage chemoattractant protein-1 (MCP-1) (Martin et al., 1997) and collagenase-3 (Mengshol et al., 2000).

Both the extensive histone-DNA contacts in individual nucleosomes and the folding of the nucleosomal chain into higher order chromatin structures limit the accessibility of the genome to DNA-binding proteins, including NF-κB. It has been proposed that stimulation with microbial products or proinflammatory cytokines may result in two distinct waves of NF-κB recruitment to target promoters: a fast recruitment to constitutively and immediately accessible promoters (e.g. the TNFα promoter) and a late recruitment to promoters requiring stimulus-dependent modifications in chromatin structure to make NF-κB sites accessible (Saccani et al., 2001). The later recruitment was shown to selectively occur on the promoters of a subset of cytokine and chemokine genes, including IL-6, IL-8 and MCP-1, through p38 MAP kinase–dependent phosphorylation and phosphoacetylation of histone H3 (Saccani et al., 2002). p38 is not a general modulator of NF-κB recruitment to chromatin as other cryptic NF-κB sites, e.g. in the macrophage inflammatory protein-1 (MIP-1) promoter, are switched into an accessible state in a p38-independent manner. A highly promoter-specific facilitation of NF-κB recruitment to selected binding sites introduces a novel mechanism of gene activation specificity during inflammatory responses.

The mechanism of gene expression optimization that has received relatively little attention so far is the effect of genetic variations on the level of response to NF-κB. A naturally occurring polymorphism at nucleotide -863 in the human

TNFα promoter region provided an opportunity to dissect the functional NF-κB interactions at a single binding site (Udalova et al., 2000). It was found that this site normally binds both p50/p65 and p50/p50 NF-κB complexes, but a single base change specifically inhibited p50/p50 binding. Reporter gene analysis indicated that the variant allele acts to elevate inducible TNFα production. This is of fundamental interest, as it raises the question of how the genetic background may influence the immune gene regulation and consequently result in balancing of the inflammatory response. For instance, patients with severe rheumatoid arthritis, the chronic inflammatory disease in which TNFα has an essential role in pathogenesis, were twice as likely to possess the TNFα-863 allele than the control group (Udalova & Kwiatkowski, 2002). Predictions of the NF-κB-binding sites that are likely to have a functional effect were done for the human chromosome 22 and identified about fifty single nucleotide polymorphisms that would alter NF-κB binding to DNA (Udalova et al., 2002b). Together with a number of other observations of functional consequences of nucleotide polymorphisms that alter protein-DNA interactions (Knight et al., 1999; Farzaneh-Far et al., 2001; Hohjoh & Tokunaga, 2001; Price et al., 2001; van Heel et al., 2002), this novel genetic mechanism of gene regulation may have important implications in disease pathogenesis.

3. NF-κB AND ITS ROLE IN DISEASE PATHOGENESIS

3.1. Scope

The profound effect that NF-κB has on the regulation of numerous genes involved in inflammation and the ability of many such genes to potently activate NF-κB themselves, led to the suggestion that NF-κB may also be essential for the development and/or progression of inflammatory diseases. Currently, an increasing amount of data on the role of NF-κB is being generated by studies in patients and patient tissue, and is being complemented by studies in relevant animal models.

3.2. Rheumatoid arthritis

The rheumatoid synovium is a site of active inflammation. It is infiltrated by a diverse spectrum of cells such as activated T-lymphocytes, macrophages, fibroblasts, endothelial cells and plasma cells, and is enriched with almost every cytokine and inflammatory mediator known (Feldmann & Maini, 2001). Cytokines such as TNFα and IL-1 that activate NF-κB are elevated in the synovium, and activated NF-κB has been detected in human synovial tissue from both early and later stage patients in macrophage- and fibroblast-like synoviocytes (Asahara et al., 1995; Handel et al., 1995; Marok et al., 1996). In particular, macrophage-like synoviocytes that localize in the synovial lining layer and the vascular endothelium have been shown to contain p65 and p50 NF-κB subunits in their nucleus (Handel et al., 1995). In animal models of arthritis, such as collagen-induced arthritis in mice or adjuvant-induced arthritis, pristane-induced arthritis and streptococcal cell wall (SCW)-induced arthritis in rats, NF-κB is also activated (Tsao et al., 1997; Han et al., 1998; Miagkov et al., 1998; Palombella et al., 1998).

The activation of NF-κB has important functional consequences. Studies that we performed in dissociated synovial membrane cultures from rheumatoid arthritis patients with active disease demonstrated that overexpression of IκBα, the natural inhibitor of NF-κB, significantly reduces the production of TNFα (Foxwell et al., 1998). As TNFα is central to the inflammatory cascade that takes place in the rheumatoid synovium, and is an already established and proven therapeutic target for the reduction of disease activity, this suggested that NF-κB-blocking agents may also result in therapeutic benefit in these patients. Interestingly, inhibition of NF-κB in these cultures also inhibited the production of a large number of other proinflammatory cytokines such as IL-1β, IL-6 and IL-8, without significantly affecting the production of the anti-inflammatory cytokines IL-10 and IL-11, or the release of soluble TNF receptors (Bondeson et al., 1999). In addition, inhibition of NF-κB blocked the secretion of matrix metalloproteinases (MMPs) 1, 3 and 13 (Bondeson et al., 1999). These observations suggested that NF-κB and NF-κB-dependent gene expression are central to the production of a large number of proinflammatory, but not of anti-inflammatory mediators in the rheumatoid synovium essential for disease pathogenesis (Fig. 2). These proinflammatory mediators in turn induce NF-κB activation, thus establishing a vicious circle that maintains NF-κB activation and perpetuates inflammation.

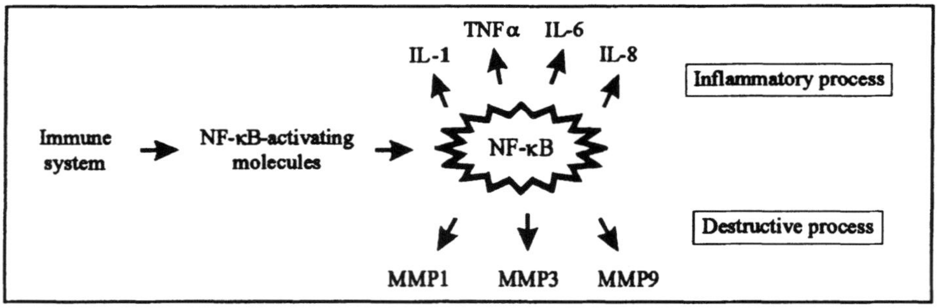

Fig. 2. NF-κB is central to the regulation of inflammatory and destructive processes in the rheumatoid synovium. Therapeutics aimed at blocking NF-κB or upstream NF-κB-activating molecules constitute a promising new approach to the treatment of rheumatoid arthritis and possibly other chronic inflammatory diseases.

NF-κB may also have a prominent role in rheumatoid arthritis joint destruction as it promotes osteoclast maturation and increased bone-resorbing activity (Iotsova et al., 1997), and inhibits at the same chondrocyte differentiation and repair of damaged cartilage tissue (de Crombrugghe et al., 2000; Murakami et al., 2000). In animal models of arthritis, the genetic inactivation of c-Rel and NF-κB1 in $rel^{-/-}$- and $nfkb1^{-/-}$-deficient mice protected the animals from the development of collagen-induced arthritis (CIA) (Campbell et al., 2000), as did the transgenic expression of (superrepressor) IκBα in cells of the T-cell lineage (Seetharaman et al., 1999). Similarly, the administration of NF-κB decoy oligonucleotides suppressed the severity of already established disease in both CIA and SCW rat arthritis by inhibiting the production of TNFα and IL-1 within the joints and by reducing paw swelling and disease activity (Miagkov et al., 1998; Tomita et al., 1999). A proteosome inhibitor that prevents IκB degradation also afforded protection in SCW-induced rat arthritis (Palombella et al., 1998). Thus, blocking NF-κB may reduce the inflammatory response and restore the cytokine equilibrium in the rheumatoid joint by reducing at the same time cartilage and bone damage.

3.3. Inflammatory bowel disease (IBD)

IBD comprises a heterogeneous group of diseases including Crohn's disease (CD) and ulcerative colitis (UC). Several genetic and environmental factors are implicated in the pathophysiology of IBD that ultimately result in a common final manifestation, which is mucosal inflammation (Fiocchi, 1998). Deregulated cytokine production and signaling mechanisms by epithelial cells, mucosal lymphocytes and macrophages seem to play a major role in the inflammatory response that takes place in the colonic mucosa. NF-κB, in particular, has been found to be increased in patients with IBD and p65 to be localized in epithelial cells and lamina propria macrophages in biopsies of patients (Neurath et al., 1996; Neurath & Pettersson, 1997; Rogler et al., 1998; Schreiber et al., 1998). Increased NF-κB DNA-binding activity has been demonstrated in nuclear extracts from both biopsy specimens and isolated lamina propria mononuclear cells from CD patients (Schreiber et al., 1998), and in vitro treatment of lamina propria macrophages with p65 antisense oligonucleotides was shown to be more effective in downregulating the expression of TNFα, IL-1β and IL-6 than 5-aminosalicylic acid or glucocorticoids (Neurath et al., 1996; Neurath & Pettersson, 1997). Collectively, these observations suggested that NF-κB activation may be important in the generation and maintenance of mucosal inflammation, although its activation may not be specific to IBD. Indeed, increased NF-κB levels have also been found in nonspecific colitis or diverticulitis biopsy speciments and were not significantly different from those found in CD or UC specimens (Rogler et al., 1998).

NF-κB was also found to be activated in several animal models that closely resemble human IBD. Thus, in TNBS-induced colitis in mice, increased NF-κB DNA-binding activity, that consisted mainly of the p65 and p50 NF-κB subunits, was detected in nuclear extracts from lamina propria macrophages (Neurath et al., 1996). Inhibition of this activity in vivo by intravenous or local colonic administration of p65 antisense oligonucleotides abrogated the clinical signs of established colitis (Neurath et al., 1996; Neurath & Pettersson, 1997). Strikingly, the efficacy of p65 antisense oligonucleotides in treating TNBS-induced colitis appeared to be superior than the administration (single or daily) of glucocorticoids. In addition, in IL-10-deficient mice, another model of chronic intestinal inflammation (Kuhn et al., 1993), increased levels of NF-κB DNA-binding activity and increased levels of p65 protein expression were found in lamina propria macrophages. Again, administration of p65 antisense oligonucleotides improved the clinical and histological signs of established colitis (Neurath et al., 1996). Finally, in DSS-induced colitis in mice, intravenous or oral administration of an inhibitor that blocks NF-κB nuclear translocation, reduced both the inflammation and injury scores of the diseased colon (Fujihara et al., 2000). These in vivo studies suggest that specific blockade of NF-κB may also have therapeutic efficacy in the human forms of colonic intestinal inflammation such as CD and UC.

3.4. Psoriasis

Psoriasis is a chronic remitting-relapsing skin disease. It is caused by increased epidermal proliferation and infiltration of inflammatory/immune cells into the epidermis that result in raised scaly plaques on the skin surface and skin lesions. This condition is often seen in conjunction with nail disease and arthritis. Numerous inflammatory mediators have been reported to be upregulated in skin biopsy speciments from psoriatic lesions, including TNFα, IL-1 and IL-15 (Danning et al., 2000). The p65 NF-κB subunit has also been shown to be activated in psoriatic lesions, although it is not known whether this is a cause or consequence of the inflammatory process taking place (Danning et al., 2000). Certainly, experiments in an animal model of psoriasis using a selective proteasome inhibitor that prevents IκB degradation reduced the severity of psoriasis, indicating that NF-κB activation may indeed be essential in that process (Zollner et al., 2002). In addition, the observation that cyclosporin, glucocorticoids and fumaric acid esters, commonly used for the treatment of psoriasis in humans (Asadullah et al., 2002), all can inhibit NF-κB activation, also supports this possibility.

3.5. Inflammatory lung diseases

Acute and chronic alveolar and/or bronchial inflammation also takes place in a number of diseases of the lung, collectively termed inflammatory lung diseases. These include acute lung injury (ALI), acute respiratory distress syndrome (ARDS), fibrosing alveolitis (FA) and asthma, and are characterized by an inflammatory response that involves the recruitment to the lungs and the activation of immune cells. This is accompanied by the upregulation of adhesion molecules and the secretion of cytokines, chemokines, oxidants and other inflammatory mediators, and results in lung injury. Airway epithelial cells contribute to that process as they are also able to produce inflammatory mediators (Thompson et al., 1995a).

NF-κB seems to be important in lung inflammation as increased NF-κB activity has been detected in bronchoalveolar lavage biopsies from ALI (Blackwell & Christman, 1997), ARDS (Schwartz et al., 1996; Christman et al., 1998), FA

(Conron et al., 2002) and asthma patients (Hart et al., 1998) compared to healthy volunteers (Farver et al., 1998). In asthma patients in particular, increased NF-κB activity is also present in the submucosal cells, in the endothelium and the bronchial epithelium (Barnes & Adcock, 1997). Few studies have been performed to evaluate the role of NF-κB in lung inflammation. In bronchoalveolar lavage biopsies from FA patients, we recently showed that overexpression of IκBα inhibits the constitutive production of both macrophage and T-cell-derived cytokines such as TNFα, IL-6, IL-8 and IFNγ, demonstrating that NF-κB regulates the expression of a number of inflammatory mediators in this system (Conron et al., 2001). The examination of putative kinases that may be involved in that process revealed that IKK2, but not NIK was required (Conron et al., 2002). Several animal models of lung inflammation have also been used. In rat and mouse models of endotoxic shock, intraperitoneal administration of endotoxin induces NF-κB activation in alveolar macrophages and lung tissue that is associated with lung neutrophilia, epithelial permeability and lipid peroxidation (Blackwell et al., 1996; Blackwell et al., 1999). This was suppressed by treatment with antioxidants that inhibit NF-κB activation (Blackwell et al., 1996). Similarly, hemorrhage-induced shock in mice resulted in rapid NF-κB activation in intraparenchymal lung mononuclear cells and in lung inflammation (Shenkar & Abraham, 1997). This may occur through stimulation of α-adrenergic receptors and the generation of reactive oxygen species. Finally, mice deficient in the p50 NF-κB subunit of NF-κB are incapable of mounting eosinophilic airway inflammation. This is due to the lack of production of IL-5 and eotaxin, MIP-1α and MIP-1β, but not to T-cell activation or adhesion molecule expression (Yang et al., 1998). The same was true for c-Rel-deficient mice that were resistant to allergen-induced pulmonary inflammation, airway hyperresponsiveness and eosinophilia (Donovan et al., 1999). As these are common features of asthma, it is likely that p50, c-rel and probably other NF-κB subunits are also important in the human disease. Although the stimuli for NF-κB activation in asthmatic lungs have not been defined, agents that correlate with asthma exacerbations, such as various allergens, ozone, rhinoviruses and respiratory syncytial viruses, have all been shown to activate NF-κB (Christman et al., 2000).

3.6. Atherosclerosis

Atherosclerosis and its clinical manifestations of heart attack, stroke and peripheral vascular insufficiency can be viewed as a multistep, chronic inflammatory disease. The development of atherosclerotic lesions can be subdivided into three phases, viz. the initiation phase (formation of small fatty streaks), the expansion phase (endothelial cell growth and coalescence of fatty streaks) and the progression phase to plaques (recruitment of smooth muscle cells, deposition of collagen and formation of a fibrous cap). This process is under the control of cytokines, chemokines, growth factors and other soluble mediators and adhesion molecules (produced within the lesion) that regulate the recruitment and proliferation of monocytes and macrophages to the vessel wall, the migration of smooth muscle cells from the media to the intima of the vessel, and the proliferation of vascular endothelial cells. As many of these inflammatory mediators can be regulated by NF-κB, it has been suggested that NF-κB may also play a major role in the pathogenesis of atherosclerosis (Collins & Cybulsky, 2001). Indeed, activated NF-κB has been detected in human atherosclerotic plaques, whereas it is almost absent in healthy vessels devoid of atherosclerosis. Studies using an antibody that recognizes p65 dissociated from IκB revealed that activated p65 localizes in the intima and media of atheromatous areas of atherosclerotic lesions in smooth muscle cells, macrophages and endothelial cells (Brand et al., 1996). This is in contrast to healthy vessels where p65 and p50 are mainly localized in the cytoplasm. Elevated NF-κB levels have also been detected in arterial smooth muscle cells after injury and have been suggested to account for the expression of several genes, including ICAM-1, VCAM-1 and MCP-1, all of which facilitate the recruitment of monocytes and macrophages to the atherosclerotic lesions (Landry et al., 1997).

Few studies have also examined the activation of NF-κB in animal models of atherosclerosis. One study demonstrated activated NF-κB in intimal cells from coronary arteries of pigs fed with a hypercholesterolemic diet (Wilson et al., 2000). Interestingly, another study in mice suggested that there is elevated NF-κB expression in vessel regions of high probability for the development of atherosclerotic plaques when compared to adjacent regions where plaques are unlikely to develop (Hajra et al., 2000). As these high probability regions coincide with areas exposed to high shear stress, the authors of the study proposed that shear stress may prime certain regions of the vessel wall by inducing steady state levels of NF-κB that may upregulate the expression of cytokines and adhesion molecules upon an additional atherogenic stimulus. In humans, many atherogenic stimuli or risk factors have been shown to activate NF-κB. Thus, oxidative stress (Li & Karin, 1999), oxidized LDL and VLDL (Brand et al., 1997; Calara et al., 1998; Dichtl et al., 1999), hypertension and angiotensin II (Han et al., 1999; Pueyo et al., 2000), hyperglycemia and elevated advanced glycation end products (Morigi et al., 1998; Schmidt et al., 1999), homocysteine and various infectious agents such as cytomegalovirus and Chlamydia pneumonia (Welch & Loscalzo, 1998) have all been shown to induce NF-κB activation in vitro or in vivo. In summary, there is good evidence that deregulated NF-κB may be involved in the development, but also some of the pathological features of atherosclerosis, although more studies are needed to demonstrate that blockade of NF-κB can result in therapeutic benefit in animal models of the disease and ultimately patients.

3.7. Diabetes

There are two forms of diabetes, viz. type 1 (often referred to as insulin-dependent diabetes mellitus or juvenile diabetes) and type 2, known as insulin-independent diabetes or adult onset diabetes. Type 1 is the most serious and is caused by the autoimmune apoptotic destruction of the insulin-producing beta cells in the islets of Langerhans. This involves the activation of T-cells and antigen-presenting cells, and the production of cytokines such as INFγ, TNFα and IL-1β (Igaz et al., 2000). Type 2 diabetes develops slowly and is probably due to alterations in fat metabolism as the body gains weight and becomes resistant to insulin. Common complications of diabetes include hypertension and cardiovascular disease.

Inappropriate NF-κB activation has recently been implicated in both types of diabetes. In type 1 diabetes, peripheral blood mononuclear cells from patients have increased NF-κB DNA-binding activity (Hofmann et al., 1998; Bierhaus et al., 2001) and increased NF-κB DNA-binding activity has been associated with more pronounced diabetic complications such as diabetic nephropathy (Hofmann et al., 1999; Mohamed et al., 1999). In human islets, the expression of a mutated nondegradable form of the IκBα or the antiapoptotic gene A20 (that also inhibits NF-κB in this system) prevents cytokine-induced beta cell death (Grey et al., 1999; Giannoukakis et al., 2000; Heimberg et al., 2001) by blocking the apoptotic pathway and probably by preventing the NF-κB-dependent upregulation of the Fas gene (Darville & Eizirik, 2001). Evidence for an important role of NF-κB in type 1 diabetes also comes from studies in animal models. In chemically induced diabetes (alloxan and steptozotocin-induced diabetes) in mice, administration of NF-κB decoy oligonucleotides (Quan et al., 2001) or other NF-κB inhibitors such as dietary zinc and Amomum xanthoides extract (Ho et al., 2001; Park & Park, 2001) inhibited the development of diabetes. In addition, nonobese diabetic (NOD) mice that are commonly used to model human type 1 diabetes exhibit elevated levels of NF-κB activation that leads to enhanced DC and T-cell function and increased autoimmune destruction of the islets of Langerhans (Hayashi et al., 1993; Weaver et al., 2001; Poligone et al., 2002). Polymorphisms in the NF-κB locus have recently been associated with human type 1 diabetes (Hegazy et al., 2001).

NF-κB has also been implicated in the development of insulin-resistance that leads to type II diabetes. The discovery that salicylates used to treat insulin-resistance potently inhibit IKKβ (Yin et al., 1998) and the observation that IKKβ-deficient mice are protected from the development of fat-induced insulin resistance (Kim et al., 2001) suggest that NF-κB activation may be essential for that process.

3.8. Neurodegenerative diseases

3.8.1. Scope

Most cells within the nervous system (including neurons, astrocytes, microglia and oligodendrocytes) express functional NF-κB (Kaltschmidt et al., 1994; O'Neill & Kaltschmidt, 1997). This can be upregulated by a wide spectrum of stimuli that include reactive oxygen species (Schreck et al., 1992), inflammatory cytokines (Barger et al., 1995), viral and bacterial products, depolarization

(Guerrini et al., 1995; Kaltschmidt et al., 1995), nerve growth factor (Carter et al., 1996), β-amyloid (Barger & Mattson, 1996; the secreted form of β-amyloid precursor protein) and dopamine (Weingarten et al., 2001). Recently, NF-κB activation was suggested to play an important role in the development of many neurodegenerative diseases by regulating survival and growth of neurons, microglia and astrocytes.

3.8.2. Multiple sclerosis (MS)

MS is a demyelinating disease where myelin in the central nervous system is lost. This is a result of an immune response that develops against oligodendrocytes (the cells that produce myelin) and effectively inhibits their function and induces their depletion (Kieseier et al., 1999). The mechanisms, however, that regulate these events remain poorly understood. Studies by immunohistochemistry have shown that NF-κB is upregulated and localized in the nuclei of microglia and oligodendrocytes in chronic active MS lesions (Gveric et al., 1998; Bonetti et al., 1999). As IL-1 and TNFα are also localized in active MS lesions, these cytokines are likely to account for the observed upregulation of NF-κB (Hofman et al., 1989; Selmaj et al., 1991; Woodroofe & Cuzner, 1993; Liu et al., 1996; Bierhaus et al., 2001). Upregulated NF-κB, in turn, may amplify the inflammatory process taking place in MS by increasing the expression of NF-κB-controlled adhesion molecules and cytokines. Ultimately, this leads to disease progression, disability and even death.

In experimental encephalomyelitis (EAE), an animal model of MS and autoimmune neural damage, NF-κB activation has also been reported (Pahan & Schmid, 2000). Interestingly, NF-κB-deficient mice are resistant to induced EAE, possibly due to a deficiency in the differentiation of myelin oligodendrocyte glycoprotein-specific T-cells to Th1- or Th2-type effector cells (Hilliard et al., 1999). However, although inflammatory factors that become upregulated by NF-κB may act to promote autoimmune disease, NF-κB may at the same time upregulate many antiapoptotic genes, thus protecting neurons and oligodendrocytes from cell death and further damage (Beg & Baltimore, 1996; Van Antwerp et al., 1996; Wang et al., 1996; Bonetti et al., 1997; Bonetti & Raine, 1997). The pathogenesis of this disease may thus rely on the balance between the 'beneficial' and 'pathological' effects of NF-κB.

3.8.3. Alzheimer's disease (AD)

AD is a neurodegenerative disease that affects the elderly. The main symptoms are memory loss and dementia, with the severity of disease being linked to the amplitude of neuronal and synaptic loss. Aggregates of amyloid beta peptides build up forming plaques along with neurofibrilary tangles in neurons that interfere with the normal function of the brain (Walsh et al., 2002). Interestingly, the areas of the brain that are affected by AD pathology also demonstrate increased NF-κB activity (Yan et al., 1995). Thus, p65 NF-κB has been localized in neurons and astrocytes next to amyloid plaques found in brain sections from AD patients (Terai et al., 1996; Kaltschmidt et al., 1997; O'Neill & Kaltschmidt, 1997; Ferrer et al., 1998). A mechanism by which NF-κB may worsen AD has been proposed. NF-κB can be activated by a number of stimuli that include soluble amyloid precursor protein (Barger & Mattson, 1996), β-amyloid peptides possibly through activation of RAGE (Kaltschmidt et al., 1997), IL-1β and glutamate (Grilli et al., 1996a). This creates a positive regulatory loop as NF-κB also upregulates the expression of the amyloid precursor protein gene (Grilli et al., 1996a), producing higher levels of β-amyloid peptide that further activates NF-κB and induces neuronal cell death. The observation, however, that NF-κB can also be beneficial in AD by upregulating the expression of genes like SOD (Mattson et al., 1997; Keller et al., 1998) and other antiapoptotic proteins that may aid neuronal survival (Barger et al., 1995; Zong et al., 1999) has complicated matters. Therefore, it is still unclear whether NF-κB is a potential therapeutic target in AD, and more studies are needed.

4. NF-κB IN THERAPY

Although there is a large amount of data that deregulated NF-κB is central to the pathogenesis of a number of inflammatory diseases, the therapeutic potential of NF-κB blockade has not yet been tested in the clinic. Evidence, however, from other immunosuppressive and anti-inflammatory drugs already used for the treatment of these diseases (Table 3) suggests that it is likely to be very beneficial, as the common property of these drugs is their ability to inhibit NF-κB activation among their multiple effects.

Table 3. Drugs used for the treatment of inflammatory diseases that have been proposed to block NF-κB.

Disease	*Treatment*
Rheumatoid arthritis	Glucocorticoids, NSAIDs, cyclosporin A and tacromilus, methotrexate, sulfasalazine, leflunomide, gold and D-penicillamine compounds, anti-TNFα blocking agents
IBD	Glucocorticoids, NSAIDs, cyclosporin A and tacromilus, methotrexate, sulfasalazine, anti-TNFα blocking agents
Psoriasis	Glucocorticoids, NSAIDs, cyclosporin A and tacromilus, methotrexate, anti-TNF blocking agents
Inflammatory lung diseases	Glucocorticoids, NSAIDs, cyclosporin A and tacromilus
Atherosclerosis	NSAIDs
Insulin resistance (diabetes)	NSAIDs
Neurodegenerative diseases	Glucocorticoids, NSAIDs, cyclosporin A and tacromilus

Thus glucocorticoids, such as dexamethasone and prednisone, which are frequently used for the treatment of rheumatoid arthritis, IBD and inflammatory lung diseases (Geier & Miner, 1992; Caesar et al., 1997; Keatings et al., 1997; Norbiato et al., 1997; Wilckens & De Rijk, 1997) owe, at least in part, their therapeutic effects to their ability to inhibit NF-κB-dependent gene expression. Glucocorticoids have been shown to upregulate the transcription and expression of IκBα, possibly by yet unidentified glucocorticoid-responsive elements in the IκBα promoter (Auphan et al., 1995; Scheinman et al., 1995a), and to physically interact with p65, thereby preventing transactivation of NF-κB (Caldenhoven et al., 1995; Scheinman et al., 1995b; De Bosscher et al., 1997; Heck et al., 1997).

In addition, NSAIDs were recently suggested to inhibit NF-κB activity, an effect that could account for their anti-inflammatory action. Thus, sodium salicylates and aspirin were shown to inhibit IκBα phosphorylation and degradation by preventing ATP-binding to IKK2 at concentrations found in the serum of patients treated for inflammatory conditions (Kopp & Ghosh, 1994; Grilli et al., 1996b; Pierce et al., 1996; Yin et al., 1998). At the same time, methotrexate, sulfasalazine and leflunomide were shown to inhibit IκBα phosphorylation and degradation, probably by interfering with the IKK complex (Egan & Sandborn, 1998; Wahl et al., 1998; Manna et al., 2000; Majumdar & Aggarwal, 2001), and demonstrating that certain NSAIDs may exert their anti-inflammatory effects by blocking the activation of NF-κB.

Similarly, gold and D-penicillamine compounds and thalidomine, that are used drugs, especially for the treatment of rheumatoid arthritis, all have inhibitory effects on NF-κB activation (Yang et al., 1995; Keifer et al., 2001). Immunosuppressive agents such as cyclosporin A and tacrolimus (FK-506), that are mainly employed in transplantation, but also in inflammatory diseases such as rheumatoid arthritis, were also able to block the activation of NF-κB in addition to their effects in other pathways such as the activation of NF-AT. Cyclosporin A has been claimed to prevent IκBα degradation and NF-κB activation by interfering with the proteasome function (Marienfeld et al., 1997; Meyer et al., 1997), whereas tacromilus seems to specifically inhibit c-Rel, but not RelB or p50 nuclear translocation (Venkataraman et al., 1995). Finally, anticytokine biologicals such as anti-TNFα blocking agents, shown to be highly successful in the treatment of a number of inflammatory diseases (Andreakos et al., 2002), may owe part of their efficacy to the prevention of cytokine-induced NF-κB activation and NF-κB-dependent gene expression. Support for that comes from a study in refractory Crohn's disease, where anti-TNFα blocking agents have been demonstrated to induce disease remission in 30-50% of the patients. It was shown that the anti-TNFα blocking agent infliximab reduced significantly nuclear p65 expression in the colonic mucosa within one week after a single infusion, and an extended downregulation of p65 seemed to be associated with a sustained induction of remission (Nikolaus et al., 2000). In contrast, reaccumulation of nuclear p65 correlated with the increase in TNFα secretion, and was usually observed before patients presented with a clinical relapse. These findings indicated that the activation of NF-κB may be secondary to the secretion of TNFα, and suggested a possible mechanism by which infliximab improves Crohn's disease.

However, as the majority of these drugs is associated with various side effects, especially in the long term, the specific inhibition of NF-κB may result in increased therapeutic efficacy and reduced toxicity. Initial attempts to identify novel NF-κB inhibitors focused on interfering with proteasome function or using natural plant products such as flavonoids to prevent IκB degradation and NF-κB activation. However, it became apparent that these inhibitors would also lack specificity and thus be subject to more severe side effects. Current attempts are focusing on the design of specific NF-κB inhibitors that block NF-κB without affecting other pathways. This requires the identification of upstream signaling components that lead to the activation of NF-κB, the determination of which of these are activated and rate-limiting in disease states, as well as the elucidation of their molecular structure and biochemical properties. In that way, some of the main worries concerning NF-κB-specific drugs, such as liver toxicity or interference with host defences in the long term (Beg et al., 1995; Li et al., 1999a-b; Tanaka et al., 1999; Lavon et al., 2000; Rudolph et al., 2000) may be overcome. Certainly, the fact that NSAIDs or glucocorticoids, that block NF-κB, can be tolerated long-term indicates that the use of NF-κB inhibitors is a valid strategy. Of course, under conditions where NF-κB inhibitors will be needed for short-term therapy, these worries are not of much concern. Ultimately, clinical trials will be needed to resolve these issues and assess the potential of NF-κB inhibitors as novel anti-inflammatory agents as well as their therapeutic efficacy.

5. REFERENCES

Agalioti T, Lomvardas S, Parekh B, Yie J, Maniatis T & Thanos D (2000) Ordered recruitment of chromatin modifying and general transcription factors to the IFN-β promoter. Cell 103: 667-678.

Ahmad M, Theofanidis P & Medford RM (1998) Role of activating protein-1 in the regulation of the vascular cell adhesion molecule-1 gene expression by tumor necrosis factor-α. J. Biol. Chem. 273: 4616-4621.

Allport VC, Slater DM, Newton R & Bennett PR (2000) NF-κB and AP-1 are required for cyclo-oxygenase 2 gene expression in amnion epithelial cell line (WISH). Mol. Hum. Reprod. 6: 561-565.

Andreakos E, Foxwell B, Brennan F, Maini R & Feldmann M (2002) Cytokines and anti-cytokine biologicals in autoimmunity: present and future. Cytokine Growth Factor Rev. 13: 299-314.

Asadullah K, Volk HD & Sterry W (2002) Novel immunotherapies for psoriasis. Trends Immunol. 23: 47-53.

Asahara H, Asanuma M, Ogawa N, Nishibayashi S & Inoue H (1995) High DNA-binding activity of transcription factor NF-κB in synovial membranes of patients with rheumatoid arthritis. Biochem. Mol. Biol. Int. 37: 827-832.

Auphan N, DiDonato JA, Rosette C, Helmberg A & Karin M (1995) Immunosuppression by glucocorticoids: inhibition of NF-κB activity through induction of IκB synthesis. Science 270: 286-290.

Baer M, Dillner A, Schwartz RC, Sedon C, Nedospasov S & Johnson PF (1998) Tumor necrosis factor α transcription in macrophages is attenuated by an autocrine factor that preferentially induces NF-κB p50. Mol. Cell. Biol. 18: 5678-5689.

Barger SW, Horster D, Furukawa K, Goodman Y, Krieglstein J & Mattson MP (1995) Tumor necrosis factors α and β protect neurons against amyloid β-peptide toxicity: evidence for involvement of a κB-binding factor and attenuation of peroxide and Ca^{2+} accumulation. Proc. Natl. Acad. Sci. USA 92: 9328-9332.

Barger SW & Mattson MP (1996) Induction of neuroprotective κB-dependent transcription by secreted forms of the Alzheimer's β-amyloid precursor. Brain Res. Mol. Brain Res. 40: 116-126.

Barnes PJ & Adcock IM (1997) NF-κB: a pivotal role in asthma and a new target for therapy. Trends

Pharmacol. Sci. 18: 46-50.
Barnes PJ & Karin M (1997) Nuclear factor-κB: a pivotal transcription factor in chronic inflammatory diseases. N. Engl. J. Med. 336: 1066-1071.
Beg AA, Sha WC, Bronson RT, Ghosh S & Baltimore D (1995) Embryonic lethality and liver degeneration in mice lacking the RelA component of NF-κB. Nature 376: 167-170.
Beg AA & Baltimore D (1996) An essential role for NF-κB in preventing TNF-α-induced cell death. Science 274: 782-784.
Belich MP, Salmeron A, Johnston LH & Ley SC (1999) TPL-2 kinase regulates the proteolysis of the NF-κB-inhibitory protein NF-κB1 p105. Nature 397: 363-368.
Bierhaus A, Schiekofer S, Schwaninger M, Andrassy M, Humpert PM, Chen J, Hong M, Luther T, Henle T, Kloting I, Morcos M, Hofmann M, Tritschler H, Weigle B, Kasper M, Smith M, Perry G, Schmidt AM, Stern DM, Haring HU, Schleicher E & Nawroth PP (2001) Diabetes-associated sustained activation of the transcription factor nuclear factor-κB. Diabetes 50: 2792-2808.
Blackwell TS, Blackwell TR, Holden EP, Christman BW & Christman JW (1996) In vivo antioxidant treatment suppresses nuclear factor-κB activation and neutrophilic lung inflammation. J. Immunol. 157: 1630-1637.
Blackwell TS & Christman JW (1997) The role of nuclear factor-κB in cytokine gene regulation. Am. J. Respir. Cell Mol. Biol. 17: 3-9.
Blackwell TS, Yull FE, Chen CL, Venkatakrishnan A, Blackwell TR, Hicks DJ, Lancaster LH, Christman JW & Kerr LD (1999) Use of genetically altered mice to investigate the role of nuclear factor-κB activation and cytokine gene expression in sepsis-induced ARDS. Chest 116: 73S-74S.
Bogdan C (2001) Nitric oxide and the immune response. Nat. Immunol. 2: 907-916.
Bondeson J, Foxwell B, Brennan F & Feldmann M (1999) Defining therapeutic targets by using adenovirus: blocking NF-κB inhibits both inflammatory and destructive mechanisms in rheumatoid synovium but spares anti-inflammatory mediators. Proc. Natl. Acad. Sci. USA 96: 5668-5673.
Bonetti B, Pohl J, Gao YL & Raine CS (1997) Cell death during autoimmune demyelination: effector but not target cells are eliminated by apoptosis. J. Immunol. 159: 5733-5741.
Bonetti B & Raine CS (1997) Multiple sclerosis: oligodendrocytes display cell death-related molecules in situ but do not undergo apoptosis. Ann. Neurol. 42: 74-84.
Bonetti B, Stegagno C, Cannella B, Rizzuto N, Moretto G & Raine CS (1999) Activation of NF-κB and c-jun transcription factors in multiple sclerosis lesionsImplications for oligodendrocyte pathology. Am. J. Pathol. 155: 1433-1438.
Bours V, Franzoso G, Azarenko V, Park S, Kanno T, Brown K & Siebenlist U (1993) The oncoprotein Bcl-3 directly transactivates through κB motifs via association with DNA-binding p50B homodimers. Cell 72: 729-739.
Brand K, Page S, Rogler G, Bartsch A, Brandl R, Knuechel R, Page M, Kaltschmidt C, Baeuerle PA & Neumeier D (1996) Activated transcription factor nuclear factor-κB is present in the atherosclerotic lesion. J. Clin. Invest. 97: 1715-1722.
Brand K, Eisele T, Kreusel U, Page M, Page S, Haas M, Gerling A, Kaltschmidt C, Neumann FJ, Mackman N, Baeurele PA, Walli AK & Neumeier D (1997) Dysregulation of monocytic nuclear factor-κB by oxidized low-density lipoprotein. Arterioscler. Thromb. Vasc. Biol. 17: 1901-1909.
Caesar I, Gross V, Roth M, Andus T, Schmidt C, Raedsch R, Weber A, Gierend M, Ewe K & Scholmerich J (1997) Treatment of active and postactive ileal and colonic Crohn's disease with oral pH-modified-release budesonide. German Budesonide Study Group. Hepatogastroenterology 44: 445-451.
Calara F, Dimayuga P, Niemann A, Thyberg J, Diczfalusy U, Witztum JL, Palinski W, Shah PK, Cercek B, Nilsson J & Regnstrom J (1998) An animal model to study local oxidation of LDL and its biological effects in the arterial wall. Arterioscler. Thromb. Vasc. Biol. 18: 884-893.
Caldenhoven E, Liden J, Wissink S, Van de Stolpe A, Raaijmakers J, Koenderman L, Okret S, Gustafsson JA & Van der Saag PT (1995) Negative cross-talk between RelA and the glucocorticoid receptor: a possible mechanism for the antiinflammatory action of glucocorticoids. Mol. Endocrinol. 9: 401-412.
Campbell IK, Gerondakis S, O'Donnell K & Wicks IP (2000) Distinct roles for the NF-κB1 (p50) and

c-Rel transcription factors in inflammatory arthritis. J. Clin. Invest. 105: 1799-1806.
Carey M (1998) The enhanceosome and transcriptional synergy. Cell 92: 5-8.
Carter BD, Kaltschmidt C, Kaltschmidt B, Offenhauser N, Bohm-Matthaei R, Baeuerle PA & Barde YA (1996) Selective activation of NF-κB by nerve growth factor through the neurotrophin receptor p75. Science 272: 542-545.
Chang L & Karin M (2001) Mammalian MAP kinase signalling cascades. Nature 410: 37-40.
Chin AI, Dempsey PW, Bruhn K, Miller JF, Xu Y & Cheng G (2002) Involvement of receptor-interacting protein 2 in innate and adaptive immune responses. Nature 416: 190-194.
Christman JW, Lancaster LH & Blackwell TS (1998) Nuclear factor κB: a pivotal role in the systemic inflammatory response syndrome and new target for therapy. Intensive Care Med. 24: 1131-1138.
Christman JW, Sadikot RT & Blackwell TS (2000) The role of nuclear factor-κB in pulmonary diseases. Chest 117: 1482-1487.
Ciechanover A, Gonen H, Bercovich B, Cohen S, Fajerman I, Israel A, Mercurio F, Kahana C, Schwartz AL, Iwai K & Orian A (2001) Mechanisms of ubiquitin-mediated limited processing of the NF-κB1 precursor protein p105. Biochimie 83: 341-349.
Collins T & Cybulsky MI (2001) NF-κB: pivotal mediator or innocent bystander in atherogenesis? J. Clin. Invest. 107: 255-264.
Conron M, Bondeson J, Pantelidis P, Beynon HL, Feldmann M, duBois RM & Foxwell BM (2001) Alveolar macrophages and T cells from sarcoid but not normal lung are permissive to adenovirus infection and allow analysis of NF-κB-dependent signaling pathways. Am. J. Respir. Cell. Mol. Biol. 25: 141-149.
Conron M, Andreakos E, Pantelidis P, Smith C, Beynon HL, Dubois RM & Foxwell BM (2002) Nuclear factor-κB activation in alveolar macrophages requires IκB kinase-β but not nuclear factor-κB inducing kinase. Am. J. Respir. Crit. Care Med. 165: 996-1004.
Couture R, Harrisson M, Vianna RM & Cloutier F (2001) Kinin receptors in pain and inflammation. Eur. J. Pharmacol. 429: 161-176.
Danning CL, Illei GG, Hitchon C, Greer MR, Boumpas DT & McInnes IB (2000) Macrophage-derived cytokine and nuclear factor κB p65 expression in synovial membrane and skin of patients with psoriatic arthritis. Arthritis Rheum. 43: 1244-1256.
Darville MI & Eizirik DL (2001) Cytokine induction of Fas gene expression in insulin-producing cells requires the transcription factors NF-κB and C/EBP. Diabetes 50: 1741-1748.
De Bosscher K, Schmitz ML, Vanden Berghe W, Plaisance S, Fiers W & Haegeman G (1997) Glucocorticoid-mediated repression of nuclear factor-κB-dependent transcription involves direct interference with transactivation. Proc. Natl. Acad. Sci. USA 94: 13504-13509.
de Crombrugghe B, Lefebvre V, Behringer RR, Bi W, Murakami S & Huang W (2000) Transcriptional mechanisms of chondrocyte differentiation. Matrix Biol. 19: 389-394.
Dichtl W, Nilsson L, Goncalves I, Ares MP, Banfi C, Calara F, Hamsten A, Eriksson P & Nilsson J (1999) Very low-density lipoprotein activates nuclear factor-κB in endothelial cells. Circ. Res. 84: 1085-1094.
Donovan CE, Mark DA, He HZ, Liou HC, Kobzik L, Wang Y, De Sanctis GT, Perkins DL & Finn PW (1999) NF-κB/Rel transcription factors: c-Rel promotes airway hyperresponsiveness and allergic pulmonary inflammation. J. Immunol. 163: 6827-6833.
Dumitru CD, Ceci JD, Tsatsanis C, Kontoyiannis D, Stamatakis K, Lin JH, Patriotis C, Jenkins NA, Copeland NG, Kollias G & Tsichlis PN (2000) TNF-α induction by LPS is regulated posttranscriptionally via a Tpl2/ERK-dependent pathway. Cell 103: 1071-1083.
Egan LJ & Sandborn WJ (1998) Inhibition of nuclear factor κB by sulfasalazine: a new target for inflammatory bowel disease therapy? Gastroenterology 115: 1295-1296.
Falvo JV, Uglialoro AM, Brinkman BM, Merika M, Parekh BS, Tsai EY, King HC, Morielli AD, Peralta EG, Maniatis T, Thanos D & Goldfeld AE (2000) Stimulus-specific assembly of enhancer complexes on the tumor necrosis factor α gene promoter. Mol. Cell. Biol. 20: 2239-2247.
Farver CF, Raychaudhuri B, Buhrow LT, Connors MJ & Thomassen MJ (1998) Constitutive NF-κB levels in human alveolar macrophages from normal volunteers. Cytokine 10: 868-871.
Farzaneh-Far A, Davies JD, Braam LA, Spronk HM, Proudfoot D, Chan SW, O'Shaughnessy KM,

Weissberg PL, Vermeer C & Shanahan CM (2001) A polymorphism of the human matrix γ-carboxyglutamic acid protein promoter alters binding of an activating protein-1 complex and is associated with altered transcription and serum levels. J. Biol. Chem. 276: 32466-32473.

Feldmann M & Maini RN (2001) Anti-TNFα therapy of rheumatoid arthritis: what have we learned? Annu. Rev. Immunol. 19: 163-196.

Ferrer I, Marti E, Lopez E & Tortosa A (1998) NF-κB immunoreactivity is observed in association with βA4 diffuse plaques in patients with Alzheimer's disease. Neuropathol. Appl. Neurobiol. 24: 271-277.

Fiocchi C (1998) Inflammatory bowel disease: etiology and pathogenesis. Gastroenterology 115: 182-205.

Foxwell B, Browne K, Bondeson J, Clarke C, de Martin R, Brennan F & Feldmann M (1998) Efficient adenoviral infection with IκBα reveals that macrophage tumor necrosis factor α production in rheumatoid arthritis is NF-κB dependent. Proc. Natl. Acad. Sci. USA 95: 8211-8215.

Franzoso G, Bours V, Park S, Tomita-Yamaguchi M, Kelly K & Siebenlist U (1992) The candidate oncoprotein Bcl-3 is an antagonist of p50/NF-κB-mediated inhibition. Nature 359: 339-342.

Franzoso G, Bours V, Azarenko V, Park S, Tomita-Yamaguchi M, Kanno T, Brown K & Siebenlist U (1993) The oncoprotein Bcl-3 can facilitate NF-κB-mediated transactivation by removing inhibiting p50 homodimers from select κB sites. EMBO J. 12: 3893-3901.

Franzoso G, Carlson L, Poljak L, Shores EW, Epstein S, Leonardi A, Grinberg A, Tran T, Scharton-Kersten T, Anver M, Love P, Brown K & Siebenlist U (1998) Mice deficient in nuclear factor (NF)-κB/p52 present with defects in humoral responses germinal center reactions and splenic microarchitecture. J. Exp. Med. 187: 147-159.

Fujihara SM, Cleaveland JS, Grosmaire LS, Berry KK, Kennedy KA, Blake JJ, Loy J, Rankin BM, Ledbetter JA & Nadler SG (2000) A D-amino acid peptide inhibitor of NF-κB nuclear localization is efficacious in models of inflammatory disease. J. Immunol. 165: 1004-1012.

Fujita T, Nolan GP, Liou HC, Scott ML & Baltimore D (1993) The candidate proto-oncogene bcl-3 encodes a transcriptional coactivator that activates through NF-κB p50 homodimers. Genes Dev. 7: 1354-1363.

Funk CD (2001) Prostaglandins and leukotrienes: advances in eicosanoid biology. Science 294: 1871-1875.

Geier DL & Miner PB (1992) New therapeutic agents in the treatment of inflammatory bowel disease. Am. J. Med. 93: 199-208.

Ghosh S, May MJ & Kopp EB (1998) NF-κB and Rel proteins: evolutionarily conserved mediators of immune responses. Annu. Rev. Immunol. 16: 225-260.

Giannoukakis N, Rudert WA, Trucco M & Robbins PD (2000) Protection of human islets from the effects of interleukin-1β by adenoviral gene transfer of an IκB repressor. J. Biol. Chem. 275: 36509-36513.

Grey ST, Arvelo MB, Hasenkamp W, Bach FH & Ferran C (1999) A20 inhibits cytokine-induced apoptosis and nuclear factor κB-dependent gene activation in islets. J. Exp. Med. 190: 1135-1146.

Grilli M, Goffi F, Memo M & Spano P (1996a) Interleukin-1β and glutamate activate the NF-κB/Rel binding site from the regulatory region of the amyloid precursor protein gene in primary neuronal cultures. J. Biol. Chem. 271: 15002-15007.

Grilli M, Pizzi M, Memo M & Spano P (1996b) Neuroprotection by aspirin and sodium salicylate through blockade of NF-κB activation. Science 274: 1383-1385.

Guerrini L, Blasi F & Denis-Donini S (1995) Synaptic activation of NF-κB by glutamate in cerebellar granule neurons in vitro. Proc. Natl. Acad. Sci. USA 92: 9077-9081.

Gveric D, Kaltschmidt C, Cuzner ML & Newcombe J (1998) Transcription factor NF-κB and inhibitor IκBα are localized in macrophages in active multiple sclerosis lesions. J. Neuropathol. Exp. Neurol. 57: 168-178.

Hajra L, Evans AI, Chen M, Hyduk SJ, Collins T & Cybulsky MI (2000) The NF-κB signal transduction pathway in aortic endothelial cells is primed for activation in regions predisposed to atherosclerotic lesion formation. Proc. Natl. Acad. Sci. USA 97: 9052-9057.

Han Y, Runge MS & Brasier AR (1999) Angiotensin II induces interleukin-6 transcription in vascular smooth muscle cells through pleiotropic activation of nuclear factor-κB transcription factors. Circ. Res. 84: 695-703.

Han Z, Boyle DL, Manning AM & Firestein GS (1998) AP-1 and NF-κB regulation in rheumatoid arthritis and murine collagen-induced arthritis. Autoimmunity 28: 197-208.

Handel ML, McMorrow LB & Gravallese EM (1995) Nuclear factor-κB in rheumatoid synovium. Localization of p50 and p65. Arthritis Rheum. 38: 1762-1770.

Hart LA, Krishnan VL, Adcock IM, Barnes PJ & Chung KF (1998) Activation and localization of transcription factor nuclear factor-κB in asthma. Am. J. Respir. Crit. Care Med. 158: 1585-1592.

Hayashi T, Sekine T & Okamoto T (1993) Identification of a new serine kinase that activates NFκB by direct phosphorylation. J. Biol. Chem. 268: 26790-26795.

Heck S, Bender K, Kullmann M, Gottlicher M, Herrlich P & Cato AC (1997) IκBα-independent downregulation of NF-κB activity by glucocorticoid receptor. EMBO J. 16: 4698-4707.

Hegazy DM, O'Reilly DA, Yang BM, Hodgkinson AD, Millward BA & Demaine AG (2001) NFκB polymorphisms and susceptibility to type 1 diabetes. Genes Immun. 2: 304-308.

Heimberg H, Heremans Y, Jobin C, Leemans R, Cardozo AK, Darville M & Eizirik DL (2001) Inhibition of cytokine-induced NF-κB activation by adenovirus-mediated expression of a NF-κB super-repressor prevents β-cell apoptosis. Diabetes 50: 2219-2224.

Hilliard B, Samoilova EB, Liu TS, Rostami A & Chen Y (1999) Experimental autoimmune encephalomyelitis in NF-κB-deficient mice: roles of NF-κB in the activation and differentiation of autoreactive T cells. J. Immunol. 163: 2937-2943.

Ho E, Quan N, Tsai YH, Lai W & Bray TM (2001) Dietary zinc supplementation inhibits NFκB activation and protects against chemically induced diabetes in CD1 mice. Exp. Biol. Med. 226: 103-111.

Hofman FM, Hinton DR, Johnson K & Merrill JE (1989) Tumor necrosis factor identified in multiple sclerosis brain. J. Exp. Med. 170: 607-612.

Hofmann MA, Schiekofer S, Kanitz M, Klevesath MS, Joswig M, Lee V, Morcos M, Tritschler H, Ziegler R, Wahl P, Bierhaus A & Nawroth PP (1998) Insufficient glycemic control increases nuclear factor-κB binding activity in peripheral blood mononuclear cells isolated from patients with type 1 diabetes. Diabetes Care 21: 1310-1316.

Hofmann MA, Schiekofer S, Isermann B, Kanitz M, Henkels M, Joswig M, Treusch A, Morcos M, Weiss T, Borcea V, Abdel Khalek AK, Amiral J, Tritschler H, Ritz E, Wahl P, Ziegler R, Bierhaus A & Nawroth PP (1999) Peripheral blood mononuclear cells isolated from patients with diabetic nephropathy show increased activation of the oxidative-stress sensitive transcription factor NF-κB. Diabetologia 42: 222-232.

Hohjoh H & Tokunaga K (2001) Allele-specific binding of the ubiquitous transcription factor OCT-1 to the functional single nucleotide polymorphism (SNP) sites in the tumor necrosis factor-α gene (TNFA) promoter. Genes Immun. 2: 105-109.

Igaz P, Falus A, Glaz E & Racz K (2000) Cytokines in diseases of the endocrine system. Cell Biol. Int 24: 663-668.

Iotsova V, Caamano J, Loy J, Yang Y, Lewin A & Bravo R (1997) Osteopetrosis in mice lacking NF-κB1 and NF-κB2. Nat. Med. 3: 1285-1289.

Kaltschmidt B, Uherek M, Volk B, Baeuerle PA & Kaltschmidt C (1997) Transcription factor NF-κB is activated in primary neurons by amyloid β peptides and in neurons surrounding early plaques from patients with Alzheimer disease. Proc. Natl. Acad. Sci. USA 94: 2642-2647.

Kaltschmidt C, Kaltschmidt B, Neumann H, Wekerle H & Baeuerle PA (1994) Constitutive NF-κB activity in neurons. Mol. Cell. Biol. 14: 3981-3992.

Kaltschmidt C, Kaltschmidt B & Baeuerle PA (1995) Stimulation of ionotropic glutamate receptors activates transcription factor NF-κB in primary neurons. Proc. Natl. Acad. Sci. USA 92: 9618-9622.

Karin M (1999) How NF-κB is activated: the role of the IκB kinase (IKK) complex. Oncogene 18: 6867-6874.

Karin M & Ben-Neriah Y (2000) Phosphorylation meets ubiquitination: the control of NFκB activity. Annu. Rev. Immunol. 18: 621-663.

Keatings VM, Jatakanon A, Worsdell YM & Barnes PJ (1997) Effects of inhaled and oral glucocorticoids on inflammatory indices in asthma and COPD. Am. J. Respir. Crit. Care Med. 155: 542-548.

Keifer JA, Guttridge DC, Ashburner BP & Baldwin AS (2001) Inhibition of NF-κB activity by thalidomide through suppression of IκB kinase activity. J. Biol. Chem. 276: 22382-22387.

Keller JN, Kindy MS, Holtsberg FW, St Clair DK, Yen HC, Germeyer A, Steiner SM, Bruce-Keller AJ, Hutchins JB & Mattson MP (1998) Mitochondrial manganese superoxide dismutase prevents neural apoptosis and reduces ischemic brain injury: suppression of peroxynitrite production lipid peroxidation and mitochondrial dysfunction. J. Neurosci. 18: 687-689.

Kelliher MA, Grimm S, Ishida Y, Kuo F, Stanger BZ & Leder P (1998) The death domain kinase RIP mediates the TNF-induced NF-κB signal. Immunity 8: 297-303.

Kieseier BC, Storch MK, Archelos JJ, Martino G & Hartung HP (1999) Effector pathways in immune mediated central nervous system demyelination. Curr. Opin. Neurol. 12: 323-336.

Kim JK, Kim YJ, Fillmore JJ, Chen Y, Moore I, Lee J, Yuan M, Li ZW, Karin M, Perret P, Shoelson SE & Shulman GI (2001) Prevention of fat-induced insulin resistance by salicylate. J. Clin. Invest. 108: 437-446.

Knight JC, Udalova I, Hill AV, Greenwood BM, Peshu N, Marsh K & Kwiatkowski D (1999) A polymorphism that affects OCT-1 binding to the TNF promoter region is associated with severe malaria. Nat. Genet. 22: 145-150.

Kobayashi K, Inohara N, Hernandez LD, Galan JE, Nunez G, Janeway CA, Medzhitov R & Flavell RA (2002) RICK/RIP2/CARDIAK mediates signalling for receptors of the innate and adaptive immune systems. Nature 416: 194-199.

Kopp E & Ghosh S (1994) Inhibition of NF-κB by sodium salicylate and aspirin. Science 265: 956-959.

Kuhn R, Lohler J, Rennick D, Rajewsky K & Muller W (1993) Interleukin-10-deficient mice develop chronic enterocolitis. Cell 75: 263-274.

Kunsch C, Ruben SM & Rosen CA (1992) Selection of optimal κB/Rel DNA-binding motifs: interaction of both subunits of NF-κB with DNA is required for transcriptional activation. Mol. Cell. Biol. 12: 4412-4421.

Landry DB, Couper LL, Bryant SR & Lindner V (1997) Activation of the NF-κB and IκB system in smooth muscle cells after rat arterial injury. Induction of vascular cell adhesion molecule-1 and monocyte chemoattractant protein-1. Am. J. Pathol. 151: 1085-1095.

Lavon I, Goldberg I, Amit S, Landsman L, Jung S, Tsuberi BZ, Barshack I, Kopolovic J, Galun E, Bujard H & Ben-Neriah Y (2000) High susceptibility to bacterial infection but no liver dysfunction in mice compromised for hepatocyte NF-κB activation. Nat. Med. 6: 573-577.

Ley K (2001) Pathways and bottlenecks in the web of inflammatory adhesion molecules and chemoattractants. Immunol. Res. 24: 87-95.

Li N & Karin M (1999) Is NF-κB the sensor of oxidative stress? FASEB J. 13: 1137-1143.

Li Q, Van Antwerp D, Mercurio F, Lee KF & Verma IM (1999a) Severe liver degeneration in mice lacking the IκB kinase 2 gene. Science 284: 321-325.

Li ZW, Chu W, Hu Y, Delhase M, Deerinck T, Ellisman M, Johnson R & Karin M (1999b) The IKKβ subunit of IκB kinase (IKK) is essential for nuclear factor κB activation and prevention of apoptosis. J. Exp. Med. 189: 1839-1845.

Liu ZG, Hsu H, Goeddel DV & Karin M (1996) Dissection of TNF receptor 1 effector functions: JNK activation is not linked to apoptosis while NF-κB activation prevents cell death. Cell 87: 565-576.

Majumdar S & Aggarwal BB (2001) Methotrexate suppresses NF-κB activation through inhibition of IκBα phosphorylation and degradation. J. Immunol. 167: 2911-2920.

Manna SK, Mukhopadhyay A & Aggarwal BB (2000) Leflunomide suppresses TNF-induced cellular responses: effects on NF-κB activator protein-1: c-Jun N-terminal protein kinase and apoptosis. J. Immunol. 165: 5962-5969.

Marienfeld R, Neumann M, Chuvpilo S, Escher C, Kneitz B, Avots A, Schimpl A & Serfling E (1997) Cyclosporin A interferes with the inducible degradation of NF-κB inhibitors but not with the processing of p105/NF-κB1 in T cells. Eur. J. Immunol. 27: 1601-1609.

Marok R, Winyard PG, Coumbe A, Kus ML, Gaffney K, Blades S, Mapp PI, Morris CJ, Blake DR, Kaltschmidt C & Baeuerle PA (1996) Activation of the transcription factor nuclear factor-κB in human inflamed synovial tissue. Arthritis Rheum. 39: 583-591.

Martin T, Cardarelli PM, Parry GC, Felts KA & Cobb RR (1997) Cytokine induction of monocyte chemoattractant protein-1 gene expression in human endothelial cells depends on the cooperative action of NF-κB and AP-1. Eur. J. Immunol. 27: 1091-1097.

Matsusaka T, Fujikawa K, Nishio Y, Mukaida N, Matsushima K, Kishimoto T & Akira S (1993) Transcription factors NF-IL6 and NF-κB synergistically activate transcription of the inflammatory cytokines interleukin 6 and interleukin 8. Proc. Natl. Acad. Sci. USA 90: 10193-10197.

Mattson MP, Goodman Y, Luo H, Fu W & Furukawa K (1997) Activation of NF-κB protects hippocampal neurons against oxidative stress-induced apoptosis: evidence for induction of manganese superoxide dismutase and suppression of peroxynitrite production and protein tyrosine nitration. J. Neurosci. Res. 49: 681-697.

Mengshol JA, Vincenti MP, Coon CI, Barchowsky A & Brinckerhoff CE (2000) Interleukin-1 induction of collagenase 3 (matrix metalloproteinase 13) gene expression in chondrocytes requires p38: c-Jun N-terminal kinase and nuclear factor κB: differential regulation of collagenase 1 and collagenase 3. Arthritis Rheum. 43: 801-811.

Meyer S, Kohler NG & Joly A (1997) Cyclosporine A is an uncompetitive inhibitor of proteasome activity and prevents NF-κB activation. FEBS Lett. 413: 354-358.

Miagkov AV, Kovalenko DV, Brown CE, Didsbury JR, Cogswell JP, Stimpson SA, Baldwin AS & Makarov SS (1998) NF-κB activation provides the potential link between inflammation and hyperplasia in the arthritic joint. Proc. Natl. Acad. Sci. USA 95: 13859-13864.

Mohamed AK, Bierhaus A, Schiekofer S, Tritschler H, Ziegler R & Nawroth PP (1999) The role of oxidative stress and NF-κB activation in late diabetic complications. Biofactors 10: 157-167.

Morigi M, Angioletti S, Imberti B, Donadelli R, Micheletti G, Figliuzzi M, Remuzzi A, Zoja C & Remuzzi G (1998) Leukocyte-endothelial interaction is augmented by high glucose concentrations and hyperglycemia in a NF-κB-dependent fashion. J. Clin. Invest. 101: 1905-1915.

Mortensen RF (2001) C-reactive protein inflammation and innate immunity. Immunol. Res. 24: 163-176.

Murakami S, Lefebvre V & de Crombrugghe B (2000) Potent inhibition of the master chondrogenic factor Sox9 gene by interleukin-1 and tumor necrosis factor-α. J. Biol. Chem. 275: 3687-3692.

Neish AS, Khachigian LM, Park A, Baichwal VR & Collins T (1995a) Sp1 is a component of the cytokine-inducible enhancer in the promoter of vascular cell adhesion molecule-1. J. Biol. Chem. 270: 28903-28909.

Neish AS, Read MA, Thanos D, Pine R, Maniatis T & Collins T (1995b) Endothelial interferon regulatory factor 1 cooperates with NF-κB as a transcriptional activator of vascular cell adhesion molecule 1. Mol. Cell. Biol. 15: 2558-2569.

Neurath MF, Pettersson S, Meyer zum Buschenfelde KH & Strober W (1996) Local administration of antisense phosphorothioate oligonucleotides to the p65 subunit of NF-κB abrogates established experimental colitis in mice. Nat. Med. 2: 998-1004.

Neurath MF & Pettersson S (1997) Predominant role of NF-κB p65 in the pathogenesis of chronic intestinal inflammation. Immunobiology 198: 91-98.

Nikolaus S, Raedler A, Kuhbacker T, Sfikas N, Folsch UR & Schreiber S (2000) Mechanisms in failure of infliximab for Crohn's disease. Lancet 356: 1475-1479.

Norbiato G, Bevilacqua M, Vago T & Clerici M (1997) Glucocorticoids and Th-1: Th-2 type cytokines in rheumatoid arthritis osteoarthritis asthma atopic dermatitis and AIDS. Clin. Exp. Rheumatol. 15: 315-323.

O'Neill LA & Kaltschmidt C (1997) NF-κB: a crucial transcription factor for glial and neuronal cell function. Trends Neurosci. 20: 252-258.

Pahan K & Schmid M (2000) Activation of nuclear factor-κB in the spinal cord of experimental allergic encephalomyelitis. Neurosci. Lett. 287: 17-20.

Pahl HL (1999) Activators and target genes of Rel/NF-κB transcription factors. Oncogene 18: 6853-6866.

Palombella VJ, Conner EM, Fuseler JW, Destree A, Davis JM, Laroux FS, Wolf RE, Huang J, Brand S, Elliott PJ, Lazarus D, McCormack T, Parent L, Stein R, Adams J & Grisham MB (1998) Role of the proteasome and NF-κB in streptococcal cell wall-induced polyarthritis. Proc. Natl. Acad. Sci. USA 95: 15671-15676.

Pan J & McEver RP (1995) Regulation of the human P-selectin promoter by Bcl-3 and specific homodimeric members of the NF-κB/Rel family. J. Biol. Chem. 270: 23077-23083.

Park BH & Park JW (2001) The protective effect of Amomum xanthoides extract against alloxan-induced diabetes through the suppression of NFκB activation. Exp. Mol. Med. 33: 64-68.

Pierce JW, Read MA, Ding H, Luscinskas FW & Collins T (1996) Salicylates inhibit IκB-α phosphorylation endothelial-leukocyte adhesion molecule expression and neutrophil transmigration. J. Immunol. 156: 3961-3969.

Poligone B, Weaver DJ, Sen P, Baldwin AS & Tisch R (2002) Elevated NF-κB activation in non-obese diabetic mouse dendritic cells results in enhanced APC function. J. Immunol. 168: 188-196.

Price SJ, Greaves DR & Watkins H (2001) Identification of novel functional genetic variants in the human matrix metalloproteinase-2 gene: role of Sp1 in allele-specific transcriptional regulation. J. Biol. Chem. 276: 7549-7558.

Pueyo ME, Gonzalez W, Nicoletti A, Savoie F, Arnal JF & Michel JB (2000) Angiotensin II stimulates endothelial vascular cell adhesion molecule-1 via nuclear factor-κB activation induced by intracellular oxidative stress. Arterioscler. Thromb. Vasc. Biol. 20: 645-651.

Quan N, Ho E, La W, Tsai YH & Bray T (2001) Administration of NF-κB decoy inhibits pancreatic activation of NF-κB and prevents diabetogenesis by alloxan in mice. FASEB J. 15: 1616-1618.

Roebuck KA (1999) Regulation of interleukin-8 gene expression. J. Interferon Cytokine Res. 19: 429-438.

Rogler G, Brand K, Vogl D, Page S, Hofmeister R, Andus T, Knuechel R, Baeuerle PA, Scholmerich J & Gross V (1998) Nuclear factor κB is activated in macrophages and epithelial cells of inflamed intestinal mucosa. Gastroenterology 115: 357-369.

Ruddle NH (1999) Lymphoid neo-organogenesis: lymphotoxin's role in inflammation and development. Immunol. Res. 19: 119-125.

Rudolph D, Yeh WC, Wakeham A, Rudolph B, Nallainathan D, Potter J, Elia AJ & Mak TW (2000) Severe liver degeneration and lack of NF-κB activation in NEMO/IKKγ-deficient mice. Genes Dev. 14: 854-862.

Saccani S, Pantano S & Natoli G (2001) Two waves of nuclear factor κB recruitment to target promoters. J. Exp. Med. 193: 1351-1359.

Saccani S, Pantano S & Natoli G (2002) p38-Dependent marking of inflammatory genes for increased NF-κB recruitment. Nat Immunol. 3: 69-75.

Scheinman RI, Cogswell PC, Lofquist AK & Baldwin AS (1995a) Role of transcriptional activation of IκBα in mediation of immunosuppression by glucocorticoids. Science 270: 283-286.

Scheinman RI, Gualberto A, Jewell CM, Cidlowski JA & Baldwin AS (1995b) Characterization of mechanisms involved in transrepression of NF-κB by activated glucocorticoid receptors. Mol. Cell. Biol. 15: 943-953.

Schmidt AM, Yan SD, Wautier JL & Stern D (1999) Activation of receptor for advanced glycation end products: a mechanism for chronic vascular dysfunction in diabetic vasculopathy and atherosclerosis. Circ. Res. 84: 489-497.

Schmitz ML & Baeuerle PA (1991) The p65 subunit is responsible for the strong transcription activating potential of NF-κB. EMBO J. 10: 3805-3817.

Schmitz ML, Bacher S & Kracht M (2001) IκB-independent control of NF-κB activity by modulatory phosphorylations. Trends Biochem. Sci. 26: 186-190.

Schreck R, Albermann K & Baeuerle PA (1992) Nuclear factor κB: an oxidative stress-responsive transcription factor of eukaryotic cells (a review). Free Radic. Res. Commun. 17: 221-237.

Schreiber S, Nikolaus S & Hampe J (1998) Activation of nuclear factor κB inflammatory bowel disease. Gut 42: 477-484.

Schwartz MD, Moore EE, Moore FA, Shenkar R, Moine P, Haenel JB & Abraham E (1996) Nuclear factor-κB is activated in alveolar macrophages from patients with acute respiratory distress syndrome. Crit. Care Med. 24: 1285-1292.

Seetharaman R, Mora AL, Nabozny G, Boothby M & Chen J (1999) Essential role of T cell NF-κB activation in collagen-induced arthritis. J. Immunol. 163: 1577-1583.

Selmaj K, Raine CS, Cannella B & Brosnan CF (1991) Identification of lymphotoxin and tumor necrosis factor in multiple sclerosis lesions. J. Clin. Invest. 87: 949-954.

Senftleben U, Cao Y, Xiao G, Greten FR, Krahn G, Bonizzi G, Chen Y, Hu Y, Fong A, Sun SC & Karin M (2001) Activation by IKKα of a second evolutionary conserved NF-κB signaling pathway. Science 293: 1495-1499.

Sha WC, Liou HC, Tuomanen EI & Baltimore D (1995) Targeted disruption of the p50 subunit of NF-κB leads to multifocal defects in immune responses. Cell 80: 321-330.

Shenkar R & Abraham E (1997) Hemorrhage induces rapid in vivo activation of CREB and NF-κB in murine intraparenchymal lung mononuclear cells. Am. J. Respir. Cell Mol. Biol. 16: 145-152.

Simeonidis S, Liang S, Chen G & Thanos D (1997) Cloning and functional characterization of mouse IκBε. Proc. Natl. Acad. Sci. USA 94: 14372-14377.

Smith C, Andreakos E, Crawley JB, Brennan FM, Feldmann M & Foxwell BM (2001) NF-κB-inducing kinase is dispensable for activation of NF-κB in inflammatory settings but essential for lymphotoxin β receptor activation of NF-κB in primary human fibroblasts. J. Immunol. 167: 5895-5903.

Stein B, Baldwin AS, Ballard DW, Greene WC, Angel P & Herrlich P (1993) Cross-coupling of the NF-κB p65 and Fos/Jun transcription factors produces potentiated biological function. EMBO J. 12: 3879-3891.

Tanaka M, Fuentes ME, Yamaguchi K, Durnin MH, Dalrymple SA, Hardy KL & Goeddel DV (1999) Embryonic lethality liver degeneration and impaired NF-κB activation in IKK-β-deficient mice. Immunity 10: 421-429.

Terai K, Matsuo A & McGeer PL (1996) Enhancement of immunoreactivity for NF-κB in the hippocampal formation and cerebral cortex of Alzheimer's disease. Brain Res. 735: 159-168.

Thanos D & Maniatis T (1995) Virus induction of human IFNβ gene expression requires the assembly of an enhanceosome. Cell 83: 1091-1100.

Thompson AB, Robbins RA, Romberger DJ, Sisson JH, Spurzem JR, Teschler H & Rennard SI (1995a) Immunological functions of the pulmonary epithelium. Eur. Respir. J. 8: 127-149.

Thompson JE, Phillips RJ, Erdjument-Bromage H, Tempst P & Ghosh S (1995b) IκB-β regulates the persistent response in a biphasic activation of NF-κB. Cell 80: 573-582.

Tomita T, Takeuchi E, Tomita N, Morishita R, Kaneko M, Yamamoto K, Nakase T, Seki H, Kato K, Kaneda Y & Ochi T (1999) Suppressed severity of collagen-induced arthritis by in vivo transfection of nuclear factor κB decoy oligodeoxynucleotides as a gene therapy. Arthritis Rheum. 42: 2532-2542.

Tsao PW, Suzuki T, Totsuka R, Murata T, Takagi T, Ohmachi Y, Fujimura H & Takata I (1997) The effect of dexamethasone on the expression of activated NF-κB in adjuvant arthritis. Clin. Immunol. Immunopathol. 83: 173-178.

Udalova IA, Richardson A, Denys A, Smith C, Ackerman H, Foxwell B & Kwiatkowski D (2000) Functional consequences of a polymorphism affecting NF-κB p50-p50 binding to the TNF promoter region. Mol. Cell. Biol. 20: 9113-9119.

Udalova IA, Mott R, Field D & Kwiatkowski D (2002a) Quantitative prediction of NF-κB DNA-protein interactions. Proc. Natl. Acad. Sci. USA 99: 8167-8172.

Udalova IA, Richardson A, Ackerman H, Wordsworth P & Kwiatkowski D (2002b) Association of accelerated erosive rheumatoid arthritis with a polymorphism that alters NF-κB binding to the TNF promoter region. Rheumatology 41: 830-831.

Van Antwerp DJ, Martin SJ, Kafri T, Green DR & Verma IM (1996) Suppression of TNF-α-induced apoptosis by NF-κB. Science 274: 787-789.

van Heel DA, Udalova IA, De Silva AP, McGovern DP, Kinouchi Y, Hull J, Lench NJ, Cardon LR, Carey AH, Jewell DP & Kwiatkowski D (2002) Inflammatory bowel disease is associated with a TNF polymorphism that affects an interaction between the OCT1 and NF-κB transcription factors. Hum. Mol. Genet. 11: 1281-1289.

Venkataraman L, Burakoff SJ & Sen R (1995) FK506 inhibits antigen receptor-mediated induction of c-rel in B and T lymphoid cells. J. Exp. Med. 181: 1091-1099.

Wahl C, Liptay S, Adler G & Schmid RM (1998) Sulfasalazine: a potent and specific inhibitor of nuclear factor κB. J. Clin. Invest. 101: 1163-1174.

Walsh DM, Klyubin I, Fadeeva JV, Cullen WK, Anwyl R, Wolfe MS, Rowan MJ & Selkoe DJ (2002) Naturally secreted oligomers of amyloid β protein potently inhibit hippocampal long-term potentiation in vivo. Nature 416: 535-539.

Wang CY, Mayo MW & Baldwin AS (1996) TNF- and cancer therapy-induced apoptosis: potentiation by inhibition of NF-κB. Science 274: 784-787.

Wang D, Westerheide SD, Hanson JL & Baldwin AS (2000) Tumor necrosis factor α-induced phosphorylation of RelA/p65 on Ser529 is controlled by casein kinase II. J. Biol. Chem. 275: 32592-32597.

Weaver DJ, Poligone B, Bui T, Abdel-Motal UM, Baldwin AS & Tisch R (2001) Dendritic cells from nonobese diabetic mice exhibit a defect in NF-κB regulation due to a hyperactive IκB kinase. J. Immunol. 167: 1461-1468.

Weih F, Durham SK, Barton DS, Sha WC, Baltimore D & Bravo R (1997) p50-NF-κB complexes partially compensate for the absence of RelB: severely increased pathology in $p50^{-/-}relB^{-/-}$ double-knockout mice. J. Exp. Med. 185: 1359-1370.

Weingarten P, Bermak J & Zhou QY (2001) Evidence for non-oxidative dopamine cytotoxicity: potent activation of NF-κB and lack of protection by anti-oxidants. J. Neurochem. 76: 1794-1804.

Welch GN & Loscalzo J (1998) Homocysteine and atherothrombosis. N. Engl. J. Med. 338: 1042-1050.

Whiteside ST, Epinat JC, Rice NR & Israel A (1997) IκBε, a novel member of the IκB family, controls RelA and cRel NF-κB activity. EMBO J. 16: 1413-1426.

Whitley MZ, Thanos D, Read MA, Maniatis T & Collins T (1994) A striking similarity in the organization of the E-selectin and β interferon gene promoters. Mol. Cell. Biol. 14: 6464-6475.

Wilckens T & De Rijk R (1997) Glucocorticoids and immune function: unknown dimensions and new frontiers. Immunol. Today 18: 418-424.

Wilson SH, Caplice NM, Simari RD, Holmes DR, Carlson PJ & Lerman A (2000) Activated nuclear factor-κB is present in the coronary vasculature in experimental hypercholesterolemia. Atherosclerosis 148: 23-30.

Woodroofe MN & Cuzner ML (1993) Cytokine mRNA expression in inflammatory multiple sclerosis lesions: detection by non-radioactive in situ hybridization. Cytokine 5: 583-588.

Wulczyn FG, Naumann M & Scheidereit C (1992) Candidate proto-oncogene bcl-3 encodes a subunit-specific inhibitor of transcription factor NF-κB. Nature 358: 597-599.

Xiao G, Harhaj EW & Sun SC (2001) NF-κB-inducing kinase regulates the processing of NF-κB2 p100. Mol. Cell 7: 401-409.

Yan SD, Yan SF, Chen X, Fu J, Chen M, Kuppusamy P, Smith MA, Perry G, Godman GC, Nawroth P, Zweiter JL & Stern D (1995) Non-enzymatically glycated tau in Alzheimer's disease induces neuronal oxidant stress resulting in cytokine gene expression and release of amyloid β-peptide. Nat. Med. 1: 693-699.

Yang JP, Merin JP, Nakano T, Kato T, Kitade Y & Okamoto T (1995) Inhibition of the DNA-binding activity of NF-κB by gold compounds in vitro. FEBS Lett. 361: 89-96.

Yang L, Cohn L, Zhang DH, Homer R, Ray A & Ray P (1998) Essential role of nuclear factor κB in the induction of eosinophilia in allergic airway inflammation. J. Exp. Med. 188: 1739-1750.

Yin MJ, Yamamoto Y & Gaynor RB (1998) The anti-inflammatory agents aspirin and salicylate inhibit the activity of IκB kinase-β. Nature 396: 77-80.

Zhong H, SuYang H, Erdjument-Bromage H, Tempst P & Ghosh S (1997) The transcriptional activity of NF-κB is regulated by the IκB-associated PKAc subunit through a cyclic AMP-independent mechanism. Cell 89: 413-424.

Zollner TM, Podda M, Pien C, Elliott PJ, Kaufmann R & Boehncke WH (2002) Proteasome inhibition reduces superantigen-mediated T cell activation and the severity of psoriasis in a SCID-hu model. J. Clin. Invest. 109: 671-679.

Zong WX, Edelstein LC, Chen C, Bash J & Gelinas C (1999) The prosurvival Bcl-2 homolog Bfl-1/A1 is a direct transcriptional target of NF-κB that blocks TNFα-induced apoptosis. Genes Dev. 13: 382-387.

6. AFFILIATIONS

Imperial College of Science, Technology and Medicine, Faculty of Medicine, Kennedy Institute of Rheumatology, 1 Aspenlea Road, Hammersmith, London W6 8LH, UK (E. Andreakos, S. Sacre & B.M. Foxwell); The John Radcliffe Hospital, University Department of Paediatrics, Headley Way, Headington, Oxford OX3 9DU, UK (I. Udalova).

G. COURTOIS & A. ISRAËL

NF-κB IN GENETIC DISORDERS

1. INTRODUCTION

The transcription factor NF-κB is kept inactive in the cytoplasm by interaction with the inhibitory molecule IκB. In response to multiple stimuli, such as inflammatory cytokines, bacterial lipopolysaccharide (LPS), viral infection or stress, IκB is phosphorylated on two critical serine residues. This modification allows its recognition and destruction via the proteasome degradation machinery. As a consequence, free NF-κB enters the nucleus and activates transcription of a variety of genes participating in immune and inflammatory response, cell adhesion, growth control or protection against apoptosis (Karin & Ben-Neriah, 2000). For many years, the kinase that phosphorylates IκB (IKK) has remained elusive. Upon biochemical fractionation, it has finally been identified as a high-molecular mass complex migrating around 700-900 kDa and containing two related catalytic subunits, viz. IKK-1 (or IKKα) and IKK-2 (or IKKβ) (Israël, 2000). An additional component, NEMO/IKKγ, has been subsequently identified by genetic complementation of an NF-κB activation-defective cell line (Yamaoka et al., 1998) and sequencing of IKK-associated polypeptides (Rothwarf et al., 1998; Mercurio et al., 1999). In contrast to IKK-1 and IKK-2, NEMO/IKKγ exhibits no catalytic properties, but acts as a regulatory subunit of IKK. Cell lines defective for NEMO/IKKγ do not activate NF-κB in response to many stimuli, demonstrating the key role of this protein in activating the NF-κB pathway. The NEMO/IKKγ gene is located on chromosome X at Xq28 (Jin & Jeang, 1999). Such a location has important consequences for human pathology.

2. LOSS OF FUNCTION MUTATIONS OF NEMO/IKKγ RESULTS IN INCONTINENTIA PIGMENTI

Incontinentia pigmenti (IP) is a severe X-linked genodermatosis, with an incidence of between 1/10,000 and 1/100,000, that presents almost exclusively in females, as male cases die in utero (Landy & Donnai, 1993). In affected females, the disorder is highly variable in presentation. Typically, IP has four dermatological stages that begin neonatally with blisters and an inflammatory response, accompanied by a massive eosinophilic granulocyte infiltration into the epidermis. Subsequently, verrucous hyperkeratotic lesions evolve and disappear over time, leaving behind areas of hyperpigmentation that follow the lines of Blaschko.

R. Beyaert (ed.), Nuclear Factor κB. Regulation and Role in Disease, 325-334.

These generally disappear by the second decade and adults may show areas of dermal scarring. In addition to the manifestations at the epidermis level, IP patients also suffer from ophthalmologic problems (abnormalities of the developing retinal vessels), odontological problems (missing or deformed teeth) and, in rare cases, neurological problems (convulsive disorders, motor or mental retardation). The gene responsible for IP was mapped to an interval of about 2 Mb distal to the color vision locus in Xq28. The gene encoding NEMO was also mapped to Xq28 and found to overlap with the G6PD locus (Jin & Jeang, 1999). This localization as well as the exquisite sensitivity to apoptosis of cells derived from IP patients suggested that NEMO might be involved in this pathology. Remarkably, the analysis of a large collection of patients showed that 85% carried the same complex rearrangement of the NEMO gene (Smahi et al., 2000). This rearrangement involves excision of the region between two MER67B repeated sequences located upstream of exon 4 and downstream of exon 10, respectively, resulting in the synthesis of a truncated 133 amino acids NEMO molecule (corresponding to exons 1-3), which is devoid of activity. In IP patients carrying this rearrangement, a lack of NF-κB activation was demonstrated by studying foetus-derived primary fibroblasts: these cells are unresponsive to all tested NF-κB-activating stimuli, they do not show degradation of the IκB molecules when stimulated, and they are very sensitive to TNF-induced apoptosis. Other mutations have been described and most of them result in more or less severe truncations of NEMO/IKKγ (Smahi et al., 2000; Aradhya et al., 2001b). NEMO/IKKγ knockout mice have been engineered by several groups (Makris et al., 2000; Rudolph et al., 2000; Schmidt-Supprian et al., 2000) and their phenotype appears very similar to the one of IP patients. Male mice die very early during embryogenesis (E12) from massive liver apoptosis. A similar phenotype has also been reported for relA and IKK-2$^{-/-}$ mice (Beg et al., 1995; Li et al., 1999a-b; Tanaka et al., 1999), but death occurs a bit later, at E15.5 and E14.5, respectively. Importantly, liver apoptosis has been demonstrated to be triggered by TNFα. Indeed, relA$^{-/-}$/TNF$\alpha^{-/-}$ (Doi et al., 1999) and IKK-2$^{-/-}$/TNF-R1$^{-/-}$ (Senftleben et al., 2001) double-knockout mice can survive until birth (but die later from bacterial infections). It remains to be determined whether liver apoptosis is also responsible for male lethality in IP. In contrast to males, NEMO/IKK$\gamma^{+/-}$ female mice develop normally, but soon after birth exhibit a transient dermatosis, characterized by patchy skin lesions with massive granulocyte infiltration, hyperproliferation and increased apoptosis of keratinocytes, reminiscent of what is observed in human IP. Hyperpigmentation, a hallmark of IP due to the presence in the dermis of phagocytes containing melanosome complexes, was also observed in skin sections from NEMO$^{+/-}$ mice. Ultrastructural analysis of these mice indicated that, contrary to what is observed in wild-type animals, basal layer keratinocytes are in loose contact with each other, forming filopodia that extend into the dilated intercellular spaces (Schmidt-Supprian et al., 2000). This phenomenon was also observed in the skin of IP patients. Interestingly, a similar picture has been described in the basal layer of the epidermis of IKK1$^{+/-}$ mice

(Hu et al., 1999). In addition, the extensive X-inactivation skewing in blood leukocytes, which is a major feature of IP patients, is also observed in NEMO/IKK$\gamma^{+/-}$ mice (Schmidt-Supprian et al., 2000). The only difference observed between IP pathology and NEMO/IKK$\gamma^{-/-}$ is the high level of mortality occurring in female mice within 6-10 days after birth, which is never observed in humans. The phenotype of IP patients is difficult to assess, due to the complex interplay that exists in several tissues between cells carrying a normal copy of NEMO/IKKγ and those carrying a defective one. IP females exhibit extremely skewed X-inactivation in blood, resulting from a counter-selection of cells carrying the mutated NEMO/IKKγ locus. For some unknown reasons, X-inactivation skewing is less complete in the skin and selective elimination of cells bearing a mutated X chromosome only starts at or around birth. This process generates the characteristic dermatosis observed in IP patients. The nature of the signal that triggers IP cell elimination at this particular time is unknown, but the cascade of events that occurs has been proposed to be the following (Makris et al., 2000). Upon an unknown kind of stress, necrosis of some NEMO/IKKγ^{-} keratinocytes may happen, triggering a local inflammation response and the secretion of cytokines, such as TNFα. As a consequence, NEMO/IKKγ^{-} cells would be extensively destroyed by TNFα-induced apoptosis. According to this scenario, if elimination of all NEMO/IKKγ^{-} cells is not complete at this stage, a second wave of elimination may occur later. Such recurrence has indeed been observed in rare cases of IP (van Leeuwen et al., 2000).

An alternative explanation, that does not involve any cell necrosis, may be proposed based on the recent generation of mice with selective IKK-2 ablation in the epidermis (Pasparakis et al., 2002). In these animals, epidermis development proceeds normally until birth. At P4-P5 their skin starts becoming hard and inflexible and by day P7-P8 a widespread scaling is observed that precedes death. At P7, mice exhibit a thickened epidermis and this hyperproliferation is accompanied with cell infiltration into the dermis and severe inflammation. Several inflammatory cytokines, such as IL-1 and TNFα, and chemokines accumulate in the epidermis or dermis during progression of the disease. IL-1 is the first to be detected in the epidermis, before the onset of the pathology. Although the signal triggering the whole process remains unknown, the skin defect is linked to keratinocyte dysfunction, resulting from impaired NF-κB activation. Importantly, purified keratinocytes do not exhibit hyperproliferation on their own, indicating that the defect is not cell-autonomous. Hyperproliferation of keratinocytes was observed in various mice with disturbed NF-κB activation, leading to the proposal that this transcription factor may participate in keratinocyte cell cycle (Seitz et al., 1998; Kaufman & Fuchs, 2000). Skin-specific IKK-$2^{-/-}$ mice rather demonstrate that hyperproliferation is a secondary event resulting from inflammation. The role of TNFα in the pathology appears essential, since crossing TNF-$R1^{-/-}$ mice with IKK-2 $skin^{-/-}$ mice completely suppresses the skin manifestations.

Skin-specific $IKK2^{-/-}$ and NEMO/IKK$\gamma^{+/-}$ mice show several similarities that provide some clues on the dermatosis observed in IP patients. Nevertheless, an important difference concerns the apoptotic component of the disease. In contrast to NEMO/IKK$\gamma^{+/-}$, apoptosis of keratinocytes in skin-specific IKK-2$^{-/-}$ does not contribute substantially to the pathology, whereas lesion clearance of IP patients directly results from NEMO/IKKγ^{-} cell apoptosis. Based on the data gathered from mouse studies (Makris et al., 2000; Schmidt-Supprian et al., 2000), a putative sequence of events for IP dermatosis can be proposed (Fig. 1).

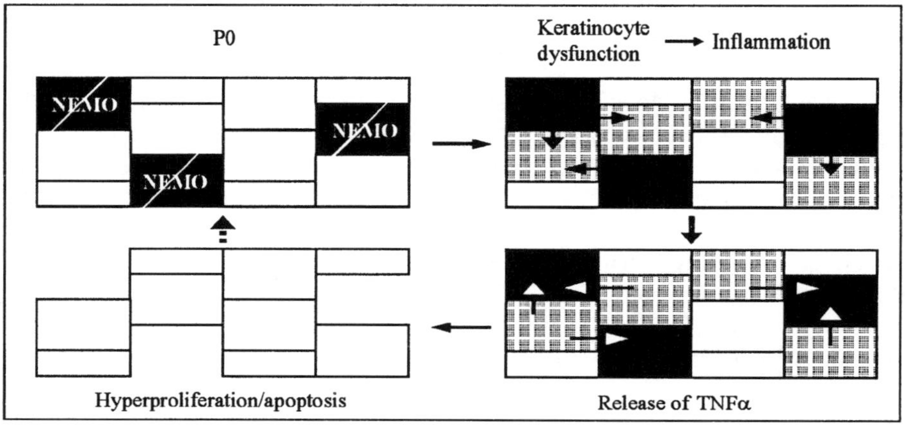

Fig. 1. Putative model for IP dermatosis. The sequence of events is based on transgenic mice analysis (Makris et al., 2000; Pasparakis et al., 2002). At birth, NEMO^{+} and NEMO^{-} cells coexist in the epidermis of IP patients/NEMO^{-} female mice. Within one or two weeks after birth, an inflammation reaction driven by NEMO^{-} keratinocytes occurs. Though the initial trigger remains unknown, NF-κB was shown to play a major role in epidermis homeostasis. Surrounding cells, from the (epi)dermis, release various inflammatory mediators (including TNFα) and chemokines. TNFα triggers hyperproliferation and apoptosis of NEMO^{-} keratinocytes as well as clearance of skin lesions. A new cycle of elimination (broken arrow) may occur later during life as observed in a few patients.

3. HYPOMORPHIC MUTATIONS OF NEMO/IKKγ RESULT IN EDA-ID

As indicated above, complete loss of NF-κB activation is lethal for males during embryogenesis, but females can survive, exhibiting a complex phenotype, mostly due to lyonization. In a sense, these results are a bit frustrating as they do not tell us much about the consequences of NF-κB inactivation in humans. In fact, hypomorphic NEMO mutations, if they can be identified, would probably tell us more by allowing male patients to survive. This type of mutation has indeed been identified recently, and several publications report a new syndrome exclusively affecting boys (Zonana et al., 2000; Aradhya et al., 2001a; Döffinger et al., 2001; Jain et al., 2001). In this case, their single X chromosome carries the mutated gene, directly allowing to observe the physiological consequences of NF-κB dysfunction in humans.

The conclusion of these analyses is that partial NF-κB deficiency in humans gives rise to a series of symptoms that vary a bit according to the severity of the mutations. The so-called "milder" forms exhibit immunodeficiency (ID) associated with a complex syndrome called anhydrotic ectodermal dysplasia (EDA). The more severe form shows these two symptoms, but in addition osteopetrosis and primary lymphedema can be observed (OL-EDA-ID) (Döffinger et al., 2001).

For years, a rare and complex X-linked syndrome associating anhydrotic ectodermal dysplasia and immunodeficiency (EDA-ID) had been described, but its genetic cause was unknown (Frix & Bronson, 1986; Sitton & Reimund, 1992; Abinun et al., 1996; Schweizer et al., 1999). Anhidrotic ectodermal dysplasia (EDA) is a well-known pathology in which the patient has no sweat glands, sparse scalp hair and rare conical teeth (Pinheiro & Freire-Maia, 1994). Three distinct loci were shown to be responsible for EDA. The first, located on X chromosome, codes for a new member of the TNF family, viz. ectodysplasin/EDA-A1 (Kere et al., 1996), a membrane-associated protein produced in cell types/tissues of ectodermal origin, such as keratinocytes, hair follicles and sweat glands (Montonen et al., 1998). Ectodysplasin is a type II transmembrane protein (Ezer et al., 1999; Mikkola et al., 1999) released as a trimer in the extracellular compartment after furin-dependent cleavage (Chen et al., 2001; Elomaa et al., 2001; Schneider et al., 2001). Allelic mutations in DL, the orthologue of mouse "downless" which encodes a putative Eda receptor (EDAR) and which is located on chromosome 2, account for both autosomal recessive and dominant types of EDA (Headon & Overbeek, 1999; Monreal et al., 1999). EDAR expression is restricted to placodes, thickenings of epithelia where the formation of epidermal appendages begins (Headon & Overbeek, 1999). The direct demonstration that EDAR is a receptor for ectodysplasin was recently provided (Tucker et al., 2000; Yan et al., 2000; Kumar et al., 2001; Laurikkala et al., 2001) and recent data indicate that ectodysplasin/EDAR interaction results in NF-κB activation (Yan et al., 2000; Döffinger et al., 2001; Kumar et al., 2001). The third locus whose disruption leads to EDA ("crinkled"), has been identified recently (Headon et al., 2001; Yan et al., 2002). It encodes a death domain-containing adaptor protein EDARADD/CR, which binds EDAR by a homotypic death domain interaction and mediates NF-κB activation. Thus mutations in three genes encoding members of a signaling cascade leading to NF-κB activation result in anhydrotic ectodermal dysplasia (Fig. 2). Although the details of the ectodysplasin/EDAR signaling pathway are not known, analysis of "downless" mice (the model of EDAR pathology) showed that this pathway is involved very early during development of hair follicle morphogenesis (Headon & Overbeek, 1999), assigning a previously unrecognized role for NF-κB in this process.

In contrast to EDA patients, those with EDA-ID also exhibit a severe immunodeficiency and the morbidity during the first years of their life is high, due to recurrent infections involving gram-negative or gram-positive bacteria, mycobacteria, etc. However, the rarity and heterogeneity of patients, and the paucity of immunological abnormalities have hampered the recognition of this syndrome as an autonomous disorder. The first cases of EDA-ID were reported in 1986 and 1992 in boys suffering from multiple life-threatening infections (Frix & Bronson, 1986; Sitton & Reimund, 1992). Additional cases were subsequently identified, but paradoxically no major immunological abnormalities were found, although some patients showed impaired antibody response to polysaccharide antigens (Abinun et al., 1996; Schweizer et al., 1999). Interestingly, all the reported cases were male, including three siblings, suggesting an X-linked recessive mode of inheritance in EDA-ID. These characteristics, together with the association with a perturbed immune response and some similarities with IP, led to analysis of the NEMO/IKKγ gene in several EDA-ID patients. Most of them indeed carry mutations in NEMO/IKKγ, but instead of leading to large truncations of the NEMO/IKKγ molecule as observed in IP, the mutations are missense mutations or small deletions only affecting the zinc finger located at the very C-terminus of the molecule (Zonana et al., 2000; Aradhya et al., 2001a; Döffinger et al., 2001; Jain et al., 2001). Interestingy, all these mutations lead to reduced but not abolished NF-κB activation, explaining why affected male patients survive. Two patients affected with the more severe form of the pathology (OL-EDA-ID) were found to carry a mutation of the stop codon, leading to the addition of 27 irrelevant amino acids at the C-terminus of the molecule (Smahi et al., 2000; Döffinger et al., 2001). This seemingly benign mutation actually turned out to be very severe as it strongly destabilizes the NEMO molecule, leading to almost undetectable levels of this protein (Döffinger et al., 2001). A more detailed analysis of the immunological status of two patients carrying this severe mutation showed that responses to several stimuli, such as LPS, IL-1β, IL-18, TNFα and CD154 were affected, accounting for the occurrence of multiple severe infections (Döffinger et al., 2001). Despite the number of pathways affected and the severity and multiplicity of infectious diseases, patients with EDA-ID have apparently normal T-cell and B-cell responses to antigen, and seem to be characterized by an impaired recognition of pathogens. This impairment of the innate immunity had to be expected: although the specific pathways activated after exposure to various classes of pathogens are still incompletely characterized, the recurrent use of toll-like receptor (TLR) molecules in such a process is clearly established as well as the requirement for NF-κB in TLR signaling. Other pathways participating in the fight against pathogens (IL-1, IL-18, TNF, CD40) are also impaired in EDA-ID; all are known to be NF-κB-dependent. Interestingly, two recently studied EDA-ID patients seemed to exhibit a slightly different immunological phenotype (Jain et al., 2001). NF-κB activation by TNF and LPS is normal, and the patients present with a hyper-IgM syndrome. The major defect lies in the CD40 pathway (which is also affected in the OL-EDA-ID patients). As both LPS and CD40 use TRAF6 as an adaptor to activate the IKK complex, this dissociation is a bit surprising. It is too early to generalize the conclusion that EDA-ID and OL-EDA-ID patients exhibit a different immunological phenotype, before more cases have been analyzed. In conclusion, both symptoms exhibited by EDA-ID patients can be reconciled with a defect in NF-κB activation: the immunological defect is due to the impairment of a series of important signaling pathways, whereas the EDA symptom is caused by an impairment of the downstream events of the ectodysplasin/EDAR/EDARADD signaling cascade.

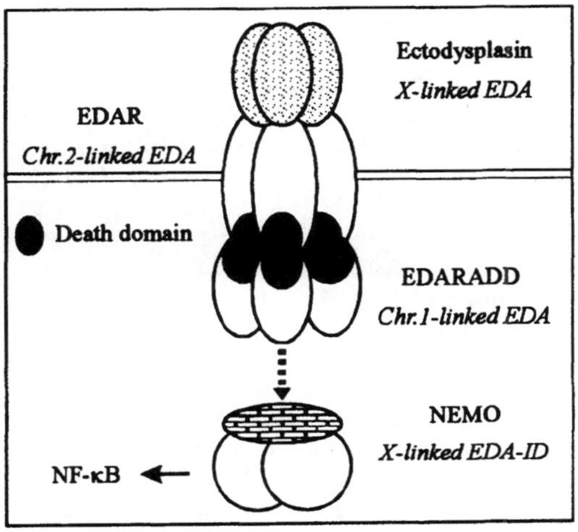

Fig. 2. The ectodysplasin/EDAR signaling pathway in skin appendage morphogenesis. Ectodysplasin, a member of the TNF superfamily, is the ligand for EDAR, a member of the TNF-R superfamily. EDAR interacts with the death domain-containing adaptor EDARADD via its own death domain. This interaction results in IKK/NF-κB activation through NEMO (broken arrow). TRAF molecules may participate in this pathway (Headon et al., 2001; Yan et al., 2002). Each component of the cascade has been genetically linked to EDA pathology (italics).

Concerning the OL-EDA-ID patients, osteopetrosis, which results from defective bone resorption by osteoclasts, can be easily tracked down to NF-κB, since it has been shown that $p50/p52^{-/-}$ mice exhibit osteopetrosis (Franzoso et al., 1997; Iotsova et al., 1997). Moreover, the RANK pathway that plays an essential role in osteoclasts function is itself NF-κB-dependent (Hsu et al., 1999). Primary lymphedema, a dysfunction specifically affecting lymphatic vessels, is still poorly understood at the genetic/biochemical level. However, the gene causing familial lymphedema, viz. V-EGFR-3, has been identified and overexpression of its product results in NF-κB activation (Karkkainen et al., 2000).

To summarize, patients with IP and EDA-ID exhibit quite distinct pathologies despite bearing mutations of the same gene. IP patients are girls who suffer from severe dermatosis after birth and are usually followed by dermatologists, whereas EDA-ID patients are boys who experience severe recurrent infections and are treated in clinical immunology departments. It seems that increasing the severity of NF-κB dysfunction leads to a continuum of syndromes, from EDA-ID for relatively mild forms to OL-EDA-ID for more severe forms (both affecting only males) to IP (where affected males cannot survive until birth) for a complete lack of NF-κB activation. The complexity of the picture is likely to be further expanded in the future when mutations affecting other components of the NF-κB-signaling pathway will be identified.

4. REFERENCES

Abinun M, Spickett G, Appleton AL, Flood T & Cant AJ (1996) Anhidrotic ectodermal dysplasia associated with specific antibody deficiency. Eur. J. Pediatr. 155: 146-147.

Aradhya S, Courtois G, Rajkovic A, Lewis RA, Levy M, Israël A & Nelson DL (2001a) Atypical forms of incontinentia pigmenti in male individuals result from mutations of a cytosine tract in exon 10 of NEMO (IKK-γ). Am. J. Hum. Genet. 68: 765-771.

Aradhya S, Woffendin H, Jakins T, Bardaro T, Esposito T, Smahi A, Shaw C, Levy M, Munnich A, D'Urso M, Lewis RA, Kenwrick S & Nelson DL (2001b) A recurrent deletion in the ubiquitously expressed NEMO (IKK-γ) gene accounts for the vast majority of incontinentia pigmenti mutations. Hum. Mol. Genet. 10: 2171-2179.

Beg AA, Sha WC, Bronson RT, Ghosh S & Baltimore D (1995) Embryonic lethality and liver degeneration in mice lacking the RelA component of NF-κB. Nature 376: 167-170.

Chen Y, Molloy SS, Thomas L, Gambee J, Bachinger HP, Ferguson B, Zonana J, Thomas G & Morris NP (2001) Mutations within a furin consensus sequence block proteolytic release of ectodysplasin-A and cause X-linked hypohidrotic ectodermal dysplasia. Proc. Natl. Acad. Sci. USA 98: 7218-7223.

Döffinger R, Smahi A, Bessia C, Geissmann F, Feinberg J, Durandy A, Bodemer C, Kenwrick S, Dupuis-Girod S, Blanche S, Wood P, Rabia SH, Headon DJ, Overbeek PA, Le Deist F, Holland SM, Belani K, Kumararatne DS, Fischer A, Shapiro R, Conley ME, Reimund E, Kalhoff H, Abinun M, Munnich A, Israël A, Courtois G & Casanova JL (2001) X-linked anhidrotic ectodermal dysplasia with immunodeficiency is caused by impaired NF-κB signaling. Nat. Genet. 27: 277-285.

Doi TS, Marino MW, Takahashi T, Yoshida T, Sakakura T, Old LJ & Obata Y (1999) Absence of tumor necrosis factor rescues RelA-deficient mice from embryonic lethality. Proc. Natl. Acad. Sci. USA 96: 2994-2999.

Elomaa O, Pulkkinen K, Hannelius U, Mikkola M, Saarialho-Kere U & Kere J (2001) Ectodysplasin is released by proteolytic shedding and binds to the EDAR protein. Hum. Mol. Genet. 10: 953-962.

Ezer S, Bayes M, Elomaa O, Schlessinger D & Kere J (1999) Ectodysplasin is a collagenous trimeric type II membrane protein with a tumor necrosis factor-like domain and co-localizes with cytoskeletal structures at lateral and apical surfaces of cells. Hum. Mol. Genet. 8: 2079-2086.

Franzoso G, Carlson L, Xing L, Poljak L, Shores EW, Brown KD, Leonardi A, Tran T, Boyce BF & Siebenlist U (1997) Requirement for NF-κB in osteoclast and B-cell development. Genes Dev. 11: 3482-3496.

Frix CD & Bronson DM (1986) Acute miliary tuberculosis in a child with anhidrotic ectodermal dysplasia. Pediatr. Dermatol. 3: 464-467.

Headon DJ & Overbeek PA (1999) Involvement of a novel Tnf receptor homologue in hair follicle induction. Nat. Genet. 22: 370-374.

Headon DJ, Emmal SA, Ferguson BM, Tucker AS, Justice MJ, Sharpe PT, Zonana J & Overbeek PA (2001) Gene defect in ectodermal dysplasia implicates a death domain adapter in development. Nature 414: 913-916.

Hsu H, Lacey DL, Dunstan CR, Solovyev I, Colombero A, Timms E, Tan HL, Elliott G, Kelley MJ, Sarosi I, Wang L, Xia XZ, Elliott R, Chiu L, Black T, Scully S, Capparelli C, Morony S, Shimamoto G, Bass MB & Boyle WJ (1999) Tumor necrosis factor receptor family member RANK mediates osteoclast differentiation and activation induced by osteoprotegerin ligand. Proc. Natl. Acad. Sci. USA 96: 3540-3545.

Hu Y, Baud V, Delhase M, Zhang P, Deerinck T, Ellisman M, Johnson R & Karin M (1999) Abnormal morphogenesis but intact IKK activation in mice lacking the IKKα subunit of IκB kinase. Science 284: 316-320.

Iotsova V, Caamano J, Loy J, Yang Y, Lewin A & Bravo R (1997) Osteopetrosis in mice lacking NF-κB1 and NF-κB2. Nat. Med. 3: 1285-1289.

Israël A (2000) The IKK complex: an integrator of all signals that activate NF-κB? Trends Cell Biol. 10: 129-133.

Jain A, Ma CA, Liu S, Brown M, Cohen J & Strober W (2001) Specific missense mutations in NEMO result in hyper-IgM syndrome with hypohydrotic ectodermal dysplasia. Nat. Immunol. 2: 223-228.

Jin DY & Jeang KT (1999) Isolation of full-length cDNA and chromosomal localization of human NF-κB modulator NEMO to Xq28. J. Biomed. Sci. 6: 115-120.

Karin M & Ben-Neriah Y (2000) Phosphorylation meets ubiquitination: the control of NF-κB activity.

Annu. Rev. Immunol. 18: 621-663.

Karkkainen MJ, Ferrell RE, Lawrence EC, Kimak MA, Levinson KL, McTigue MA, Alitalo K & Finegold DN (2000) Missense mutations interfere with VEGFR-3 signalling in primary lymphoedema. Nat. Genet. 25: 153-159.

Kaufman CK & Fuchs E (2000) It's got you covered. NF-κB in the epidermis. J. Cell Biol. 149: 999-1004.

Kere J, Srivastava AK, Montonen O, Zonana J, Thomas N, Ferguson B, Munoz F, Morgan D, Clarke A, Baybayan P, Chen EY, Ezer S, Saarialho-Kere U, de la Chapelle A & Schlessinger D (1996) X-linked anhidrotic (hypohidrotic) ectodermal dysplasia is caused by mutation in a novel transmembrane protein. Nat. Genet. 13: 409-416.

Kumar A, Eby MT, Sinha S, Jasmin A & Chaudhary PM (2001) The ectodermal dysplasia receptor activates the nuclear factor-κB, JNK, and cell death pathways and binds to ectodysplasin A. J. Biol. Chem. 276: 2668-2677.

Landy SJ & Donnai D (1993) Incontinentia pigmenti (Bloch-Sulzberger syndrome). J. Med. Genet. 30: 53-59.

Laurikkala J, Mikkola M, Mustonen T, Aberg T, Koppinen P, Pispa J, Nieminen P, Galceran J, Grosschedl R & Thesleff I (2001) TNF signaling via the ligand-receptor pair ectodysplasin and edar controls the function of epithelial signaling centers and is regulated by Wnt and activin during tooth organogenesis. Dev. Biol. 229: 443-455.

Li Q, Van Antwerp D, Mercurio F, Lee KF & Verma IM (1999a) Severe liver degeneration in mice lacking the IκB kinase 2 gene. Science 284: 321-325.

Li ZW, Chu W, Hu Y, Delhase M, Deerinck T, Ellisman M, Johnson R & Karin M (1999b) The IKKβ subunit of IκB kinase (IKK) is essential for nuclear factor κB activation and prevention of apoptosis. J. Exp. Med. 189: 1839-1845.

Makris C, Godfrey VL, Krahn-Senftleben G, Takahashi T, Roberts JL, Schwarz T, Feng L, Johnson RS & Karin M (2000) Female mice heterozygous for IKKγ/NEMO deficiencies develop a dermatopathy similar to the human X-linked disorder incontinentia pigmenti. Mol. Cell 5: 969-979.

Mercurio F, Murray BW, Shevchenko A, Bennett BL, Young DB, Li JW, Pascual G, Motiwala A, Zhu H, Mann M & Manning AM (1999) IκB kinase (IKK)-associated protein 1, a common component of the heterogeneous IKK complex. Mol. Cell. Biol. 19: 1526-1538.

Mikkola ML, Pispa J, Pekkanen M, Paulin L, Nieminen P, Kere J & Thesleff I (1999) Ectodysplasin, a protein required for epithelial morphogenesis, is a novel TNF homologue and promotes cell-matrix adhesion. Mech. Dev. 88: 133-146.

Monreal AW, Ferguson BM, Headon DJ, Street SL, Overbeek PA & Zonana J (1999) Mutations in the human homologue of mouse dl cause autosomal recessive and dominant hypohidrotic ectodermal dysplasia. Nat. Genet. 22: 366-369.

Montonen O, Ezer S, Saarialho-Kere UK, Herva R, Karjalainen-Lindsberg ML, Kaitila I, Schlessinger D, Srivastava AK, Thesleff I & Kere J (1998) The gene defective in anhidrotic ectodermal dysplasia is expressed in the developing epithelium, neuroectoderm, thymus, and bone. J. Histochem. Cytochem. 46: 281-289.

Pasparakis M, Courtois G, Hafner M, Schmidt-Supprian M, Nenci A, Toksoy A, Krampert M, Goebeler M, Gillitzer R, Israël A, Krieg T, Rajewsky K & Haase I (2002) TNF-mediated inflammatory skin disease in mice with epidermis-specific deletion of IKK2. Nature 417: 861-866.

Pinheiro M & Freire-Maia N (1994) Ectodermal dysplasias: a clinical classification and a causal review. Am. J. Med. Genet. 53: 153-162.

Rothwarf DM, Zandi E, Natoli G & Karin M (1998) IKK-γ is an essential regulatory subunit of the IκB kinase complex. Nature 395: 297-300.

Rudolph D, Yeh WC, Wakeham A, Rudolph B, Nallainathan D, Potter J, Elia AJ & Mak TW (2000) Severe liver degeneration and lack of NF-κB activation in NEMO/IKKγ-deficient mice. Genes Dev. 14: 854-862.

Schmidt-Supprian M, Bloch W, Courtois G, Addicks K, Israël A, Rajewsky K & Pasparakis M (2000) NEMO/IKKγ-deficient mice model incontinentia pigmenti. Mol. Cell 5: 981-992.

Schneider P, Street SL, Gaide O, Hertig S, Tardivel A, Tschopp J, Runkel L, Alevizopoulos K,

Ferguson BM & Zonana J (2001) Mutations leading to X-linked hypohidrotic ectodermal dysplasia affect three major functional domains in the tumor necrosis factor family member ectodysplasin-A. J. Biol. Chem. 276: 18819-18827.

Schweizer P, Kalhoff H, Horneff G, Wahn V & Diekmann L (1999) Polysaccharide specific humoral immunodeficiency in ectodermal dysplasia. Case report of a boy with two affected brothers. Klin. Pädiatr. 211: 459-461.

Seitz CS, Lin Q, Deng H & Khavari PA (1998) Alterations in NF-κB function in transgenic epithelial tissue demonstrate a growth inhibitory role for NF-κB. Proc. Natl. Acad. Sci. USA 95: 2307-2312.

Senftleben U, Li ZW, Baud V & Karin M (2001) IKKβ is essential for protecting T cells from TNFα-induced apoptosis. Immunity 14: 217-230.

Sitton JE & Reimund EL (1992) Extramedullary hematopoiesis of the cranial dura and anhidrotic ectodermal dysplasia. Neuropediatrics 23: 108-110.

Smahi A, Courtois G, Vabres P, Yamaoka S, Heuertz S, Munnich A, Israël A, Heiss NS, Klauck SM, Kioschis P, Wiemann S, Poustka A, Esposito T, Bardaro T, Gianfrancesco F, Ciccodicola A, D'Urso M, Woffendin H, Jakins T, Donnai D, Stewart H, Kenwrick SJ, Aradhya S, Yamagata T, Levy M, Lewis RA & Nelson DL (2000) Genomic rearrangement in NEMO impairs NF-κB activation and is a cause of incontinentia pigmenti. The International Incontinentia Pigmenti (IP) Consortium. Nature 405: 466-472.

Tanaka M, Fuentes ME, Yamaguchi K, Durnin MH, Dalrymple SA, Hardy KL & Goeddel DV (1999) Embryonic lethality, liver degeneration, and impaired NF-κB activation in IKK-β-deficient mice. Immunity 10: 421-429.

Tucker AS, Headon DJ, Schneider P, Ferguson BM, Overbeek P, Tschopp J & Sharpe PT (2000) Edar/Eda interactions regulate enamel knot formation in tooth morphogenesis. Development 127: 4691-4700.

van Leeuwen RL, Wintzen M & van Praag MC (2000) Incontinentia pigmenti: an extensive second episode of a "first-stage" vesicobullous eruption. Pediatr. Dermatol. 17: 70.

Yamaoka S, Courtois G, Bessia C, Whiteside ST, Weil R, Agou F, Kirk HE, Kay RJ & Israël A (1998) Complementation cloning of NEMO, a component of the IκB kinase complex essential for NF-κB activation. Cell 93: 1231-1240.

Yan M, Wang LC, Hymowitz SG, Schilbach S, Lee J, Goddard A, de Vos AM, Gao WQ & Dixit VM (2000) Two-amino acid molecular switch in an epithelial morphogen that regulates binding to two distinct receptors. Science 290: 523-527.

Yan M, Zhang Z, Brady JR, Schilbach S, Fairbrother WJ & Dixit VM (2002) Identification of a novel death domain-containing adaptor molecule for ectodysplasin-A receptor that is mutated in crinkled mice. Curr. Biol. 12: 409-413.

Zonana J, Elder ME, Schneider LC, Orlow SJ, Moss C, Golabi M, Shapira SK, Farndon PA, Wara DW, Emmal SA & Ferguson BM (2000) A novel X-linked disorder of immune deficiency and hypohidrotic ectodermal dysplasia is allelic to incontinentia pigmenti and due to mutations in IKK-γ (NEMO). Am. J. Hum. Genet. 67: 1555-1562.

5. AFFILIATION

Institut Pasteur, Unité de Biologie Moléculaire de l'Expression Génique, FRE2364 CNRS, 26 rue du Dr. Roux, 75724 Paris, France.

R. DE MARTIN

NF-κB IN THE VASCULAR SYSTEM

1. INTRODUCTION

From the contributions up to now it should have become very clear that NF-κB plays a central role in diverse aspects of innate and adaptive immune and, more generally, stress-response reactions. In this regard, the vascular wall that represents one of the main barriers in the organism, namely the interface between blood and underlying tissue, is of outstanding importance. The vascular wall consists mainly of endothelial cells (EC) and smooth muscle cells (SMC) and, not unexpectedly, NF-κB has multiple functions in these cells. NF-κB deregulation is not only associated, but in many instances functionally involved in a variety of vascular disorders; it is the purpose of this chapter to discuss some of them in more detail.

2. THE INFLAMMATORY RESPONSE OF THE ENDOTHELIUM

Inflammation is one of the first actions in host defence, and the endothelium plays a strategic role due to the need of immune cells to adhere to and pass through the vessel wall in order to fight, for example, invading microorganisms. However, it should be noted that the term "inflammation" is used in a much broader sense than in the above meaning, since it appears to be an underlying mechanism or at least a common feature of a quite diverse set of diseases including, for example, diabetic microangiopathies, atopic dermatitis, inflammatory bowel disease, rheumatoid arthritis, psoriasis and arteriosclerosis. Historically, the inflammatory response of the endothelium has been designated "activation" (Willms-Kretschmer et al., 1967) and was later redefined (Pober, 1988). It means a "quantitative change in the level of expression of certain gene products that, in turn, endow endothelial cells with the new capacities that cumulatively allow them to perform new functions" and follows the course of events that happen when EC encounter inflammatory noxae. Soon, the definition was further refined to distinguish between type I and II activation, with type I activation relating to very rapid and transcription-independent events such as secretion of von Willebrand factor and platelet-activating factor (PAF), as well as the early release of P-selectin (although there is transcriptional upregulation of P-selectin at later times as well). In contrast, type II activation includes progressive induction that is essentially characterized by gene expression of cytokines (IL-1, IL-6, IL-

R. Beyaert (ed.), Nuclear Factor κB. Regulation and Role in Disease, 335-351.

8), cell adhesion molecules (E-selectin, ICAM-1, VCAM-1) and procoagulatory molecules (e.g. tissue factor).

The inflammatory response of EC is elicited not only, as the name illustrates, by the "classical" inflammatory cytokines and polypeptides such as IL-1 and TNFα, but also by microorganisms and their cell wall components such as Candida albicans, Plasmodium falciparum, Chlamydia pneumoniae and bacterial lipopolysaccharide (LPS). They all lead to activation of NF-κB, as this is the case for several others. Agents that are relevant in the context of vascular biology include oxidized LDL, oxidized phospholipids, advanced glycation end products (AGE), thrombin, certain complement factors, heat shock, shear stress and, last but not least, cold shock and hypoxia reperfusion, as it may occur during transplantation. Homocysteine has been reported to activate NF-κB in SMC, but to suppress its activity in EC (Desai et al., 2001; Stangl et al., 2001). With the exception of the oxidized lipids, which may rather act through interaction with lipid rafts, and the physical stimuli, most of these agents bind to specific cell surface receptors. EC express TNF-R1 and TNF-R2 (Schmid et al., 1995), as well as the related CD40 (Wagner et al., 2002), and the IL-1 receptor system including the high-affinity IL-1R and the type II decoy receptor (Coyne et al., 2001). Microbial products bind to different Toll-like receptors (TLR) of which nine are currently known. LPS binds to TLR-4, which is constitutively expressed on EC, whereas TLR-2, that acts as a receptor for gram-positive and fungal cell wall components and lipoproteins, is expressed mainly after stimulation with LPS or IFNγ (Faure et al., 2001). AGE bind to receptors for AGE (RAGE), thereby activating EC, an effect that appears to be of special importance in diseases like diabetic angiopathies (Wautier et al., 1996), nondiabetic uremia and occlusive vascular and renal disease (Bierhaus et al., 1996).

The signaling pathways leading to NF-κB activation have been reviewed in detail in previous chapters. Briefly, signals originating from distinct receptors are first propagated further towards the nucleus through interaction with partially receptor-specific adapter molecules and kinases, such as TRAF1 and TRAF2 for TNF receptors, TRAF6 for IL-1R, and TRAF6 and MyD88 for TLRs, before they appear to converge at the level of the signalsome. This multi-protein complex contains, among others, the IκB kinases IKKγ/NEMO, IKK1 and IKK2. The latter is crucial for transmitting the inflammatory signal, as evidenced by genetic (knockout mice) models, and phosphorylates IκBα on Ser32 and Ser36, which is in turn a signal for IκBα ubiquitination and proteasome-mediated degradation. Therefore, this signaling pathway appears to follow the general scheme of NF-κB activation; however, an alternative pathway resulting in IκBα degradation has been described that could be of relevance in the context of reperfusion injury, namely the proteolysis-independent activation of NF-κB through IκBα phosphorylation on tyrosine (Imbert et al., 1996). This phosphorylation event was first described as a (nonphysiological) response to the tyrosine phosphorylation inhibitor pervanadate, but has subsequently been shown to occur also in response to hypoxia/reoxygenation. Moreover, a recent study demonstrates IκBα tyrosine phosphorylation also in response to erythropoietin in nerve cells; however, whether this pathway is of relevance also in EC remains to be established.

Thus, the main regulatory step for NF-κB activation is its translocation to the nucleus; however, other aspects should be kept in mind. First, phosphorylation of NF-κB RelA at several serine residues, mostly Ser276 and Ser529, are crucial for its transactivating properties. This serine phosphorylation may involve protein kinase A and/or members of the p38 MAP kinase family, and mediate the interaction of RelA and of NF-κB1/p50 with histone acetyltransferases that serve as coactivators (Zhong et al., 1998; Zhong et al., 2002). They do play an important role in the transcriptional activity of NF-κB towards certain genes such as IL-6, as evidenced by the stimulatory effect of the histone deacetylase inhibitor trichostatin A, but may be less important for others. Second, nucleo-cytoplasmic shuttling has been observed for several molecules of the NF-κB signaling pathway, and occurs constitutively in resting cells; treatment with the nuclear export inhibitor leptomycin B leads to nuclear accumulation of NF-κB within 90 min in the absence of additional stimulation. Moreover, IκBα accumulates in the nucleus, suggesting that NF-κB is inactive in the nucleus under these conditions. The findings that both NIK and IKK1, but not IKK2, shuttle between cytoplasm and nucleus as well have suggested the possibility that activation of NF-κB can also occur in the nucleus (Johnson et al., 1999; Birbach et al., 2002).

One of the most striking features that distinguishes NF-κB from many other transcription factors and that is the basis for its role as a stress response factor is the transient nature of its activation. Obviously, a transient mode of activation implies appropriate mechanisms for downregulation, and suggests that in chronic diseases these mechanisms could be impaired. Indeed, several mechanisms have been described that act to shut down NF-κB activity within up to a few hours after activation, and that occur on different levels. On the transcriptional level, IκBα is expressed in an NF-κB-dependent manner, involving distinct NF-κB-binding sites in the IκBα promoter (de Martin et al., 1993). A20, an NF-κB-dependent gene originally described to have antiapoptotic functions, has also NF-κB-inhibitory properties (Cooper et al., 1996). Furthermore, hyperphosphorylation of IKK2 in the HLH domain following stimulation leads to shut down of IKK2 kinase activity and termination of further activating signals (Delhase et al., 1999). Last but not least, NF-κB-dependent expression of COX-2 leads at later stages of activation to the generation of cyclopentenone prostaglandins that display anti-inflammatory activity (Rossi et al., 2000). However, it is not clear whether these described mechanisms are sufficient for the very efficient termination of NF-κB activity, or whether additional mechanisms exist that contribute to this phenomenon.

Multiple interactions of NF-κB with other signaling pathways have been reported, either through direct interaction with other transcription factors, or at the level of cytoplasmic signaling molecules, or both. Direct interactions include the transcription factors SP1, AP-1 and CREB, as well as transcriptional coactivators such as p300/CBP (Zhong et al., 1998). In the context of vascular cells, relevant signaling pathways known to communicate with NF-κB include the glucocorticoid receptor (GR) (although direct interaction has been reported as well), egr-1 and peroxisome proliferator-activated receptors (PPARs) (Scheinman et al., 1995b; Marx et al., 1999; Mechtcheriakova et al., 2001; Rival et al., 2002).

Glucocorticoids such as dexamethasone are potent anti-inflammatory drugs and highly relevant for clinical use. The direct interaction between the GR and the NF-κB subunits RelA and NF-κB1 has been reported, and more recently the surface of the GR has been mapped (Scheinman et al., 1995b; Tao et al., 2001). Such data suggest that binding to NF-κB occurs through distinct protein-protein interactions, arguing against the previous hypothesis that transrepression occurs through competition for a single coactivator protein. Moreover, the ability of RelA, NF-κB1 and c-Rel subunits to bind DNA was inhibited by dexamethasone. In a subsequent study, the same group found that GR also induces expression of IκBα, suggesting an additional mechanism for downregulation of NF-κB; however, no GR-responsive element in the IκBα promoter was characterized (Scheinman et al., 1995a). In EC, induction of IκBα by dexamethasone could not be demonstrated, indicating that in different cell types the mechanism(s) of GR action may be different and, in general, that they are more complex than originally estimated (Brostjan et al., 1996).

PPARs are a family of ligand-activated transcriptional regulators that are activated by fibrates, a class of substances known as lipid-lowering agents (Chinetti et al., 2000). Thus, they are of clinical relevance for the prevention of arteriosclerosis. In addition, an anti-inflammatory role for fibrates has been shown. EC express PPARα, PPARδ and PPARγ, and treatment of EC with the PPARα activator fenofibrate inhibited the TNFα-induced expression of VCAM-1 (Marx et al., 1999). Moreover, PPARα-deficient mice displayed a prolonged response to inflammatory stimuli (Devchand et al., 1996). Glutathione S-transferase pull-down experiments demonstrated that PPARα physically interacts with RelA, but also with c-Jun and CBP (Delerive et al., 1999). In addition, as in the case of the GR, induction of IκBα has been reported for PPARα, however, in the absence of functional PPAR response elements in the IκBα promoter. Taken together, PPAR activators can downregulate the activity of NF-κB at at least two distinct levels.

The final outcome of NF-κB activation in EC is the induced expression of genes that can be classified into chemoattractants that function to attract circulating immune cells, cell adhesion molecules that mediate the interaction of these cells with the vascular wall, genes that protect EC from undergoing apoptosis, and several others. Some are involved in coagulation, the first feature of an onset of repair mechanisms, and in angiogenesis, a process that involves multiple aspects of EC biology, including proliferation, migration as well as expression of proteolytic enzymes. Chemoattractants include IL-1, MCP-1 and RANTES. Firm adhesion to the vascular wall involves E-selectin, VCAM-1 and ICAM-1. Procoa gulant proteins include tissue factor and PAI-1(Table1). It appears that at this stage of the inflammatory response, there are only few events on the side of the endothelium that are independent of NF-κB, such as the loss of heparan sulfate, the expression of integrins that mediate the initial weak adhesion (rolling) of lymphocytes, the secretion of von Willebrand factor from Weibel-Palade bodies, and the initial rapid expression of P-selectin. Experiments utilizing adenovirus-mediated overexpression of either IκBα or dominant-negative IKK2 have demonstrated that the majority of the functions of the endothelium is NF-κB-dependent and thus underlines the importance of this transcription factor in the inflammatory response.

Table 1. Genes regulated by NF-κB in vascular cells.

Interleukins	IL-1, IL-6, IL-8, MCP-1, RANTES
Growth factors, receptors	VEGF, Flk/KDR
Cell adhesion molecules	E-selectin, ICAM-1, VCAM-1
Transcription factors/inhibitors	IκBα, Nur77/NAK-1, IRF-1
Apoptosis-related	cIAP-1, cIAP-2, XIAP, A20
Procoagulant	Tissue factor, PAI-1
Others	iNOS, COX-2, MMP2, MMP9, PTX-3

3. ARTERIOSCLEROSIS

Arteriosclerosis is a multi-factorial disease that involves both genetic and environmental components. Its initiation and progression develop in several stages and involve the interplay of a variety of cell types including EC, SMC and monocytes. It is therefore a complex disease that is still far away from being fully understood; however, the general consensus is to regard arteriosclerosis at least in part as an inflammatory disease. In addition, the molecular mechanisms that regulate distinct steps, such as EC activation, SMC proliferation and migration, or generation of monocyte-derived mediators, have been studied in detail. As an outcome, again NF-κB has been identified as an important regulator in several of the above aspects (Brand et al., 1996; Lindner & Collins, 1996).

At an early stage, arteriosclerosis develops as a fatty streak that progresses further to affect larger regions. Certain areas of the vasculature appear to be preferentially affected, including bifurcations and major curvatures. In these parts, disturbances of the blood flow may be expected, leading to the notion that hemodynamic forces including shear stress may dispose certain regions of the vasculature for lesion formation. The endothelium in the area of the fatty streak is activated and expresses cell adhesion molecules and chemokines that attract monocytes. They invade the intima and transform into foam cells by taking up (oxidized) lipids, and secrete a variety of cytokines and growth factors. SMC are recruited from the media, migrate into the intima and deposit collagen, resulting in the formation of the arteriosclerotic plaque that is ultimately covered by a fibrous cap. A further step is the calcification of the affected region that may involve the interplay of osteoblast- and osteoclast-like cells. Finally, instability of the plaque can result in rupture and occlusion of arterioles in the periphery, the coronary tree or the brain, as well as exposure of procoagulant mediators such as tissue factor, resulting in coagulation and occlusion at the site of rupture.

What is the role of NF-κB in the initiation and progression of arteriosclerosis? From early on it has been described that NF-κB is present in cells within the area of arteriosclerotic lesions in an active form, including EC, SMC and macrophages (Collins, 1993; Berliner et al., 1995). Likewise, the influence of dietary constituents on NF-κB has been noted (Liao et al., 1993). To address the question of functional importance, we have to look at the individual steps of atherogenesis as described above. First, at the stage of initiation, shear stress has been shown to activate NF-κB in EC and to result in monocyte adhesion (Lan et al., 1994; Weber et al., 1999). Increased amounts of NF-κB have been found in regions of perturbed blood flow, for example bifurcations. These areas have been designated high-probability areas (for the development of arteriosclerosis) in mice (Hajra et al., 2000). Both NF-κB and IκBα are expressed in these regions at higher levels as compared to adjacent sites that have a low probability to develop into arteriosclerotic lesions. Interestingly, NF-κB is localized in the cytoplasm, indicating, apart from its higher abundance, that it is still in an inactive state. However, treatment with LPS or feeding an atherogenic diet to $LDLR^{-/-}$ mice results in activation of NF-κB as assayed by translocation to the nucleus. This is associated with expression of the cell adhesion molecule VCAM-1 and E-selectin, and secretion of chemoattractants such as MCP-1, classical indicators of inflammation in EC; expression of all these is highly dependent on NF-κB. It is conceivable that a stimulatory effect, although systematic, has more pronounced consequences in cells that have high levels of NF-κB as compared to those in adjacent regions that display normal levels (Hajra et al., 2000).

Which agents may activate NF-κB at that stage? From the cognate risk factors, high serum levels of LDL or its oxidation products have been reported to activate NF-κB (Parhami et al., 1993). However, this issue remains controversial, since also the opposite has been found (Heermeier et al., 2001). Whereas short-term exposure to oxidized LDL induces NF-κB, prolonged exposure of SMC to (low levels) of the same agent renders cells refractory to stimulation (Brand et al., 1997). Furthermore, differences between heavily and mildly oxidized LDL preparations may account for several of the differences observed in the literature. Besides, a number of dietary constituents and substances found in serum, including linoleic acid (Toborek & Hennig, 1998) and oxidized 1-palmitoyl-2-arachidonoyl-sn-glycero-3-phosphorylcholine (Yeh et al., 2001), can induce NF-κB in EC, resulting in the expression of inflammatory markers, and could therefore be responsible for the initiation of arteriosclerosis. Vice versa, HDL, that has long been known to have a protective effect, was recently shown to interrupt NF-κB signaling by interfering with the sphingosine kinase signaling pathway that activates NF-κB via the generation of sphingosine 1-phosphate (Xia et al., 1999). In addition, high levels of homocysteine are known as a risk factor for arterioscle rosis; they have been shown to activate NF-κB in human vascular SMC in vitro

(Desai et al., 2001). Taken together, at the stage of initiation, the ability to activate or inhibit NF-κB appears to correlate with the cognate effects of dietary constituents. Apart from these, the renin-angiotensin system has been implicated in the development of arteriosclerosis. Besides its effect on blood pressure, which is a risk factor by itself, and the stimulation of SMC proliferation, angiotensin II has been reported to activate NF-κB in EC (Kranzhofer et al., 1999).

At the stage of the fatty streak, monocytes adhere to the activated endothelium and migrate into the intima. Uptake of lipids through the scavenger receptor or CD36 leads to their activation and transformation into foam cells. Activated monocytes secrete, among others, TNFα, that leads to further activation of EC, and to proliferation and migration of SMC. TNFα transcription in monocytes is dependent on NF-κB (Janabi et al., 2000).

SMC of the media of healthy vessels express detectable amounts of NF-κB restricted to the cytoplasm. In the arteriosclerotic lesion and after experimental balloon injury NF-κB is highly expressed in an active form. Without stimulation, SMC in culture require NF-κB for proliferation and undergo apoptosis when NF-κB is inhibited, e.g. by adenoviral overexpression of IκBα (Bellas et al., 1995; Erl et al., 1999). It is not clear whether this reflects the requirement of small amounts of undetectable NF-κB for SMC survival in the described experiments, or whether SMC in culture are more susceptible to apoptosis compared to the in vivo situation. In any case, increased numbers of apoptotic SMC have been found in arteries of rabbits transduced with an adenoviral vector expressing IκBα after balloon injury, suggesting that NF-κB is necessary for their survival, possibly by regulating the expression of antiapoptotic genes of the IAP family (Erl et al., 1999). It can thus be concluded that at least in the diseased state in the presence of activating cytokines and growth factors, NF-κB is necessary for SMC survival.

Plaque rupture is a major complication once an arteriosclerotic lesion has formed. Matrix metalloproteinases (MMP) contribute to the destruction of the extracellular matrix at the shoulder regions of arteriosclerotic plaques and thus contribute to plaque instability. The major MMP1, MMP3 and MMP9 are secreted by vascular SMC in response to cytokines and growth factors produced mainly by macrophages within the arteriosclerotic plaque. Using adenoviral gene transfer of IκBα, expression of all three MMP has been demonstrated to be NF-κB-dependent (Bond et al., 2001).

Bone metabolism is thought to be delicately balanced by mineral deposition and resorption, two functions exerted by osteoblasts and osteoclasts, respectively. These cells are derived from hematopoietic precursors of the mononuclear phagocytic lineage. Similar mechanisms have been proposed for the calcification of the arteriosclerotic plaque. Calcifying vascular cells that differentiate into osteoblast-like cells, on the one hand, and osteoclast-like cells (OLC) generated by differentiation from monocytic precursors, on the other hand, are thought to balance mineral deposition on the arterial wall (Watson et al., 1994; Shalhoub et al., 2000; Doherty et al., 2002). Metabolic disturbances in the microenvironment of the arteriosclerotic plaque may tilt this balance and impair the generation of

OLC, resulting in lack of mineral resorption. Genetic evidence from knockout and mutant mice points towards a role of osteoprotegerin and of M-CSF in OLC differentiation, and these mice show increased arterial calcification (Qiao et al., 1997; Bucay et al., 1998). Osteoprotegerin regulates the expression of the receptor activator of NF-κB (RANK) on monocytic precursers and its ligand RANKL (also known as ODF and TRANCE), expressed, for example, on EC (McGowan et al., 2001). RANK was identified as the intrinsic hematopoietic cell surface receptor that controls osteoclastogenesis (Li et al., 2000). It is a member of the TNFα receptor superfamily and couples via TRAF2, TRAF5 and TRAF6 to NF-κB and (TRAF-independently) to the JNK-signaling pathways (Darnay et al., 1998). In osteoclast precursors, stimulation of RANK by RANKL leads to activation of the NF-κB RelA/NF-κB1 complex (Wei et al., 2001). This fits well to the observation that NF-κB1/NF-κB2 double-knockout mice fail to develop mature osteoclasts (and also B-cells; Franzoso et al., 1997). Thus there is an emerging concept of the molecular and cellular mechanisms that mediate calcification of arteries, and NF-κB appears to play a central role in this process (Doherty et al., 2002).

4. RESTENOSIS

Treatment of obstructed arteriosclerotic carotid or coronary arteries by balloon angioplasty leads in a significant portion of patients to re-occlusion and therefore represents a serious clinical complication. Though the mechanisms leading to re-occlusion are still poorly understood, they involve SMC proliferation, migration and matrix deposition, and thus resemble in many parameters the later stages of the arteriosclerotic process. Moreover, activated NF-κB was found in sections of arteries after balloon injury (Lindner & Collins, 1996).

Several groups have investigated the role of NF-κB in this restenosis. Various agents associated with arterial injury were demonstrated to activate NF-κB and to cause proliferation of SMC, including IGF-1, thrombin and oxidative stress (Nakajima et al., 1994; Balaram et al., 1999; Azevedo et al., 2000). Vice versa, inhibitors of NF-κB such as N-acetylcysteine, aspirin, antisense oligonucleotides and proteasome inhibitors block SMC proliferation (Lee et al., 1998; Speir et al., 1998; Voisard et al., 2001; Meiners et al., 2002). In rat coronary arteries, antisense RelA oligonucleotides inhibited SMC proliferation and neointima formation (Autieri et al., 1995). Recently, our group has demonstrated that adenoviral gene transfer of IκBα can substantially inhibit neointima formation in a rabbit iliac artery balloon injury model (Breuss et al., 2002). Part of the NF-κB effects on SMC could also be due to the prevention of SMC apoptosis (Erl et al., 1999).

5. DIABETIC ANGIOPATHIES

A common complication of diabetes is the development of micro- and macroangiopathies. They are characterized by endothelial dysfunction, which leads to platelet and neutrophil adhesion as well as to accelerated arteriosclerosis. Stimulation of extravascular matrix protein formation becomes manifest as the main morphological characteristics of diabetic microangiopathy, namely capillary basement membrane thickening. This microvascular sclerosis limits the capacity of the microvessels to dilate, which reduces its maximum perfusion capacity and impairs autoregulation. Although basic differences exist between type I and type II diabetes, common major clinical implications are the diabetic retinopathy and nephropathy.

Concerning the cause of the described alterations, part of them can be attributed to the hemodynamic insults caused by increased microvascular pressure. However, at least two additional agents have been associated with diabetic angiopathies: high glucose levels and increased amounts of AGE. High glucose and hyperglycemia have been reported to increase leukocyte-endothelial interactions in HUVEC in an NF-κB- and PKC-dependent manner (Pieper & Riaz-ul-Haq, 1997; Morigi et al., 1998). Likewise, AGE have been demonstrated to activate NF-κB: they are generated by nonenzymatic and irreversible glycation of proteins and occur during aging and, at an advanced rate, during diabetes. They bind to several cell surface receptors including p60, p90, lactoferrin, galectin-3 and RAGE, the latter being the predominant AGE receptor on EC. Experimentally, binding of AGE that can be presented either as glycated human serum albumin or on the surface of erythrocytes, stimulated NF-κB binding activity and resulted, among others, in the induction of endothelin and tissue factor (Wautier et al., 1994; Bierhaus et al., 1997; Quehenberger et al., 2000). Thus, several lines of evidence point towards the involvement of NF-κB in the development of diabetic angiopathies; however, its precise contribution remains to be established.

6. ANGIOGENESIS

The formation of blood vessels is mainly accomplished by two processes, namely vasculogenesis and angiogenesis. Vasculogenesis is defined as the formation of new vessels from angioblasts and takes place during embryonic development of the major vascular structures of the yolk sac and of the embryo. In contrast, the term angiogenesis relates to the sprouting of new capillaries from pre-existing vessels and takes place in the adult organism. Angiogenesis is important for the (neo)vascularization of solid tumors, where supply with oxygen and nutrients becomes a limiting factor, for wound healing and during the female reproductive cycle. In general, the formation of new blood vessels involves distinct steps of proliferation, migration including matrix degradation, as well as protection from apoptosis of EC and, at a later stage, SMC and pericytes (Ronicke et al., 1996; Flamme et al., 1997; Ferrara, 2001).

Vascular endothelial growth factor (VEGF), of which four members have been identified, is regarded as the major proangiogenic factor and binds to three high-affinity tyrosine kinase receptors on EC, VEGF-R1 (Flt-1), VEGF-R2 (Flk-

1/KDR), and VEGF-R3/Flt4 (de Vries et al., 1992; Terman et al., 1992; Kukk et al., 1996). All receptors are expressed already in the mesoderm at an early stage of vascular development; however, expression of Flk is lost at later stages, except for tumor blood vessels. VEGF-R3/Flt4 appears to play a role in the development of lymphatic vessels.

In contrast to the situation in the adult, there is only little evidence for an involvement of NF-κB in the process of developmental vasculogenesis. First, both RelA- and IKK2-deficient mice died around E13 due to liver hemorrhage in an TNFα-dependent manner; however, this was reported to be primarily due to apoptosis of hepatocytes (Li et al., 1999). It is not clear whether also EC undergo apoptosis. Secondly, expression of Flk/KDR has been reported to be regulated transcriptionally involving NF-κB (Ronicke et al., 1996); however, Flk/KDR knockout mice die already between E8.5 and E9.5 (Shalaby et al., 1995), which is well before death of RelA and IKK2 knockouts, suggesting that in vivo and at least at this developmental stage NF-κB cannot be a limiting factor for Flk/KDR expression.

On the contrary, several lines of evidence suggest that NF-κB plays a critical role in angiogenesis in the adult. Angiogenesis can be assessed in several models, including in vitro tube formation assays in matrigel or collagen gels, sprouting assays, in vivo matrigel implantation models in the mouse, and retinal neovascularization models in rabbits. Expression of dominant-negative IKK2 using a recombinant adenovirus completely inhibited tube formation of HUVEC in a matrigel assay (Oitzinger et al., 2001). Likewise, in a fibrin matrix assay, hypoxia in combination with FGF-2 led to tube formation and to activation of NF-κB and Erk1/2 in human microvascular EC (Kroon et al., 2001). In a similar context in microvascular EC, TNFα (a known angiogenesis inducer) has been shown to exert its function through the stimulation of IL-8, VEGF and basic FGF, all of which are important (as shown by blocking with the respective antibodies). In addition, an antisense RelA oligonucleotide almost completely inhibited TNFα-dependent IL-8 production and partially abrogated TNFα-dependent VEGF production, suggesting that NF-κB-dependent IL-8 and VEGF expression is important for tube formation (Yoshida et al., 1997). This confirmed earlier studies on the involvement of IL-8 in in vitro angiogenesis assays (Shono et al., 1996).

In vivo, several studies have demonstrated the inhibition of angiogenesis in cancer models. Transfection of human prostatic cancer cells with IκBα inhibited the expression of VEGF, IL-8, MMP-9 as well as collagenase activity and, when injected into nude mice, resulted in slower growing tumors with lower metastatic potential as compared to the nontransfected control cells (Huang et al., 2001). A similar situation was previously found in human ovarian cancer cells (Huang et al., 2000).

In noncancer models, pyrrolidine dithiocarbamate, a widely used inhibitor of NF-κB, was found to suppress retinal neovascularization in a murine model of ischemic retinopathy (Yoshida et al., 1999). In a mouse matrigel implantation model, PAF induced expression of the NF-κB-dependent angiogenic factors TNFα, IL-1α, basic FGF and VEGF. The expression was inhibited by RelA antisense oligonucleotides or by antioxidants (Ko et al., 2002). Taken together, these studies suggest that NF-κB is activated during angiogenesis and that it leads to the expression of several proangiogenic factors, including VEGF and IL-8, both of which are of critical importance in the angiogenic process.

However, in addition MPP play an important role in the formation of new blood vessels, since they are important for matrix degradation in order to allow invasion of the newly formed endothelial sprouts into the surrounding tissue. MMP2, a type IV collagenase, is a major candidate for performing this function, as it is secreted by EC in its proform. MMP2 expression has been shown to be NF-κB-dependent (Kim & Koh, 2000). In addition, the activation of MMP2 from its precursor occurs through cleavage by a membrane-type MMP, namely MT1-MMP, the expression of which is again NF-κB-dependent (Han et al., 2001).

One important aspect of proliferation and migration of any adherent cell is prevention of apoptosis. During cell division, cells loosen their contacts to each other and to the extracellular matrix, and tend to undergo apoptosis, also termed "anoikis" (Frisch & Francis, 1994). In EC, αvβ3 integrin plays a fundamental role in the anchorage-dependent survival and prevention of apoptosis. Ligation of αvβ3 integrin leads to NF-κB activation and, vice versa, its blockade by neutralizing antibodies inhibits NF-κB activity and leads to EC death. Studies with overexpression of IκBα revealed NF-κB dependence of EC survival when plated on osteopontin and vitronectin, but not on fibronectin, laminin or collagen I (Scatena et al., 1998; Scatena & Giachelli, 2002). NF-κB then leads to the expression of antiapoptotic genes such as cIAP-1, cIAP-1, XIAP and A20, which have been described to be expressed in EC and which in some cases have also been demonstrated to be of functional importance (Krikos et al., 1992; Wrighton et al., 1996; Stehlik et al., 1998; Erl et al., 1999).

7. CONCLUSIONS

In cells of the vessel wall, ample evidence is accumulating that NF-κB plays key regulatory roles in several different pathophysiological settings. In diseases, the beneficial role of NF-κB as a stress and immune regulator is often enhanced and exaggerated to the point where adverse effects are taking place. Inhibition of NF-κB by low-molecular weight compounds or gene therapy approaches holds the promise to open new opportunities for the treatment of vascular diseases. However, a major challenge will be the development of compounds that are sufficiently specific to target NF-κB precisely in the context of a certain disease, without affecting its multiple other functions in different cells or organs. Clearly, these new putative inhibitors will therefore not target NF-κB itself, but parts of specific signaling pathways leading to its activation. Delineation of these pathways during the last years has provided evidence that indeed a diversity of such activation pathways, depending on the cell type and/or stimulus, may exist.

8. ACKNOWLEDGMENTS

The author thanks R. Hofer-Warbinek, G. Winsauer, C. Wiesner and M. Hoeth for valuable discussions and reading of the manuscript. Research was supported by grants from the Austrian Science Foundation (Sonderforschungsbereich "Microvascular Injury and Repair", SFB 5-12).

9. REFERENCES

Autieri MV, Yue TL, Ferstein GZ & Ohlstein E (1995) Antisense oligonucleotides to the p65 subunit of NF-κB inhibit human vascular smooth muscle cell adherence and proliferation and prevent neointima formation in rat carotid arteries. Biochem. Biophys. Res. Commun. 213: 827-836.

Azevedo LC, Pedro MA, Souza LC, de Souza HP, Janiszewski M, da Luz PL & Laurindo FR (2000) Oxidative stress as a signaling mechanism of the vascular response to injury: the redox hypothesis of restenosis. Cardiovasc. Res. 47: 436-445.

Balaram SK, Agrawal DK & Edwards JD (1999) Insulin like growth factor-1 activates nuclear factor-κB and increases transcription of the intercellular adhesion molecule-1 gene in endothelial cells. Cardiovasc. Surg. 7: 91-97.

Bellas RE, Lee JS & Sonenshein GE (1995) Expression of a constitutive NF-κB-like activity is essential for proliferation of cultured bovine vascular smooth muscle cells. J. Clin. Invest. 96: 2521-2527.

Berliner JA, Navab M, Fogelman AM, Frank JS, Demer LL, Edwards PA, Watson AD & Lusis AJ (1995) Atherosclerosis: basic mechanisms. Oxidation, inflammation, and genetics. Circulation 91: 2488-2496.

Bierhaus A, Ritz E & Nawroth PP (1996) Expression of receptors for advanced glycation end-products in occlusive vascular and renal disease. Nephrol. Dial. Transplant. 11: 87-90.

Bierhaus A, Chevion S, Chevion M, Hofmann M, Quehenberger P, Illmer T, Luther T, Berentshtein E, Tritschler H, Muller M, Wahl P, Ziegler R & Nawroth PP (1997) Advanced glycation end product-induced activation of NF-κB is suppressed by α-lipoic acid in cultured endothelial cells. Diabetes 46: 1481-1490.

Birbach A, Gold P, Binder BR, Hofer E, de Martin R & Schmid JA (2002) Signaling molecules of the NF-κB pathway shuttle constitutively between cytoplasm and nucleus. J. Biol. Chem. 277: 10842-10851.

Bond M, Chase AJ, Baker AH & Newby AC (2001) Inhibition of transcription factor NF-κB reduces matrix metalloproteinase-1, -3 and -9 production by vascular smooth muscle cells. Cardiovasc. Res. 50: 556-565.

Brand K, Page S, Rogler G, Bartsch A, Brandl R, Knuechel R, Page M, Kaltschmidt C, Baeuerle PA & Neumeier D (1996) Activated transcription factor nuclear factor-κB is present in the atherosclerotic lesion. J. Clin. Invest. 97: 1715-1722.

Brand K, Eisele T, Kreusel U, Page M, Page S, Haas M, Gerling A, Kaltschmidt C, Neumann FJ, Mackman N, Baeurele PA, Walli AK & Neumeier D (1997) Dysregulation of monocytic nuclear factor-κB by oxidized low-density lipoprotein. Arterioscler. Thromb. Vasc. Biol. 17: 1901-1909.

Breuss JM, Cejna M, Bergmeister H, Kadl A, Baumgartl G, Steurer S, Xu Z, Koshelnick Y, Lipp J, De Martin R, Losert U, Lammer J & Binder BR (2002) Activation of nuclear factor-κB significantly contributes to lumen loss in a rabbit iliac artery balloon angioplasty model. Circulation 105: 633-638.

Brostjan C, Anrather J, Csizmadia V, Stroka D, Soares M, Bach FH & Winkler H (1996) Glucocorticoid-mediated repression of NFκB activity in endothelial cells does not involve induction of IκBα synthesis. J. Biol. Chem. 271: 19612-19616.

Bucay N, Sarosi I, Dunstan CR, Morony S, Tarpley J, Capparelli C, Scully S, Tan HL, Xu W, Lacey DL, Boyle WJ & Simonet WS (1998) osteoprotegerin-deficient mice develop early onset osteoporosis and arterial calcification. Genes Dev. 12: 1260-1268.

Chinetti G, Fruchart JC & Staels B (2000) Peroxisome proliferator-activated receptors (PPARs): nuclear receptors at the crossroads between lipid metabolism and inflammation. Inflamm. Res. 49: 497-505.

Collins T (1993) Endothelial nuclear factor-κB and the initiation of the atherosclerotic lesion. Lab. Invest. 68: 499-508.

Cooper JT, Stroka DM, Brostjan C, Palmetshofer A, Bach FH & Ferran C (1996) A20 blocks endothelial cell activation through a NF-κB-dependent mechanism. J. Biol. Chem. 271: 18068-18073.

Coyne C, Baravick J, Howell T, 3rd, Baravick E, Willetto C & Fenwick BW (2001) Biochemical mechanisms that interact with membrane-associated IL-1 RII (60-kDa decoy) receptors in populations of adherent macrophages and vascular endothelium. Cell. Signal. 13: 765-776.

Darnay BG, Haridas V, Ni J, Moore PA & Aggarwal BB (1998) Characterization of the intracellular domain of receptor activator of NF-κB (RANK). Interaction with tumor necrosis factor receptor-associated factors and activation of NF-κB and c-Jun N-terminal kinase. J. Biol. Chem. 273: 20551-20555.

de Martin R, Vanhove B, Cheng Q, Hofer E, Csizmadia V, Winkler H & Bach FH (1993) Cytokine-inducible expression in endothelial cells of an IκBα-like gene is regulated by NF-κB. EMBO J. 12: 2773-2779.

de Vries C, Escobedo JA, Ueno H, Houck K, Ferrara N & Williams LT (1992) The fms-like tyrosine kinase, a receptor for vascular endothelial growth factor. Science 255: 989-991.

Delerive P, De Bosscher K, Besnard S, Vanden Berghe W, Peters JM, Gonzalez FJ, Fruchart JC, Tedgui A, Haegeman G & Staels B (1999) Peroxisome proliferator-activated receptor α negatively regulates the vascular inflammatory gene response by negative cross-talk with transcription factors NF-κB and AP-1. J. Biol. Chem. 274: 32048-32054.

Delhase M, Hayakawa M, Chen Y & Karin M (1999) Positive and negative regulation of IκB kinase activity through IKKβ subunit phosphorylation. Science 284: 309-313.

Desai A, Lankford HA & Warren JS (2001) Homocysteine augments cytokine-induced chemokine expression in human vascular smooth muscle cells: implications for atherogenesis. Inflammation 25: 179-186.

Devchand PR, Keller H, Peters JM, Vazquez M, Gonzalez FJ & Wahli W (1996) The PPARα-leukotriene B4 pathway to inflammation control. Nature 384: 39-43.

Doherty TM, Uzui H, Fitzpatrick LA, Tripathi PV, Dunstan CR, Asotra K & Rajavashisth TB (2002) Rationale for the role of osteoclast-like cells in arterial calcification. FASEB J. 16: 577-582.

Erl W, Hansson GK, de Martin R, Draude G, Weber KS & Weber C (1999) Nuclear factor-κB regulates induction of apoptosis and inhibitor of apoptosis protein-1 expression in vascular smooth muscle cells. Circ. Res. 84: 668-677.

Faure E, Thomas L, Xu H, Medvedev A, Equils O & Arditi M (2001) Bacterial lipopolysaccharide and IFN-γ induce Toll-like receptor 2 and Toll-like receptor 4 expression in human endothelial cells: role of NF-κB activation. J. Immunol. 166: 2018-2024.

Ferrara N (2001) Role of vascular endothelial growth factor in regulation of physiological angiogenesis. Am. J. Physiol. Cell. Physiol. 280: C1358-1366.

Flamme I, Frolich T & Risau W (1997) Molecular mechanisms of vasculogenesis and embryonic angiogenesis. J. Cell. Physiol. 173: 206-210.

Franzoso G, Carlson L, Xing L, Poljak L, Shores EW, Brown KD, Leonardi A, Tran T, Boyce BF & Siebenlist U (1997) Requirement for NF-κB in osteoclast and B-cell development. Genes Dev. 11: 3482-3496.

Frisch SM & Francis H (1994) Disruption of epithelial cell-matrix interactions induces apoptosis. J. Cell Biol. 124: 619-626.

Hajra L, Evans AI, Chen M, Hyduk SJ, Collins T & Cybulsky MI (2000) The NF-κB signal transduction pathway in aortic endothelial cells is primed for activation in regions predisposed to atherosclerotic lesion formation. Proc. Natl. Acad. Sci. USA 97: 9052-9057.

Han YP, Tuan TL, Wu H, Hughes M & Garner WL (2001) TNF-α stimulates activation of pro-MMP2 in human skin through NF-κB mediated induction of MT1-MMP. J. Cell Sci. 114: 131-139.

Heermeier K, Leicht W, Palmetshofer A, Ullrich M, Wanner C & Galle J (2001) Oxidized LDL suppresses NF-κB and overcomes protection from apoptosis in activated endothelial cells. J. Am. Soc. Nephrol. 12: 456-463.

Huang S, Robinson JB, Deguzman A, Bucana CD & Fidler IJ (2000) Blockade of nuclear factor-κB signaling inhibits angiogenesis and tumorigenicity of human ovarian cancer cells by suppressing expression of vascular endothelial growth factor and interleukin 8. Cancer Res. 60: 5334-5339.

Huang S, Pettaway CA, Uehara H, Bucana CD & Fidler IJ (2001) Blockade of NF-κB activity in human prostate cancer cells is associated with suppression of angiogenesis, invasion, and metastasis. Oncogene 20: 4188-4197.

Imbert V, Rupec RA, Livolsi A, Pahl HL, Traenckner EB, Mueller-Dieckmann C, Farahifar D, Rossi B, Auberger P, Baeuerle PA & Peyron JF (1996) Tyrosine phosphorylation of IκB-α activates NF-κB without proteolytic degradation of IκB-α. Cell 86: 787-798.

Janabi M, Yamashita S, Hirano K, Sakai N, Hiraoka H, Matsumoto K, Zhang Z, Nozaki S & Matsuzawa Y (2000) Oxidized LDL-induced NF-κB activation and subsequent expression of pro-inflammatory genes are defective in monocyte-derived macrophages from CD36-deficient patients. Arterioscler. Thromb. Vasc. Biol. 20: 1953-1960.

Johnson C, Van Antwerp D & Hope TJ (1999) An N-terminal nuclear export signal is required for the nucleocytoplasmic shuttling of IκBα. EMBO J. 18: 6682-6693.

Kim H & Koh G (2000) Lipopolysaccharide activates matrix metalloproteinase-2 in endothelial cells through an NF-κB-dependent pathway. Biochem. Biophys. Res. Commun. 269: 401-405.

Ko HM, Seo KH, Han SJ, Ahn KY, Choi IH, Koh GY, Lee HK, Ra MS & Im SY (2002) Nuclear factor κB dependency of platelet-activating factor-induced angiogenesis. Cancer Res. 62: 1809-1814.

Kranzhofer R, Schmidt J, Pfeiffer CA, Hagl S, Libby P & Kubler W (1999) Angiotensin induces inflammatory activation of human vascular smooth muscle cells. Arterioscler. Thromb. Vasc. Biol. 19: 1623-1629.

Krikos A, Laherty CD & Dixit VM (1992) Transcriptional activation of the tumor necrosis factor α-inducible zinc finger protein, A20, is mediated by κB elements. J. Biol. Chem. 267: 17971-17976.

Kroon ME, Koolwijk P, van der Vecht B & van Hinsbergh VW (2001) Hypoxia in combination with FGF-2 induces tube formation by human microvascular endothelial cells in a fibrin matrix: involvement of at least two signal transduction pathways. J. Cell Sci. 114: 825-833.

Kukk E, Lymboussaki A, Taira S, Kaipainen A, Jeltsch M, Joukov V & Alitalo K (1996) VEGF-C receptor binding and pattern of expression with VEGFR-3 suggests a role in lymphatic vascular development. Development 122: 3829-3837.

Lan Q, Mercurius KO & Davies PF (1994) Stimulation of transcription factors NF-κB and AP1 in endothelial cells subjected to shear stress. Biochem. Biophys. Res. Commun. 201: 950-956.

Lee JS, Kypreos KE & Sonenshein GE (1998) Synchronization of cultured vascular smooth muscle cells following reversal of quiescence induced by treatment with the antioxidant N-acetylcysteine. Exp. Cell Res. 239: 447-453.

Li J, Sarosi I, Yan XQ, Morony S, Capparelli C, Tan HL, McCabe S, Elliott R, Scully S, Van G, Kaufman S, Juan SC, Sun Y, Tarpley J, Martin L, Christensen K, McCabe J, Kostenuik P, Hsu H, Fletcher F, Dunstan CR, Lacey DL & Boyle WJ (2000) RANK is the intrinsic hematopoietic cell surface receptor that controls osteoclastogenesis and regulation of bone mass and calcium metabolism. Proc. Natl. Acad. Sci. USA 97: 1566-1571.

Li Q, Van Antwerp D, Mercurio F, Lee KF & Verma IM (1999) Severe liver degeneration in mice lacking the IκB kinase 2 gene. Science 284: 321-325.

Liao F, Andalibi A, deBeer FC, Fogelman AM & Lusis AJ (1993) Genetic control of inflammatory gene induction and NF-κB-like transcription factor activation in response to an atherogenic diet in mice. J. Clin. Invest. 91: 2572-2579.

Lindner V & Collins T (1996) Expression of NF-κB and IκB-α by aortic endothelium in an arterial injury model. Am. J. Pathol. 148: 427-438.

Marx N, Sukhova GK, Collins T, Libby P & Plutzky J (1999) PPARα activators inhibit cytokine-induced vascular cell adhesion molecule-1 expression in human endothelial cells. Circulation 99: 3125-3131.

McGowan NW, Walker EJ, Macpherson H, Ralston SH & Helfrich MH (2001) Cytokine-activated endothelium recruits osteoclast precursors. Endocrinology 142: 1678-1681.
Mechtcheriakova D, Schabbauer G, Lucerna M, Clauss M, De Martin R, Binder BR & Hofer E (2001) Specificity, diversity, and convergence in VEGF and TNF-α signaling events leading to tissue factor up-regulation via EGR-1 in endothelial cells. FASEB J. 15: 230-242.
Meiners S, Laule M, Rother W, Guenther C, Prauka I, Muschick P, Baumann G, Kloetzel PM & Stangl K (2002) Ubiquitin-proteasome pathway as a new target for the prevention of restenosis. Circulation 105: 483-489.
Morigi M, Angioletti S, Imberti B, Donadelli R, Micheletti G, Figliuzzi M, Remuzzi A, Zoja C & Remuzzi G (1998) Leukocyte-endothelial interaction is augmented by high glucose concentrations and hyperglycemia in a NF-κB-dependent fashion. J. Clin. Invest. 101: 1905-1915.
Nakajima T, Kitajima I, Shin H, Takasaki I, Shigeta K, Abeyama K, Yamashita Y, Tokioka T, Soejima Y & Maruyama I (1994) Involvement of NF-κB activation in thrombin-induced human vascular smooth muscle cell proliferation. Biochem. Biophys. Res. Commun. 204: 950-958.
Oitzinger W, Hofer-Warbinek R, Schmid JA, Koshelnick Y, Binder BR & de Martin R (2001) Adenovirus-mediated expression of a mutant IκB kinase 2 inhibits the response of endothelial cells to inflammatory stimuli. Blood 97: 1611-1617.
Parhami F, Fang ZT, Fogelman AM, Andalibi A, Territo MC & Berliner JA (1993) Minimally modified low density lipoprotein-induced inflammatory responses in endothelial cells are mediated by cyclic adenosine monophosphate. J. Clin. Invest. 92: 471-478.
Pieper GM & Riaz-ul-Haq (1997) Activation of nuclear factor-κB in cultured endothelial cells by increased glucose concentration: prevention by calphostin C. J. Cardiovasc. Pharmacol. 30: 528-532.
Pober JS (1988) Cytokine-mediated activation of vascular endothelium. Physiology and pathology. Am. J. Pathol. 133: 426-433.
Qiao JH, Tripathi J, Mishra NK, Cai Y, Tripathi S, Wang XP, Imes S, Fishbein MC, Clinton SK, Libby P, Lusis AJ & Rajavashisth TB (1997) Role of macrophage colony-stimulating factor in atherosclerosis: studies of osteopetrotic mice. Am. J. Pathol. 150: 1687-1699.
Quehenberger P, Bierhaus A, Fasching P, Muellner C, Klevesath M, Hong M, Stier G, Sattler M, Schleicher E, Speiser W & Nawroth PP (2000) Endothelin 1 transcription is controlled by nuclear factor-κB in AGE-stimulated cultured endothelial cells. Diabetes 49: 1561-1570.
Rival Y, Beneteau N, Taillandier T, Pezet M, Dupont-Passelaigue E, Patoiseau JF, Junquero D, Colpaert FC & Delhon A (2002) PPARα and PPARδ activators inhibit cytokine-induced nuclear translocation of NF-κB and expression of VCAM-1 in EAhy926 endothelial cells. Eur. J. Pharmacol. 435: 143-151.
Ronicke V, Risau W & Breier G (1996) Characterization of the endothelium-specific murine vascular endothelial growth factor receptor-2 (Flk-1) promoter. Circ. Res. 79: 277-285.
Rossi A, Kapahi P, Natoli G, Takahashi T, Chen Y, Karin M & Santoro MG (2000) Anti-inflammatory cyclopentenone prostaglandins are direct inhibitors of IκB kinase. Nature 403: 103-108.
Scatena M, Almeida M, Chaisson ML, Fausto N, Nicosia RF & Giachelli CM (1998) NF-κB mediates αvβ3 integrin-induced endothelial cell survival. J. Cell Biol. 141: 1083-1093.
Scatena M & Giachelli C (2002) The αvβ3 integrin, NF-κB, osteoprotegerin endothelial cell survival pathway. Potential role in angiogenesis. Trends Cardiovasc. Med. 12: 83-88.
Scheinman RI, Cogswell PC, Lofquist AK & Baldwin AS (1995a) Role of transcriptional activation of IκBα in mediation of immunosuppression by glucocorticoids. Science 270: 283-286.
Scheinman RI, Gualberto A, Jewell CM, Cidlowski JA & Baldwin AS (1995b) Characterization of mechanisms involved in transrepression of NF-κB by activated glucocorticoid receptors. Mol. Cell. Biol. 15: 943-953.
Schmid EF, Binder K, Grell M, Scheurich P & Pfizenmaier K (1995) Both tumor necrosis factor receptors, TNFR60 and TNFR80, are involved in signaling endothelial tissue factor expression by juxtacrine tumor necrosis factor α. Blood 86: 1836-1841.
Shalaby F, Rossant J, Yamaguchi TP, Gertsenstein M, Wu XF, Breitman ML & Schuh AC (1995) Failure of blood-island formation and vasculogenesis in Flk-1-deficient mice. Nature 376: 62-66.

Shalhoub V, Elliott G, Chiu L, Manoukian R, Kelley M, Hawkins N, Davy E, Shimamoto G, Beck J, Kaufman SA, Van G, Scully S, Qi M, Grisanti M, Dunstan C, Boyle WJ & Lacey DL (2000) Characterization of osteoclast precursors in human blood. Br. J. Haematol. 111: 501-512.

Shono T, Ono M, Izumi H, Jimi SI, Matsushima K, Okamoto T, Kohno K & Kuwano M (1996) Involvement of the transcription factor NF-κB in tubular morphogenesis of human microvascular endothelial cells by oxidative stress. Mol. Cell. Biol. 16: 4231-4239.

Speir E, Yu ZX, Ferrans VJ, Huang ES & Epstein SE (1998) Aspirin attenuates cytomegalovirus infectivity and gene expression mediated by cyclooxygenase-2 in coronary artery smooth muscle cells. Circ. Res. 83: 210-216.

Stangl V, Gunther C, Jarrin A, Bramlage P, Moobed M, Staudt A, Baumann G, Stangl K & Felix SB (2001) Homocysteine inhibits TNF-α-induced endothelial adhesion molecule expression and monocyte adhesion via nuclear factor-κB dependent pathway. Biochem. Biophys. Res. Commun. 280: 1093-1100.

Stehlik C, de Martin R, Kumabashiri I, Schmid JA, Binder BR & Lipp J (1998) Nuclear factor (NF)-κB-regulated X-chromosome-linked iap gene expression protects endothelial cells from tumor necrosis factor α-induced apoptosis. J. Exp. Med. 188: 211-216.

Tao Y, Williams-Skipp C & Scheinman RI (2001) Mapping of glucocorticoid receptor DNA binding domain surfaces contributing to transrepression of NF-κB and induction of apoptosis. J. Biol. Chem. 276: 2329-2332.

Terman BI, Dougher-Vermazen M, Carrion ME, Dimitrov D, Armellino DC, Gospodarowicz D & Bohlen P (1992) Identification of the KDR tyrosine kinase as a receptor for vascular endothelial cell growth factor. Biochem. Biophys. Res. Commun. 187: 1579-1586.

Toborek M & Hennig B (1998) The role of linoleic acid in endothelial cell gene expression. Relationship to atherosclerosis. Subcell. Biochem. 30: 415-436.

Voisard R, Huber N, Baur R, Susa M, Ickrath O, Both A, Koenig W & Hombach V (2001) Different effects of antisense RelA p65 and NF-κB1 p50 oligonucleotides on the nuclear factor-κB mediated expression of ICAM-1 in human coronary endothelial and smooth muscle cells. BMC Mol. Biol. 2: 7.

Wagner AH, Gebauer M, Pollok-Kopp B & Hecker M (2002) Cytokine-inducible CD40 expression in human endothelial cells is mediated by interferon regulatory factor-1. Blood 99: 520-525.

Watson KE, Bostrom K, Ravindranath R, Lam T, Norton B & Demer LL (1994) TGF-β1 and 25-hydroxycholesterol stimulate osteoblast-like vascular cells to calcify. J. Clin. Invest. 93: 2106-2113.

Wautier JL, Wautier MP, Schmidt AM, Anderson GM, Hori O, Zoukourian C, Capron L, Chappey O, Yan SD, Guillausseau P & Stern D (1994) Advanced glycation end products (AGEs) on the surface of diabetic erythrocytes bind to the vessel wall via a specific receptor inducing oxidant stress in the vasculature: a link between surface-associated AGEs and diabetic complications. Proc. Natl. Acad. Sci. USA 91: 7742-7746.

Wautier JL, Zoukourian C, Chappey O, Wautier MP, Guillausseau PJ, Cao R, Hori O, Stern D & Schmidt AM (1996) Receptor-mediated endothelial cell dysfunction in diabetic vasculopathy. Soluble receptor for advanced glycation end products blocks hyperpermeability in diabetic rats. J. Clin. Invest. 97: 238-243.

Weber KS, Draude G, Erl W, de Martin R & Weber C (1999) Monocyte arrest and transmigration on inflamed endothelium in shear flow is inhibited by adenovirus-mediated gene transfer of IκB-α. Blood 93: 3685-3693.

Wei S, Teitelbaum SL, Wang MW & Ross FP (2001) Receptor activator of nuclear factor-κB ligand activates nuclear factor-κB in osteoclast precursors. Endocrinology 142: 1290-1295.

Willms-Kretschmer K, Flax MH & Cotran RS (1967) The fine structure of the vascular response in hapten-specific delayed hypersensitivity and contact dermatitis. Lab. Invest. 17: 334-349.

Wrighton CJ, Hofer-Warbinek R, Moll T, Eytner R, Bach FH & de Martin R (1996) Inhibition of endothelial cell activation by adenovirus-mediated expression of IκBα, an inhibitor of the transcription factor NF-κB. J. Exp. Med. 183: 1013-1022.

Xia P, Vadas MA, Rye KA, Barter PJ & Gamble JR (1999) High density lipoproteins (HDL) interrupt the sphingosine kinase signaling pathway. A possible mechanism for protection against

atherosclerosis by HDL. J. Biol. Chem. 274: 33143-33147.
Yeh M, Leitinger N, de Martin R, Onai N, Matsushima K, Vora DK, Berliner JA & Reddy ST (2001) Increased transcription of IL-8 in endothelial cells is differentially regulated by TNF-α and oxidized phospholipids. Arterioscler. Thromb. Vasc. Biol. 21: 1585-1591.
Yoshida A, Yoshida S, Ishibashi T, Kuwano M & Inomata H (1999) Suppression of retinal neovascularization by the NF-κB inhibitor pyrrolidine dithiocarbamate in mice. Invest. Ophthalmol. Vis. Sci. 40: 1624-1629.
Yoshida S, Ono M, Shono T, Izumi H, Ishibashi T, Suzuki H & Kuwano M (1997) Involvement of interleukin-8, vascular endothelial growth factor, and basic fibroblast growth factor in tumor necrosis factor α-dependent angiogenesis. Mol. Cell. Biol. 17: 4015-4023.
Zhong H, Voll RE & Ghosh S (1998) Phosphorylation of NF-κB p65 by PKA stimulates transcriptional activity by promoting a novel bivalent interaction with the coactivator CBP/p300. Mol. Cell 1: 661-671.
Zhong H, May MJ, Jimi E & Ghosh S (2002) The phosphorylation status of nuclear NF-κB determines its association with CBP/p300 or HDAC-1. Mol. Cell 9: 625-636.

10. AFFILIATION

University of Vienna, Department of Vascular Biology and Thrombosis Research, Brunnerstraße 59, 1235 Vienna, Austria.

O. DIENZ, S. BACHER & M.L. SCHMITZ

NF-κB IN T-LYMPHOCYTE BIOLOGY

1. THE FUNCTION OF T-LYMPHOCYTES IN THE IMMUNE SYSTEM

NF-κB plays an important and evolutionarily conserved role in innate and adaptive immunity. Innate immunity allows immediate defence against invading pathogens, and the more specific adaptive immunity leads to generation of an immunological memory. Adaptive immune response is mediated by T- and B-cells, which are derived from the same pluripotent hematopoietic stem cell precursor in the bone marrow (Hardy & Hayakawa, 2001; Shortman & Wu, 2001). While NF-κB is essential for the maturation and proliferation of both B- and T-cells, this review will focus on the function of NF-κB in T-lymphocyte biology. T-cells translocate to the thymus, where they are selected and differentiated to mature peripheral lymphocytes (Moroy & Karsunky, 2000). There are two subtypes of T-cells, expressing the CD4 or CD8 receptor on their cell surface. Cytotoxic $CD8^+$ T-cells bind to the major histocompatibility class I (MHC I) complexes on infected cells and kill them. $CD4^+$ T-cells become activated by antigen bound to the MHC II complexes of antigen-presenting cells (APCs). This initiates T-cell proliferation and release of different cytokines, leading to activation of other T- and B-cells. As the latter strictly depend on T-cell help for activation, $CD4^+$ T-cells are also called T-helper (Th) cells.

Each T-cell specifically recognizes one given antigen. This diversity is achieved by a high recombination rate in the genes encoding the T-cell receptor (TCR) (Nemazee, 2000). To avoid the generation of self-reactive immune cells, they are selected during maturation (Jameson, 1998; Hardy & Hayakawa, 2001). In the thymus, many self-peptides are presented to a preliminary TCR (pre-TCR) for negative selection, which prevents the generation of self-reactive T-cells. On the other hand, positive selection avoids maturation of activation-defective T-cells by selection for an inducible pre-TCR. At this stage of development, T-cells are CD4 and CD8 double-positive (DP). After these processes, they generate a mature TCR and differentiate in either $CD4^+$ or $CD8^+$ cells, and travel through the blood and the lymphatic system (Ellmeier et al., 1999).

2. ACTIVATION OF T-CELLS

The antigen presented by an APC on its MHC molecule is recognized by the TCR complex at the surface of the T-cell (Garcia et al., 1999). The TCR is

R. Beyaert (ed.), Nuclear Factor κB. Regulation and Role in Disease, 353-371.

composed of an extracellular α and ß chain, which are linked by a disulfide bond near the membrane. Both chains contain a transmembrane region and a short cytoplasmic tail. The TCR is complexed with other polypeptide chains, collectively called CD3. CD3 is composed of one γ, one δ and two ϵ chains, which are also transmembrane proteins. Moreover, the TCR heterodimer is complexed with a homodimer of ζ chains, which contributes to the transmission of most of the activating signals (van Oers, 1999). Interaction of the TCR on a $CD4^+$ T-cell with a dendritic cell presenting a peptide-loaded MHC II complex leads to the initiation of many intracellular signaling cascades and the initiation of T-cell proliferation (Acuto & Cantrell, 2000). But triggering of the TCR/CD3 complex alone is not sufficient for full activation and rather causes a state of unresponsiveness. This state is termed anergy and characterized by the absence of proliferation, even if the cell is fully restimulated later (Schwartz, 1997).

According to the two-signal theory of self/nonself-discrimination, the full activation of T-lymphocytes necessarily requires two signals (Bretscher & Cohn, 1970): one is provided by stimulation of the TCR, while the second signal is mediated by the occupancy of auxiliary receptors such as CD28. The two signaling pathways derived either from TCR or CD28 lead to T-cell proliferation and cytokine release (Fig. 1). T-cell costimulation is characterized by the synergistic activation of several pathways including JNK and NF-κB. In the physiological context, costimulation serves as a control mechanism to avoid undesired activation of the immune response. Accordingly, the ligands of CD28 (B7-1/CD80 and B7-2/CD86) are only expressed by activated and mature dendritic cells, which previously encountered foreign antigens (Keller, 2001).

The biochemical mechanisms underlying the difference of T-cell costimulation versus TCR stimulation are still a matter of debate, and the nature and function of CD28-derived signals are still unclear. It might well be that CD28 activates signaling cascades that are specific for the coreceptor and not triggered by the TCR. Alternatively, costimulation may lower the threshold for T-cell activation and prolong the time of T-cell activation (Viola et al., 1999). A further possibility comes from the identification of several negative regulators in T-cell signaling. In this scenario, the CD28 costimulatory receptor might exert its function by inhibiting such negative signals. Indeed, mice deficient for the negative regulator casitas B-lineage lymphoma protein-b (Cbl-b), a component of the ubiquitin system, are affected by severe autoimmune diseases, because their T-cells proliferate independently of costimulatory signals (Bachmaier et al., 2000; Chiang et al., 2000).

The interaction between a T-lymphocyte and an APC results in a temporal and spatial reorganization of multiple cellular elements at the contact region between both cells, a specialized region referred to as the immunological synapse (Grakoui et al., 1999). A large active multi-component complex, termed supramolecular activation complex (SMAC), is formed at the immunological synapse (Dustin & Cooper, 2000; Bromley et al., 2001). This contact area is also highly enriched in cholesterol- and sphingolipid-rich membrane microdomains, termed lipid rafts. These are used as platforms for the assembly of the signaling complex and contain a number of constitutive or recruitable proteins (Sedwick & Altman, 2002).

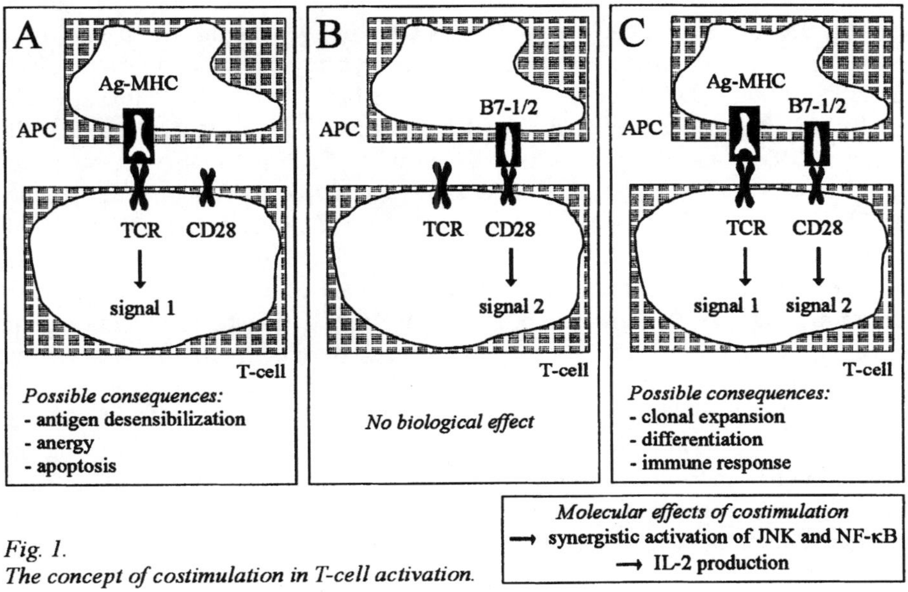

Fig. 1.
The concept of costimulation in T-cell activation.

3. THE ROLE OF NF-κB IN T-LYMPHOCYTE DEVELOPMENT

Development of T-lymphocytes is regulated by and dependent on the expression of antigen receptor genes (von Boehmer, 1997). Diversity of the antigen receptor repertoire is generated by somatic gene rearrangements of the genes encoding the two receptor subunits. Lymphocytes without productive gene rearrangements are eliminated by apoptosis, whereas lymphocytes expressing functional antigen receptors undergo further selection and differentiation before exciting the thymus or bone marrow (Melchers, 1997). The separate developmental stages of T-cells can be conveniently distinguished by the differential expression of the coreceptors CD4 and CD8. Immature thymocytes lack expression of CD4 and CD8 ($CD4^-$ $CD8^-$ double-negative (DN) cells). These cells develop into an intermediate $CD4^+$ $CD8^+$ DP stage before maturing into $CD4^+$ or $CD8^+$ single-positive (SP) cells and leaving the thymus. DN thymocytes are a very heterogeneous population that can be further subdivided into distinct developmental stages by using the differential expression of additional cell surface markers. DN cells expressing the heat-stable antigen undergo a consecutive development from $CD44^+$ $CD25^-$ to $CD44^+$ $CD25^+$ followed by $CD44^-$ $CD25^+$ and finally leading to $CD44^-$ $CD25^-$ cells. A role for NF-κB in thymocyte development was proposed by the finding that NF-κB family members are differentially expressed within the thymic compartments. RelB and c-Rel are expressed in the medulla, whereas RelA is predominant in the cortex (Weih et al., 1994; Sen et al., 1995; Denk et al., 2000). In the developing thymocyte, the highest levels of NF-κB are found in $CD44^- CD25^+$ and $CD4^- CD8^-$ DN cells, which correspond to a population that expresses the pre-T-cell receptor (Voll & Ghosh, 1999). The role of NF-κB for the thymocyte development was investigated by generating transgenic mice expressing a non-degradable form of IκBα in thymocytes (Boothby et al., 1997; Esslinger et al., 1997; Attar et al., 1998; Ferreira et al., 1999; Hettmann et al., 1999). All studies showed a loss of $CD8^+$ cells in the periphery. Among the various $CD8^+$ populations, only thymus-dependent CD8αß T-cells were impaired. In addition, mice overexpressing IκBα show a selective reduction in the number of $CD44^- CD25^+$ and $CD44^- CD25^-$ cells (Voll & Ghosh, 1999). IκBε knockout mice exhibit a 50% decrease in DN $CD44^- CD25^+$ cells (Memet et al., 1999). The defects in T-cell development can be attributed to a decreased cell survival (Ferreira et al., 1999; Hettmann et al., 1999; Voll & Ghosh, 1999), but also to impaired cell proliferation. In support of this, a role of NF-κB for positive selection was shown in vitro, where retroviral-mediated expression of IκBα blocked MHC peptide-induced differentiation of a $CD4^+/CD8^+$ cell line (Ochoa-Garay et al., 1998). Another NF-κB-regulated developmental step is the maturation of dendritic cells (Denk et al., 2000). Accordingly, mice lacking RelB do not contain dendritic cells in both thymus and spleen (Weih et al., 1995). In summary, there is clear evidence for a crucial role of NF-κB in T-cell development and maturation.

4. LESSONS FROM MOUSE MODELS: THE ROLE OF NF-κB FOR THE ACTIVATION OF T-LYMPHOCYTES

NF-κB plays not only a pivotal role for the maturation of T-cells, it is also important for various T-cell effector functions. Some clues for the function of different NF-κB/Rel family members in the immune response have come from analyses of protein expression in lymphoid organs. While p50, c-Rel, and RelA are highly expressed in all hematopoietic-derived cells including lymphocytes, the p52 and RelB protein expression pattern is more restricted (Sha et al., 1998). Nuclear p52 is only found in accessory cells of the immune system, including (follicular) dendritic cells and macrophages in the T-cell zone (Feuillard et al., 2000). In adult lymphoid tissues, high expression of RelB is restricted to dendritic cells found in the deep cortex of the lymph nodes, the periarterial lymphatic sheaths of the spleen and thymic medulla.

Mice lacking the p52 protein have an altered splenic marginal zone architecture (Iotsova et al., 1997). Immunization of these mice with T-cell-dependent antigens results in an impaired antibody response that is characterized by the inability to form follicular dendritic cell networks and germinal centers. Interestingly, this defect to generate germinal centers is not a lymphocyte-intrinsic property, as adoptively transferred p52-deficient lymphocytes are able to form these cellular structures. RelB is a critical regulator of dendritic cell differentiation. These cells act as antigen-presenting cells which enter resting tissues as precursors and, after antigenic exposure, differentiate and migrate to draining lymph nodes. Therefore, RelB knockout mice display a strongly reduced number of dendritic cells (Burkly et al., 1995; Weih et al., 1995). In addition, $RelB^{-/-}$ mice develop T-cell-dependent skin lesions similar to human atopic dermatitis (Barton et al., 2000). RelB reduces thymocyte apoptosis and regulates terminal thymocyte activation and maturation (Guerin et al., 2002). In thymuses from $RelB^{-/-}$ mice, thymocyte differentiation of SP T-cells is partially impaired. RelB-deficient mice are unable to form follicular dendritic cell networks and germinal centers induced by antigen challenge in the spleen. Reciprocal bone marrow transfer experiments show that RelB expression is required for proper formation of follicular dendritic cell networks and germinal centers in radiation-resistant stromal cells, but not in bone marrow-derived hematopoietic cells (Weih, 2001). Besides its relevance for the proper development of spleen microarchitecture, RelB is also required for expression of homing chemokines (Weih et al., 2001).

T-cells derived from mice lacking the p50 subunit also show a reduced proliferation rate, but these animals have only a mild phenotype and display an impaired ability to clear infections with Listeria monocytogenes (Sha et al., 1998) and are unable to mount airway eosinophilic inflammation (Das et al., 2001). Mice expressing the p50 protein, but not p105, have an inflammatory phenotype and suffer from lymphocytic infiltrations in lungs and liver and an increased susceptibility to infections (Ishikawa et al., 1998). These animals also exhibit splenomegalies due to erythrocytic extramedullary hematopoiesis and lymph node enlargement, although the T-cell functions are only slightly impaired. p50-deficient mice are unable to induce expression of transcription factor GATA-3 and show no Th2 cytokine production in developing, but not committed, Th2 cells. Therefore, p50 contributes to Th2 differentiation in allergic airway inflammation (Das et al., 2001). A role for p50 in the differentiation of Th cells is supported by the analysis of p50-deficient mice which are resistant to experimental autoimmune encephalomyelitis. This resistance is associated with a deficiency of myelin oligodendrocyte glycoprotein-specific T-cells to differentiate into either Th1- or Th2-type effector cells in vivo (Hilliard et al., 1999).

Whereas p52 and RelB are not important for the direct process of T-cell activation, T-lymphocytes from c-Rel knockout mice are unresponsive to most mitogenic stimuli, including concanavalin A and T-cell costimulation (Köntgen et al., 1995). The ability of these c-Rel knockout T-cells to proliferate can be restored after addition of exogenous IL-2, showing an important role for c-Rel in the regulation of IL-2 transcription. The c-Rel subunit transactivates transcription of the IL-2 promoter from an NF-κB-binding site and from the so-called CD28RE/AP composite element within this promoter (Shapiro et al., 1997). This element, which was previously defined as critical for transcriptional upregulation of IL-2 by CD28 receptor activation in T-cells, binds NF-κB and further incompletely defined transcription factors. Mice lacking c-Rel also display a markedly reduced T-cell proliferation in response to influenza virus infection (Harling-

McNabb et al., 1999). It remains to be seen in future studies whether T-cell proliferation can be restored in these animals upon exogenous administration of IL-2. The importance of the p65 subunit is only revealed by indirect approaches, as p65$^{-/-}$ animals die prenatally due to massive liver degeneration (Beg et al., 1995). p65$^{-/-}$ CD4^{+} T-cells can be obtained after transplantation of fetal liver cells from p65$^{-/-}$ mice into lethally irradiated C57BL/6 mice. These cells show a strongly decreased proliferative response after stimulation with IL-2 and αTCR/CD28 antibodies (Zheng et al., 2001), revealing the importance of p65 for T-cell activation. Alternatively, p65$^{-/-}$ T-cells can be obtained by breeding to a TNF receptor-deficient background. These double knockout mice lack secondary lymphoid organs including lymph nodes, Peyer's patches and an organized splenic microarchitecture (Alcamo et al., 2002). Therefore, p65 seems to play an important role for T-cell activation and the development of secondary lymphoid organs. Since some subunits, such as p50 and RelB, can partially compensate each other's function (Weih et al., 1997), more complex phenotypes arise in double or triple knockout animals (Gerondakis et al., 1998).

Targeted disruption of IKKα or IKKβ is prenatally lethal (Senftleben & Karin, 2002). Therefore, the role of IKKβ was studied by adoptive transfer experiments by transplanting IKKß$^{-/-}$ and IKK$\alpha^{-/-}$ hematopoietic stem cells into lethally irradiated mice. IKK$\beta^{-/-}$ radiation chimeras lack T-cells and display aberrant lymphoid organs and thymic rudiments. The cotransfer of wild-type bone marrow allows rescue of T-lymphopoiesis (Senftleben et al., 2001), raising the possibility that the lymphopoietic function of IKKβ may be mediated by extrinsic factors. Mice lacking IKKβ and the TNF receptor show an almost normal development of T-cells. Therefore, the lack of T-cell development in IKK$^{-/-}$ cells can be explained by an increased sensitivity to TNFα-induced cell death (Senftleben et al., 2001).

In summary, these experiments show that NF-κB is required for mounting of an effective immune response and the proper induction of T-cell activation by several mechanisms:

(i) Members of the NF-κB/Rel family of transcription factors are essential for the induced expression of inflammatory cytokines such as IL-2.

(ii) NF-κB/Rel family members contribute to the cytokine-independent proliferation by activation of cell cycle regulators including cyclin D1 (Joyce et al., 2001).

(iii) NF-κB is essential for the maturation of T-lymphocytes.

(iv) NF-κB is an important regulator of apoptosis, as its inhibition protects DP thymocytes from αTCR-mediated apoptosis (Hettmann et al., 1999). Besides its proapoptotic role, NF-κB exerts an important antiapoptotic function. For example, activation-induced cell death of T-lymphocytes can be prevented by adjuvants which allow T-cell survival via induction of Bcl-3 expression and concomitant induction of NF-κB transcription (Mitchell et al., 2001).

(v) NF-κB target genes contribute to the development and function of secondary lymphoid organs.

5. COMPONENTS OF THE NF-κB-INDUCING SIGNALING CASCADE TRIGGERED BY T-CELL COSTIMULATION

5.1. Protein tyrosine kinases (PTKs)

Efficient activation of a T-cell by an APC occurs by simultaneous triggering of the T-cell receptor and costimulatory receptors such as CD28, and immediately activates cytoplasmic PTKs. Triggering of these kinases is the earliest detectable molecular event after T-cell costimulation (Wange & Samelson, 1996). Cytoplasmic PTKs include members of the Syk (ZAP70 and Syk) and Src (Lck and Fyn) families. PTKs phosphorylate a plethora of substrate proteins, thus contributing to the activation of intrinsic enzyme activities and allowing the dynamic disruption or formation of multi-protein signaling complexes (Krawczyk & Penninger, 2001). Interaction of proteins is enabled by PTK-mediated phosphorylation of substrate proteins, which then allows binding to other proteins containing Src homology 2 (SH2) domains or other phosphotyrosine-binding domains. PTKs contribute to the formation of the multi-component SMAC complex formed at the immunological synapse (Dustin & Cooper, 2000; Bromley et al., 2001). Pharmacological evidence points to the importance of PTKs for TCR/CD28-induced activation of NF-κB, as inhibition of PTK activity by herbimycin A prevents NF-κB activation (Anderson et al., 1994). Along this line, Jurkat T-cell leukemia cells lacking PTK ZAP-70 fail to activate NF-κB in response to engagement of TCR and CD28 (Herndon et al., 2001). The importance of further PTKs for the activation of NF-κB in T-cells has to be revealed in future studies. One substrate for PTKs is the lymphocyte-specific signaling protein Vav1. Tyrosine-phosphorylated Vav1 is incorporated into an inducible multi-protein complex, the so-called T-cell activation signaling complex (TASC) (Hehner et al., 2000c). The TASC contains a variety of proteins including the adapter src homology 2 domain–containing leukocyte phosphoprotein of 76 kDa (SLP-76) and phospholipase C (PLC) γ1.

5.2. Vav1

The 95-kDa product of the Vav1 proto-oncogene displays a unique arrangement of signaling motifs, including a calponin homology domain, an acidic domain, a DBL homology domain with guanine nucleotide exchange factor (GEF) activity for small GTPases of the Rac/Rho family, a pleckstrin homology domain, a cysteine-rich domain, an SH2 domain flanked by two proline-binding SH3 domains and a bipartite putative nuclear localization signal (Bustelo, 2000). In vitro experiments show that Vav1, once activated by phosphatidylinositol 3,4,5-trisphosphate (PIP3) binding and Lck phosphorylation, stimulates the GDP/GTP exchange activity of Rac through its Dbl homology domain (Crespo et al., 1997; Han et al., 1998). The Vav1 protein controls several early events in T-cell signaling, including Ca^{2+} homeostasis and cytoskeletal reorganization (Fischer et al., 1998; Holsinger et al., 1998), but there is also evidence for the importance of

Vav1 for NF-κB activation. Overexpression of Vav1 in T-cells results in activation of NF-κB (Hehner et al., 2000c) via induction of IKK activity and NF-κB DNA binding (Dienz et al., 2000). Accordingly, mice lacking the Vav1 protein fail to induce NF-κB activity after T-cell costimulation (Costello et al., 1999). Addition of ionomycin cannot rescue the defect in IκBα degradation in Vav1 $T^{-/-}$-cells, suggesting that Vav1-mediated NF-κB activation is independent of calcium signals (Costello et al., 1999).

5.3. PKCθ

Recent studies showed compelling evidence for an important role of PKCθ, a Ca^{2+}-independent PKC isoform selectively expressed in T-cells and skeletal muscle, for activation of NF-κB (Isakov & Altman, 2002). T-lymphocytes from mice lacking PKCθ fail to induce NF-κB activity in response to T-cell costimulation, while activation of NF-κB still occurs in response to TNFα and IL-1 (Sun et al., 2000). Similarly, TCR/CD28-induced activation of NF-κB can be inhibited by rottlerin, a PKC inhibitor with some specificity for PKCθ (Coudronniere et al., 2000; Lin et al., 2000). Expression of a constitutively active form of PKCθ, but not of any other PKC isoform, potently activates NF-κB in T-cells. After T-cell stimulation, only PKCθ, but not any of the other PKC isoforms, is selectively transported to lipid rafts contained within the immunological synapse (Monks et al., 1998; Bi et al., 2001; Isakov & Altman, 2002). Lck-dependent raft translocation of PKCθ is required for its NF-κB-activating function (Bi et al., 2001). These experiments identify PKCθ as an important member of the NF-κB activation cascade in response to T-cell activation. PKCθ exerts its activating effects on NF-κB upon triggering of IKK activity (Dienz et al., 2000; Lin et al., 2000), but it also stimulates the transactivating function of NF-κB by physical and functional cooperation with Akt1 (Bauer et al., 2001). The activating function of PKCθ for NF-κB strictly relies on its kinase activity, but the direct phosphorylation substrate(s) on the road down to NF-κB activation are not clearly defined. PKCθ and Vav1 synergize for the activation of T-cells (Villalba et al., 2000) and NF-κB (Dienz et al., 2000). This synergism still occurs in Jurkat cells lacking SLP-76, but strictly depends on the presence of PLCγ1 (Dienz et al., 2003). In unstimulated T-cells, PKCθ is likely to form a heterodimer with Vav1 (Möller et al., 2001). Only the ligation of the costimulatory CD28 receptor, but not of the TCR, leads to the transient dissociation of the Vav1/PKCθ complex. Several experimental approaches revealed that PKCθ is located downstream from Vav1 in the control of the pathway leading to synergistic NF-κB activation.

5.4. SLP-76 and PLCγ1

The importance of some TASC proteins for the activation of NF-κB is evident from genetic experiments. Jurkat cells lacking the adapter protein SLP-76 fail to activate NF-κB (Herndon et al., 2001). Also Jurkat T-cells lacking PLCγ1 do not activate NF-κB in response to T-cell costimulation. The importance of PLCγ1 for NF-κB activation relies on its enzymatic function, as it cleaves the membrane phospholipid phosphatidylinositol 4,5-bisphosphate (PIP2) into inositol 1,4,5-trisphosphate (IP3) and diacylglycerol (DAG). While IP3 mobilizes Ca^{2+} from intracellular stores, DAG mediates activation of PKC family members (Berridge et al., 1997). Accordingly, treatment of PLCγ1-deficient cells with the phorbol ester and PKC activator PMA restores the deficient activation of NF-κB. In line with this, PKCθ activates NF-κB in SLP-76- and PLCγ1-deficient cells, showing that PKCθ is acting further downstream. In contrast, Vav1-induced NF-κB activation is normal in SLP-76⁻ cells, but absent in PLCγ1⁻ cells. TCR/CD28-stimulated recruitment of PKCθ and IKKγ to lipid rafts is lost in SLP-76⁻ or PLCγ1⁻ cells, while translocation of Vav1 remains unaffected (Dienz et al., 2003). Accordingly, recruitment of PKCθ to the immunological synapse strictly relies on the presence of SLP-76 and PLCγ1, but synapse translocation of Vav1 is independent of both proteins. Given the importance of Vav1 and the TASC proteins SLP-76 and PLCγ1 for the activation of NF-κB, it will be interesting to identify further TASC proteins and to examine their functional role for the activation of NF-κB.

5.5. *Ca^{2+} signals and reactive oxygen intermediates (ROIs)*

T-cell activation is accompanied by a release of Ca^{2+} from intracellular stores, a process important for several functions such as the activation of transcription factor NF-AT (Northrop et al., 1994). Although Ca^{2+} fluxes may contribute to NF-κB activation in response to ER overload (Pahl & Baeuerle, 1996), Ca^{2+} signals are dispensable for NF-κB activation in response to T-cell costimulation. T-cells from $Vav1^{-/-}$ mice show a lack of Ca^{2+} mobilization and NF-κB activation, but the restoration of intracellular Ca^{2+} fluxes by ionomycin rescues only the activation of NF-AT, but not of NF-κB (Costello et al., 1999). TCR/CD28 stimulation is also concomitant with an increase in ROIs, which facilitate NF-κB activation, at least in physiological concentrations. There are several ROI species, such as the short-lived superoxide anion (O_2^-) and the more stable hydrogen peroxide (H_2O_2). Hydrogen peroxide and O_2^- can form the highly reactive hydroxyl radicals in the presence of transition metal ions (Sies, 1993). ROIs are scavenged by reduced glutathione (GSH) which can be oxidized to glutathione disulfide (GSSG). T-cell costimulation induces an increase of ROIs and causes therefore a decrease of the GSH/GSSG ratio. This process facilitates TCR/CD28-triggered NF-κB activation and lowers the threshold for the activation of this transcription factor (Hehner et al., 2000a). Such signal amplification events may be important in the early phase of infection, when only low concentrations of pathogens are present. However, the molecular targets mediating this enhancing effect still remain incompletely defined. Selective pharmacological inhibitors reveal that the GSH/GSSG-derived signals are independent of lipid peroxidation processes, 5-lipoxygenases, PI3 kinase or Ca^{2+}-dependent signaling steps, but strictly rely on the activation of PTKs (Hehner et al., 2000a). Accordingly, the activity of PTKs such as Lck and Fyn can be augmented by pro-oxidative conditions (Dröge, 2002). Another possible candidate for redox regulation in the NF-κB activation pathway is PICOT, a PKCθ-interacting protein with sequence homology to thioredoxin (Witte et al. 2000).

5.6. IKK-activating kinase(s)

Several kinases, including NIK, MEKK1 and NAK, were found to phosphorylate one or both IKKs in vitro, to activate the IκB kinase complex (IKC) in vivo and to induce NF-κB-dependent gene expression, suggesting that these kinases act as IKK kinases (IKKKs) (Karin & Ben-Neriah, 2000). In the mean time, knockout models revealed a role of NIK and NAK for NF-κB transactivation, but not for the activation of the IKC (Bonnard et al., 2000; Yin et al., 2001). Similarly, $MEKK1^{-/-}$ embryonic fibroblasts show no defects in IKK activation (Xia et al., 2000). The stimulatory effect of Akt on NF-κB in T-cells (Kane et al., 1999) seems to rely primarily on Akt-mediated enhancement of NF-κB-dependent transactivation (Bauer et al., 2001).

In T-cells, two MAP3Ks have been shown to be important for T-cell costimulation-induced NF-κB activity, namely MLK3 and the proto-oncogene Cot/Tpl-2 (Lin et al., 1999; Hehner et al., 2000b). Expression of the wild-type form of MLK3 induces nuclear immigration, DNA binding and transcriptional activity of NF-κB. While MLK3 efficiently activates IKK activity and nuclear immigration of NF-κB, MLK3-induced transactivation is relatively weak (Hehner et al., 2000b). MLK3 directly phosphorylates and thus activates IKKα and IKKß, revealing its function as an IKKK. MLK3 binds, dependent on the central leucine zipper region, to components of the IKC in vivo. A kinase-deficient version of MLK3 impairs NF-κB-dependent transcription induced by T-cell costimulation, but not in response to TNFα or IL-1 (Hehner et al., 2000b). Endogenous MLK3 is phosphorylated only in response to T-cell costimulation, but not by treatment with TNFα or IL-1. In support to a signal-specific role of MLK3 in T-cells, the Tax protein encoded by the human T-lymphotropic virus type 1 (HTLV-1) upregulates MLK3 levels and thus activates NF-κB (Ng et al., 2001). TCR/CD28-mediated MLK3 activation pathway is mediated by the Rho family GTPases Rac and Cdc42 (Hehner et al., 2000b; Bock et al., 2000). The second MAP3K involved in TCR/CD28-induced activation of NF-κB is Cot. Overexpression of Cot triggers transcription from an NF-κB-dependent reporter gene and stimulates endogenous IKK activity (Lin et al., 1999). A dominant negative form of Cot prevents only TCR/CD28- but not TNFα-mediated activation of NF-κB. Cot physically assembles with NIK and IKKα in vivo, but the pathway triggered by Cot is distinct from that activated by MLK3. Whereas Cot leads to the activation of NIK (Lin et al., 1999), MLK3 does not act upstream from NIK and directly phosphorylates both IKKs. While the downstream targets for Cot are relatively well defined, it remains to be seen whether endogenous Cot is activated by T-cell costimulation.

Another possible IKKK in T-cells is PKCθ. Costimulation of human $CD4^+$ T-cells leads to the physical association of PKCθ with IKKß, and immunofluorescence studies show colocalization of both proteins in stimulated T-cells (Khoshnan et al., 2000). Both proteins are found within lipid rafts, but future studies must clarify whether PKCθ directly transphosphorylates IKKß at the conserved Ser177 and Ser181 contained in the T-loop, and thereby causes IKK activation.

5.7. CARD domain proteins

NF-κB activation in T-cells also involves several caspase recruitment domain (CARD)-containing proteins. One of these is Bcl-10, which was identified from a breakpoint in mucosa-associated lymphoid tissue (MALT) (Willis et al., 1999). NF-κB activation induced by transient overexpression of wild-type Bcl-10 in cell lines is dependent on the CARD (Thome et al., 2001). T-lymphocytes from Bcl-$10^{-/-}$ mice show no activation of NF-κB in response to T-cell costimulation (Ruland et al., 2001), thereby revealing Bcl-10 as an essential component of this signaling pathway. Also, treatment of these knockout cells with PMA and ionomycin fails to induce NF-κB activation. Since these agents bypass the early signaling steps and directly elevate Ca^{2+} fluxes and PKC activity, this finding indicates that Bcl-10 is located at the level of or downstream from PKCθ. Accordingly, early tyrosine phosphorylation, mobilization of Ca^{2+} and activation of JNK/AP-1 is normal in Bcl-$10^{-/-}$ cells (Ruland et al., 2001). How can Bcl-10 activate NF-κB in the absence of any intrinsic enzymatic activity? A possible clue comes from the finding that its viral homologue vCLAP/E10 from equine herpesvirus persistently activates NF-κB by direct binding to IKKγ. Because CARDs are known for their ability to form homo- and heteromultimers, it is conceivable that Bcl-10 multimerization leads to the activation of NF-κB. In support of this, there is evidence that Bcl-10-mediated NF-κB activation requires the arrangement of Bcl-10 in cytoplasmic filaments (Guiet & Vito, 2000). Thus, the organized assembly of proteins in ordered structures linked to the cytoskeleton is important for this NF-κB activation pathway. It was recently shown that oligomerization of IKKγ strongly activates NF-κB. Accordingly, forced oligomerization of Bcl-10 results in upregulated IKK activity. Induced protein/protein interactions are also important for NF-κB activation in response to TNFα, as dimerization of the RIP kinase activates the IKC and leads to a twofold up-shift in the molecular mass of the kinase complex (Poyet et al., 2000). These findings raise the possibility of IKC activation by oligomerization.

Bcl-10 is found in association with several other proteins. One of these is MALT1, another target of chromosomal translocation in MALT lymphoma. MALT1 is composed of a death domain, two immunoglobulin-like domains and a C-terminal caspase-like domain, and cooperates with Bcl-10 for the induction of

NF-κB (Uren et al., 2000). Other Bcl-10-interacting CARD proteins are Carma1/CARD11 (Bertin et al., 2001; Gaide et al., 2001) and CARD9 (Bertin et al., 2000). Jurkat T-cells lacking Carma1/CARD11 fail to activate NF-κB following TCR/CD28 costimulation (Wang et al., 2002), showing an essential role of this CARD protein for NF-κB activation in T-cells. However, the upstream signals regulating the NF-κB-inducing function of CARD proteins are not yet known. Since many CARD proteins, such as Bcl-10, are phosphorylated at multiple sites, CARD proteins may be regulated in a kinase-dependent way. However, neither the exact phosphorylation sites nor the involved kinases are known so far.

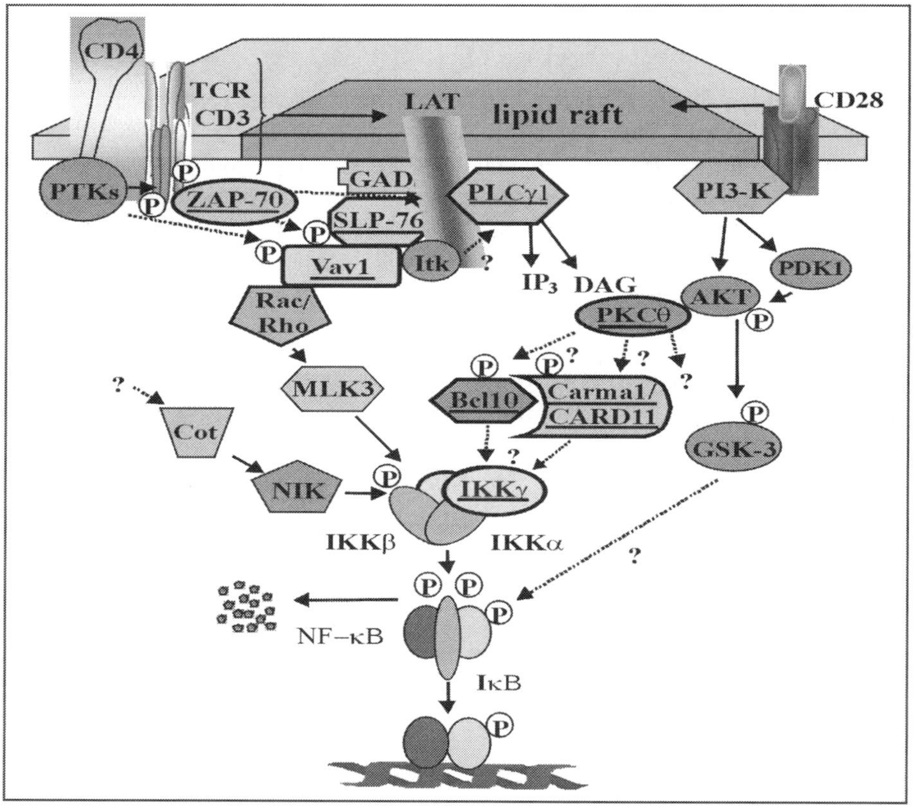

Fig. 2. Members of the signaling cascade mediating NF-κB activation induced by T-cell costimulation (essential components are underlined).

5.8. IKKs

T-cell IKKs are contained within heterogeneous complexes. Stimulation of human $CD4^+$ T-cells activates two distinct IKK complexes: a heterodimeric IKKα/IKKß complex and a homodimeric IKKß complex (Khoshnan et al., 1999). In addition to IKKα and IKKß, T-cells contain IKKε (also termed IKKi; Shimada et al.,

1999), which has approximately 30% sequence identity to IKKα/ß and which phosphorylates Ser36 of IκBα (Peters et al., 2000). However, it remains to be seen whether IκBα is the main physiologically important substrate for this kinase and whether IKKϵ might function to phosphorylate IKKα/ß (Peters & Maniatis, 2001). Although IKKϵ itself only phosphorylates Ser36, the endogenous IKKϵ complex phosphorylates both critical serine residues of IκBα. This unknown IKKϵ-associated Ser32-phosphorylating kinase can be dissociated from the IKKϵ complex upon addition of urea, but biochemical data indicate that this kinase activity is neither IKKα nor IKKß (Peters & Maniatis, 2001). IKKϵ activity can be induced upon T-cell costimulation, but not by IL-1 or TNFα. These data suggest that this kinase may be of importance in the process of TCR-derived NF-κB activation (Fig. 2).

6. FUTURE DIRECTIONS

Given the importance of NF-κB for the development and activation of T-cells, it is conceivable that T-cells utilize — dependent on the developmental stage — different NF-κB activation pathways. A good example is provided by PKCθ, as this kinase is only required for NF-κB activation in mature T-lymphocytes but not in thymocytes (Sun et al., 2000). Therefore, NF-κB activation in pre-T-cells (Aifantis et al., 2001) may use activation pathways distinct from the cascades employed in mature T-cells. Since most of the biochemical experiments have been performed with T-cell leukemia cells, such as Jurkat or CEM, much remains to be learned on the signaling routes and effector molecules used in the various developmental stages of the T-cell. On top of that, there is evidence that NF-κB is differentially regulated in Th1 and Th2 subtypes. Th1 cells, which are important during the cell-mediated immune response against intracellular pathogens, rapidly induce NF-κB elicited by T-cell costimulation. In contrast, the Th2 subset (which is involved in humoral immunity and the allergenic response) shows a very delayed and long-lasting induction of NF-κB (Dorado et al., 1998). The molecular basis underlying this slow activation process is completely unknown, as almost all NF-κB activators trigger the activity of this transcription factor within only a few minutes. The molecular mechanism of IKK activation by T-cell costimulation is currently not clear. It remains to be shown whether IKKs are directly phosphorylated by upstream kinases or whether they are activated by induced oligomerization, e.g. by CARD proteins. Most NF-κB activators are found, at least transiently, in lipid rafts. It will be interesting to learn whether the assembly of NF-κB activators in lipid rafts is required for NF-κB activation and whether the IKKs located in these membrane microdomains are in their enzymatically active state. Undesired activation of T-cells would cause autoimmunity and is prevented by negative regulators such as Cbl or SLAP/Fyb (Bachmaier et al., 2000; Chiang et al., 2000). The central role of NF-κB in T-lymphocyte biology tempts to predict also the existence of negative NF-κB regulators, which await their identification in future studies. The use of genetic models in combination with new tools in biochemistry (e.g. tandem-affinity purification technology), molecular biology (e.g. knockdown by RNA interference) and cell biology (visualization of protein migration in living cells using GFP fusion proteins) will allow to address many of these important questions.

7. NOTES

Research was supported by grants from the EU (QLK3-CT-2000-00463), the Deutsche Forschungsgemeinschaft (Schm 1417/3-1), the Fonds der chemischen Industrie, the Schweizerischer Nationalfonds, Oncosuisse and the Association for International Cancer Research. We apologize to those who made also important contributions to the issues discussed and who could not be cited due to space limits.

8. REFERENCES

Acuto O & Cantrell D (2000) T cell activation and the cytoskeleton. Annu. Rev. Immunol. 18: 165-184.

Aifantis I, Gounari F, Scorrano L, Borowski C & von Boehmer H (2001) Constitutive pre-TCR signaling promotes differentiation through Ca^{2+} mobilization and activation of NF-κB and NFAT. Nat. Immunol. 2: 403-409.

Alcamo E, Hacohen N, Schulte LC, Rennert PD, Hynes RO & Baltimore D (2002) Requirement for the NF-κB family member RelA in the development of secondary lymphoid organs. J. Exp. Med. 195: 233-244.

Anderson MT, Staal FJ, Gitler C, Herzenberg LA & Herzenberg LA (1994) Separation of oxidant-initiated and redox-regulated steps in the NF-κB signal transduction pathway. Proc. Natl. Acad. Sci. USA 91: 11527-11531.

Attar RM, Macdonald-Bravo H, Raventos-Suarez C, Durham SK & Bravo R (1998) Expression of constitutively active IκBβ in T cells of transgenic mice: persistent NF-κB activity is required for T-cell immune responses. Mol. Cell. Biol. 18: 477-487.

Bachmaier K, Krawczyk C, Kozieradzki I, Kong YY, Sasaki T, Oliveira-dos-Santos A, Mariathasan S, Bouchard D, Wakeham A, Itie A, Le J, Ohashi PS, Sarosi I, Nishina H, Lipkowitz S & Penninger JM (2000) Negative regulation of lymphocyte activation and autoimmunity by the molecular adaptor Cbl-b. Nature 403: 211-216.

Barton D, HogenEsch H & Weih F (2000) Mice lacking the transcription factor RelB develop T cell-dependent skin lesions similar to human atopic dermatitis. Eur. J. Immunol. 30: 2323-2332.

Bauer B, Krumbock N, Fresser F, Hochholdinger F, Spitaler M, Simm A, Uberall F, Schraven B & Baier G (2001) Complex formation and cooperation of protein kinase Cθ and Akt1/protein kinase Bα in the NF-κB transactivation cascade in Jurkat T cells. J. Biol. Chem. 276: 31627-3134.

Beg AA, Sha WC, Bronson RT, Ghosh S & Baltimore D (1995) Embryonic lethality and liver degeneration in mice lacking the RelA component of NF-κB. Nature 376: 167-170.

Berridge MJ (1997) Lymphocyte activation in health and disease. Crit. Rev. Immunol. 17: 155-178.

Bertin J, Wang L, Guo Y, Jacobson MD, Poyet JL, Srinivasula SM, Merriam S, DiStefano PS & Alnemri ES (2001) CARD11 and CARD14 are novel caspase recruitment domain (CARD)/membrane-associated guanylate kinase (MAGUK) family members that interact with BCL10 and activate NF-κB. J. Biol. Chem. 276: 11877-11882.

Bertin J, Guo Y, Wang L, Srinivasula SM, Jacobson MD, Poyet JL, Merriam S, Du MQ, Dyer MJ, Robison KE, DiStefano PS & Alnemri ES (2000) CARD9 is a novel caspase recruitment domain-containing protein that interacts with BCL10/CLAP and activates NF-κB. J. Biol. Chem. 275: 41082-41086.

Bi K, Tanaka Y, Coudronniere N, Sugie K, Hong S, van Stipdonk MJB & Altman A (2001) Antigen-induced translocation of PKC-θ to membrane rafts is required for T cell activation. Nat. Immunol. 2: 556-563.

Bock BC, Vacratsis PO, Qamirani E & Gallo KA (2000) Cdc42-induced activation of the mixed-lineage kinase SPRK in vivo. Requirement of the Cdc42/Rac interactive binding motif and changes in phosphorylation. J. Biol. Chem. 275: 14231-14241.

Bonnard M, Mirtsos C, Suzuki S, Graham K, Huang J, Ng M, Itie A, Wakeham A, Shahinian A, Henzel WJ, Elia AJ, Shillinglaw W, Mak TW, Cao Z & Yeh WC (2000) Deficiency of T2K leads to apoptotic liver degeneration and impaired NF-κB-dependent gene transcription. EMBO J. 19: 4976-4985.

Boothby MR, Mora AL, Scherer DC, Brockman JA & Ballard DW (1997) Perturbation of the T lymphocyte lineage in transgenic mice expressing a constitutive repressor of nuclear factor (NF)-κB. J. Exp. Med. 185: 1897-1907.

Bretscher P & Cohn M (1970) A theory of self-nonself discrimination. Science 169: 1042-1049.

Bromley SK, Burack WR, Johnson KG, Somersalo K, Sims TN, Sumen C, Davis MM, Shaw AS, Allen PM & Dustin ML (2001) The immunological synapse. Annu. Rev. Immunol. 19: 375-396.

Burkly L, Hession C, Ogata L, Reilly C, Marconi LA, Olson D, Tizard R, Cate R & Lo D (1995) Expression of relB is required for the development of thymic medulla and dendritic cells. Nature 373: 531-536.

Bustelo, XR (2000) Regulatory and signaling properties of the Vav family. Mol. Cell. Biol. 20: 1461-1477.

Chiang YJ, Kole HK, Brown K, Naramura M, Fukuhara S, Hu RJ, Jang IK, Gutkind JS, Shevach E & Gu H (2000) Cbl-b regulates the CD28 dependence of T-cell activation. Nature 403: 216-220.

Costello PS, Walters AE, Mee PJ, Turner M, Reynolds LF, Prisco A, Sarner N, Zamoyska R & Tybulewicz VLJ (1999) The Rho-family GTP exchange factor Vav is a critical transducer of T cell receptor signals to the calcium, ERK, and NF-κB pathways. Proc. Natl. Acad. Sci. USA 96: 3035-3040.

Coudronniere N, Villalba M, Englund N & Altman A (2000) NF-κB activation induced by T cell receptor/CD28 costimulation is mediated by protein kinase C-θ. Proc. Natl. Acad. Sci. USA 97: 3394-3399.

Crespo P, Schuebel KE, Ostrom AA, Gutkind JS & Bustelo XR (1997) Phosphotyrosine-dependent activation of Rac-1 GDP/GTP exchange by the Vav proto-oncogene product. Nature 385: 169-172.

Das J, Chen CH, Yang L, Cohn L, Ray P & Ray A (2001) A critical role for NF-κB in GATA3 expression and TH2 differentiation in allergic airway inflammation. Nat. Immunol. 2: 45-50.

Denk A, Wirth T & Baumann B (2000) NF-κB transcription factors: critical regulators of hematopoiesis and neuronal survival. Cytokine Growth Factor Rev. 11: 303-320.

Dienz O, Hehner S, Dröge W & Schmitz ML (2000) Synergistic activation of NF-κB by functional cooperation between Vav and PKCθ in T lymphocytes. J. Biol. Chem. 275: 24547-24551.

Dienz O, Möller A, Strecker A, Stephan N, Krammer PH, Dröge W & Schmitz ML (2003) Src homology 2 domain-containing leukocyte phosphoprotein of 76 kDa and phospholipase Cγ1 are required for NF-κB activation and lipid raft recruitment of protein kinase Cθ induced by T cell costimulation. J. Immunol. 170: 365-372.

Dorado B, Portoles P & Ballester S (1998) NF-κB in Th2 cells: delayed and long-lasting induction through the TCR complex. Eur. J. Immunol. 28: 2234-2244.

Dröge W (2002) Free radicals in the physiological control of cell function. Physiol. Rev. 82: 47-95.

Dustin ML & Cooper JA (2000) The immunological synapse and the actin cytoskeleton: molecular hardware for T cell signaling. Nat. Immunol. 11: 23-29.

Ellmeier W, Sawada S & Littman DR (1999) The regulation of CD4 and CD8 coreceptor gene expression during T cell development. Annu. Rev. Immunol. 17: 523-554.

Esslinger CW, Wilson A, Sordat B, Beermann F & Jongeneel CV (1997) Abnormal T lymphocyte development induced by targeted overexpression of IκBα. J. Immunol. 158: 5075-5078.

Ferreira V, Sidenius N, Tarantino N, Hubert P, Chatenoud L, Blasi F & Korner M (1999) In vivo inhibition of NF-κB in T-lineage cells leads to a dramatic decrease in cell proliferation and cytokine production and to increased cell apoptosis in response to mitogenic stimuli, but not to abnormal thymopoiesis. J. Immunol. 162: 6442-6450.

Feuillard J, Memet S, Goudeau B, Lilienbaum A, Schmidt-Ullrich R, Raphael M & Israel A (2000) In vivo identification of lymphocyte subsets exhibiting transcriptionally active NF-κB/Rel complexes. Int. Immunol. 12: 613-621.

Fischer KD, Kong YY, Nishina H, Tedford K, Marengere LE, Kozieradzki I, Sasaki T, Starr M, Chan G, Gardener S, Nghiem MP, Bouchard D, Barbacid M, Bernstein A & Penninger JM (1998) Vav is a regulator of cytoskeletal reorganization mediated by the T-cell receptor. Curr. Biol. 8: 554-562.

Gaide O, Martinon F, Micheau O, Bonnet D, Thome M & Tschopp J (2001) Carmal, a CARD-containing binding partner of Bcl10, induces Bcl10 phosphorylation and NF-κB activation. FEBS Lett. 496: 121-127.

Garcia KC, Degano M, Speir JA & Wilson IA (1999) Emerging principles for T cell receptor recognition of antigen in cellular immunity. Rev. Immunogenet. 1: 75-90.

Gerondakis S, Grumont R, Rourke I & Grossmann M (1998) The regulation and roles of Rel/NF-κB transcription factors during lymphocyte activation. Curr. Opin. Immunol. 10: 353-359.

Grakoui A, Bromley SK, Sumen C, Davis MM, Shaw AS, Allen PM & Dustin ML (1999) The immunological synapse: a molecular machine controlling T cell activation. Science 285: 221-217.

Guerin S, Baron ML, Valero R, Herrant M, Auberger P & Naquet P (2002) RelB reduces thymocyte apoptosis and regulates terminal thymocyte maturation. Eur. J. Immunol. 32: 1-9.

Guiet C & Vito P (2000) Caspase recruitment domain (CARD)-dependent cytoplasmic filaments mediate bcl10-induced NF-κB activation. J. Cell. Biol. 148: 1131-1140.

Han J, Luby Phelps K, Das B, Shu X, Xia Y, Mosteller RD, Krishna UM, Falck JR, White MA & Broek D (1998) Role of substrates and products of PI 3-kinase in regulating activation of Rac-related guanosine triphosphatases by Vav. Science 279: 558-560.

Hardy RR & Hayakawa K (2001) B cell development pathways. Annu. Rev. Immunol. 19: 595-621.

Harling-McNabb L, Deliyannis G, Jackson DC, Gerondakis S, Grigoriadis G & Brown LE (1999) Mice lacking the transcription factor subunit Rel can clear an influenza infection and have functional anti-viral cytotoxic T cells but do not develop an optimal antibody response. Int. Immunol. 11: 1431-1439.

Hehner SP, Breitkreutz R, Shubinsky G, Unsoeld H, Schulze-Osthoff K, Schmitz ML & Dröge, W (2000a) Enhancement of T cell receptor signaling by a mild oxidative shift in the intracellular thiol pool. J. Immunol. 165: 4319-4328.

Hehner SP, Hofmann TG, Ushmorov A, Dienz O, Wing-Lan Leung I, Lassam N, Scheidereit C, Dröge W & Schmitz ML (2000b) Mixed-lineage kinase 3 delivers CD3/CD28-derived signals into the IκB kinase complex. Mol. Cell. Biol. 20: 2556-2568.

Hehner SP, Hofmann TG, Dienz O, Dröge W & Schmitz ML (2000c) Tyrosine-phosphorylated Vav1 as a point of integration for T-cell receptor and CD28-mediated activation of JNK, p38, and interleukin-2 transcription. J. Biol. Chem. 275: 18160-18171.

Herndon TM, Shan XC, Tsokos GC & Wange RL (2001) ZAP-70 and SLP-76 regulate protein kinase Cθ and NF-κB activation in response to engagement of CD3 and CD28. J. Immunol. 166: 5654-5664.

Hettmann T, DiDonato J, Karin M & Leiden JM (1999) An essential role for nuclear factor κB in promoting double positive thymocyte apoptosis. J. Exp. Med. 189: 145-158.

Hilliard B, Samoilova EB, Liu TS, Rostami A & Chen Y (1999) Experimental autoimmune encephalomyelitis in NF-κB-deficient mice:roles of NF-κB in the activation and differentiation of autoreactive T cells. J. Immunol. 163: 2937-2943.

Holsinger LJ, Graef IA, Swat W, Chi T, Bautista DM, Davidson L, Lewis RS, Alt FW & Crabtree GR (1998) Defects in actin-cap formation in Vav-deficient mice implicate an actin requirement for lymphocyte signal transduction. Curr. Biol. 8: 563-572.

Iotsova V, Caamano J, Loy J, Yang Y, Lewin A & Bravo R (1997) Osteopetrosis in mice lacking NF-κB1 and NF-κB2. Nat. Med. 3: 1285-1289.

Isakov N & Altman A (2002) Protein kinase Cθ in T cell activation. Annu. Rev. Immunol. 20: 761-

794.

Ishikawa H, Claudio E, Dambach D, Raventos-Suarez C, Ryan C & Bravo R (1998) Chronic inflammation and susceptibility to bacterial infections in mice lacking the polypeptide (p)105 precursor (NF-κB1) but expressing p50. J. Exp. Med. 187: 985-996.

Jameson SC (1998) T cell receptor antagonism in vivo, at last. Proc. Natl. Acad. Sci. USA 95 14001-14002.

Joyce D, Albanese C, Steer J, Fu M, Bouzahzah B & Pestell RG (2001) NF-κB and cell-cycle regulation: the cyclin connection. Cytokine Growth Factor Rev. 12: 73-90.

Kane LP, Shapiro VS, Stokoe D & Weiss A (1999) Induction of NF-κB by the Akt/PKB kinase. Curr. Biol. 9: 601-604.

Karin M & Ben-Neriah Y (2000) Phosphorylation meets ubiquitination: the control of NF-κB activity. Annu. Rev. Immunol. 18: 621-663.

Keller R (2001) Dendritic cells: their significance in health and disease. Immunol. Lett. 78: 113-122.

Khoshnan A, Bae D, Tindell C-A & Nel A-E (2000) The physical association of protein kinase C θ with a lipid raft-associated inhibitor of κB factor kinase (IKK) complex plays a role in the activation of the NF-κB cascade by TCR and CD28. J. Immunol. 165: 6933-6940.

Khoshnan A, Kempiak SJ, Bennett BL, Bae D, Xu W, Manning AM, June CH & Nel AE (1999) Primary human $CD4^+$ T cells contain heterogeneous IκB kinase complexes: role in activation of the IL-2 promoter. J. Immunol. 163: 5444-5452.

Köntgen F, Grumont RJ, Strasser A, Metcalf D, Li R, Tarlinton D & Gerondakis S (1995) Mice lacking the c-rel proto-oncogene exhibit defects in lymphocyte proliferation, humoral immunity, and interleukin-2 expression. Genes Dev. 9: 1965-1977.

Krawczyk C & Penninger JM (2001) Molecular controls of antigen receptor clustering and autoimmunity. Trends Cell Biol. 11: 212-220.

Lin X, O'Mahony A, Mu Y, Geleziunas R & Greene WC (2000) Protein kinase C-θ participates in NF-κB activation induced by CD3-CD28 costimulation through selective activation of IκB kinase β. Mol. Cell. Biol. 20: 2933-2940.

Lin X, Cunningham ET, Mu Y, Geleziunas R & Greene WC (1999) The proto-oncogene Cot kinase participates in CD3/CD28 induction of NF-κB acting through the NF-κB-inducing kinase and IκB kinases. Immunity 10: 271-280.

Melchers F (1997) Lymphoid development. Res. Immunol. 148: 426-428.

Memet S, Laouini D, Epinat JC, Whiteside ST, Goudeau B, Philpott D, Kayal S, Sansonetti PJ, Berche P, Kanellopoulos J & Israel A (1999) IκBε-deficient mice: reduction of one T cell precursor subspecies and enhanced Ig isotype switching and cytokine synthesis. J. Immunol. 163: 5994-6005.

Mitchell TC, Hildeman D, Kedl RM, Teague TK, Schaefer BC, White J, Zhu Y, Kappler J & Marrack P (2001) Immunological adjuvants promote activated T cell survival via induction of Bcl-3. Nat. Immunol. 2: 397-402.

Möller A, Dienz O, Hehner SP, Dröge W & Schmitz ML (2001) Protein kinase Cθ cooperates with Vav1 to induce JNK activity in T-cells. J. Biol. Chem. 276: 20022-20028.

Monks CR, Freiberg BA, Kupfer H, Sciaky N & Kupfer A (1998) Three-dimensional segregation of supramolecular activation clusters in T cells. Nature 395: 82-86.

Moroy T & Karsunky H (2000) Regulation of pre-T-cell development. Cell Mol. Life Sci. 57: 957-975.

Nemazee D (2000) Receptor selection in B and T lymphocytes. Annu. Rev. Immunol. 18: 19-51.

Northrop JP, Ho SN, Chen L, Thomas DJ, Timmerman LA, Nolan GP, Admon A & Crabtree, GR (1994) NF-AT components define a family of transcription factors targeted in T-cell activation. Nature 369: 497-502.

Ochoa-Garay J, Kaye J & Coligan JE (1998) Nuclear factor κB is required for peptide antigen-induced differentiation of a $CD4^+CD8^+$ thymocyte line. J. Immunol. 160: 3835-3843.

Pahl HL & Baeuerle PA (1996) Activation of NF-κB by ER stress requires both Ca^{2+} and reactive oxygen intermediates as messengers. FEBS lett. 392: 129-136.

Peters RT & Maniatis T (2001) A new family of IKK-related kinases may function as IκB kinase kinases. Biochim. Biophys. Acta 1471: M57-M62.

Peters RT, Liao SM & Maniatis T (2000) IKKε is part of a novel PMA-inducible IκB kinase complex. Mol. Cell 5: 513-522.

Poyet JL, Srinivasula SM, Lin JH, Fernandes-Alnemri T, Yamaoka S, Tsichlis PN & Alnemri ES (2000) Activation of the IκB kinases by RIP via IKKγ/NEMO-mediated oligomerization. J. Biol. Chem. 275: 37966-37977.

Ruland J, Duncan GS, Elia A, del-Barco-Barrantes I, Nguyen L, Plyte S, Millar DG, Bouchard D, Wakeham A, Ohashi PS & Mak TW (2001) Bcl10 is a positive regulator of antigen receptor-induced activation of NF-κB and neural tube closure. Cell 104: 33-42.

Schwartz RH (1997) T cell clonal anergy. Curr. Opin. Immunol. 9: 351-357.

Sedwick CE & Altman A (2002) Ordered just so: lipid rafts and lymphocyte function. Sci STKE (122):RE2.

Sen J, Venkataraman L, Shinkai Y, Pierce JW, Alt FW, Burakoff SJ & Sen R (1995) Expression and induction of nuclear factor-κB-related proteins in thymocytes. J. Immunol. 154: 3213-3221.

Senftleben U & Karin M (2002) The IKK/NF-κB pathway. Crit. Care Med. 30: S18-S26.

Senftleben U, Li ZW, Baud V & Karin M (2001) IKKβ is essential for protecting T cells from TNFα-induced apoptosis. Immunity 14: 217-230.

Sha WC (1998) Regulation of immune responses by NF-κB/Rel transcription factor. J. Exp. Med. 187: 143-146.

Shapiro VS, Truitt KE, Imboden JB & Weiss A (1997) CD28 mediates transcriptional upregulation of the interleukin-2 (IL-2) promoter through a composite element containing the CD28RE and NF-IL-2B AP-1 sites. Mol. Cell. Biol. 17: 4051-4058.

Shimada T, Kawai T, Takeda K, Matsumoto M, Inoue J, Tatsumi Y, Kanamaru A & Akira S (1999) IKK-i, a novel lipopolysaccharide-inducible kinase that is related to IκB kinases. Int. Immunol. 11: 1357-1362.

Shortman K & Wu L (2001) Parentage and heritage of dendritic cells. Blood 97: 3325.

Sies H (1993) Strategies of antioxidant defense. Eur. J. Biochem. 215: 213-219.

Sun Z, Arendt CW, Ellmeier W, Schaeffer EM, Sunshine MJ, Gandhi L, Annes J, Petrzilka D, Kupfer A, Schwartzberg PL & Littman DR (2000) PKC-θ is required for TCR-induced NF-κB activation in mature but not immature T lymphocytes. Nature 404: 402-407.

Thome M, Gaide O, Micheau O, Martinon F, Bonnet D, Gonzalez M & Tschopp J (2001) Equine herpesvirus protein E10 induces membrane recruitment and phosphorylation of its cellular homologue, bcl-10. J. Cell. Biol. 152: 1115-1122.

Uren G-A, O'Rourke K, Aravind L, Pisabarro T-M, Seshagiri S, Koonin VE & Dixit MV (2000) Identification of paracaspases and metacaspases: two ancient families of caspase-like proteins, one of which plays a key role in MALT lymphoma. Mol. Cell 6: 961-967.

van Oers NS (1999) T cell receptor-mediated signs and signals governing T cell development. Semin. Immunol. 11: 227-237.

Villalba M, Coudronniere N, Deckert M, Teixeiro E, Mas P & Altman A (2000) A novel functional interaction between Vav and PKCθ is required for TCR-induced T cell activation. Immunity 12: 151-160.

Viola A, Schroeder S, Sakakibara Y & Lanzavecchia A (1999) T lymphocyte costimulation mediated by reorganization of membrane microdomains. Science 283: 680-682.

Voll RE & Ghosh S (1999) Role of NF-κB in T-lymphocyte development. Cold Spring. Harb. Symp. Quant. Biol. 64: 485-490.

von Boehmer H (1997) Aspects of lymphocyte developmental biology. Immunol. Today 18: 260-262.

Wang D, Case SM, You Y, Bertin J, Nunez G & Lin X (2002) CARD11 is required for T cell receptor (TCR)-induced NF-κB activation. Keystone Symposium NF-κB: Bench to Bedside, Abstract book, p. 84.

Wange RL & Samelson LE (1996) Complex complexes: signaling at the TCR. Immunity 5: 197-205.

Weih DS, Yilmaz ZB & Weih F (2001) Essential role of RelB in germinal center and marginal zone formation and proper expression of homing chemokines. J. Immunol. 167: 1909-1919.

Weih F, Carrasco D & Bravo R (1994) Constitutive and inducible Rel/NF-κB activities in mouse thymus and spleen. Oncogene 9: 3289-3297.

Weih F, Carrasco D, Durham SK, Barton DS, Rizzo CA, Ryseck RP, Lira SA & Bravo R (1995) Multiorgan inflammation and hematopoietic abnormalities in mice with a targeted disruption of RelB, a member of the NF-κB/Rel family. Cell 80: 331-340.

Weih F, Durham SK, Barton DS, Sha WC, Baltimore D & Bravo R (1997) p50-NF-κB complexes partially compensate for the absence of RelB: severely increased pathology in $p50^{-/-}relB^{-/-}$ double-knockout mice. J. Exp. Med. 185: 1359-1370.

Willis TG, Jadayel DM, Du MQ, Peng H, Perry AR, Abdul-Rauf M, Price H, Karran L, Majekodunmi O, Wlodarska I, Pan L, Crook T, Hamoudi R, Isaacson PG & Dyer MJ (1999) Bcl10 is involved in t(1;14)(p22;q32) of MALT B cell lymphoma and mutated in multiple tumor types. Cell 96: 35-45.

Witte S, Villalba M, Bi K, Liu Y, Isakov N & Altman A (2000) Inhibition of the c-Jun N-terminal kinase/AP-1 and NF-κB pathways by PICOT, a novel protein kinase C-interacting protein with a thioredoxin homology domain. J. Biol. Chem. 275: 1902-1909.

Xia Y, Makris C, Su B, Li E, Yang J, Nemerow GR & Karin M (2000) MEK kinase 1 is critically required for c-Jun N-terminal kinase activation by proinflammatory stimuli and growth factor-induced cell migration. Proc. Natl. Acad. Sci. USA 97: 5243-5248.

Yin L, Wu L, Wesche H, Arthur CD, White JM, Goeddel DV & Schreiber RD (2001) Defective lymphotoxin-β receptor-induced NF-κB transcriptional activity in NIK-deficient mice. Science 291: 2162-2165.

Zheng Y, Ouaaz F, Bruzzo P, Singh V, Gerondakis S & Beg AA (2001) NF-κB RelA (p65) is essential for TNFα-induced Fas expression but dispensable for both TCR-induced expression and activation-induced cell death. J. Immunol. 166: 4949-4957.

9. AFFILIATIONS

German Cancer Research Center, Division of Immunochemistry, Im Neuenheimer Feld 280, 69120 Heidelberg, Germany (O. Dienz & S. Bacher). University of Bern, Department for Chemistry and Biochemistry, Freiestrasse 3, 3012 Bern, Switzerland (M.L. Schmitz).

B. KALTSCHMIDT & C. KALTSCHMIDT

NF-κB IN THE NERVOUS SYSTEM

1. INTRODUCTION

NF-κB is a transcription factor with inducible activity present in all neuronal cell types (O'Neill & Kaltschmidt, 1997). Bona fide NF-κB in the nervous system was first described in 1993, consisting of inducible DNA-binding complexes composed of p50 and p65 subunits (Kaltschmidt et al., 1993b). This evidence and the already well-characterized role of NF-κB in the immune system and in inflammation prompted us to propose potential roles of NF-κB in many different neurological diseases (Kaltschmidt et al., 1993a). Now about ten years after, we are happy to summarize the large body of evidence in literature, which proves many of the initial work and extended the field in so many novel directions.

2. SIGNAL TRANSDUCTION

2.1. *Subunit composition*

Annotated DNA sequence information suggests that there are still the previously biochemically characterized five DNA-binding subunits: p50, p52, RelA (p65), c-Rel and RelB. Well-characterized inhibitory subunits are IκBα, IκBβ, IκBγ (p105), IκBδ (p100) and IκBϵ (Whiteside & Israël, 1997). Within the nervous system, inducible NF-κB is most frequently composed of two DNA-binding subunits (e.g. p50 or p65). NF-κB is either constitutively active or forms a complex with the inhibitory subunit IκBα (Bakalkin et al., 1993; Kaltschmidt et al., 1993a; Rattner et al., 1993; Guerrini et al., 1995; Kaltschmidt et al., 1994b; Kaltschmidt et al., 1995a-b; Kaltschmidt et al., 1997). There are reports of other κB-binding activities, such as brain-specific transcription factor (BETA), specifically detected in gray matter extracts (Körner et al., 1989), developing brain factors (DBFs), which were reported to be highly enriched in developing cortex (Cauley & Verma, 1994), and neuronal κB-binding factor (NKBF) with different target sequence requirements (Moerman et al., 1999). So far, these other κB-binding factors were not assigned to specific genes nor could be tested directly in reporter gene assays. It seems to be granted that there is an additional level of complexity added by overlapping mutually exclusive or synergistically acting binding sites for other transcription factors. In vitro, in glia and neurons a sustained NF-κB activation for up to 72 h was observed (O'Neill & Kelly, 1995;

R. Beyaert (ed.), Nuclear Factor κB. Regulation and Role in Disease, 373-392.

Kaltschmidt et al., 1999) after stimulation with IL-1 or TNF. In vivo an induced activity for up to 72 h was reported in CA1 neurons after ischemia (Clemens et al., 1997). Traumatic brain injury resulted first in activation of axonal NF-κB and later activated NF-κB was detected in neuronal nuclei, where activation could be detected as long as one year after brain injury (Nonaka et al., 1999). Some explain the sustained NF-κB activation as a differential use of IκB isoforms α and β. For glial cells, it was reported that sustained IL-1-induced NF-κB activity was still present even after the IκBα protein amount had returned to the level measured before stimulation, whereas IκBβ protein levels remained low, suggesting that IκBβ is the negative regulator for sustained NF-κB activation (Bourke et al., 2000). A biphasic response, which is repressed by IκBα, was reported for TNF-stimulated neural cells (Kemler & Fontana, 1999).

2.2. *Activators*

Many activators of NF-κB were identified (Table 1), some with specificity for the nervous system such as the neurotransmitter glutamate acting as an NF-κB activator via the main ionotropic glutamate receptors or the neurotrophins, which display an amazing specificity: NGF is activating NF-κB via the p75 receptor (Carter et al., 1996b), whereas other cytokines (such as NT-3 or NT4/5) do not activate. In microglia, all neurotrophins activate NF-κB (Nakajima et al., 1998). There is much complexity in the system: one molecule (e.g. TNF) can activate or repress NF-κB in neurons (Kaltschmidt et al., 1999), and there are cell type-specific effects.

2.3. *Repressors*

Anti-inflammatory cytokines from the immune system, such as TGFβ or IL-10, are known to also inhibit NF-κB in the nervous system (Table 2). Mechanistically it is less clear how these molecules act in the nervous system. One possibility might be the induction of IκB transcription (Arsura et al., 1996). Some molecules are known as activators, but they seem to act, at higher concentrations, as repressors. Interestingly, the lipid peroxidation product 4-hydroxy-2,3-nonenal inhibits both constitutive and inducible NF-κB activity (Camandola et al., 2000).

2.4. *Retrograde signal transduction in neurons*

Synaptic plasticity leading to information storage has been proposed to result from a synaptic dialogue (Routtenberg, 2000). A first step would be glutamate release from the presynaptic site, followed by a modification of the presynaptic release process. The latter is thought to require a retrograde messenger, which travels along the axon to switch on gene expression in order to replenish presynaptic protein supply. We and others think that NF-κB might be crucially involved in this important process of synaptic plasticity.

Table 1. Examples of molecules activating NF-κB in the nervous system.

Molecules	*Cell types*	*References*
Glutamate	neurons	Guerrini et al., 1995
Kainate	neurons	Kaltschmidt et al., 1995a
N-methyl-D-aspartate	neurons	Lipsky et al., 2001
TNF	neuroblastoma	Drew et al., 1993
TNF	neurons	Barger et al., 1995 Kaltschmidt et al., 1999
TNF	astrocytes	Sparacio et al., 1992
TNF	microglia	Lee et al., 2000
IL-1	glia cells	O'Neill & Kelly, 1995
IL-1	neurons	Grilli et al., 1996a
Secreted β-amyloid precursor	neurons	Barger & Mattson, 1996
Secreted β-amyloid precursor	microglia	Barger & Harmon, 1997
β-Amyloid	neurons	Behl et al., 1994 Kaltschmidt et al., 1997
β-Amyloid	astrocytes	Kaltschmidt et al., 1997
β-Amyloid/IFNγ	microglia	Bonaiuto et al., 1997
ATP	microglia	Ferrari et al., 1997)
ATP/IL-1	astrocytes	John et al., 2001
Adenosine	neurons	Basheer et al., 2001
LPS	microglia	Bauer et al., 1997
LPS	astrocytes	Pistritto et al., 1999
EPO	neurons	Digicaylioglu & Lipton, 2001
Bradykinin	neurons	Schwaninger et al., 1999
Stromal-derived cell factor-1α	astrocytes	Han et al., 2001
Epidermal growth factor	astrocytes	Zelenaia et al., 2000
Vascular endothelial growth factor	neuroblastoma	Jin et al., 2000
Nerve growth factor	neurons	Wood, 1995
Nerve growth factor	Schwann cells	Carter et al., 1996a
Nerve growth factor	oligodendrocytes	Yoon et al., 1998
Ciliary neurotrophic factor	PNS neurons	Middleton et al., 2000
Leukemia inhibitory factor	PNS neurons	Middleton et al., 2000
IL-6	PNS neurons	Middleton et al., 2000
Cardiotrophin 1	PNS neurons	Middleton et al., 2000
Brain-derived neurotrophic factor	microglia	Nakajima et al., 1998
Neurotrophin 3	microglia	Nakajima et al., 1998
Neurotrophin 4/5	microglia	Nakajima et al., 1998
Insulin-like growth factor	neuroblastoma	Heck et al., 1999
Activity-dependent neurotrophic factor	neuroblastoma	Glazner et al., 2000
Pigment epithelium-derived factor	neurons	Yabe et al., 2001
H_2O_2	neurons	Kaltschmidt et al., 1997
H_2O_2	oligodendrocytes	Vollgraf et al., 1999
Gangliosides	neurons	Pyo et al., 1999
Glutaredoxin	neurons	Daily et al., 2001

Table 2. Examples of molecules repressing NF-κB in the nervous system.

Molecules	*Cell types*	*References*
IL-4	astrocytes	Pousset et al., 2000
IL-10	astrocytes	Pousset et al., 2000
IL-10	neurons	Bachis et al., 2001
IL-10	microglia	Ehrlich et al., 1998
Transforming growth factor-β	microglia	Hu et al., 1999
Transforming growth factor-β	neurons	Kaltschmidt & Kaltschmidt, 2001
TNF/H_2O_2	neurons	Ginis et al., 2000
TNF	neurons	Kaltschmidt et al., 1999
NO	neurons	Togashi et al., 1997
NO	microglia	Colasanti et al., 1995
Hydroxy-nonenal	neurons	Camandola et al., 2000
Glucocorticoids	neurons	Braun et al., 2000
β-Amyloid peptide	neurons	Kaltschmidt et al., 1997
Melatonin	neurons	Lezoualc'h et al., 1998
Corticotropin-releasing hormone	neurons	Lezoualc'h et al., 2000
Aspirin	neurons	Grilli et al., 1996b
Triflusal	glia	Acarin et al., 2001
LY341122	neurons	Stephenson et al., 2000
Ganglioside	astrocyte	Massa, 1993
Pyrrolidinedithiocarbamate	neurons	Kaltschmidt et al., 1995a
Pyrrolidinedithiocarbamate	microglia	Bauer et al., 1997
Vitamin E	neurons	Post et al., 2002

The capability of NF-κB in transmitting information from, for example, active synapses to the nucleus is supported by some studies demonstrating the presence of NF-κB in synaptic regions. Synaptosomes contain presynaptic proteins that are sealed and stabilized by the postsynaptic density, and inducible forms of NF-κB have been found in synaptosomal preparations (Kaltschmidt et al., 1993b). Low-salt extracts prepared from synaptosomes contain NF-κB proteins, such as p50 and p65, together with IκBα. Synaptophysin cofractionates with NF-κB proteins during purification. Co-localization of synaptophysin and NF-κB proteins has also been detected in rat cerebral cortex (Kaltschmidt et al., 1993b). In cortical extracts, NF-κB could be activated with the detergent desoxycholate (DOC), which resulted in two specific DNA-binding complexes with different sensitivities for DOC. Supershifting and inhibition with recombinant IκBα showed a bona fide DNA-binding complex that includes the RelA and p50 subunit. Similar complexes have been detected using hippocampal synaptosomal preparations (Meberg et al., 1996). In addition, these authors reported a robust increase in p65 mRNA after long-term potentiation in vivo. It is possible that this is part of a feed-forward mechanism leading to increased DNA-binding to κB elements during long-term potentiation. Recently also an important influence of NF-κB on long-term suppression of synaptic transmission was reported (Albensi & Mattson, 2000). Purkinje cell synapses were analyzed using light microscopy and en passant synapses were found to contain NF-κB (Guerrini et al., 1995). In addition, an electron microscopy study found NF-κB-like and IκBα-like immunoreactivities within dendrites, including dendritic spines and postsynaptic densities, of neurons in the hippocampus and the cerebral cortex (Suzuki et al., 1997). Using an activity-specific anti-p65 antibody we could detect activated NF-κB in granule cell dendrites within the fascia dentata of rat hippocampus (Kaltschmidt et al., 2000). NF-κB is activated in neurons by glutamate and depolarization (Guerrini et al., 1995; Kaltschmidt et al., 1995a). In Drosophila melanogaster the NF-κB homologue Dorsal co-localizes with the IκB homologue Cactus within the nervous system. Both proteins are detected at high levels in postsynaptic sites of glutamatergic neuromuscular junctions (Cantera et al., 1999). Thus NF-κB might be utilized as a retrograde messenger in both presynaptic and postsynaptic compartments (Fig. 1).

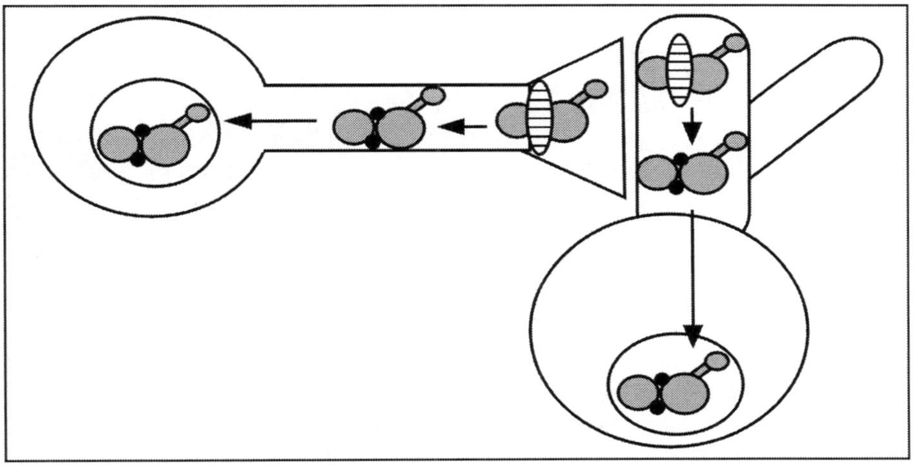

Fig. 1. Representation of NF-κB localization in neurons. DNA-binding complexes (p50 and p65 with additional transactivating domain) are gray, nuclear localization signals are black and IκB is striped. In addition to the localization of inducible NF-κB in the cytoplasm, there are inducible complexes at both presynaptic and postsynaptic sites. Upon activation by glutamatergic transmission or synaptotoxic agents such as Aβ, NF-κB is activated and transported, dependent on the nuclear localization signals to the nucleus, where replenishing protein expression is initiated.

Memory consolidation in a crab also involves activation of NF-κB-like activity (Freudenthal et al., 1998). Activated NF-κB was detected in Aplysia axons (Povelones et al., 1997) and in rat (Sulejczak & Skup, 2000). In the rat traumatic brain injury resulted first in an activation of axonal NF-κB and later activated NF-κB was detected in neuronal nuclei. Activation could be detected as long as one year after brain injury (Nonaka et al., 1999). As an injury signal sensed by synpases Aβ from diffuse plaques is able to activate synaptic NF-κB (Ferrer et al., 1998). Recently we directly analyzed the transport of NF-κB in living neurons (Wellmann et al., 2001). To enable analysis of translocation of GFP-tagged p65 in living hippocampal neurons we fused a GFP tag to the p65 subunit of NF-κB and confirmed that this fusion protein (GFP-p65) retained its functionality as a transcription factor. GFP-p65 was present in the nuclei of neurons, but after overexpression together with IκBα the distribution of the protein changed from nuclear to neuritic (in dendrites and axons). In glutamate-stimulated hippocampal neurons a return of GFP-p65 from a neuritic to a nuclear distribution was observed. Interestingly, the redistribution of GFP-p65 was dependent on a functional nuclear localization sequence (NLS). This NLS-dependent retrograde transport has already been described for Aplysia axons (Schmied & Ambron, 1997). We conclude that NF-κB is capable of being a signal transducer, transmitting information from, for example, active synapses to the nucleus, in addition to its well-known role of a transcription factor. In this way NF-κB might be involved in translating short-term signals from distant sites in neurites into long-term changes in gene expression, which may have a key role in plasticity, development and survival.

2.5. Cross-coupling with other signaling pathways

Much novel information is emerging on the intracellular interactors of different signaling cascades (Fig. 2). Two pathways seem to be characterized in great detail, more particularly erythropoietin (EPO) signaling and neurotrophin signaling. Neurotrophins, such as nerve growth factor (NGF), are well-characterized in their ability to support survival and differentiation of neurons; further functions include induction of apoptosis (Frade et al., 1996). There are two receptor systems for NGF: on the one hand, a tyrosine kinase receptor (TrkA) with high selectivity for NGF; on the other hand, a member of the TNF receptor superfamily with an intracellular death domain, namely the 75-kDa neurotrophin receptor (p75NTR). In contrast to TrkA, p75NTR can bind three neurotrophins: NGF, brain-derived neurotrophic factor (BDNF) and neurotrophin-3 (NT-3) with similar affinity (Carter et al., 1996a). Despite the fact that p75NTR was among the first growth factor receptors to be cloned, the function of this receptor was obscure and all NGF effects were initially thought to be related with the activation of the high-affinity NGF receptor TrkA. Two populations of neural cells seem to exist: one with p75NTR only (e.g. Schwann cells and oligodendrocytes) and another with both NGF receptors. We identified NF-κB activation after stimulation of p75NTR with NGF (Carter et al., 1996b) in Schwann cells. Schwann cells were used since they are devoid of detectable TrkA expression. NF-κB activation via NGF was abolished in Schwann cell cultures derived from p75NTR-deficient animals. In contrast to BDNF and NT-3, only NGF was capable to activate NF-κB in Schwann cells.

Evidence from many nonneural systems has shown that many TNF receptor family members interact with TNF receptor–associated factors (TRAFs) to activate NF-κB. TRAF6 is expressed in Schwann cells and can induce in these cells a profound activation of an NF-κB-dependent reporter gene (Khursigara et al.,

1999). Furthermore, TRAF6 associates within 1 min with p75NTR in an NGF-dependent manner. Interaction of TRAF6 with p75NTR was mapped to a juxta-membrane region not requiring the cytoplasmic death domain. In proofing a physiological significance of the biochemical data, transdominant-negative TRAF6 (devoid of the N-terminal RING finger, viz. amino acids 289-522) inhibited NGF-mediated activation of NF-κB in Schwann cells.

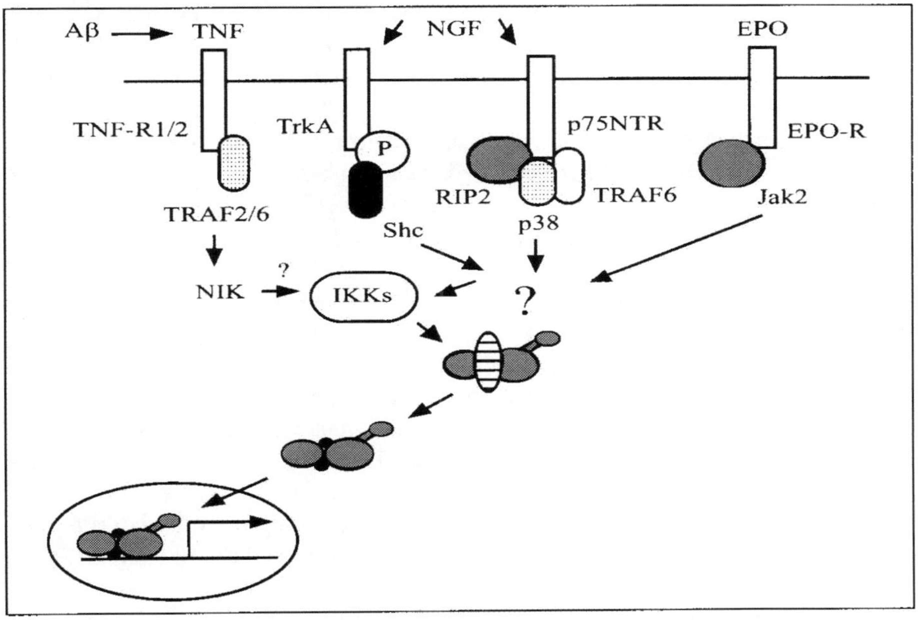

Fig. 2. Proposed scheme for different signaling pathways converging on NF-κB activation in neural cells.

Yeast two-hybrid screening identified the MAP kinase p38β2 as an interactor with the cytoplasmic domain of p75NTR (Wang et al., 2000). Phosphorylation by p38β2 could initiate p75NTR phosphorylation. Interestingly, activation of NF-κB by p38β2 was accompanied with a downmodulation of AP-1 activity, suggesting how p75NTR discriminates between cell death and survival. Activation of NF-κB by NGF was also observed in other cell lines containing only the p75NTR, such as PCNA (Carter et al., 1996a) and oligodendrocytes (Ladiwala et al., 1998). Activation of NF-κB through NGF shows a slow kinetic (Carter et al., 1996b), which is slower than the activation by TNF (Foehr et al., 2000) and depends on the age of primary culture (Khursigara et al., 1999). Additional synergistic effects, such as cross-talk with TNF signaling, could be considered as well (Bhakar et al., 1999). It has become clear that the observed lack of p75NTR-mediated NF-κB activation in HEK293 cells (Bhakar et al., 1999) is due to a lack of sufficient protein kinase RIP2 in these cells (Khursigara et al., 2001). RIP2, also named RICK or CARDIAK, is a protein kinase crucially involved in NGF-mediated NF-κB in Schwann cells. TNF-mediated signaling was identified as transmitter of Aβ-mediated signal transduction in astrocytes (Akama & Van Eldik, 2000). Expression of transdominant-negative mutants of TRAF2, or TRAF6 and NIK, abrogated the expression of the NF-κB target gene, namely inducible NO synthase (iNOS).

Pheochromocytoma 12 (PC12) cells can be differentiated by NGF in sympathetic neuron-like cells, which are protected from stress-induced apoptosis, depending on NF-κB activation (Taglialatela et al., 1997). PC12 cells express both NGF receptors, namely TrkA and p75NTR. Recently, chimeric receptors containing the extracellular domain of platelet-derived growth factor receptor (PDGFR) fused to the transmembrane and cytoplasmic domains of either TrkA or p75NTR were used to investigate NGF-mediated NF-κB activation in PC12 cells. Because PC12 cells lack endogenous PDGFR, addition of PDGF permits selective activation of either TrkA or p75NTR responses. Both TrkA and p75NTR activate NF-κB (Foehr et al., 2000) in PC12 cells.

In line with the results in Schwann cells, a transdominant-negative (DN) mutant of TRAF6 blocked p75NTR-mediated NF-κB activation, whereas a DN mutant of the phosphotyrosine-recognizing interactor Shc inhibited only TrkA-mediated NF-κB activation. Coimmunoprecipitation showed a direct interaction of Shc with TrkA and of TRAF6 and p75NTR, but not vice versa. Both pathways converge in activation of IκB kinases. Two kinases IKKα and IKKβ within a signalosome also containing NEMO can be responsible for phosphorylation of IκB on Ser32 and Ser36. Overexpression of DN IκBα suppressed both TrkA- and p75NTR-mediated NF-κB activation. Similarly, a kinase-dead mutant of IKKα interfered with both pathways.

At odds with these data in wild-type PC12 cells, an NF-κB-dependent expression of Bcl-x_L, independently of IκB serine phosphorylation, was reported in primary hippocampal neurons and SH-SY5Y cells (Bui et al., 2001). No IκB degradation after NGF treatment was detected. This NGF effect has been described as IκB tyrosine phosphorylation (Bui et al., 2001). So far, it is not clear why this investigation disagrees with other findings in the same cell type. Tyrosine phosphorylation of IκB seems to be involved in another recently described neuroprotective pathway activated by EPO (Digicaylioglu & Lipton, 2001). EPO was shown to be produced also within the brain as part of an anti-hypoxic answer. EPO receptors were detected on cerebrocortical neurons. A preconditioning of cerebrocortical neurons with EPO resulted in neuroprotection. The EPO receptor-associated tyrosine-specific kinase Janus kinase-2 (Jak2) has been shown to be responsible for NF-κB activation mediated by EPO. In vitro kinase assays displayed a Jak2-catalyzed tyrosine phosphorylation of recombinant IκBα. Surprisingly, EPO induced both a serine and a tyrosine phosphorylation of IκBα. Transdominant IκBα repressed EPO-mediated neuroprotection, arguing for an important role of serine phosphorylation. In nonneuronal cells, transdominant-negative Jak2 did not suppress NF-κB activation mediated by EPO (Bittorf et al., 2001). EPO has also been shown to act through induction of two NF-κB target genes, namely XIAP and cIAP2 (Digicaylioglu & Lipton, 2001), both containing the baculovirus IAP repeat and known to act as apoptosis suppressors.

3. SYSTEM BIOLOGY

3.1. Role in glia

There is much evidence that NF-κB activation is crucial for glial cell activation, since many proinflammatory stimuli act on glia (Table 1). The question remains whether this NF-κB-mediated transcription of target genes acts proapoptotically or antiapoptotically on neuronal survival. We think there is evidence for both, depending on the system. Cytotoxic agents such as nitric oxide (NO) could be produced in microglia after NF-κB-mediated induction of iNOS (Colasanti et al., 1995). On the other hand, high concentrations of NO repress NF-κB activation (Colasanti et al., 1995; Togashi et al., 1997). Similarly, co-cultured astrocytes repress, by a yet unknown mechanism, NF-κB activation by excitotoxic stimulation (Kaltschmidt & Kaltschmidt, 2000). Thus the production of proinflammatory reactions in glia is tightly controlled at the level of NF-κB activation. But also neurotrophic factors, such as NGF, BDNF or IL-6, can be produced by glial cells in an NF-κB-dependent manner. Also this cytokine response seems to be tightly controlled by NF-κB-inhibiting cytokines such as TGFβ, IL-4 and IL-10 (Table 2).

3.2. Role in neurons

There have been controversial theories about a role of NF-κB in neuroprotection or neurodegeneration. Following the observation on the combinatorial role of transcription factors in promoters/enhancers it might be very unlikely that one factor has only one physiological function. It is more likely that depending on the stimulus the team of activated transcription factors specify target gene expression, which might decide whether NF-κB is involved in life or death. In this respect it might be interesting that nearly all of the proteins activating NF-κB in neurons (Table 1) act as neuroprotective agents. Examples include NGF, ADNF, PEDF, TNF, EPO, LIF, CNTF and CT-1. Cytokine-dependent survival of peripheral neurons was mediated by p65 as shown in p65-deficient mice (Middleton et al., 2000). Trigeminal neurons from p65-deficient mice were impaired in NGF-dependent survival in comparison to wild-type. A nine amino-acids peptide derived from the activity-dependent neurotrophic factor (ADNF) activated NF-κB in hippocampal neurons. ADNF-mediated protection against Fe^{2+} was abolished by κB decoys (Glazner et al., 2000). TNF protected hippocampal and cerebellar granule neurons against Aβ in an NF-κB activation-dependent manner (Barger et al., 1995; Kaltschmidt et al., 1999). Knockouts of TNF receptor enhanced lesion size in a cortical impact paradigm and overexpression of the NF-κB target gene MnSOD reduced lesion size (Sullivan et al., 1999). Adenoviral overexpression of

IκB superrepressor inhibited Bcl-2 and Bcl-x after TNF stimulation in hippocampal neurons (Tamatani et al., 1999). Further antiapoptotic NF-κB target genes include cIAP and XIAP which were identified after EPO-mediated NF-κB activation (Digicaylioglu & Lipton, 2001). Knockout of the p50 subunit reduced protection against excitotoxic lesions (Yu et al., 1999) and exacerbated neuronal apoptosis in an experimental model of Huntington disease (Yu et al., 2000). Pigment epithelium-derived factor (PEDF) protected cerebellar granule cells against excitotoxic lesions (Yabe et al., 2001) in an NF-κB-dependent manner.

Overexpression of dominant-negative IκBα suppressed both TrkA- and p75NTR-mediated activation of NF-κB (Foehr et al., 2000). Similarly a kinase dead-mutant of IKKα interfered with both pathways. Interestingly, the physiological outcome of both pathways was different. Repression of p75NTR-mediated NF-κB significantly enhanced apoptosis, similar to the enhanced apoptosis rate in primary neurons after Aβ treatment together with IκBα superrepressor expression (Kaltschmidt et al., 1999). In contrast TrkA ligation does not lead to increased apoptosis when NF-κB is inhibited. Neurite outgrowth was dependent on NF-κB activation mediated by TrkA, but NF-κB activation seems not to be absolutely required for neurite outgrowth, since total inhibition of NF-κB activation results in about 65% reduction of neurite formation. Also IκB expression inhibited neuronal differentiation and Bcl-2 gene expression in SH-SY5Y cells (Feng & Porter, 1999). Taken together it might be reasonable to assume that in the context of p75NTR signaling NF-κB could play an antiapoptotic role counterbalancing the activation of a caspase cascade as described for the TNF receptor. The composite of NGF signaling via both receptors might determine the biological outcome; however, NF-κB activation by both p75NTR or TrkA results in neuroprotection. Using cDNA arrays the following NF-κB target genes for both receptors in PC12 cells were identified (Foehr et al., 2000):

— TrkA: Bcl-x (4.1x up); GSTp (5.5x up); synapsin II (3.1x up); neuromodulin (5.2x up);

— p75: Bcl-x (4.5x up); GSTp (5x up).

Further evidence for a role of NF-κB in neuroprotection is provided by the description of an Akt-dependent activation of NF-κB in neuronal cells (Jin et al., 2000) and by the inhibition of protective NF-κB activation by the proapoptotic prostate apoptosis response-4 (Par-4) (Camandola & Mattson, 2000).

Preconditioning, a term coined for an exposure to a low dose of noxious stimulus, which in turn protects against the noxious dose, has long been known. We have recently identified NF-κB activation as a crucial mechanism in Aβ-mediated preconditioning (Kaltschmidt et al., 1999). A nanomolar dose of Aβ activated NF-κB in cerebellar granule cells and protected against neurotoxic micromolar concentrations. This neuroprotection was abrogated when IκB was overexpressed. Now there are novel studies suggesting that NF-κB is a crucial factor inducing neuroprotection in many different neuroprotection paradigms (Blondeau et al., 2001; Digicaylioglu & Lipton, 2001; Ravati et al., 2001; Ginis et al., 2002). Finally it has been described that long-lasting NF-κB activation is neuroprotective (Lezoualc'h et al., 1998) and that the neuroprotective expression of MnSOD might not be mediated by p300 (Ginis et al., 2002).

In contrast to the described role for NF-κB in neuroprotection, some studies suggest a neurodegenerative role in ischemia. Chemical preconditioning efficiently reduced neuronal damage after glucose deprivation in neuronal cultures; however, this protective effect was not associated with NF-κB activation (Weih et al., 1999). An NF-κB-dependent reporter gene was activated during cerebral ischemia in vivo. In p50-deficient animals, the infarct size was significantly reduced in the ischemic hemisphere in comparison to wild-type animals (Schneider et al., 1999). Since hypoxic damage in vivo is a complex phenomenon involving glia and neurons, it might very well be that glial NF-κB activation is the driving force for neuronal degeneration in this paradigm.

It is noteworthy to mention that recent data presented evidence for a role of NF-κB as cellular response against hypoxia-induced apoptosis (Digicaylioglu & Lipton, 2001). EPO was induced after hypoxia and acted in turn as neuroprotective NF-κB activator. There are also reports of a neurodegenerative role of NF-κB in cultured neurons. It has previously been reported that aspirin-mediated neuroprotection is correlated with inhibition of NF-κB (Grilli et al., 1996b). Similarly, oxidative stress-mediated toxicity in clonal hippocampal cell lines could be repressed by antioxidative NF-κB inhibition (Post et al., 1998). It has been proposed that like in nonneuronal cells NF-κB might act upstream of p53 transcription (Uberti et al., 2000).

In summary, actually there is more information on molecular mechanisms for a neuroprotective role of NF-κB, but activation of NF-κB in glia might also be involved in neurodegeneration.

3.3. Role in development

A role of NF-κB in neuronal development is described in IKK1 and IKK2 double-knockout mice, which have an increased apoptosis rate in the developing neuroectoderm and suffer from a defect in hind brain closure, albeit not with a 100% penetrance (Li et al., 2000). In this line a developmental regulation of the neuronal expression of NF-κB-dependent transgenes has been described (Schmidt-Ullrich et al., 1996). Induction of strong reporter gene activity was described in the same phase when cortex and cerebellum reach functional maturity. It has been suggested that NF-κB might be responsible for tissue maintenance, for example as an inhibitor of apoptosis. Furthermore, it has been shown that neuronal differentiation of the human neuroblastoma cell line SH-SY5Y by retinoic acid or phorbol ester is associated with Bcl-2 transcription (Feng & Porter, 1999). Neuronal differentiation and increase in Bcl-2 mRNA was inhibited in SH-SY5Y cells expressing the superrepressor IκBα. Radiation activated NF-κB and induced differentiation. Along this line, it has been shown that the NF-κB target gene IL-6 is essential for neuronal differentiation in PC12 cells induced by radiation (Abeyama et al., 1995).

3.4. Role in neurodegenerative diseases

There is genetic evidence from early onset Alzheimer patients, linking this disease to the transmembrane protein amyloid precursor protein (APP). Interestingly the APP might be an NF-κB target gene (Grilli et al., 1996a). During the progression of Alzheimer's disease APP is aberrantly preoteolytically processed, resulting in the production of neurotoxic Aβ peptides which are major components of senile plaques found in Alzheimer patients' brains. Activated NF-κB was detected in post-mortem brains of Alzheimer disease patients (Terai et al., 1996; Boissiere et al., 1997; Kaltschmidt et al., 1997; Lukiw & Bazan, 1998). NF-κB was activated in neurons in the vicinity of diffuse Aβ deposits (Kaltschmidt et al., 1997; Ferrer et al., 1998). A decrease of activated NF-κB was observed in neurons in the vicinity of later plaque stages (Kaltschmidt et al., 1999). In vitro NF-κB is activated by Aβ in cultured neurons in an inverted-U-shaped dose response curve (Kaltschmidt et al., 1995b). In primary cultured neurons, a low dose of Aβ activated NF-κB, while a high dose did not (Kaltschmidt et al., 1999). A similar dose response was detected with TNF, which might reflect the different affinities of the two receptors TNF-R1 and TNF-R2. Recent data analyzing TNF-R1- and TNF-R2-deficient animals showed that TNF in neurons act via the high-affinity TNF-R1 as an NF-κB activator (Yang et al., 2002). Furthermore the low dose of Aβ, which is NF-κB activating, elicits protection against a subsequent high toxic dose of Aβ (Kaltschmidt et al., 1999). Similarly a protection against neurotoxic Aβ concentrations was achieved in cultures pretreated with an NF-κB-activating dose of Aβ. Finally overexpression of super-repressor IκB abbrogated the protective effect of preconditioning. Therefore we put forward that activated NF-κB in neurons is protecting against Aβ from diffuse deposits. This might also explain the low rate of neurodegeneration around diffuse plaques, which were frequently found in elderly patients without cognitive impairments (Kaltschmidt et al., 1999). One of the crucial markers for Alzheimer is the increase in the frequency of more mature plaques. In accordance with the neuroprotective hypothesis, around these plaques activated NF-κB was strongly reduced (Kaltschmidt et al., 1999). NF-κB might be seen as a regulator of neuroprotective cell function which must be activated to preserve neuronal cell function in the aged brain (Toliver-Kinsky et al., 1997). Inhibition of NF-κB by proteosome inhibitors results in a strong induction of neuronal apoptosis (Taglialatela et al., 1998). Another peptide derived from proteolytic processing of Aβ is the secreted form of APP (sAPPα). Also this protein is an NF-κB-dependent neuroprotective agent (Guo et al., 1998). sAPPα can counteract the proapoptotic actions of mutated presenilin-1. Interestingly, overexpression of mutated presenilin-1 resulted in an enhanced activation of NF-κB, followed by a prolonged inhibition (Guo et al., 1998). This second wave of inhibited NF-κB is similar to the inhibition of NF-κB activation by high doses of Aβ (Kaltschmidt et al., 1997; Kaltschmidt et al., 1999) and might be the consequence of an oxidative modification. Signaling of Aβ might involve the receptor of advanced glycation end products (RAGE) (Yan et al., 1995). RAGE ligation activates NF-κB via a Ras-dependent pathway (Huttunen et al., 1999).

Other chronic diseases of the nervous system with NF-κB involvement include multiple sclerosis (MS), Parkinson disease, Huntington disease and amyotropic lateral sclerosis (ALS). MS is characterized by a degeneration of oligodendrocytes. In experimental autoimmune encephalomyelitis, a model for MS, it has been shown that NF-κB is activated at the height of the disease in microglia (Kaltschmidt et al., 1994a). A similar activation was described in post-mortem brains of MS patients (Gveric et al., 1998; Bonetti et al., 1999). Based on this observation an amelioration of the disease was achieved using a therapy with the NF-κB-inhibiting antioxidant PDTC (Pahan & Schmid, 2000). In this line it was reported that p50-deficient mice were significantly resistant against the induction of EAE (Hilliard et al., 1999). Since MS is mainly a disease of glial cells and NF-κB activation is confined to glia and macrophages, it seems reasonable to speculate that this is a good example showing that NF-κB activation within glia might be a suitable therapeutic target. There is evidence that oxidative damage to motor neurons is a prime reason of ALS. Surprisingly, no NF-κB activation in motor neurons of ALS patients was detected (Migheli et al., 1997), but NF-κB activation was confined to astrocytes. Thus it might be that the increased astrocyte inflammation, together with the lack of neuroprotective NF-κB activation, is one of the disease mechanisms. In Parkinson patients' brains, a strong NF-κB activation in substantia nigra neurons was detected in comparison to healthy controls (Hunot et al., 1997). Furthermore, hydroxydopamine-induced neurodegeneration, one of the model lesions of Parkinson disease, was ameliorated with NF-κB inhibition via tea extracts (Levites et al., 2002). On the other hand, dopamine-induced neuronal apoptosis was accompanied by inhibition of NF-κB and could be rescued by activation of NF-κB via glutaredoxin (Daily et al., 2001). Finally, p50-deficient mice showed increased damage of striatal neurons and enhanced motor disfunctions after mitochondrial toxin 3-nitropropionic treatment, an experimental model for Huntington disease (Yu et al., 2000). Further studies defining the NF-κB target genes in each of the relevant cell types are under way and might shed additional light on the role of NF-κB in neuroprotection and disease.

4. NOTES

Due to space constraints many of important references could not be included and we apologize to those whose important work was not cited.

5. REFERENCES

Abeyama K, Kawano K, Nakajima T, Takasaki I, Kitajima I & Maruyama I (1995) Interleukin 6 mediated differentiation and rescue of cell redox in PC12 cells exposed to ionizing radiation. FEBS Lett. 364: 298-300.

Acarin L, Gonzalez B & Castellano B (2001) Triflusal posttreatment inhibits glial nuclear factor-κB, downregulates the glial response, and is neuroprotective in an excitotoxic injury model in postnatal brain. Stroke 32: 2394-2402.

Akama KT & Van Eldik LJ (2000) β-Amyloid stimulation of inducible nitric-oxide synthase in astrocytes is interleukin-1β- and tumor necrosis factor-α (TNFα)-dependent, and involves a TNFα receptor-associated factor- and NFκB-inducing kinase-dependent signaling mechanism. J. Biol. Chem. 275: 7918-7924.

Albensi BC & Mattson MP (2000) Evidence for the involvement of TNF and NF-κB in hippocampal synaptic plasticity. Synapse 35: 151-159.

Arsura M, Wu M & Sonenshein GE (1996) TGFβ1 inhibits NF-κB/Rel activity inducing apoptosis of B cells: transcriptional activation of IκBα. Immunity 5: 31-40.

Bachis A, Colangelo AM, Vicini S, Doe PP, De Bernardi MA, Brooker G & Mocchetti I (2001) Interleukin-10 prevents glutamate-mediated cerebellar granule cell death by blocking caspase-3-like activity. J. Neurosci. 21: 3104-3112.

Bakalkin G, Yakovleva T & Terenius L (1993) NF-κB-like factors in the murine brain. Developmentally-regulated and tissue-specific expression. Brain Res. Mol. Brain Res. 20: 137-146.

Barger SW, Horster D, Furukawa K, Goodman Y, Krieglstein J & Mattson MP (1995) Tumor necrosis factors α and β protect neurons against amyloid β-peptide toxicity: evidence for involvement of a κB-binding factor and attenuation of peroxide and Ca^{2+} accumulation. Proc. Natl. Acad. Sci. USA 92: 9328-9332.

Barger SW & Mattson MP (1996) Induction of neuroprotective κB-dependent transcription by secreted forms of the Alzheimer's β-amyloid precursor. Brain Res. Mol. Brain Res. 40: 116-126.

Barger SW & Harmon AD (1997) Microglial activation by Alzheimer amyloid precursor protein and modulation by apolipoprotein E. Nature 388: 878-881.

Basheer R, Rainnie DG, Porkka-Heiskanen T, Ramesh V & McCarley RW (2001) Adenosine, prolonged wakefulness, and A1-activated NF-κB DNA binding in the basal forebrain of the rat. Neuroscience 104: 731-739.

Bauer MK, Lieb K, Schulze-Osthoff K, Berger M, Gebicke-Haerter PJ, Bauer J & Fiebich BL (1997) Expression and regulation of cyclooxygenase-2 in rat microglia. Eur. J. Biochem. 243: 726-731.

Behl C, Davis JB, Lesley R & Schubert D (1994) Hydrogen peroxide mediates amyloid β protein toxicity. Cell 77: 817-827.

Bhakar AL, Roux PP, Lachance C, Kryl D, Zeindler C & Barker PA (1999) The p75 neurotrophin receptor (p75NTR) alters tumor necrosis factor-mediated NF-κB activity under physiological conditions, but direct p75NTR-mediated NF-κB activation requires cell stress. J. Biol. Chem. 274: 21443-21449.

Bittorf T, Buchse T, Sasse T, Jaster R & Brock J (2001) Activation of the transcription factor NF-κB by the erythropoietin receptor: structural requirements and biological significance. Cell Signal. 13: 673-681.

Blondeau N, Widmann C, Lazdunski M & Heurteaux C (2001) Activation of the nuclear factor-κB is a key event in brain tolerance. J. Neurosci. 21: 4668-4677.

Boissiere F, Hunot S, Faucheux B, Duyckaerts C, Hauw JJ, Agid Y & Hirsch EC (1997) Nuclear translocation of NF-κB in cholinergic neurons of patients with Alzheimer's disease. Neuroreport 8: 2849-2852.

Bonaiuto C, McDonald PP, Rossi F & Cassatella MA (1997) Activation of nuclear factor-κB by β-amyloid peptides and interferon-γ in murine microglia. J. Neuroimmunol. 77: 51-56.

Bonetti B, Stegagno C, Cannella B, Rizzuto N, Moretto G & Raine CS (1999) Activation of NF-κB and c-jun transcription factors in multiple sclerosis lesions. Implications for oligodendrocyte pathology. Am. J. Pathol. 155: 1433-1438.

Bourke E, Kennedy EJ & Moynagh PN (2000) Loss of IκB-β is associated with prolonged NF-κB activity in human glial cells. J. Biol. Chem. 275: 39996-40002.

Braun S, Liebetrau W, Berning B & Behl C (2000) Dexamethasone-enhanced sensitivity of mouse hippocampal HT22 cells for oxidative stress is associated with the suppression of nuclear factor-κB. Neurosci. Lett. 295: 101-104.

Bui NT, Livolsi A, Peyron JF & Prehn JH (2001) Activation of nuclear factor κB and Bcl-x survival gene expression by nerve growth factor requires tyrosine phosphorylation of IκBα. J. Cell Biol. 152: 753-764.

Camandola S & Mattson MP (2000) Pro-apoptotic action of PAR-4 involves inhibition of NF-κB activity and suppression of BCL-2 expression. J. Neurosci. Res. 61: 134-139.

Camandola S, Poli G & Mattson MP (2000) The lipid peroxidation product 4-hydroxy-2,3-nonenal inhibits constitutive and inducible activity of nuclear factor κB in neurons. Brain Res. Mol. Brain Res. 85: 53-60.

Cantera R, Kozlova T, Barillas-Mury C & Kafatos FC (1999) Muscle structure and innervation are affected by loss of Dorsal in the fruit fly, Drosophila melanogaster. Mol. Cell. Neurosci. 13: 131-141.

Carter BD, Dechant G, Frade JM, Kaltschmidt C & Barde YA (1996a) Neurotrophins and their p75 receptor. Cold Spring Harbor Symp. Quant. Biol. 61: 407-415.

Carter BD, Kaltschmidt C, Kaltschmidt B, Offenhauser N, Bohm-Matthaei R, Baeuerle PA & Barde YA (1996b) Selective activation of NF-κB by nerve growth factor through the neurotrophin receptor p75. Science 272: 542-545.

Cauley K & Verma IM (1994) κB enhancer-binding complexes that do not contain NF-κB are developmentally regulated in mammalian brain. Proc. Natl. Acad. Sci. USA 91: 390-394.

Clemens JA, Stephenson DT, Dixon EP, Smalstig EB, Mincy RE, Rash KS & Little SP (1997) Global cerebral ischemia activates nuclear factor-κB prior to evidence of DNA fragmentation. Brain Res. Mol. Brain Res. 48: 187-196.

Colasanti M, Persichini T, Menegazzi M, Mariotto S, Giordano E, Caldarera CM, Sogos V, Lauro GM & Suzuki H (1995) Induction of nitric oxide synthase mRNA expression. Suppression by exogenous nitric oxide. J. Biol. Chem. 270: 26731-26733.

Daily D, Vlamis-Gardikas A, Offen D, Mittelman L, Melamed E, Holmgren A & Barzilai A (2001) Glutaredoxin protects cerebellar granule neurons from dopamine-induced apoptosis by dual activation of the ras-phosphoinositide 3-kinase and jun n-terminal kinase pathways. J. Biol. Chem. 276: 21618-21626.

Digicaylioglu M & Lipton SA (2001) Erythropoietin-mediated neuroprotection involves cross-talk between Jak2 and NF-κB signalling cascades. Nature 412: 641-647.

Drew PD, Lonergan M, Goldstein ME, Lampson LA, Ozato K & McFarlin DE (1993) Regulation of MHC class I and β2-microglobulin gene expression in human neuronal cells. Factor binding to conserved cis-acting regulatory sequences correlates with expression of the genes. J. Immunol. 150: 3300-3310.

Ehrlich LC, Hu S, Peterson PK & Chao CC (1998) IL-10 down-regulates human microglial IL-8 by inhibition of NF-κB activation. Neuroreport 9: 1723-1726.

Feng Z & Porter AG (1999) NF-κB/Rel proteins are required for neuronal differentiation of SH-SY5Y neuroblastoma cells. J. Biol. Chem. 274: 30341-30344.

Ferrari D, Wesselborg S, Bauer MK & Schulze-Osthoff K (1997) Extracellular ATP activates transcription factor NF-κB through the P2Z purinoreceptor by selectively targeting NF-κB p65. J. Cell Biol. 139: 1635-1643.

Ferrer I, Marti E, Lopez E & Tortosa A (1998) NF-κB immunoreactivity is observed in association with βA4 diffuse plaques in patients with Alzheimer's disease. Neuropathol. Appl. Neurobiol. 24: 271-277.

Foehr ED, Lin X, O'Mahony A, Geleziunas R, Bradshaw RA & Greene WC (2000) NF-κB signaling promotes both cell survival and neurite process formation in nerve growth factor-stimulated PC12 cells. J. Neurosci. 20: 7556-7563.

Frade JM, Rodriguez-Tebar A & Barde YA (1996) Induction of cell death by endogenous nerve growth factor through its p75 receptor. Nature 383: 166-168.

Freudenthal R, Locatelli F, Hermitte G, Maldonado H, Lafourcade C, Delorenzi A & Romano A (1998) κB like DNA-binding activity is enhanced after spaced training that induces long-term memory in the crab Chasmagnathus. Neurosci. Lett. 242: 143-146.

Ginis I, Hallenbeck JM, Liu J, Spatz M, Jaiswal R & Shohami E (2000) Tumor necrosis factor and reactive oxygen species cooperative cytotoxicity is mediated via inhibition of NF-κB. Mol. Med. 6:

1028-1041.

Ginis I, Jaiswal R, Klimanis D, Liu J, Greenspon J & Hallenbeck JM (2002) TNF-α-induced tolerance to ischemic injury involves differential control of NF-κB transactivation: the role of NF-κB association with p300 adaptor. J. Cereb. Blood Flow Metab. 22: 142-152.

Glazner GW, Camandola S & Mattson MP (2000) Nuclear factor-κB mediates the cell survival-promoting action of activity-dependent neurotrophic factor peptide-9. J. Neurochem. 75: 101-108.

Grilli M, Goffi F, Memo M & Spano P (1996a) Interleukin-1β and glutamate activate the NF-κB/Rel binding site from the regulatory region of the amyloid precursor protein gene in primary neuronal cultures. J. Biol. Chem. 271: 15002-15007.

Grilli M, Pizzi M, Memo M & Spano P (1996b) Neuroprotection by aspirin and sodium salicylate through blockade of NF-κB activation. Science 274: 1383-1385.

Guerrini L, Blasi F & Denis-Donini S (1995) Synaptic activation of NF-κB by glutamate in cerebellar granule neurons in vitro. Proc. Natl. Acad. Sci. USA 92: 9077-9081.

Guo Q, Robinson N & Mattson MP (1998) Secreted β-amyloid precursor protein counteracts the pro-apoptotic action of mutant presenilin-1 by activation of NF-κB and stabilization of calcium homeostasis. J. Biol. Chem. 273: 12341-12351.

Gveric D, Kaltschmidt C, Cuzner ML & Newcombe J (1998) Transcription factor NF-κB and inhibitor IκBα are localized in macrophages in active multiple sclerosis lesions. J. Neuropathol. Exp. Neurol. 57: 168-178.

Han Y, He T, Huang DR, Pardo CA & Ransohoff RM (2001) TNF-α mediates SDF-1α-induced NF-κB activation and cytotoxic effects in primary astrocytes. J. Clin. Invest. 108: 425-435.

Heck S, Lezoualc'h F, Engert S & Behl C (1999) Insulin-like growth factor-1-mediated neuroprotection against oxidative stress is associated with activation of nuclear factor κB. J. Biol. Chem. 274: 9828-9835.

Hilliard B, Samoilova EB, Liu TS, Rostami A & Chen Y (1999) Experimental autoimmune encephalomyelitis in NF-κB-deficient mice:roles of NF-κB in the activation and differentiation of autoreactive T cells. J. Immunol. 163: 2937-2943.

Hu S, Chao CC, Ehrlich LC, Sheng WS, Sutton RL, Rockswold GL & Peterson PK (1999) Inhibition of microglial cell RANTES production by IL-10 and TGF-β. J. Leukoc. Biol. 65: 815-821.

Hunot S, Brugg B, Ricard D, Michel PP, Muriel MP, Ruberg M, Faucheux BA, Agid Y & Hirsch EC (1997) Nuclear translocation of NF-κB is increased in dopaminergic neurons of patients with Parkinson disease. Proc. Natl. Acad. Sci. USA 94: 7531-7536.

Huttunen HJ, Fages C & Rauvala H (1999) Receptor for advanced glycation end products (RAGE)-mediated neurite outgrowth and activation of NF-κB require the cytoplasmic domain of the receptor but different downstream signaling pathways. J. Biol. Chem. 274: 19919-19924.

Jin KL, Mao XO & Greenberg DA (2000) Vascular endothelial growth factor rescues HN33 neural cells from death induced by serum withdrawal. J. Mol. Neurosci. 14: 197-203.

John GR, Simpson JE, Woodroofe MN, Lee SC & Brosnan CF (2001) Extracellular nucleotides differentially regulate interleukin-1β signaling in primary human astrocytes: implications for inflammatory gene expression. J. Neurosci. 21: 4134-4142.

Kaltschmidt B, Baeuerle PA & Kaltschmidt C (1993a) Potential involvement of the transcription factor NF-κB in neurological disorders. Mol. Aspects Med. 14: 171-190.

Kaltschmidt C, Kaltschmidt B & Baeuerle PA (1993b) Brain synapses contain inducible forms of the transcription factor NF-κB. Mech. Dev. 43: 135-147.

Kaltschmidt C, Kaltschmidt B, Lannes-Vieira J, Kreutzberg GW, Wekerle H, Baeuerle PA & Gehrmann J (1994a) Transcription factor NF-κB is activated in microglia during experimental autoimmune encephalomyelitis. J. Neuroimmunol. 55: 99-106.

Kaltschmidt C, Kaltschmidt B, Neumann H, Wekerle H & Baeuerle PA (1994b) Constitutive NF-κB activity in neurons. Mol. Cell. Biol. 14: 3981-3992.

Kaltschmidt C, Kaltschmidt B & Baeuerle PA (1995a) Stimulation of ionotropic glutamate receptors activates transcription factor NF-κB in primary neurons. Proc. Natl. Acad. Sci. USA 92: 9618-9622.

Kaltschmidt C, Kaltschmidt B, Henkel T, Stockinger H & Baeuerle PA (1995b) Selective recognition of the activated form of transcription factor NF-κB by a monoclonal antibody. Biol. Chem. Hoppe Seyler 376: 9-16.

Kaltschmidt B, Uherek M, Volk B, Baeuerle PA & Kaltschmidt C (1997) Transcription factor NF-κB is activated in primary neurons by amyloid β peptides and in neurons surrounding early plaques from patients with Alzheimer disease. Proc. Natl. Acad. Sci. USA 94: 2642-2647.

Kaltschmidt B, Uherek M, Wellmann H, Volk B & Kaltschmidt C (1999) Inhibition of NF-κB potentiates amyloid β-mediated neuronal apoptosis. Proc. Natl. Acad. Sci. USA 96: 9409-9414.

Kaltschmidt B, Deller T, Frotscher M & Kaltschmidt C (2000) Ultrastructural localization of activated NF-κB in granule cells of the rat fascia dentata. Neuroreport 11: 839-844.

Kaltschmidt B & Kaltschmidt C (2000) Constitutive NF-κB activity is modulated via neuron-astroglia interaction. Exp. Brain Res. 130: 100-104.

Kaltschmidt B & Kaltschmidt C (2001) DNA array analysis of the developing rat cerebellum: transforming growth factor-β2 inhibits constitutively activated NF-κB in granule neurons. Mech. Dev. 101: 11-19.

Kemler I & Fontana A (1999) Role of IκBα and IκBβ in the biphasic nuclear translocation of NF-κB in TNFα-stimulated astrocytes and in neuroblastoma cells. Glia 26: 212-220.

Khursigara G, Orlinick JR & Chao MV (1999) Association of the p75 neurotrophin receptor with TRAF6. J. Biol. Chem. 274: 2597-2600.

Khursigara G, Bertin J, Yano H, Moffett H, DiStefano PS & Chao MV (2001) A prosurvival function for the p75 receptor death domain mediated via the caspase recruitment domain receptor-interacting protein 2. J. Neurosci. 21: 5854-5863.

Körner M, Rattner A, Mauxion F, Sen R & Citri Y (1989) A brain-specific transcription activator. Neuron 3: 563-572.

Ladiwala U, Lachance C, Simoneau SJ, Bhakar A, Barker PA & Antel JP (1998) p75 neurotrophin receptor expression on adult human oligodendrocytes: signaling without cell death in response to NGF. J. Neurosci. 18: 1297-1304.

Lee SJ, Zhou T, Choi C, Wang Z & Benveniste EN (2000) Differential regulation and function of Fas expression on glial cells. J. Immunol. 164: 1277-1285.

Levites Y, Youdim MB, Maor G & Mandel S (2002) Attenuation of 6-hydroxydopamine (6-OHDA)-induced nuclear factor-κB (NF-κB) activation and cell death by tea extracts in neuronal cultures. Biochem. Pharmacol. 63: 21-29.

Lezoualc'h F, Sparapani M & Behl C (1998) N-acetyl-serotonin (normelatonin) and melatonin protect neurons against oxidative challenges and suppress the activity of the transcription factor NF-κB. J. Pineal Res. 24: 168-178.

Lezoualc'h F, Engert S, Berning B & Behl C (2000) Corticotropin-releasing hormone-mediated neuroprotection against oxidative stress is associated with the increased release of non-amyloidogenic amyloid β precursor protein and with the suppression of nuclear factor-κB. Mol. Endocrinol. 14: 147-159.

Li Q, Estepa G, Memet S, Israël A & Verma IM (2000) Complete lack of NF-κB activity in IKK1 and IKK2 double-deficient mice: additional defect in neurulation. Genes Dev. 14: 1729-1733.

Lipsky RH, Xu K, Zhu D, Kelly C, Terhakopian A, Novelli A & Marini AM (2001) Nuclear factor κB is a critical determinant in N-methyl-D-aspartate receptor-mediated neuroprotection. J. Neurochem. 78: 254-264.

Lukiw WJ & Bazan NG (1998) Strong nuclear factor-κB-DNA binding parallels cyclooxygenase-2 gene transcription in aging and in sporadic Alzheimer's disease superior temporal lobe neocortex. J. Neurosci. Res. 53: 583-592.

Massa PT (1993) Specific suppression of major histocompatibility complex class I and class II genes in astrocytes by brain-enriched gangliosides. J. Exp. Med. 178: 1357-1363.

Meberg PJ, Kinney WR, Valcourt EG & Routtenberg A (1996) Gene expression of the transcription factor NF-κB in hippocampus: regulation by synaptic activity. Brain Res. Mol. Brain Res. 38: 179-190.

Middleton G, Hamanoue M, Enokido Y, Wyatt S, Pennica D, Jaffray E, Hay RT & Davies AM (2000) Cytokine-induced nuclear factor κB activation promotes the survival of developing neurons.

J. Cell Biol. 148: 325-332.

Migheli A, Piva R, Atzori C, Troost D & Schiffer D (1997) c-Jun, JNK/SAPK kinases and transcription factor NF-κB are selectively activated in astrocytes, but not motor neurons, in amyotrophic lateral sclerosis. J. Neuropathol. Exp. Neurol. 56: 1314-1322.

Moerman AM, Mao X, Lucas MM & Barger SW (1999) Characterization of a neuronal κB-binding factor distinct from NF-κB. Brain Res. Mol. Brain Res. 67: 303-315.

Nakajima K, Kikuchi Y, Ikoma E, Honda S, Ishikawa M, Liu Y & Kohsaka S (1998) Neurotrophins regulate the function of cultured microglia. Glia 24: 272-289.

Nonaka M, Chen XH, Pierce JE, Leoni MJ, McIntosh TK, Wolf JA & Smith DH (1999) Prolonged activation of NF-κB following traumatic brain injury in rats. J. Neurotrauma 16: 1023-1034.

O'Neill LA & Kelly VP (1995) Sustained activation of NFκB and transient IκBα degradation induced by tumour necrosis factor in 1321N1 human astrocytoma. Biochem. Soc. Trans. 23: 597S.

O'Neill LA & Kaltschmidt C (1997) NF-κB: a crucial transcription factor for glial and neuronal cell function. Trends Neurosci. 20: 252-258.

Pahan K & Schmid M (2000) Activation of nuclear factor-κB in the spinal cord of experimental allergic encephalomyelitis. Neurosci. Lett. 287: 17-20.

Pistritto G, Franzese O, Pozzoli G, Mancuso C, Tringali G, Preziosi P & Navarra P (1999) Bacterial lipopolysaccharide increases prostaglandin production by rat astrocytes via inducible cyclo-oxygenase: evidence for the involvement of nuclear factor κB. Biochem. Biophys. Res. Commun. 263: 570-574.

Post A, Holsboer F & Behl C (1998) Induction of NF-κB activity during haloperidol-induced oxidative toxicity in clonal hippocampal cells: suppression of NF-κB and neuroprotection by antioxidants. J. Neurosci. 18: 8236-8246.

Post A, Rucker M, Ohl F, Uhr M, Holsboer F, Almeida OF & Michaelidis TM (2002) Mechanisms underlying the protective potential of α-tocopherol (vitamin E) against haloperidol-associated neurotoxicity. Neuropsychopharmacology 26: 397-407.

Pousset F, Dantzer R, Kelley KW & Parnet P (2000) Interleukin-1 signaling in mouse astrocytes involves Akt: a study with interleukin-4 and IL-10. Eur. Cytokine Netw. 11: 427-434.

Povelones M, Tran K, Thanos D & Ambron RT (1997) An NF-κB-like transcription factor in axoplasm is rapidly inactivated after nerve injury in Aplysia. J. Neurosci. 17: 4915-4920.

Pyo H, Joe E, Jung S, Lee SH & Jou I (1999) Gangliosides activate cultured rat brain microglia. J. Biol. Chem. 274: 34584-34589.

Rattner A, Körner M, Walker MD & Citri Y (1993) NF-κB activates the HIV promoter in neurons. EMBO J. 12: 4261-4267.

Ravati A, Ahlemeyer B, Becker A, Klumpp S & Krieglstein J (2001) Preconditioning-induced neuroprotection is mediated by reactive oxygen species and activation of the transcription factor nuclear factor-κB. J. Neurochem. 78: 909-919.

Routtenberg A (2000) It's about time. Am. Psychol. Assoc. 2: 17-34.

Schmidt-Ullrich R, Memet S, Lilienbaum A, Feuillard J, Raphael M & Israël A (1996) NF-κB activity in transgenic mice: developmental regulation and tissue specificity. Development 122: 2117-2128.

Schmied R & Ambron RT (1997) A nuclear localization signal targets proteins to the retrograde transport system, thereby evading uptake into organelles in aplysia axons. J. Neurobiol. 33: 151-160.

Schneider A, Martin-Villalba A, Weih F, Vogel J, Wirth T & Schwaninger M (1999) NF-κB is activated and promotes cell death in focal cerebral ischemia. Nat. Med. 5: 554-559.

Schwaninger M, Sallmann S, Petersen N, Schneider A, Prinz S, Libermann TA & Spranger M (1999) Bradykinin induces interleukin-6 expression in astrocytes through activation of nuclear factor-κB. J. Neurochem. 73: 1461-1466.

Sparacio SM, Zhang Y, Vilcek J & Benveniste EN (1992) Cytokine regulation of interleukin-6 gene expression in astrocytes involves activation of an NF-κB-like nuclear protein. J. Neuroimmunol. 39: 231-242.

Stephenson D, Yin T, Smalstig EB, Hsu MA, Panetta J, Little S & Clemens J (2000) Transcription factor nuclear factor-κB is activated in neurons after focal cerebral ischemia. J. Cereb. Blood Flow

Metab. 20: 592-603.
Sulejczak D & Skup M (2000) Axoplasmic localisation of the NFκB p65 subunit in the rat brain. Acta Neurobiol. Exp. 60: 217.
Sullivan PG, Bruce-Keller AJ, Rabchevsky AG, Christakos S, Clair DK, Mattson MP & Scheff SW (1999) Exacerbation of damage and altered NF-κB activation in mice lacking tumor necrosis factor receptors after traumatic brain injury. J. Neurosci. 19: 6248-6256.
Suzuki T, Mitake S, Okumura-Noji K, Yang JP, Fujii T & Okamoto T (1997) Presence of NF-κB-like and IκB-like immunoreactivities in postsynaptic densities. Neuroreport 8: 2931-2935.
Taglialatela G, Robinson R & Perez-Polo JR (1997) Inhibition of nuclear factor κB (NFκB) activity induces nerve growth factor-resistant apoptosis in PC12 cells. J. Neurosci. Res. 47: 155-162.
Taglialatela G, Kaufmann JA, Trevino A & Perez-Polo JR (1998) Central nervous system DNA fragmentation induced by the inhibition of nuclear factor κB. Neuroreport 9: 489-493.
Tamatani M, Che YH, Matsuzaki H, Ogawa S, Okado H, Miyake S, Mizuno T & Tohyama M (1999) Tumor necrosis factor induces Bcl-2 and Bcl-x expression through NFκB activation in primary hippocampal neurons. J. Biol. Chem. 274: 8531-8538.
Terai K, Matsuo A & McGeer PL (1996) Enhancement of immunoreactivity for NF-κB in the hippocampal formation and cerebral cortex of Alzheimer's disease. Brain Res. 735: 159-168.
Togashi H, Sasaki M, Frohman E, Taira E, Ratan RR, Dawson TM & Dawson VL (1997) Neuronal (type I) nitric oxide synthase regulates nuclear factor κB activity and immunologic (type II) nitric oxide synthase expression. Proc. Natl. Acad. Sci. USA 94: 2676-2680.
Toliver-Kinsky T, Papaconstantinou J & Perez-Polo JR (1997) Age-associated alterations in hippocampal and basal forebrain nuclear factor κB activity. J. Neurosci. Res. 48: 580-587.
Uberti D, Grilli M & Memo M (2000) Contribution of NF-κB and p53 in the glutamate-induced apoptosis. Int. J. Dev. Neurosci. 18: 447-454.
Vollgraf U, Wegner M & Richter-Landsberg C (1999) Activation of AP-1 and nuclear factor-κB transcription factors is involved in hydrogen peroxide-induced apoptotic cell death of oligodendrocytes. J. Neurochem. 73: 2501-2509.
Wang JJ, Tasinato A, Ethell DW, Testa MP & Bredesen DE (2000) Phosphorylation of the common neurotrophin receptor p75 by p38β2 kinase affects NF-κB and AP-1 activities. J. Mol. Neurosci. 15: 19-29.
Weih M, Bergk A, Isaev NK, Ruscher K, Megow D, Riepe M, Meisel A, Victorov IV, Dirnagl U & Dirnagi U (1999) Induction of ischemic tolerance in rat cortical neurons by 3-nitropropionic acid: chemical preconditioning. Neurosci. Lett. 272: 207-210.
Wellmann H, Kaltschmidt B & Kaltschmidt C (2001) Retrograde transport of transcription factor NF-κB in living neurons. J. Biol. Chem. 276: 11821-11829.
Whiteside ST & Israël A (1997) IκB proteins: structure, function and regulation. Semin. Cancer Biol. 8: 75-82.
Wood JN (1995) Regulation of NF-κB activity in rat dorsal root ganglia and PC12 cells by tumour necrosis factor and nerve growth factor. Neurosci. Lett. 192: 41-44.
Yabe T, Wilson D & Schwartz JP (2001) NFκB activation is required for the neuroprotective effects of pigment epithelium-derived factor (PEDF) on cerebellar granule neurons. J. Biol. Chem. 276: 43313-43319.
Yan SD, Yan SF, Chen X, Fu J, Chen M, Kuppusamy P, Smith MA, Perry G, Godman GC, Nawroth P, Zweiter JL & Stern D (1995) Non-enzymatically glycated tau in Alzheimer's disease induces neuronal oxidant stress resulting in cytokine gene expression and release of amyloid β-peptide. Nat. Med. 1: 693-699.
Yang L, Lindholm K, Konishi Y, Li R & Shen Y (2002) Target depletion of distinct tumor necrosis factor receptor subtypes reveals hippocampal neuron death and survival through different signal transduction pathways. J. Neurosci. 22: 3025-3032.
Yoon SO, Casaccia-Bonnefil P, Carter B & Chao MV (1998) Competitive signaling between TrkA and p75 nerve growth factor receptors determines cell survival. J. Neurosci. 18: 3273-3281.
Yu Z, Zhou D, Bruce-Keller AJ, Kindy MS & Mattson MP (1999) Lack of the p50 subunit of nuclear factor-κB increases the vulnerability of hippocampal neurons to excitotoxic injury. J.

Neurosci. 19: 8856-8865.
Yu Z, Zhou D, Cheng G & Mattson MP (2000) Neuroprotective role for the p50 subunit of NF-κB in an experimental model of Huntington's disease. J. Mol. Neurosci. 15: 31-44.
Zelenaia O, Schlag BD, Gochenauer GE, Ganel R, Song W, Beesley JS, Grinspan JB, Rothstein JD & Robinson MB (2000) Epidermal growth factor receptor agonists increase expression of glutamate transporter GLT-1 in astrocytes through pathways dependent on phosphatidylinositol 3-kinase and transcription factor NF-κB. Mol. Pharmacol. 57: 667-678.

6. AFFILIATION

University of Witten/Herdecke, Institute of Neurobiochemistry, Stockumer Straße 10, 58448 Witten, Germany.

A.S. BALDWIN Jr.

NF-κB IN ONCOGENESIS AND AS A TARGET FOR CANCER THERAPY

1. INTRODUCTION

Given the breadth of biological mechanisms associated with NF-κB, it is not surprising that NF-κB has been discovered to be involved with basic processes in cancer. The roles for NF-κB in cancer appear to be complex and likely involve the ability of this transcription factor to control apoptosis, cell-cycle progression, cell differentiation, angiogenesis and cell migration. Consistent with its role in oncogenesis, NF-κB is activated by most oncoproteins and there is evidence that NF-κB interacts with key proteins involved in controlling oncogenesis. Importantly, it has been reported that NF-κB is activated in cancer cells by several chemotherapies and by radiation, and that in many cases this response inhibits the ability of the cancer therapy to induce cell death. Based on these findings, NF-κB is proposed to be a new target for cancer therapy.

2. NF-κB AND ITS ASSOCIATION WITH ONCOGENESIS

The history of NF-κB and cancer dates to studies on viral-induced oncogenesis. v-Rel, the viral homologue of c-Rel, was identified as the transforming gene of the avian retrovirus RevT, which is highly oncogenic and causes aggressive leukemia and lymphomas in chickens. Studies on v-Rel (Ghosh et al., 1998; Gilmore, 1999; Rayet & Gelinas, 1999; Baldwin, 2001; Karin et al., 2001) have set the stage for findings on NF-κB and IκB family members in human cancer. Relevant to human cancer, the c-Rel, NF-κB2 (p100/p52), and Bcl-3 genes are amplified and/or rearranged in certain cancers. For example, the Bcl-3 gene was localized to a [t(14,19) (q32;q13.1)] chromosomal translocation found in some B-cell chronic lymphocytic leukemias and is overexpressed in certain B-cell neoplasms (Rayet & Gelinas, 1999; Baldwin, 2001; Karin et al., 2002). Mutations in the IκBα gene have been detected in Hodgkin's lymphoma (Cabannes et al., 1999) and are suggested to contribute to constitutively active NF-κB found in Hodgkin's cells. This implies a tumor suppressor function for IκB as well as an involvement of NF-κB in this disease. Genetic rearrangements involved in cancer and not involving NF-κB or IκB genes may impinge on the NF-κB regulatory pathway. For example, a translocation placing the Bcl-10 gene under the control

R. Beyaert (ed.), Nuclear Factor κB. Regulation and Role in Disease, 393-408.

of the IgG promoter results in chronic Bcl-10 expression, NF-κB activation and lymphoma (Willis et al., 1999).

Extensive evidence links NF-κB with oncogenic transformation. For example, the Tax oncoprotein from the human T-cell leukemia virus-I (HTLV-I) induces NF-κB activity, and activation of NF-κB was shown to be required for transformation of rat fibroblasts by the HTLV-I Tax protein (Baldwin, 2001; Karin et al., 2002). Similarly, most viral transforming proteins stimulate NF-κB transcriptional activity, either through the induction of nuclear accumulation or through the stimulation of transactivation potential (Mosialos, 1997). For example, the LMP1 protein of Epstein-Barr virus activates NF-κB and NF-κB appears to be required for LMP1-induced cellular transformation (Mosialos, 1997; He et al., 2000). Consistent with a possible role for NF-κB in transformation and tumorigenesis, many human derived solid tumor cell lines display increased nuclear NF-κB (p50-p65 heterodimer) levels and activity (Baldwin, 2001; Karin et al., 2002), although this may not always reflect the natural state of the tumor.

Further data support the involvement of NF-κB in oncogenesis. Thus, it was shown that inhibition of NF-κB by expression of a modified form of IκBα (super-repressor IκBα) blocked focus formation induced by oncogenic H-Ras in NIH3T3 cells (Finco et al., 1997). Consistent with this, Ras-induced transformation is blocked by dominant negative forms of IKK which block NF-κB (Arsura et al., 2000). The oncogenic fusion protein BCR-ABL associated with chronic myelogenous leukemia and some cases of acute lymphoblastic leukemia was shown to activate NF-κB, and tumorigenesis driven by BCR-ABL could be blocked by expression of superrepressor IκBα (Reuther et al., 1998). Additionally, NF-κB was shown to be activated in Hodgkin's lymphoma and inhibition of NF-κB blocked growth of these cells (Bargou et al., 1997). It has been shown that inhibition of NF-κB in head and neck squamous cell carcinoma inhibits cell survival and tumor growth (Duffey et al., 1999). As mentioned earlier, there are numerous reports of NF-κB being activated in a variety of tumor cell lines grown in vitro (Baldwin, 2001). For example, the classic form of NF-κB (p50-p65) has been shown to be activated (i.e. nuclear) in breast cancer cell lines and in breast tumors (Sovak et al., 1997). Our observations agree that NF-κB appears to be dysregulated in breast cancer, but we find that human breast tumors typically display an accumulation of nuclear p52/NF-κB2 and Bcl-3 along with c-Rel rather than p65 (Cogswell et al., 2000). NF-κB has been reported to activated (i.e. nuclear) in cancer cells from other solid tumors (e.g. prostate and pancreatic cancers) (Baldwin, 2001) and in C-cell carcinoma where the activation is associated with expression of the Ret tyrosine kinase (Ludwig et al., 2001). Interestingly, NF-κB was found not to be activated in colorectal tumor cells but rather in the tumor-associated stromal cells (Hardwick et al., 2001).

Much of the functional evidence linking NF-κB with oncogenesis has been obtained through the use of the superrepressor form of IκBα to block NF-κB. Thus, experiments typically have utilized cell lines stably expressing the modified, phosphorylation/degradation resistant IκBα or have utilized adenoviral delivery of the inhibitor (Baldwin, 2001). If IκB can inhibit NF-κB-independent pathways, then the proposed link between NF-κB and oncogenesis can be called into question. This caveat holds for any inhibitor that is not specific for NF-κB.

3. MECHANISMS OF REGULATION OF NF-κB IN TUMOR CELLS

3.1. Activation by oncoproteins and suppression by tumor suppressor proteins

NF-κB is activated by most oncoproteins, but the mechanisms associated with this process are not always clear. The mechanism of activation of NF-κB by certain viral oncoproteins is probably the best understood. For example, the LMP1 protein of Epstein Barr virus promotes NF-κB activation through the recruitment of TRAF2 and the formation of a complex that includes NIK and IKK and through the downstream activation of the kinase Tpl-2/Cot (Cahir-McFarland et al., 1999; Busch & Bishop, 2001; Eliopoulous et al., 2002). TRAF6 is also required for LMP1 to activate NF-κB (Schultheiss et al., 2001). The HTLV-I Tax oncoprotein interacts directly with IKK leading to phosphorylation of the IKKβ and IKKγ subunits and to its constitutive activation (Carter et al., 2001). The activation of NF-κB in Hodgkin's cells is at least partly based on ligand-independent signaling derived from overexpressed CD30 (Horie et al., 2002). Her-2/neu, the EGF receptor family member overexpressed in some breast cancers, has been shown to activate NF-κB through a PI3-kinase-dependent mechanism (Pianetti et al., 2001). Oncogenic Ras expression in epithelial cells is associated with enhanced IKK activity that is dependent on PI3-kinase activity (Arsura et al., 2000). The exact contribution of IKK in these processes is unclear since stable Ras-expressing fibroblasts have elevated NF-κB transcriptional activity without significantly increased nuclear levels of NF-κB (Finco et al., 1997). In that regard, Ras and other oncoproteins are able to strongly activate the transcriptional activation domain of the p65 subunit as assayed through fusion with the Gal4 DNA binding domain (Finco et al., 1997). Several cancer cell lines have been shown to exhibit elevated IKK activity associated with the activation of NF-κB (Gasparian et al., 2002; Suh et al., 2002) and elevated IKK activity has been reported in breast tumors (Romieu-Mourez et al., 2001). Interestingly, it has been reported that IKKα is a critical intermediate in a signaling pathway that promotes mammary epithelial cell proliferation through a mechanism involving cyclin D1 expression (Cao et al., 2001) potentially linking this pathway with the development of breast cancer. In melanoma cells, constitutive activation of NF-κB is apparently controlled by elevated levels of reactive oxygen species generated from NADPH-quinone oxidoreductase (Brar et al., 2001). In pancreatic cancer cell lines, autocrine production of IL-1 confers constitutive activation of NF-κB (Arlt et al., 2002). Thus, multiple signaling pathways, potentially all dependent on IKK activity, lead to the activation of NF-κB in cancer cells.

Given the proposed importance of NF-κB in oncogenesis, it is not surprising that certain tumor suppressor pathways have been shown to negatively regulate

NF-κB. For example, there is evidence that the tumor suppressor protein p53 can negatively regulate NF-κB through cross-competition for transcriptional coactivators (Webster & Perkins, 1999). Recently it was shown that the tumor suppressor protein menin, the loss of which is associated with multiple endocrine neoplasia, physically interacts with and inhibits NF-κB function (Heppner et al., 2001). PTEN is a lipid phosphatase responsible for dephosphorylating the PI3-kinase product phosphatidylinositol 3,4,5-triphosphate, which is involved in regulating the serine/threonine kinase Akt/PKB, and Akt is known to regulate NF-κB activity. PTEN has tumor suppressor function and, correspondingly, is found mutated in prostate cancer and other neoplasms. Several groups have shown that PTEN can block NF-κB activation, although there is disagreement as to the exact mechanism whereby this functions, with groups indicating that PTEN can block nuclear accumulation of NF-κB (Gustin et al., 2001; Pianetti et al., 2001), while others indicate that PTEN can block NF-κB transactivation potential and DNA binding without affecting IκBα degradation (Koul et al., 2001; Mayo et al., 2002). Recently it was shown that E2F1 interacts with and inhibits NF-κB and that the ability of E2F1 to potentiate apoptosis occurs at least partly through a mechanism of NF-κB inhibition (Tanaka et al., 2002). Though the involvement of E2F1 in cancer is likely to be complex with described roles in controlling cell proliferation and apoptosis, loss of E2F1 in mice leads to tumor formation (Yamasaki et al., 1996). Based on this, E2F1 can be considered a tumor suppressor protein and the ability to negatively regulate NF-κB may be an important aspect of E2F1 function. Given the importance of NF-κB in oncogenesis, it is likely that other tumor suppressor pathways will be shown to modulate NF-κB either directly or through modulation of specific signaling components which ultimately target NF-κB function (Fig. 1).

3.2. Apparent roles for NF-κB in oncogenesis involve suppression of apoptosis and induction of proliferation, cell invasion, angiogenesis and inflammatory responses

Suppression of apoptosis is considered a hallmark of cancer and a number of groups have demonstrated that NF-κB activation can block cell death pathways (Baldwin, 2001). NF-κB has been shown to activate TRAF1 and TRAF2, and c-IAP1 and c-IAP2 to block caspase-8 activation induced by TNF. Other anti-apoptotic genes have been shown to be activated by NF-κB and include Bcl-2, the Bcl-2 homologues A1/Bfl-1 and Bcl-x_L, IEX-1 and XIAP (Baldwin, 2001). Of significant interest regarding the suppression of apoptosis is the observation that NF-κB can antagonize p53 function, possibly through the cross-competition for transcriptional coactivators (Webster & Perkins, 1999), although others have found a dependence on NF-κB for p53-induced apoptosis (Ryan et al., 2000).

Consistent with an antiapoptotic role for NF-κB in oncogenesis, inhibition of NF-κB leads to the induction of apoptosis after induction of expression of an oncogenic allele of H-Ras (RasV12) (Mayo et al., 1997). Other evidence was presented that inhibition of NF-κB in transformed cells can induce apoptosis (Baldwin, 2001). It has also been shown that the activation of NF-κB by growth factors suppresses the apoptotic response induced by c-Myc (Romashkova & Makarov, 1999). As discussed above, E2F-1 expression leads to apoptosis by a mechanism at least partly associated with the ability of E2F-1 to interact with and inhibit NF-κB (Tanaka et al., 2002). Antiapoptotic genes potentially regulated by NF-κB are known to be expressed in a variety of cancers. For example, IAP proteins and Bcl-x_L are upregulated in different tumors (LaCasse et al., 1998; Baldwin, 2001). An additional mechanism whereby NF-κB may be involved in oncogenesis, is by promoting cell proliferation. It has been shown that NF-κB (the p50 and p65 heterodimer) can promote cell growth by binding to the cyclin D1 promoter and by stimulating cyclin D1 gene transcription (Guttridget et al., 1999; Hinz et al., 1999). Upregulation of cyclin D1 gene expression by NF-κB is associated with enhanced G1/S transition and Rb hyperphosphorylation. We have found that Bcl-3 expression in association with p52 potently induces cyclin D1 gene expression (Westerheide et al., 2001). Interestingly, whereas Bcl-3 promoted G1/S, it correspondingly suppressed G2/M transition.

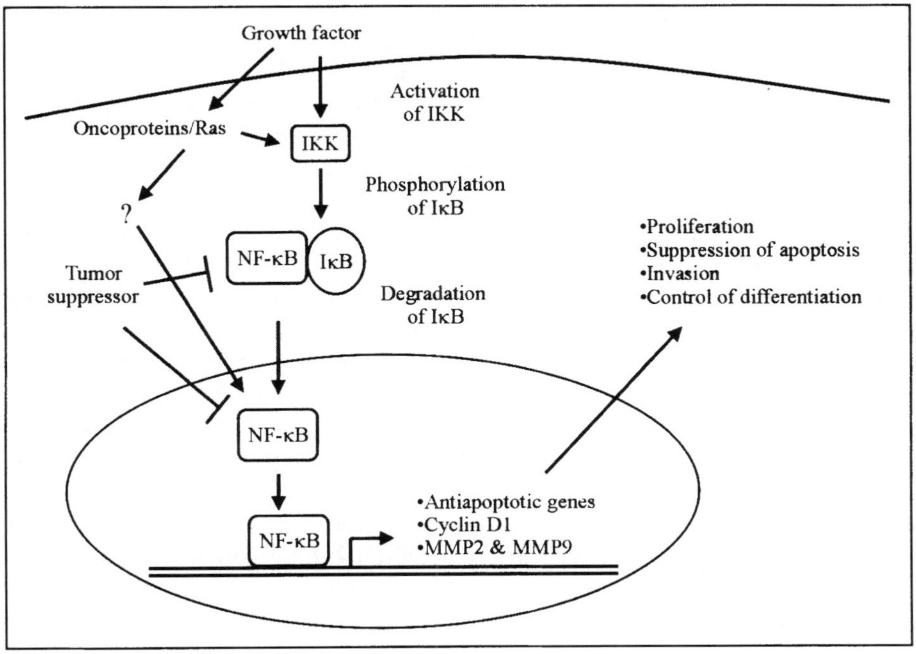

Fig. 1. Ability of oncoproteins and tumor suppressor proteins to modulate NF-κB activity. In the nucleus, NF-κB is proposed to regulate the expression of genes that control apoptosis, differentiation, metastasis and proliferation.

What other NF-κB-regulated genes or processes may contribute to oncogenesis? COX-2, a protein involved in inflammation, is found upregulated in more aggressive forms of colorectal cancer (Tsujii et al., 1998), is known to be transcriptionally activated by NF-κB (Baldwin, 2001) and is known to promote angiogenesis (Tsujii et al., 1998). Inhibition of NF-κB suppressed angiogenesis along with VEGF and IL-8 expression in one model (Huang et al., 2000), but it is unclear whether this may be indirect since smaller tumors due to inhibition of NF-κB could be less hypoxic and therefore have less angiogenesis. Relative to other aspects of oncogenic control, NF-κB is known to regulate expression of cell adhesion molecules and cell surface proteases (such as MMP9 and MMP2; Baldwin, 2001). These observations are consistent with reports showing that NF-κB can promote metastasis (Andela et al., 2000; Huang et al., 2001). Another mechanism whereby NF-κB may regulate tumor progression and tumor cell migration is by controlling chemokine/chemokine receptor gene expression. For example, NF-κB is known to regulate RANTES gene expression and RANTES has been demonstrated to be expressed in breast tumors (Azenshtein et al., 2002). Furthermore, the chemokine receptors CCR7 and CXCR4 are upregulated in breast tumors and are predicted to be involved in metastasis to different tissue/organ sites (Muller et al., 2002) and NF-κB is proposed to regulate expression of CCR7 (Hopken et al., 2002). It was shown that $TNF^{-/-}$ animals are inhibited in their ability to undergo skin carcinogenesis (Moore et al., 1999). Since NF-κB positively regulates TNF gene expression and since TNF activates NF-κB, these results may correlate with a role for NF-κB in skin carcinogenesis. Consistent with the involvement of cytokines in cancer, it has been suggested that the upregulation of NF-κB in head and neck squamous cell carcinoma promotes pro-inflammatory cytokine production and possibly metastasis (Wolf et al., 2001). Similarly, TNFα plays an important role in multiple myeloma by the NF-κB-dependent upregulation of IL-6 in bone marrow stromal cells (Hideshima et al., 2001a). IL-6 functions to promote growth and survival of multiple myeloma cells. IL-6 is also associated with bone metastasis and enhancement of osteoclast activity associated with bone resorption (Pederson et al., 1999). NF-κB has been shown to block differentiation of myoblasts into myofibers through the induced posttranscriptional loss of MyoD mRNA (Guttridge et al., 2000) and NF-κB activation leads to a similar loss of Sox9 mRNA and a block on differentiation of chondrocytes (R. Sitcheran & A.S. Baldwin, in preparation). Thus it is possible that NF-κB activation promotes the loss of differentiation associated with certain hematological malignancies and solid tumors. NF-κB has also been reported to regulate expression of the murine telomerase catalytic subunit (TERT) gene (Yin et al., 2000), implicating NF-κB in control of telomerase activity and its associated immortalization functions.

It is interesting to consider the role of NF-κB in inflammatory responses (Ghosh et al., 1998) associated with oncogenesis. Accumulating evidence indicates that inflammation can promote a number of cancers. For example, it is known that both hereditary and sporadic forms of chronic pancreatitis are associated with an increased risk of developing pancreatic cancer (Farrow & Evers, 2000). Inflammatory bowel disease (IBD) is also associated with the development of colorectal cancer and NF-κB appears to be also involved with the development of IBD (Neurath et al., 1996). Thus, it has been suggested that cytokines, reactive oxygen species and proinflammatory signal transduction pathways associated with inflammation may play a key role in the development of pancreatic cancer or other malignancies (Farrow & Evers, 2000). Nucleotide polymorphisms found in the IL-1 gene, associated with enhanced expression levels, are correlated with enhanced risk for gastric cancer (El-Omar et al., 2001). Given the ability of NF-κB to both regulate cytokine gene expression and to function downstream of cytokine receptors to control expression of genes, such as COX-2 or IL-8, strongly implicate this transcription factor in inflammatory mechanisms associated with cancer. The fact that a number of anti-inflammatory compounds not only function as preventive agents in cancer but also block NF-κB activation is further evidence for the potential role of NF-κB in inflammation-associated oncogenesis.

4. NF-κB AND TUMOR SUPPRESSOR POTENTIAL

Whereas evidence generally supports the pro-oncogenic potential for NF-κB family members, some emerging evidence suggests a tumor suppressor function for these proteins. Probably the best example of this exists from studies which indicate that transgenic expression of the superrepressor form of IκBα in skin leads to skin cancer (van Hogerlinden et al., 1999). In these studies, inhibition of NF-κB signaling resulted in an increase in frequency of apoptotic cells and the spontaneous development of squamous cell carcinoma. It was reported that p65 null fibroblasts have properties associated with transformation (e.g. formation of colonies in soft agar) and re-expression of p65 in these cells suppressed these properties (Gapuzan et al., 2002). c-Myc expression was found to block NF-κB transactivation potential (Tanaka et al., 2002; You et al., 2002), suggesting that the oncoprotein c-Myc can function to inhibit NF-κB activity. Evidence that NF-κB functions downstream of p53 to promote apoptosis is also consistent with an antioncogenic role for NF-κB (Ryan et al., 2000). The human papillomavirus oncoprotein E7 has been reported to block NF-κB activation by interacting with IKK (Spitkovsky et al., 2002). Moreover, the papillomavirus E6 protein inhibited NF-κB-dependent transcriptional activity. Clearly, cell or signaling context may utilize NF-κB subunits in different ways to regulate oncogenesis.

5. APPROACHES TO INHIBIT NF-κB AND EVIDENCE THAT NF-κB IS THE TARGET OF VARIOUS CHEMOPREVENTIVE COMPOUNDS

A number of relatively effective inhibitors of NF-κB activation have been identified. Recently, proteasome inhibitors have been used to block NF-κB nuclear localization through the inhibition of IκBα degradation and the processing of the p105 NF-κB precursor (Adams, 1999; Karin & Ben-Neriah, 2000; Yamamoto & Gaynor, 2001). Several studies have shown that proteasome inhibitors have significant antitumor activity in murine xenograft models (Orlowski et al., 1998; Sunwoo et al., 2001), and one of the proteasome inhibitors, PS-341, has been shown to inhibit NF-κB and, correspondingly, inhibit cell survival, tumor growth and angiogenesis in squamous cell carcinoma tumor models (Sunwoo et al., 2001). As the major proteolytic system in eukaryotic cells (Orlowski, 1999), however, the ubiquitin-proteasome pathway controls the breakdown of most cellular proteins, and thereby influences a variety of physiologic processes. Obviously, a proteasome inhibitor will affect the stability of numerous proteins normally controlled by the ubiquitin/proteasome degradation pathway and specificity, regarding NF-κB as a unique target, is therefore an issue with this type of agent.

Other molecules have been shown to inhibit NF-κB, including a peptide which blocks the interactions of IKKγ with the catalytic core of IKK, the NSAID sulindac, cyclopentenone prostaglandins, arsenic trioxide, thalidomide, a variety of antioxidants, natural products such as parthenolide, curcumin and resveratrol, and synthetic small molecule inhibitors of the IKK complex developed by pharmaceutical companies (Kapahi et al., 2000; Baldwin, 2001; Keifer & Baldwin, 2001; Yamamoto & Gaynor, 2001; Orlowski & Baldwin, 2002). Relevant to cancer, resveratrol was shown to induce apoptosis of Ras-transformed cells and sulindac also induced cell death in certain transformed cells (Yamamoto & Gaynor, 2001). Arsenic trioxide, which directly modifies IKK on a critical cysteine (Kapahi et al., 2000), has shown efficacy in several hematological malignancies (Waxman & Anderson, 2001). More recently, a drug (PS-1145) was identified (Hideshima et al., 2002) which blocks the IκB kinase (IKK) and has potent NF-κB inhibitory properties. PS-1145 blocks multiple myeloma cell proliferation but not as effectively as the proteasome inhibitor PS-341. Overall, these data underscore the potential relevance of targeting NF-κB in cancer through monotherapies that block different constituents in the NF-κB signal transduction pathway. They also suggest that the ability of certain chemopreventive compounds (resveratrol, NSAIDs, curcumin, etc.) to suppress the initiation of cancer may be related to their ability to block NF-κB (Yamamoto & Gaynor, 2001).

6. NF-κB AND CHEMO- AND RADIORESISTANCE

The realization that NF-κB can inhibit apoptosis led us to examine whether NF-κB is induced by chemotherapy and whether it plays a role in blocking the efficacy of cancer chemotherapies and radiation. Thus, we showed that HT1080 fibrosarcoma cells exposed to ionizing radiation or to the chemotherapy daunorubicin exhibited enhanced activation of NF-κB (Baldwin, 2001). Inhibition of NF-κB under these experimental conditions led to a dramatically improved apoptotic response to ionizing radiation or daunorubicin treatment as compared to the control cells. HT1080 fibrosarcoma tumors grown in nude mice were induced to undergo apoptosis when infected with an adenovirus expressing a modified form of IκBα along with systemic delivery of the chemotherapeutic agent CPT-11 (Wang et al., 1999). Other tumors (for example, those derived from the colorectal tumor cell line LOVO) showed basically identical responses to the combined treatment. In fact, LOVO tumors could be eliminated with CPT-11 systemic treatment and with adenoviral delivery of IκBα (Cusack et al., 2000). The use of the proteasome inhibitor PS-341 as a systemic inhibitor of NF-κB in conjunction with CPT-11 treatment also led to significantly improved chemotherapeutic responses through enhanced apoptosis (Cusack et al., 2001). Lactacystin, a structurally unrelated proteasome inhibitor, has also been reported to enhance apoptosis induced by drugs such as etoposide and doxorubicin (Ogiso et al., 2000). It should be noted that wild-type p53 is apparently not needed for the enhanced apoptotic response observed with chemotherapy plus NF-κB inhibition since the response occurs in cancer cells that have mutated or deleted p53 (Cusack et al., 2000). Similar to the data with chemotherapy, inhibition of NF-κB sensitized LOVO cells to radiation with an approximate 80% reduction in tumor volume using xenografts receiving radiation and PS-341 (Russo et al., 2001).

The form of chemotherapy or stimulus used may determine whether NF-κB is pro- or antiapoptotic (Kaltschmidt et al., 2000; Baldwin, 2001). For example, the majority of chemotherapies that we have studied induce some form of DNA damage and inhibition of NF-κB activation promotes cell death. However, it was proposed recently that NF-κB may be required for Taxol-induced cell death (Huang et al., 2000). This is consistent with studies that suggest that NF-κB can function proapoptotically based on the type of stimulus to which cells are exposed. For example, cell death induced by hydrogen peroxide has been reported to require NF-κB and there is a report that NF-κB is required for p53-dependent cell death (Dumont et al., 1999; Ryan et al., 2000). More studies are needed to determine chemotherapy specificity and potentially tumor specificity regarding the role of NF-κB in controlling cancer therapy efficacy.

7. NF-κB INHIBITION AND CANCER THERAPY CLINICAL TRIALS

Agents that inhibit NF-κB function are currently in clinical use as cancer chemotherapeutics, while others are undergoing clinical development. One of these drugs is arsenic trioxide (ATO) which inhibits IKK activity and which was reported to have activity against promyelocytic leukemia in studies from China (Waxman & Anderson, 2001). A multicenter trial in the United States in patients with relapsed acute promyelocytic leukemia (APL) found a complete response rate of 85% (Niu et al., 1999). Studies have also found that ATO has comparable activity in patients with newly diagnosed APL (Orlowski & Baldwin, 2002), and this agent is being evaluated in ongoing clinical trials in patients with other hematologic and solid tumors. Indeed, preliminary evidence of some activity in patients with multiple myeloma has recently been reported (Orlowski & Baldwin, 2001). It is presently unclear whether the ability of arsenic trioxide to inhibit IKK activity is related to its efficacy as a cancer therapy.

Thalidomide is another agent in clinical use that inhibits NF-κB, which it does at least partly by inhibiting IκB kinase activity (Keifer et al., 2001). In patients with refractory multiple myeloma this drug has substantial activity, with a response rate of up to 32% (Singhal et al., 1999). Recent studies using combinations with other agents such as dexamethasone suggest that such regimens may have even greater efficacy (Orlowski & Baldwin, 2002). Interestingly, dexamethasone can inhibit NF-κB through mechanisms distinct from that of thalidomide (Yamamoto & Gaynor, 2001). These encouraging findings have prompted trials in other hematologic malignancies that have documented activity in Hodgkin's and non-Hodgkin's lymphomas (Orlowski & Baldwin, 2001). In contrast, while there is a response rate in AIDS-related Kaposi's sarcoma (Little et al., 2000), the results with single agent thalidomide in other solid tumors have been more disappointing. Occasional responses have been noted in patients with gliomas (Fine et al., 2000) and androgen-independent prostate cancer (Figg et al., 2001a), but of those patients who have derived any benefit most have exhibited stable disease rather than regression. As is the case in myeloma there is the possibility that combination of thalidomide with standard chemotherapeutic agents may prove more efficacious, and such studies are currently underway (Figg et al., 2001b).

While ATO and thalidomide were already in clinical use before their abilities to inhibit NF-κB were identified, several newer classes of chemotherapeutics are in development that were intended, at least in part, to target NF-κB. The best example of agents in this category are inhibitors of the proteasome, and specifically PS-341 (Adams, 1999). Several phase I studies of this drug have been completed, and excitement has been generated in the area of multiple myeloma since significant activity was seen in that patient population including one complete response, prompting a phase II trial which is currently underway. In patients with solid tumors the drug has been well-tolerated and an occasional response has been seen (Orlowski & Baldwin, 2002). Based on some of the preclinical data described earlier, combination phase I trials of PS-341 with standard chemotherapy drugs such as CPT-11 and gemcitabine are ongoing, and if successful will lead to phase II studies where their promise will be better evaluated. Since proteasome inhibition also augments radiation sensitivity clinical trials evaluating this approach are also underway.

With respect to the available clinical data it is interesting to note that drugs that target NF-κB such as thalidomide and PS-341 have to date shown the greatest success as single agents in the therapy of hematologic malignancies. This is perhaps reflective of the important role of NF-κB in the biology of diseases such as multiple myeloma where adherence of these cells to bone marrow stroma triggers synthesis of the myeloma survival and growth factor interleukin-6 through an NF-κB-dependent mechanism (Chauhan et al., 1996). The role of NF-κB in antiapoptotic signaling has also raised new hope that inhibitors will enhance the efficacy of conventional chemotherapeutic drugs. Since the NF-κB inhibitors in current clinical use may also work through other mechanisms, however, such as interruption of growth signaling pathways (Hideshima et al., 2001b), it will be difficult to determine how much of their activity is due to their impact on NF-κB. The application of future agents such as IKK inhibitors that more specifically target the NF-κB signaling cascade will be helpful in this regard. From a clinical perspective a phase III trial randomizing patients to, for example, either chemotherapy (or radiation) alone or chemotherapy (or radiation) in conjunction with a specific NF-κB inhibitor would be required to prove that the preclinical xenograft data can be reproduced in a human population. While such results are likely to be several years away, at this time it appears to be one of the more promising new approaches to cancer therapy.

8. ACKNOWLEDGMENTS

Research was supported by grants from the Leukemia and Lymphoma Society, the National Cancer Institute and the National Cancer Institute Breast Cancer SPORE program.

9. REFERENCES

Adams J (1999) Proteasome inhibitors: a novel class of potent and effective antitumor agents. Cancer Res. 59: 2615-2622.

Arlt A, Vorndamann J, Muerkoster S, Yu H, Schmidt W, Folsch U & Schafer H (2002) Autocrine production of IL-1β confers constitutive activation of NF-κB activity and chemoresistance in pancreatic carcinoma cell lines. Cancer Res. 62: 910-916.

Andela V, Schwarz E, Puzas J, O'Keefe R & Rosier R (2000) Tumor metastasis and the reciprocal regulation of prometastatic and antimetastatic factors by NF-κB. Cancer Res. 60: 6557-6562.

Arsura M, Mercurio F, Oliver A, Thorgeirsson S & Sonenshein G (2000) Role of the IκB kinase complex in oncogenic Ras and Raf-mediated transformation of rat liver epithelial cells. Mol. Cell. Biol. 20: 5381-5391.

Azenshtein E, Luboshits G, Shina S, Neumark E, Shahbazian D, Weil M, Wigler N, Keydar I & Ben-Baruch A (2002) The CC chemokine RANTES in breast carcinoma progression: regulation of expression and potential mechanisms of promalignant activity. Cancer Res. 15: 1093-1102.

Baldwin A (2001) Control of oncogenesis and cancer therapy resistance by the transcription factor NF-κB. J. Clin. Invest. 107: 241-246.

Bargou R, Emmerich F, Krpammann D, Bommert K, Mapara M, Arnold W, Royer H, Grinstein E, Greiner A, Scheidereit C & Dorken B (1997) Constitutive NF-κB/RelA activation is required for proliferation and survival of Hodgkin's disease tumor cells. J. Clin. Invest. 100: 2961-2969.

Brar S, Kennedy T, Whorton A, Sturrock A, Hecksteadt T, Ghio A & Hoidal J (2001) Reactive oxygen species from NADPH: quinone oxidorectase constitutively active NF-κB in malignant melanoma cells. Am. J. Physiol. 280: 659-676.

Busch L & Bishop G (2001) Multiple C-terminal regions of the EBV oncoprotein LMP1 cooperatively regulate signaling to B lymphocytes via TNF receptor-associated (TRAF)-dependent and TRAF-independent mechanisms. J. Immunol. 167: 5805-5813.

Cabannes, E, Khan G, Aillet F, Jarret, R & Hay R (1999) Mutations in the IκBα gene in Hodgkin's disease suggest a tumor suppressor role for IκBα. Oncogene 18: 3063-3070.

Cahir-McFarland E, Izumi K & Mosialos G (1999) Epstein-Barr virus transformation: involvement of LMP1-mediated activation of NF-κB. Oncogene 18: 6959-6964.

Cao Y, Bonizzi G, Seagroves T, Greten F, Johnson R, Schmidt E & Karin M (2001) IKKα provides an essential link between RANK signaling and cyclin D1 expression during mammary gland development. Cell 107: 763-775.

Carter R, Geyer B, Xie M, Acevedo-Suarez C & Ballard D (2001) Persistent activation of NF-κB by tax transforming protein involves chronic phosphorylation of IκB kinase subunits IKKβ and IKKγ. J. Biol. Chem. 276: 24445-24448.

Chauhan D, Uchiyama H, Akbarali Y, Urashima M, Yamamoto K, Libermann T & Anderson K (1996) Multiple myeloma cell adhesion-induced interleukin-6 expression in bone marrow stromal cells involves activation of NF-κB. Blood 87: 1104-1112.

Cogswell P, Guttridge D, Funkhouser W & Baldwin A (2000) Selective activation of NF-κB subunits in human breast cancer: potential roles for NF-κB2/p52 and Bcl-3. Oncogene 19: 1123-1131.

Cusack J, Liu R & Baldwin A (2000) Inducible chemoresistance to CPT-11 in colorectal cancer cells and a xenograft model is overcome by inhibition of NF-κB activation. Cancer Res. 60: 2323-2330.

Cusack J, Liu R, Houston M, Abendroth K, Elliott P, Adams J & Baldwin A (2001) Enhanced chemosensitivity to CPT-11 with proteasome inhibitor PS-341: implications for systemic NF-κB inhibition. Cancer Res. 61: 3535-3540.

Duffey D, Chen Z, Dong G, Ondrey F, Wolf J, Brown K, Siebenlist U & Van Waes C (1999) Expression of a dominant mutant IκBα in human head and neck squamous cell carcinoma inhibits survival, proinflammatory cytokine expression, and tumor growth in vivo. Cancer Res. 59: 3468-3474.

Dumont A, Hehner S, Hofmann T, Ueffing M, Dröge W & Schmitz M (1999) Hydrogen peroxide-induced apoptosis is CD95-independent, requires the release of mitochondria-derived reactive oxygen species and the activation of NF-κB. Oncogene 18: 747-757.

Eliopoulos A, Davies C, Blake S, Murray P, Najafipour S, Tsichlis P & Young L (2002) The oncogenic protein kinase Tpl-2/Cot contributes to EBV-encoded LMP1-induced NF-κB signaling downstream of TRAF2. J. Virol. 76: 4567-4579.

El-Omar M, Carrington M, Chow W, McColl K, Bream J, Young H, Herrera J, Lissowska J, Yuan C, Rothman N, Lanyon G, Martin M, Fraumeni J & Rabkin C (2001) The role of IL-1 polymorphisms in the pathogenesis of gastric cancer. Nature 412: 99.

Farrow B & Evers B (2002) Inflammation and the development of pancreatic cancer. Surg. Oncol. 10: 153-169.

Figg WD, Dahut W, Duray P, Hamilton M, Tompkins A, Steinberg S, Jones E, Premkumar A, Linehan W, Floeter M, Chen C, Dixon S, Kohler K, Kruger E, Gubish E, Pluda J & Reed E (2001a) A randomized phase II trial of thalidomide, an angiogenesis inhibitor, in patients with androgen-independent prostate cancer. Clin. Cancer Res. 7: 1888-93.

Figg WD, Arlen P, Gulley J, Fernandez P, Noone M, Fedenko K, Hamilton M, Parker C, Kruger E, Pluda A & Dahut W (2001b) A randomized phase II trial of docetaxel (taxotere) plus thalidomide in androgen-independent prostate cancer. Sem. Oncol. 28: 62-66.

Finco T, Westwick J, Norris J, Beg A, Der C & Baldwin A (1997) Oncogenic H-Ras-induced signaling activates NF-κB transcriptional activity, which is required for cellular transformation. J. Biol. Chem. 272: 24113-24116.

Fine HA, Figg W, Jaeckle K, Wen P, Kyritsis A, Loeffler J, Levin V, Black P, Kaplan R, Pluda J & Yung W (2000) Phase II trial of the antiangiogenic agent thalidomide in patients with recurrent high-grade gliomas. J. Clin. Oncol. 18: 708-715.

Gapuzan M, Yufit P & Gilmore T (2002) Immortalized embryonic mouse fibroblasts lacking the RelA subunit of transcription factor NF-κB have a malignantly transformed phenotype. Oncogene 21: 2484-2492.

Gasparian A, Yao Y, Kowalczyk D, Lyakh L, Karseladze A, Slaga T & Budunova I (2002) The role of IKK in constitutive activation of NF-κB transcription factor in prostate carcinoma cells. J. Cell Sci. 115: 141-151.

Ghosh S, May M & Kopp E (1998) NF-κB and Rel proteins: evolutionarily conserved mediators of the immune response. Annu. Rev. Immunol. 16: 225-260.

Gilmore T (1999) Multiple mutations contribute to the oncogenicity of the retroviral protein v-Rel. Oncogene 18: 6925-6937.

Gustin J, Maehama T, Dixon J & Donner D (2001) The PTEN tumor suppressor protein inhibits TNF-induced NF-κB activity. J. Biol. Chem. 276: 27740-27744.

Guttridge D, Albanese C, Reuther J, Pestell R & Baldwin A (1999) NF-κB controls cell growth and differentiation through transcriptional regulation of cyclin D1. Mol. Cell. Biol. 198: 5785-5799.

Guttridge D, Mayo M, Madrid L, Wang C-Y & Baldwin A (2000) NF-κB-induced loss of MyoD mRNA: possible role in muscle decay and cachexia. Science 289: 2363-2366.

Hardwick J, van den Brink G, Offerhaus G, van Deventer S & Peppelenbosch M (2001) NF-κB, p38 MAPK, and JNK are highly expressed and active in the stroma of human colonic adenomatous polyps. Oncogene 20: 819-827.

He Z, Xin B, Yang X, Chan C & Cao L (2000) NF-κB activation is involved in LMP1-mediated transformation and tumorigenesis of rat-1 fibroblasts. Cancer Res. 60: 1845-1848.

Heppner C, Bilimoria K, Agarwal S, Kester M, Whitty L, Guru S, Chandrasekharappa S, Collins F, Spiegel A, Marx S & Burns A (2001) The tumor suppressor protein menin interacts with NF-κB and inhibits NF-κB-mediated transactivation. Oncogene 20: 4917-4925.

Hideshima T, Chauhan D, Schlossman R, Richardson P & Anderson K (2001a) The role of TNFα in the pathophysiology of human multiple myeloma: therapeutic applications. Oncogene 20: 4519-4527.

Hideshima T, Richardson P, Chauhan D, Palombella V, Elliott P, Adams J & Anderson K (2001b) The proteasome inhibitor PS-341 inhibits growth, induces apoptosis, and overcomes drug resistance in human multiple myeloma cells. Cancer Res 61: 3071-3076.

Hideshima T, Chauhan D, Richardson P, Mitsiades C, Mitsiades N, Hayashi T, Munshi N, Dang L, Castro A, Palombella V, Adams J & Anderson K (2002) NF-κB as a therapeutic target in multiple myeloma. J. Biol. Chem. 277: 16639-16647.

Hinz M, Krappmann D, Eichten A, Heder A, Scheidereit C & Strauss M (1999) NF-κB function in growth control: regulation of cyclin D1 expression and G0/G1-to-S-phase transition. Mol. Cell. Biol. 19: 2690-2698.

Hopken U, Foss H, Meyer D, Hinz M, Leder K, Stein H & Lipp M (2002) Upregulation of chemokine receptor CCR7 in classic but not lymphocyte-predominant Hodgkin disease correlates with distinct dissemination of neoplastic cells in lymphoid organs. Blood 99: 1109-1126.

Horie R, Watanabe T, Morishita, Ito K, Ishida T, Kanegae Y, Saito L, Higashihara M, Mori S, Kadin M & Watanabe T (2002) Ligand-independent signaling by overexpressed CD30 drives NF-κB activation in Hodgkin-Reed-Sternberg cells. Oncogene 21: 2493-2503.

Huang S, DeGuzman A, Bucana C & Fidler I (2000) NF-κB activity correlates with growth, angiogenesis and metastasis of human melanoma cells in nude mice. Clin. Cancer Res. 6: 2573-2581.

Huang S, Pettaway C, Uehara H, Bucana D & Fidler I (2001) Blockade of NF-κB activity in human prostate cancer cells is associated with suppression of angiogenesis, invasion and metastasis. Oncogene 20: 4188-4197.

Huang Y, Johnson K, Norris J & Fan W (2000) NF-κB/IκB signaling pathways may contribute to the mediation of paclitaxel-induced apoptosis in solid tumor cells. Cancer Res. 60: 4426-4432.

Kaltschmidt B, Kaltschmidt C, Hofmann T, Hehner S, Dröge W & Schmitz M (2000) The pro- or anti-apoptotic function of NF-κB is determined by the nature of the apoptotic stimulus. Eur. J. Biochem. 267: 3828-3835.

Kapahi P, Takahashi T, Natoli G, Adams S, Chen Y, Tsien R & Karin M (2000) Inhibition of NF-κB activation by arsenite through reaction with a critical cysteine in the activation loop of IKK. J. Biol. Chem. 275: 36062-36066.

Karin M & Ben-Neriah Y (2000) Phosphorylation meets ubiquitination: the control of NF-κB activity. Annu. Rev. Immunol. 18: 621-663.

Karin M, Cao Y, Greten F & Li Z-W (2002) NF-κB in cancer: from innocent bystander to major culprit. Nat. Rev. Cancer 2: 301-310.

Keifer J & Baldwin A (2001) Inhibition of NF-κB activity by thalidomide through suppression of IκB kinase activity. J. Biol. Chem. 276: 22382-22387.

Koul D, Yao Y, Abbruzzese J, Yung W & Reddy S (2001) Tumor suppress MMAC/PTEN inhibits cytokine-induced NF-κB activation without interfering with the IκB degradation pathway. J. Biol. Chem. 276: 11402-11408.

LaCasse E, Baird S, Korneluk R & MacKenzie A (1998) The inhibitors of apoptosis (IAPs) and their emerging role in cancer. Oncogene 17: 3247-3259.

Little R, Wyvill K, Pluda J, Welles L, Marshall V, Figg W, Newcomb F, Tosato G, Feigal E, Steinberg S, Whitby D, Goedert J & Yarchoan R (2000) Activity of thalidomide in AIDS-related Kaposi's sarcoma. J. Clin. Oncol. 18: 2593-602.

Ludwig L, Kessler H, Wagner M, Hoang-Vu C, Dralle H, Adler G, Bohm B & Schmid R (2001) NF-κB is constitutively active in C-cell carcinoma and required for RET-induced transformation. Cancer Res. 61: 4526-4535.

Mayo M, Wang C-Y, Cogswell P, Rogers-Graham K, Lowe S, Der C & Baldwin A (1997) Requirement of NF-κB activation to suppress p53-independent apoptosis induced by oncogenic Ras. Science 278: 1812-1815.

Mayo M, Madrid L, Westerheide S, Jones D, Yuan X, Baldwin A & Whang Y (2002) PTEN blocks TNF-induced NF-κB-dependent transcription by inhibiting the transactivation potential of the p65 subunit. J. Biol. Chem. 277: 11116-11125.

Moore R, Owens D, Stamp G, Arnott C, Burke F, East N, Holdsworth H, Turner L, Rollins B, Pasparakis M, Kollias G & Baldwill F (1999) Mice deficient in TNFα are resistant to skin carcinogenesis. Nat. Med. 5: 828-831.

Mosialos, G. 1997. The role of Rel/NF-κB proteins in viral oncogenesis and the regulation of viral transcription. Sem. Cancer Biol. 8: 121-129.

Mujumdar S, Lamothe B & Aggarwal B (2002) Thalidomide suppresses NF-κB activation induced by TNF and H_2O_2, but not that activated by ceramide, lipopolysaccharides, or phorbol ester. J. Immunol. 168: 2644-2651.

Muller A, Homey B, Soto H, Ge N, Catron D, Buchanan M, McClanahan T, Murphy E, Yuan W, Wagner S, Barrera J, Monar A, Verastegui E & Zlotnik A (2001) Involvement of chemokine receptors in breast cancer metastasis. Nature 410: 50-56.

Neurath M, Pettersson S, Mey zum Buschenfeld K & Strober W (1996) Local administration of antisense phosphorothioate oligonucleotides to the p65 subunit of NF-κB abrogates established experimental colitis in mice. Nat. Med. 2: 998-1004.

Niu C, Yan H, Yu T, Sun HP, Liu JX, Li XS, Wu W, Zhang FQ, Chen Y, Zhou L, Li JM, Zeng XY, Yang RR, Yuan MM, Ren MY, Gu FY, Cao Q, Gu BW, Su XY, Chen GQ, Xiong SM, Zhang T, Waxman S, Wang ZY, Chen Z, Hu J, Shen Z-X & Chen SJ (1999) Studies on treatment of acute promyelocytic leukemia with arsenic trioxide: remission induction, follow-up, and molecular monitoring in 11 newly diagnosed and 47 relapsed acute promyelocytic leukemia patients. Blood 94: 3315-3324.

Ogiso Y, Tomida A, Lei S, Omura S & Tsuro T (2000) Proteasome inhibition circumvents solid tumor resistance to topoisomerase II-directed drugs. Cancer Res. 60: 2429-2434.

Orlowski RZ, Eswara J, Lafond-Walker A, Grever M, Orlowski M & Dang CV (1998) Tumor growth inhibition in a murine model of human Burkitt's lymphoma by a proteasome inhibitor. Cancer Res. 58: 4342-4348.

Orlowski RZ (1999) The role of the ubiquitin-proteasome pathway in apoptosis. Cell Death Differ. 6: 303-313.

Orlowski RZ & Baldwin AS (2002) NF-κB as a therapeutic target in cancer. Trends Mol. Med. 8: 385-389.

Pederson L, Winding B, Foged N, Spelsberg T & Oursler M (1999) Identification of breast cancer cell line-derived paracrine factors that stimulate osteoclast activity. Cancer Res. 59: 5849-5855.

Pianetti S, Arsura M, Romieu-Mourez R, Coffey R & Sonenshein G (2001) Her-2/neu overexpression induces NF-κB via a PI3-kinase/Akt pathway involving calpain-mediated degradation of IκBα that can be inhibited by the tumor suppressor PTEN. Oncogene 20: 1287-1299.

Rayet B & Gelinas C (1999) Aberrant Rel/NF-κB genes and activity in human cancer. Oncogene 18: 6938-6947.

Reuther J, Reuther G, Cortez D, Prendergast A & Baldwin A (1998) A requirement for NF-κB activation in Bcr-Abl-mediated transformation. Genes Dev. 12: 968-981.

Romieu-Mourez R, Landesman-Bollag E, Seldin D, Traish A, Mercurio F & Sonenshein G (2001) Roles of IKK and protein kinase CKII in activation of NF-κB in breast cancer. Cancer Res. 61: 3810-3818.

Romashkova J & Makarov S (1999) NF-κB is a target of AKT in anti-apoptotic PDGF signaling. Nature 401: 86-90.

Russo S, Tepper J, Baldwin A, Liu R, Adams J, Elliott P & Cusack J (2001) Enhancement of radiosensitivity by proteasome inhibition: implications for a role of NF-κB. Int. J. Rad. Oncol. Biol. Phys. 50: 183-193.

Ryan K, Ernst M, Rice N & Vousden N (2000) Role of NF-κB in p53-mediated programmed cell death. Nature 404: 892-897.

Schultheiss U, Puschner S, Kremmer E, Mak T, Engelmann H, Hammerschmidt W & Keiser A (2001) TRAF6 is a critical mediator of signal transduction by the viral oncogene LMP1. EMBO J. 20: 5678-5691.

Singhal S, Mehta J, Desikan R, Ayers D, Roberson P, Eddlemon P, Munshi N, Anaissie E, Wilson C, Dhodapkar, Zeddis J & Barlogie B (1999) Antitumor activity of thalidomide in refractory multiple myeloma N. Engl. J. Med. 341: 1565-1571.

Sovak M, Bellas R, Kim D, Zanieski G, Rogers A, Traish A & Sonenshein G (1997) Aberrant NF-κB/Rel expression and the pathogenesis of breast cancer. J. Clin. Invest. 100: 2952-2960.

Spitkovsky D, Hehner S, Hofmann T, Moller A & Schmitz M (2002) The human papillomavirus oncoprotein E7 attenuates NF-κB activation by targeting the IKK complex. J. Biol. Chem. 277: 25576-25582.

Suh J, Payvandi F, Edelstein L, Amenta P, Zong W, Gelinas C & Rabson A (2002) Mechanisms of constitutive NF-κB activation in human prostate cancer cells. Prostate 52: 183-200.

Sunwoo J, Chen Z, Dong G, Yeh N, Crowl-Bancroft C, Sausville E, Adams J, Elliott P & Van Waes C (2001) Novel proteasome inhibitor PS-341 inhibits activation of NF-κB, cell survival, tumor growth, and angiogenesis in squamous cell carcinoma. Clin. Cancer Res. 7: 1419-1428.

Tanaka H, Matsumara I, Ezoe S, Satoh Y, Sakamaki T, Albanese C, Machi T, Pestell R & Kanakura Y (2002) E2F1 and c-Myc potentiate apoptosis through inhibition of NF-κB activity that facilitates MnSOD-mediated ROS elimination. Mol. Cell 9: 1017-1029.

Tsujii M, Kawano S, Tsuji S, Sawaoka H, Hori M & DuBois R (1998) Cyclo-oxygenase regulates angiogenesis induced by colon carcinoma cells. Cell 93: 705-716.

van Hogerlinden M, Rozell B, Ahrlund-Richter L & Toftgard R (1999) Squamous cell carcinoma and increased apoptosis in skin with inhibited Rel/NF-κB signaling. Cancer Res. 59: 3299-3303.

Wang C-Y, Cusack J, Liu R & Baldwin A (1999) Control of inducible chemoresistance: enhanced anti-tumor therapy through increased apoptosis by inhibition of NF-κB. Nat. Med. 5: 412-417.

Waxman S & Anderson K (2001) History of development of arsenic derivatives in cancer therapy. Oncologist 6: 3-10.

Webster G & Perkins N (1999) Transcriptional cross-talk between NF-κB and p53. Mol. Cell. Biol. 19: 3485-3495.

Westerheide S, Mayo M, Anest V, Hanson J & Baldwin A (2001) The putative oncoprotein Bcl-3 induces cyclin D1 to stimulate G1 transition. Mol. Cell. Biol. 21: 8428-8436.

Willis T, Jadayel D, Du M, Peng H, Perry A, Abdul-Rauf M, Price H, Karran L, Majekodunmi O, Wlodarska I, Pan L, Crook T, Hamoudi R, Isaacson P & Dyer M (1999) Bcl10 is involved in t(1;14)(p22;q32) of MALT B cell lymphoma and mutated in multiple tumor types. Cell 96: 35-45.

Wolf J, Chen Z, Dong G, Sunwoo J, Bancroft C, Capo D, Yeh N, Mukaida N & Van Waes C (2001) IL-1α promotes NF-κB and AP-1-induced IL-8 expression, cell survival and proliferation in head and neck squamous cell carcinomas. Clin. Cancer Res. 7: 1812-1820.

Yamamoto Y & Gaynor R (2001) Therapeutic potential of inhibition of the NF-κB pathway in the treatment of inflammation and cancer. J. Clin. Invest. 107: 135-142.

Yamasaki L, Jacks T, Bronson R, Goillot E, Harlow E & Dyson N (1996) Tumor induction and tissue atrophy in mice lacking E2F1. Cell 85: 537-548.

Yin L, Hubbard A & Giardina C (2000) NF-κB regulates transcription of the mouse telomerase catalytic subunit. J. Biol. Chem. 275: 36671-36675.

You Z, Madrid LV, Saims D, Sedivy J & Wang C-Y (2002) c-Myc sensitizes cells to tumor necrosis factor-mediated apoptosis by inhibiting nuclear factor κB transactivation. J. Biol. Chem. 277: 36671-36677.

10. AFFILIATION

University of North Carolina, Lineberger Comprehensive Cancer Center, School of Medicine, Chapel Hill, NC 27599, USA.

P. GEORGEL

CONTROL OF IMMUNITY BY Rel PROTEINS IN DROSOPHILA

1. INTRODUCTION

To fight off invading microorganisms, Drosophila uses an impressive and efficient arsenal that is able to eliminate different types of pathogens by three main strategies (Tzou et al., 2002). The first is activation of proteolytic cascades that lead to coagulation and melanization once the cuticle has been injured. This reaction, during which blood clotting occurs and phenoloxydase converts dopamine to melanin, is thought to limit the pathogen dissemination. However, this aspect of the immune defence is still poorly studied in Drosophila and few molecular data are yet available. The second facet of the fly innate immune system is the cellular reaction. Among the different types of blood cells, plasmatocytes have phagocytotic activities and lamellocytes are responsible for the encapsulation of large particles, such as egg wasp during parasitic infections (Lanot et al., 2001). Several markers of the hematopoietic lineage have been recently described in Drosophila; they exhibit strong conservation with molecules involved in mammalian hematopoiesis (Lebestky et al., 2000). Even if these studies have demonstrated the role of the Toll pathway for proper hemocyte proliferation (Qiu et al., 1998), the involvement of Rel/NF-κB factors in this process is still an open question and will not be discussed in this review. The third weapon, which is used by the fly to eliminate microbes, consists of a cocktail of seven lytic peptide families, which are rapidly produced by the fat body a few hours following a septic injury (Table 1). Their combined concentration reaches 300 μM in the hemolymph (Bulet et al., 1999; Meister et al., 2000); this high level of induction prompted several laboratories to investigate their transcriptional regulation during the immune reaction. The accumulation of biochemical and genetic data over the past ten years undoubtedly shows that the genes encoding the antimicrobial peptides of Drosophila are controlled by two Rel/NF-κB proteins, namely dorsal-related immune factor (Dif) and Relish, which turn on antifungal and antibacterial peptide-encoding genes, respectively (Tzou et al., 2002). Furthermore, Dif and Relish are the ultimate actors of two independent signaling cascades that are activated after an infection with fungi or Gram-positive bacteria (in the case of the Toll/Dif pathway), or after an infection with Gram-negative bacteria (in the case of the imd/Relish pathway). Many partners of

R. Beyaert (ed.), Nuclear Factor κB. Regulation and Role in Disease, 409-421.

these signal transduction pathways have recently been identified by genetic screens. This allowed to show that both the Toll and imd pathways exhibit striking similarities with mammalian cascades that also regulate NF-κB activity, namely the Toll-like receptors (TLR) and the tumor necrosis factor (TNF)-signaling pathways. These data pointed to common evolutionary origins of insect and mammalian immune systems (Hoffmann & Reichhart, 2002).

Table 1. Drosophila-inducible antimicrobial peptides and their regulation.

Peptide type	*Main biological activity*	*Concentration (μM) after septic injury*	*Mainly regulated by*
Diptericin	anti-Gram-negative	0.5	imd/Relish
Attacin	anti-Gram-negative	ND	imd/Relish
Drosocin	anti-Gram-negative	40	imd/Relish
Drosomycin	antifungal	100	Toll/Dif
Metchnikowin	antifungal	40	imd/Relish and Toll/Dif
Cecropin	anti-Gram-negative and antifungal	50	imd/Relish and Toll/Dif
Defensin	anti-Gram-positive	1	imd/Relish and Toll/Dif

Activity and concentration of the peptides in the blood after septic injury are based on Bulet et al., 1999. The pathway controlling their transcription in the fat body results from compilation of several studies in which mutant flies were used to monitor the immune-inducible expression (Petersen et al., 1995; Lemaitre et al., 1996; Levashina et al., 1998; Leulier et al., 2000; Rutschmann et al., 2000a; Georgel et al., 2001; Vidal et al., 2001).

2. Rel/NF-κB PROTEINS IN DROSOPHILA

The first Drosophila protein containing a Rel homology domain (RHD) similar to that present in NF-κB was identified on the basis of its role in the establishment of the dorso-ventral axis of the embryo (Steward, 1987). Mutant females that lack a functional dorsal gene lay dorsalized embryos, hence the name dorsal. The Dorsal protein contains a 300-amino acid long RHD and an activation domain (Fig. 1). It was therefore classified, with the mammalian cRel and RelA, in the first subfamily of Rel/NF-κB factors (Huguet et al., 1997).

The dorsal phenotype was shown to be shared by other mutants; this allowed the definition of the "dorsal group" genes which are all necessary for the dorso-ventral axis specification (Anderson, 1987). Among these genes, those encoding the Toll receptor, the kinase Pelle and the inhibitor Cactus have homologues which are involved in the NF-κB-dependent activation of inflammatory genes in mammals (Belvin & Anderson, 1996). By analogy with the mammalian system, it was therefore very tempting to speculate that Dorsal could also participate in the fly immune defence. This hypothesis, which was also supported by the presence of regulatory motifs similar to the NF-κB-binding site in the promoters of the antimicrobial peptide genes (Engström et al., 1993; Georgel et al., 1993; Kappler et al., 1993), prompted the first genetic analysis of the humoral immune reaction in Drosophila (Lemaitre et al., 1996). The results obtained provided the conceptual basis of the insect host defence mechanism. In essence, the data indicated that the cytokine-like protein Spaetzle, Toll, Pelle, Tube and Cactus are required for the antifungal defence of Drosophila. However, the nature of the Rel protein responsible for induction of the drosomycin gene remained unknown. Indeed, mutants lacking a functional copy of the dorsal gene do not exhibit an increased sensitivity to fungal and bacterial infections. This was interpreted as a possible redundancy with another Rel/NF-κB protein (Reichhart et al., 1993), which was simultaneously identified and cloned by its homology to dorsal. This second Rel protein was named Dorsal-related immune factor (Dif) (Ip et al., 1993) and is most similar to Dorsal in its RHD. However, both proteins exhibit different DNA-binding characteristics (Gross et al., 1996) and largely differ in their C-terminal activation domain (Ip & Levine, 1994).

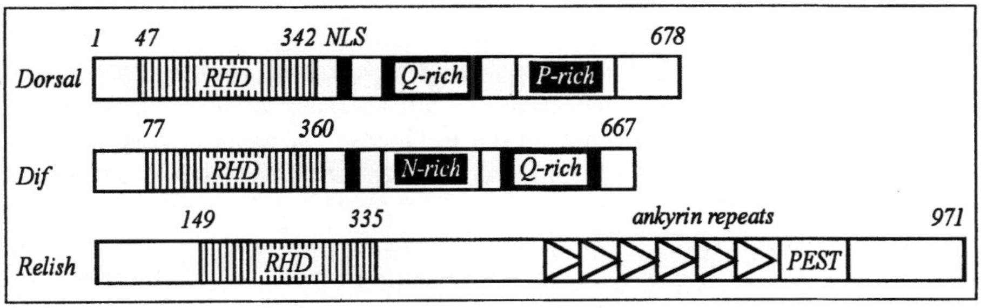

Fig. 1. The Drosophila Rel/NF-κB transactivators.

Furthermore, the expression of the two proteins was shown to be induced and both factors are transported into the nucleus upon an immune challenge. As shown by immuno-histochemistry, Dif and Dorsal are present in the nuclei of fat body cells in $Toll^{10B}$ flies, in which the receptor is constitutively activated. This favored their potential involvement in the Drosophila immune reaction. In addition, cell transfection assays had shown that Dorsal and Dif are able to induce the transcription of reporter genes controlled by κB motifs present in the promoters of antimicrobial peptide genes (Petersen et al., 1995; Gross et al., 1996). As Dif mutants were not available, and since Dorsal and Dif could heterodimerize in vitro (Gross et al., 1996), the model proposed at that time implied that the optimal induction of immunity genes in Drosophila involved Dif/Dorsal complexes (Ip & Levine, 1994).

The situation became more complex with the identification of a third RHD, containing proteins encoded by the relish gene in a differential display screen (Dushay et al., 1996). This search for immune-induced transcripts in flies injected with the Gram-negative bacterium Enterobacter cloacae revealed the existence of a compound protein containing six C-terminal ankyrin repeats and a PEST domain in addition to the RHD (Fig. 1). These features place Relish in the same subgroup of the Rel/NF-κB as the mammalian p100 and p105, which are the precursors of p52 and p50, respectively (Ghosh & Karin, 2002). Similarly to Dorsal and Dif, Relish overexpression in the tumorous blood cell line mbn-2 induced antimicrobial peptide expression; this effect was enhanced with a truncated protein in which the ankyrin repeats had been deleted.

A clear picture of the role of the three Rel/NF-κB factors of Drosophila emerged with the production and analysis of mutants. Adult flies carrying an EMS-induced mutation in the dif gene fail to express the drosomycin gene in response to an infection with the fungus Beauveria bassiana and are therefore very sensitive to this pathogen (Rutschmann et al., 2000a). In addition, this mutation blocks the constitutive drosomycin expression observed in Toll gain-of-function mutants, thus indicating that Dif exerts its effects downstream of the previously identified Toll receptor. At the larval stage, Dif and Dorsal function redundantly in the transcriptional regulation of the drosomycin gene (Manfruelli et al., 1999; Meng et al., 1999; Rutschmann et al., 2000a). Dif mutants show no resistance defects after an infection with Gram-negative bacteria. On the other hand, relish mutants were generated by P-element excision. Homozygous mutant adult flies exhibit a dramatic reduction of antibacterial peptide gene expression and a marked susceptibility to infections with Gram-negative bacteria (Hedengren et al., 1999). These genetic data conferred a prominent function to Dif and Relish in the Drosophila innate response and attenuated somewhat that of Dorsal, whose role in larvae and adults is still unclear. As previously mentioned, Dorsal can substitute for Dif in drosomycin induction during larval stages. Furthermore, overexpression of Dorsal induces lamellocyte differentiation and melanotic capsule formation. This indicates that the Rel/NF-κB protein Dorsal, in addition to its role during embryogenesis, may participate in the cellular defence of Drosophila (Govind, 1999). Additionally, Dorsal/Dif and Dorsal/Relish heterodimers have been observed in transfected S2 cells (Gross et al., 1996; Han & Ip, 1999), which could also indicate that Dorsal may serve as a cofactor mediating the specific induction of target genes encoding antimicrobial peptides by different inducers. Nevertheless, the genetic studies described above clearly indicate that Dif is the primary mediator of antifungal infections and infections with anti-Gram-positive bacteria, whereas Relish is crucial for the fly to mount an efficient anti-Gram-negative response.

3. THE Toll PATHWAY AND Dif ACTIVATION

The initial discovery of the role of the Toll pathway (Fig. 2) in the Drosophila antifungal response (Lemaitre et al., 1996), which is closely followed by the identification of TLR in mammals (Medzhitov et al., 1997; Poltorak et al., 1998), raised the question of the nature of the extracellular molecules able to sense the pathogens in the hemolymph. This question was reinforced by the observation that the Toll pathway is also necessary for the fly to fight off Gram-positive bacteria (Rutschmann et al., 2002). Unlike its mammalian counterparts TLR, the Drosophila Toll receptor cannot be considered a pathogen recognition receptor as previously defined (Janeway, 1989), since the Toll ligand is believed to be the product of the spaetzle gene. Spaetzle is synthesized as an inactive precursor, which is proteolytically processed to generate a C-terminal fragment related to cysteine knot-growth factors. This cleavage is the end point of a proteolytic cascade, which has been extensively studied during embryonic development (Anderson & Beddington, 1997).

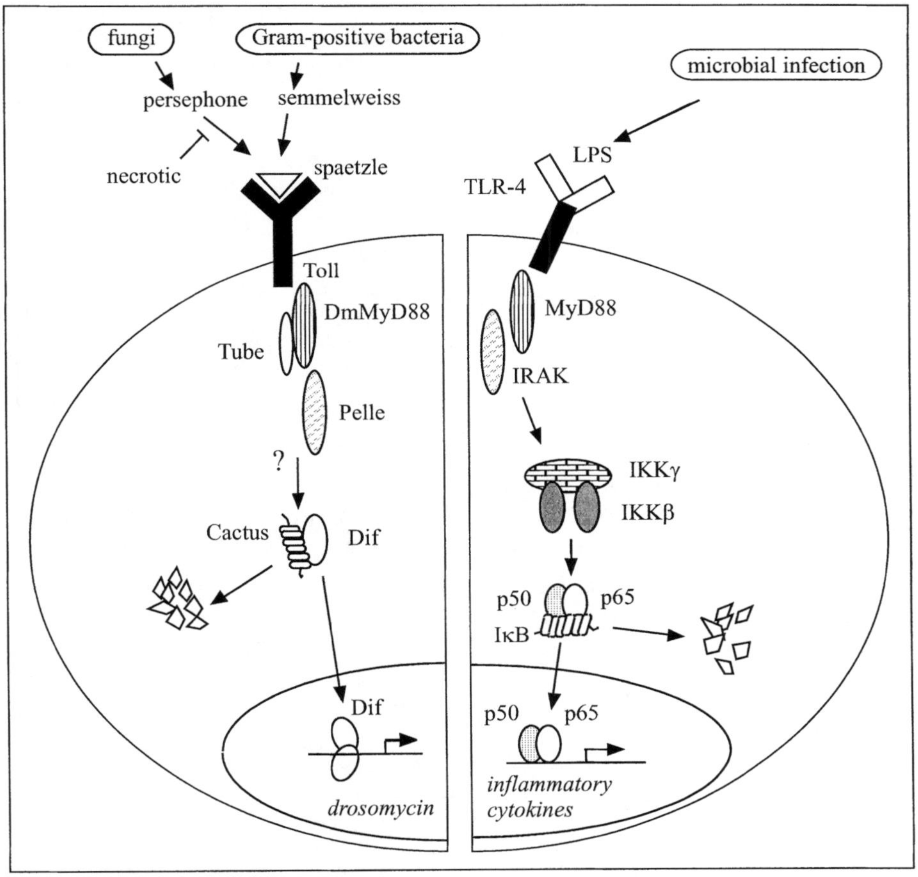

Fig. 2. The Drosophila and mammalian Toll pathways. TLR-4 is activated by lipopolysaccharide (LPS). The other TLR (Akira et al., 2001), sensing different pathogen-associated molecular patterns, are not shown. Homologous proteins between the two transduction cascades are represented by identical shading. A more detailed description of the mammalian TLR pathway has previously been given (Silverman & Maniatis, 2001).

Since mutants for the genes encoding proteases of the embryonic cascade showed no immune phenotype, the identity of the enzymes responsible for Spaetzle cleavage during the immune reaction remained unknown. The identification of the serine protease inhibitor Spn43Ac encoded by the necrotic (nec) gene was an important step to solve this question (Levashina et al., 1999). In nec mutants, the drosomycin gene is expressed in the absence of immune challenge and Spaetzle is constitutively cleaved, which indicated that an upstream protease controlled by Spn43Ac was specifically involved in the Drosophila immune defence. This hypothesis was recently confirmed by the identification of a new gene (persephone) during a genetic screen designed to isolate mutations that could suppress the nec phenotype (Ligoxygakis et al., 2002). Persephone encodes a serine protease; mutant flies are highly sensitive to fungal infections, but not to infections with Gram-positive bacteria. This effect is correlated with the lack of induction of the drosomycin gene after a fungal infection and shows that two distinct extracellular pathways converge on the Toll receptor in response to infections with fungi or Gram-positive bacteria.

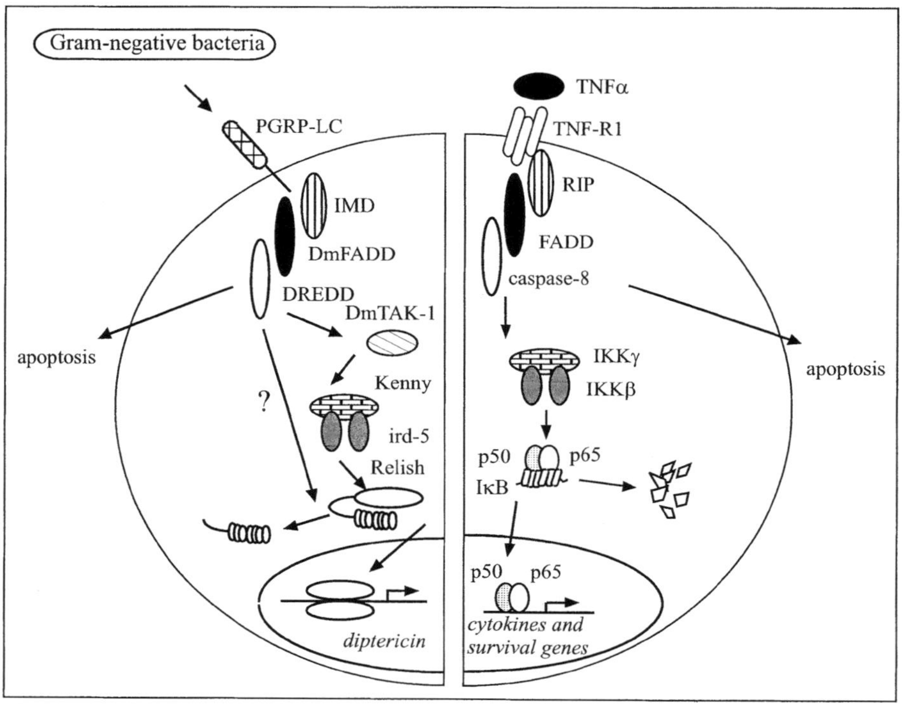

Fig. 3. The Drosophila imd/Relish cascade and the mammalian TNFα pathway. Identical decoration represents several components which have a similar structure in the two signaling cascades. An exhaustive description of the mammalian TNFα-signaling cascade has previously been given (Silverman & Maniatis, 2001).

The activation of this pathway after an infection with Gram-positive bacteria has also been partially unraveled by the discovery of the semmelweiss (seml) gene, which encodes a peptidoglycan recognition protein (PGRP) (Michel et al., 2001). Mutations in the semmelweiss gene reduce the survival rate of flies challenged with Gram-positive bacteria, without affecting the susceptibility to fungal infections. This suggests that the PGRP-SA protein functions as a pattern recognition receptor (PRR) in Drosophila to activate the Toll pathway upon interaction with peptidoglycan from Gram-positive bacteria. In spite of these progresses, it is obvious that many components of the extracellular cascades, initiated by the presence of fungi or Gram-positive bacteria in the hemolymph and leading to the production of active Spaetzle, are still missing. Once activated by cleaved Spaetzle, the Toll receptor then transduces the signal by interactions between its intracellular domain and several partners, all of which contain a death domain. These are the adaptor proteins DmMyD88 and Tube, as well as the serine-threonine kinase Pelle (Imler & Hoffmann, 2001; Tauszig-Delamasure et al., 2002). This large complex signals the phosphorylation of the ankyrin-repeat inhibitor Cactus, which is then degraded by the ubiquitin/proteasome pathway. As Pelle is not directly responsible for the signal-dependent Cactus phosphorylation (Silverman & Maniatis, 2001), the nature of the Cactus kinase is still unknown. After Cactus degradation, the Rel/NF-κB protein Dif is transported into the nucleus by a mechanism involving the nuclear transport factor (NTF-2), a component of the nuclear import machinery, which adds an additional step to control the immune response (Bhattacharya & Steward, 2002). Nuclear Dif activates the expression of numerous genes, among which the antifungal peptide drosomycin is a prototypical marker, as demonstrated by genome-wide analysis of transcripts in wild-type and mutant flies (De Gregorio et al., 2001; Irving et al., 2001; De Gregorio et al., 2002).

4. THE imd PATHWAY AND Relish ACTIVATION

The resistance to Gram-negative bacteria was shown to be independent from the Toll pathway by the discovery of the immune deficiency (imd) mutation, in which the immune inducibility of the diptericin gene is impaired (Lemaitre et al., 1995; Corbo & Levine, 1996). In 1996, imd was the sole mutation known to affect the process, but since then significant progress has been made to unravel the signaling cascade that leads to the resistance against Gram–negative bacteria (Fig. 3). The most upstream actor identified so far is a transmembrane PGRP encoded by the PGRP-LC gene (Choe et al., 2002; Gottar et al., 2002; Ramet et al., 2002). However, the exact modalities by which Gram-negative bacteria present in the hemolymph are sensed and how this activates the imd pathway are not known; they may involve, as is the case in the Toll pathway, an extracellular proteolytic cascade. Signal transduction then induces interactions between two cytoplasmic components containing a death domain, Imd (Georgel et al., 2001) and DmFadd (Naitza et al., 2002). This complex also contains the death effector domain (DED) protein Dredd, a caspase-8 homologue (Chen et al., 1998; Leulier et al., 2000). Downstream of Imd, Dredd and DmFadd, a mitogen-activated protein 3 kinase (MAP3K) homologous to the mammalian transforming growth factor-activated kinase (TAK-1) is required for full diptericin transcriptional induction (Vidal et al., 2001). DmTAK-1 is thought to activate a signalosome-equivalent complex composed of an IKKβ homologue, encoded by the ird-5 gene (Lu et al., 2001), and Kenny, which is similar to the IKKγ/NEMO factor (Elrod-Erickson et al., 2000; Rutschmann et al., 2000b). The Drosophila genome contains also an IKKϵ-like protein, but the latter does not play any apparent role in the insect host defence (Liao et al., 2002). The Drosophila IKKβ/γ complex phosphorylates Relish, whose cleavage by an unidentified protease achieves its activation by removing the C-terminal ankyrin-domain part of the protein (Silverman et al., 2000; Stöven et al., 2000). Relish is then free to move into the nucleus to transactivate target genes such as diptericin, attacin, drosocin and cecropin. As mentioned above in the case of Dif, Relish nuclear translocation is also subjected to regulation by NTF-2 (Bhattacharya & Steward, 2002).

This scheme, however, suffers from several uncertainties. First, the PGRP-LC protein possesses no apparent sequence susceptible to transduce the signal that induces Imd and DmFadd oligomerization. The existence of a yet unidentified receptor, with which PGRP-LC could interact, is therefore suggested. Second, the position of Dredd is somewhat controversial. Overexpression of Dredd was used to perform epistatic analysis, which indicated that the dredd gene acts downstream of the IKK signalosome (Vidal et al., 2001). Consequently, this caspase-8 homologue was proposed to cleave Relish, a process shown to be proteasome-independent (Silverman & Maniatis, 2001). However, biochemical experiments, which indicate that purified Dredd cannot cleave Relish in vitro, do not support this hypothesis. Furthermore, the presence of Dredd in the complex containing Imd and DmFadd also favors the role of Dredd as an apical caspase rather than an effector protease (Hu & Yang, 2000; Horng & Medzhitov, 2001). Third, this leaves open the possibility that one or more additional caspases are necessary in this signal transduction pathway. Their involvement was suggested by overexpression of the viral inhibitor of apoptosis p35 in flies, which blocks the signal-induced diptericin expression (Georgel et al., 2001).

5. CONCLUSION

During the past five years, our understanding of the Drosophila immune defence against fungi or bacteria has greatly improved by genetic studies, which allowed the study of this response in vivo and the description of mutant phenotypes. More recently, sequencing of the Drosophila genome, combined with microarray analysis, has facilitated the identification of genes that might play a role in the host defence (Khush & Lemaitre, 2000). Altogether, these approaches have permitted to draw a clear picture of the activation pathways that turn on antimicrobial peptide genes (Tzou et al., 2002). One of the most interesting aspects of the fly humoral reaction is the detection of the invading microorganisms by pathogen recognition receptors of the PGRP family (Werner et al., 2000). Some of these are circulating molecules, as demonstrated by hemolymph transfer experiments (Michel et al., 2001) and others are predicted to be membrane-bound (Choe et al., 2002; Gottar et al., 2002; Ramet et al., 2002). Their discovery will raise new questions regarding the role of the homologous proteins in innate immunity in humans and mice. This situation is evocative of the demonstration of the role of the Drosophila Toll receptor in the antifungal defence, a finding that paved the way for the search of TLR in mammals and which led to their discovery as PRR (Akira et al., 2001; Kimbrell & Beutler, 2001). Despite the fact that Drosophila Toll is not a PRR, the Toll/Dif pathway is highly similar to the mammalian cascade that leads to TLR-mediated NF-κB activation. Furthermore, the imd/Relish pathway, which resembles the TNFα-activated cascade, also points to a common ancestry between insects and mammals in the immune reaction (Hoffmann & Reichhart, 2002).

However, the question of the mechanisms that lead to activation of the Rel/NF-κB proteins Dif and Relish still remains unsolved. In particular, the kinase that mediates Cactus immune-inducible degradation (Nicolas et al., 1998) constitutes a missing piece in the Toll pathway puzzle. The Relish cleavage that allows the RHD to be translocated into the nucleus also awaits to be unraveled. A caspase cleavage site has been identified between the RHD and the ankyrin repeats (Silverman & Maniatis, 2001), but the identity of the enzyme is not yet known. Furthermore, even if the heterogeneity of the complexes containing Rel/NF-κB proteins is largely proved in Drosophila, the in vivo relevance of these data is still poorly documented. This is illustrated by the lack of information regarding the physiological role of the different potential heterodimeric combinations of the Rel/NF-κB proteins Dorsal, Dif and Relish. In addition, studies performed in mammalian systems have taught us that the activity of Rel/NF-κB proteins largely depends on their association with cofactors (Ghosh & Karin, 2002). Except for the case of the GATA factor Serpent, which binds to sequence motifs located close to the κB-response element of the cecropin promoter and which is required for gene expression (Petersen et al., 1999), no additional cofactor for Dif or Relish has been described. In this context, the role of HMG1/Y in the formation of the enhanceasome on the IFNβ (Yie et al., 1999) or nitric-oxyde synthase (Perrella et al., 1999) promoters has to be mentioned. The roles in Drosophila immunity of homologous proteins, which facilitate NF-κB binding in mammals, have never been investigated. Finally, to complete the list of the Drosophila Rel/NF-κB factors, the existence of a homologue of mammalian NF-AT has to be mentioned. In mammals, this factor exerts pleiotropic roles (Graef et al., 2001), among which T-cell-mediated immune reactions have been studied in knockout mice. Our unpublished observations do not favor a role for this factor in the humoral defence of the fly, as mutants have no defect in the transcription of antimicrobial peptide genes after a septic injury. However, a function for the Drosophila NF-AT in the cellular immune response cannot be excluded.

This review has highlighted the power of the genetic approach, which allowed precise dissection of the Toll and imd pathways and identification of many genes involved in the Drosophila immune defence. However, it also demonstrates that it will not be possible to get away from biochemical investigations to unravel the exact mechanisms of Dif and Relish activation, which remain largely unanswered. Since the catalogue of the different partners involved in the Drosophila host defence is now well described, it is time to study their intimate relationships at the molecular level.

6. ACKNOWLEDGMENTS

The author is grateful to Prof. J. Hoffmann for his continuous support and longstanding encouragement. He thanks Prof. J.-L. Imler for critical reading of the manuscript and all the members of the laboratory for invaluable discussions.

7. REFERENCES

Akira S, Takeda K & Kaisho T (2001) Toll-like receptors: critical proteins linking innate and acquired immunity. Nat. Immunol. 2: 675-680.

Anderson K & Beddington R (1997) Pattern formation and developmental mechanisms. Curr. Opin. Genet. Dev. 7: 455-458.

Anderson KV (1987) Dorsal-ventral embryonic pattern genes of Drosophila. Trends Genet. 3: 91-97.

Belvin MP & Anderson KV (1996) A conserved signaling pathway: the Drosophila Toll-dorsal pathway. Annu. Rev. Cell Dev. Biol. 12: 393-416.

Bhattacharya A & Steward R (2002) The Drosophila homolog of NTF-2, the nuclear transport factor-2, is essential for immune response. EMBO Rep. 3: 378-383.

Bulet P, Hetru C, Dimarcq JL & Hoffmann D (1999) Antimicrobial peptides in insects; structure and function. Dev. Comp. Immunol. 23: 329-344.

Chen P, Rodriguez A, Erskine R, Thach T & Abrams JM (1998) Dredd, a novel effector of the apoptosis activators reaper, grim, and hid in Drosophila. Dev. Biol. 201: 202-216.

Choe KM, Werner T, Stoven S, Hultmark D & Anderson KV (2002) Requirement for a peptidoglycan recognition protein (PGRP) in Relish activation and antibacterial immune responses in Drosophila. Science 296: 359-362.

Corbo JC & Levine M (1996) Characterization of an immunodeficiency mutant in Drosophila. Mech. Dev. 55: 211-220.

De Gregorio E, Spellman PT, Rubin GM & Lemaitre B (2001) Genome-wide analysis of the Drosophila immune response by using oligonucleotide microarrays. Proc. Natl. Acad. Sci. USA 98: 12590-12595.

De Gregorio E, Spellman PT, Tzou P, Rubin GM & Lemaitre B (2002) The Toll and Imd pathways are the major regulators of the immune response in Drosophila. EMBO J. 21: 2568-2579.

Dushay MS, Asling B & Hultmark D (1996) Origins of immunity: Relish: a compound Rel-like gene in the antibacterial defense of Drosophila. Proc. Natl. Acad. Sci. USA 93: 10343-10347.

Elrod-Erickson M, Mishra S & Schneider D (2000) Interactions between the cellular and humoral immune responses in Drosophila. Curr. Biol. 10: 781-784.

Engström Y, Kadalayil L, Sun SC, Samakovlis C, Hultmark D & Faye I (1993) κB-like motifs regulate the induction of immune genes in Drosophila. J. Mol. Biol. 232: 327-333.

Georgel P, Meister M, Kappler C, Lemaitre B, Reichhart JM & Hoffmann JA (1993) Insect immunity: the diptericin promoter contains multiple functional regulatory sequences homologous to mammalian acute-phase response elements. Biochem. Biophys. Res. Com. 197: 508-517.

Georgel P, Naitza S, Kappler C, Ferrandon D, Zachary D, Swimmer C, Kopczynski C, Duyk G, Reichhart JM & Hoffmann JA (2001) Drosophila immune deficiency (IMD) is a death domain protein that activates antibacterial defense and can promote apoptosis. Dev. Cell 1: 503-514.

Ghosh S & Karin M (2002) Missing pieces in the NF-κB puzzle. Cell 109: S81-96.

Gottar M, Gobert V, Michel T, Belvin M, Duyk G, Hoffmann JA, Ferrandon D & Royet J (2002) The Drosophila immune response against Gram-negative bacteria is mediated by a peptidoglycan recognition protein. Nature 416: 640-644.

Govind S (1999) Control of development and immunity by rel transcription factors in Drosophila. Oncogene 18: 6875-6887.

Graef IA, Chen F & Crabtree GR (2001) NFAT signaling in vertebrate development. Curr. Opin. Genet. Dev. 11: 505-512.

Gross I, Georgel P, Kappler C, Reichhart JM & Hoffmann JA (1996) Drosophila immunity: a comparative analysis of the Rel proteins dorsal and Dif in the induction of the genes encoding diptericin and cecropin. Nucleic Acids Res. 24: 1238-1245.

Han ZS & Ip YT (1999) Interaction and specificity of Rel-related proteins in regulating Drosophila immunity gene expression. J. Biol. Chem. 274: 21355-21361.

Hedengren M, Asling B, Dushay MS, Ando I, Ekengren S, Wihlborg M & Hultmark D (1999) Relish: a central factor in the control of humoral but not cellular immunity in Drosophila. Mol. Cell 4: 1-20.

Hoffmann JA & Reichhart JM (2002) Drosophila innate immunity: an evolutionary perspective. Nat. Immunol. 3: 121-126.

Horng T & Medzhitov R (2001) Drosophila MyD88 is an adapter in the Toll signaling pathway. Proc. Natl. Acad. Sci. USA 98: 12654-12658.

Hu S & Yang X (2000) dFADD, a novel death domain-containing adapter protein for the Drosophila caspase DREDD. J. Biol. Chem. 275: 30761-30764.

Huguet C, Crepieux P & Laudet V (1997) Rel/NF-κB transcription factors and IκB inhibitors: evolution from a unique common ancestor. Oncogene 15: 2965-2974.

Imler J & Hoffmann JA (2001) Toll receptors in innate immunity. Trends Cell Biol. 11: 304-311.

Ip YT & Levine M (1994) Molecular genetics of Drosophila immunity. Curr. Opin. Genet. Dev. 4: 672-677.

Ip YT, Reach M, Engstrom Y, Kadalayil L, Cai H, Gonzalez-Crespo S, Tatei K & Levine M (1993) Dif, a dorsal-related gene that mediates an immune-response in Drosophila. Cell 75: 753-763.

Irving P, Troxler L, Heuer TS, Belvin M, Kopczynski C, Reichhart J, Hoffmann JA & Hetru C (2001) A genome-wide analysis of immune responses in Drosophila. Proc. Natl. Acad. Sci. USA 98: 15119-15124.

Janeway CA (1989) Evolution and revolution in immunology. Cold Spring Harbor Symp. Quant. Biol. 54: 1-13.

Kappler C, Meister M, Lagueux M, Gateff E, Hoffmannn JA & Reichhart JM (1993) Insect immunity.Two 17-bp repeats nesting a κB-related sequence confer inducibility to the diptericin gene and bind a polypeptide in bacteria-challenged Drosophila. EMBO J. 12: 1561-1568.

Khush RS & Lemaitre B (2000) Genes that fight infection: what the Drosophila genome says about animal immunity. Trends Genet. 16: 442-449.

Kimbrell DA & Beutler B (2001) The evolution and genetics of innate immunity. Nat. Rev. Genet. 2: 256-267.

Lanot R, Zachary D, Holder F & Meister M (2001) Postembryonic hematopoiesis in Drosophila. Dev. Biol. 230: 243-257.

Lebestky T, Chang T, Hartenstein V & Banerjee U (2000) Specification of Drosophila hematopoietic lineage by conserved transcription factors. Science 288: 146-149.

Lemaitre B, Kromer-Metzger E, Michaut L, Nicolas E, Meister M, Georgel P, Reichhart JM & Hoffmann JA (1995) A recessive mutation, immune deficiency (imd), defines two distinct control pathways in the Drosophila host defence. Proc. Natl. Acad. Sci. USA 92: 9465-9469.

Lemaitre B, Nicolas E, Michaut L, Reichhart JM & Hoffmann JA (1996) The dorsoventral regulatory gene cassette spätzle/Toll/cactus controls the potent antifungal response in Drosophila adults. Cell 86: 973-983.

Leulier F, Rodriguez A, Khush RS, Abrams JM & Lemaitre B (2000) The Drosophila caspase Dredd is required to resist Gram-negative bacterial infections. EMBO Rep. 1: 353-358.

Levashina EA, Langley E, Green C, Gubb D, Ashburner M, Hoffmann JA & Reichhart JM (1999) Constitutive activation of Toll-mediated antifungal defense in serpin-deficient Drosophila. Science 285: 1917-1919.

Levashina EA, Ohresser S, Lemaitre B & Imler JL (1998) Two distinct pathways can control expression of the gene encoding the Drosophila antimicrobial peptide metchnikowin. J. Mol. Biol. 278: 515-527.

Liao D, Silverman N, Zhou R & Maniatis T (2002) DmIKKϵ is not required for the Drosophila antifungal immune response. Paper presented at A. Dros. Res. Conf. 43.

Ligoxygakis P Pelte N Hoffmann JA & Reichhart JM (2002) Activation of Drosophila Toll during fungal infection by a blood serine protease. Science 297: 114-116.

Lu Y, Wu LP & Anderson KV (2001) The antibacterial arm of the Drosophila innate immune response requires an IκB kinase. Genes Dev. 15: 104-110.

Manfruelli P, Reichhart JM, Steward R, Hoffmann JA & Lemaitre B (1999) A mosaic analysis in Drosophila fat body cells of the control of antimicrobial peptide genes by the Rel proteins Dorsal and DIF. EMBO J. 18: 3380-3391.

Medzhitov R, Preston-Hurlburt P & Janeway CA (1997) A human homologue of the Drosophila Toll protein signals activation of adaptive immunity. Nature 388: 394-397.

Meister M, Hetru C & Hoffmann JA (2000) The antimicrobial host defense of Drosophila. Curr. Top. Microbiol. Immunol. 248: 17-36.

Meng X, Khanuja BS & Ip YT (1999) Toll receptor-mediated Drosophila immune response requires Dif, an NF-κB factor. Genes Dev. 13: 792-797.

Michel T, Reichhart J, Hoffmann JA & Royet J (2001) Drosophila Toll is activated by Gram-positive bacteria through a circulating peptidoglycan recognition protein. Nature 414: 756-759.

Naitza S Rosse C Kappler C Georgel P Belvin M Gubb D Camonis J Hoffmann JA & Reichhart JM (2002) The Drosophila immune defense against gram-negative infection requires the death protein dFADD. Immunity 17: 575-581.

Nicolas E, Reichhart JM, Hoffmann JA & Lemaitre B (1998) In vivo regulation of the IκB homologue cactus during the immune response of Drosophila. J. Biol. Chem. 273: 10463-10469.

Perrella MA, Pellacani A, Wiesel P, Chin MT, Foster LC, Ibanez M, Hsieh CM, Reeves R, Yet SF & Lee ME (1999) High mobility group-I(Y) protein facilitates nuclear factor-κB binding and transactivation of the inducible nitric-oxide synthase promoter/enhancer. J. Biol. Chem. 274: 9045-9052.

Petersen UM, Bjorklund G, Ip YT & Engstrom Y (1995) The dorsal-related immunity factor, Dif, is a sequence-specific trans-activator of Drosophila cecropin gene expression. EMBO J. 14: 3146-3158.

Petersen UM, Kadalayil L, Rehorn KP, Hoshizaki DK, Reuter R & Engstrom Y (1999) Serpent regulates Drosophila immunity genes in the larval fat body through an essential GATA motif. EMBO J. 18: 4013-4022.

Poltorak A, He X, Smirnova I, Liu M, Van Huffel C, Du X, Birdwell D, Alejos E, Silva M, Galanos C, Freudenberg M, Ricciardi-Castagnoli P, Layton B & Beutler B (1998) Defective LPS signaling in C3H/HeJ and C57BL/10ScCr mice: mutations in Tlr4 gene. Science 282: 2085-2088.

Qiu P, Pan PC & Govind S (1998) A role for the Drosophila Toll/Cactus pathway in larval hematopoiesis. Development 125: 1909-1920.

Ramet M, Manfruelli P, Pearson A, Mathey-Prevot B & Ezekowitz RA (2002) Functional genomic analysis of phagocytosis and identification of a Drosophila receptor for E. coli. Nature 416: 644-648.

Reichhart JM, Georgel P, Meister M, Lemaitre B, Kappler C & Hoffmann JA (1993) Expression and nuclear translocation of the Rel/NF-κB-related morphogen dorsal during the immune-response of Drosophila. Comptes Rendus de l'Academie des Sciences. III. Sciences de la Vie-Life Sciences 316: 1218-1224.

Rutschmann S, Jung AC, Hetru C, Reichhart JM, Hoffmann JA and Ferrandon D (2000a) The Rel protein DIF mediates the antifungal, but not the antibacterial, response in Drosophila. Immunity 12: 569-580.

Rutschmann S, Jung AC, Rui Z, Silverman N, Hoffmann JA & Ferrandon D (2000b) Role of Drosophila IKKγ in a Toll-independent antibacterial immune response. Nat. Immunol. 1: 342-347.

Rutschmann S Kilinc A & Ferrandon D (2002) The Toll pathway is required for resistance to gram-positive bacterial infections in Drosophila. J. Immunol. 168: 1542-1546.

Silverman N & Maniatis T (2001) NF-κB signaling pathways in mammalian and insect innate immunity. Genes Dev. 15: 2321-2342.

Silverman N, Zhou R, Stoven S, Pandey N, Hultmark D & Maniatis T (2000) A Drosophila IκB kinase complex required for Relish cleavage and antibacterial immunity. Genes Dev. 14: 2461-2471.

Steward R (1987) Dorsal, an embryonic polarity gene in Drosophila, is homologous to the vertebrate proto-oncogene, c-rel. Science 238: 692-694.

Stöven S, Ando I, Kadalayil L, Engström Y & Hultmark D (2000) Activation of the Drosophila NF-κB factor Relish by rapid endoproteolytic cleavage. EMBO Rep. 1: 347-352.

Tauszig-Delamasure S, Bilak H, Capovilla M, Hoffmann JA & Imler JL (2002) Drosophila MyD88 is required for the response to fungal and Gram-positive bacterial infections. Nat. Immunol. 3: 91-97.

Tzou P, De Gregorio E & Lemaitre B (2002) How Drosophila combats microbial infection: a model to study innate immunity and host-pathogen interactions. Curr. Opin. Microbiol. 5: 102-110.

Vidal S, Khush RS, Leulier F, Tzou P, Nakamura M & Lemaitre B (2001) Mutations in the Drosophila dTAK1 gene reveal a conserved function for MAPKKKs in the control of rel/NF-κB dependent innate immune responses. Genes Dev. 15: 1900-1912.

Werner T, Liu G, Kang D, Ekengren S, Steiner H & Hultmark D (2000) A family of peptidoglycan recognition proteins in the fruit fly Drosophila melanogaster. Proc. Natl. Acad. Sci. USA 97: 13772-13777.

Yie J, Merika M, Munshi N, Chen G & Thanos D (1999) The role of HMG I(Y) in the assembly and function of the IFN-β enhanceosome. EMBO J. 18: 3074-3089.

8. AFFILIATION

Centre National de la Recherche Scientifique, Institut de Biologie Moléculaire et Cellulaire, UPR 9022, 15 rue Descartes, 67000 Strasbourg, France.

INDEX

FSC
www.fsc.org
MIX
Papier | Fördert
gute Waldnutzung
FSC® C083411

Kolibri 360 GmbH
Ferdinand-Jühlke-Straße 7
99095 Erfurt, Deutschland
produktsicherheit@kolibri360.de